Advances in Global Change Research

Volume 83

This book series is indexed in SCOPUS.

Advances in Global Change Research

Aims and Scopes

This book series is aimed at addressing a range of environmental issues through state-of-the-art and/or interdisciplinary approaches. The books in the series can either be monographs or edited volumes based, for example, on the outcomes of conferences and workshops, or by invitation of experts. The topics that the series can consider publishing include, but are not limited to:

Physical and biological elements of earth system science, in particular

- Climate change
- Biodiversity
- Sea-level rise
- Paleo-climates and paleo-environments

Social aspects of global change

- Environmentally-triggered migrations
- Environmental change and health
- Food security and water availability
- Access to essential resources in a changing world

Economic and policy aspects of global change

- Economic impacts
- Cost-benefit analyses
- Environmental governance
- Energy transition

Methodologies for addressing environmental issues

- Planetary data analysis
- Earth observations from space
- Proxy data analyses
- Numerical modeling
- Statistical analyses

Solutions to global environmental problems

- Sustainability
- Ecosystem services
- Climate services
- Technological/engineering solutions

Books in the series should be at least 200 pages in length, and include a table of contents.

Images supplied in colour will be reproduced in both the print and electronic versions of the book at no cost to the author/editor.

Manuscripts should be provided as a Word document and will be professionally typeset at no cost to the authors/editors.

All contributions to an edited volume should undergo standard peer review to ensure high scientific quality, while monographs should also be reviewed by at least two experts in the field.

Manuscripts that have undergone successful review should then be prepared according to the Publisher's guidelines manuscripts: https://www.springer.com/gp/authors-editors/book-authors-editors/book-manuscript-guidelines

Pritha Datta · Pulak Mishra · Bhagirath Behera
Editors

Climate Change and Human Well-being

Bridging Subjective and Objective Perspectives

Springer

Editors
Pritha Datta
Rekhi Centre of Excellence for the Science of Happiness
Indian Institute of Technology Kharagpur
Kharagpur, West Bengal, India

Bhagirath Behera
Department of Humanities and Social Sciences
Indian Institute of Technology Kharagpur
Kharagpur, West Bengal, India

Pulak Mishra
Department of Humanities and Social Sciences
Rekhi Centre of Excellence for the Science of Happiness
Indian Institute of Technology Kharagpur
Kharagpur, West Bengal, India

ISSN 1574-0919 ISSN 2215-1621 (electronic)
Advances in Global Change Research
ISBN 978-3-032-25478-8 ISBN 978-3-032-25479-5 (eBook)
https://doi.org/10.1007/978-3-032-25479-5

This Springer imprint is published by the registered company Springer Nature Switzerland AG
The registered company address is: Gewerbestrasse 11, 6330 Cham, Switzerland

To all of us who now have to endure the realities of climate change, and to all the future generations who will be heirs to its burden may we develop a wisdom that will take care of human well-being and of the earth as a whole.

Preface

Climate change is commonly discussed in relation to increasing temperatures, altering precipitations, severe weather conditions, and economical damages. But behind these statistics, there is something more personal and immediate, i.e., how it influences the manner in which people live, feel and how they construct their identities in this world. Human well-being, in terms of health, livelihoods, culture or subjective feelings of happiness and security are inextricably linked to the current climate crisis.

The book, *Climate Change and Human Well-being: Bridging Subjective and Objective Perspectives*, is an attempt to unite varied views to contemplate on this relationship. It poses the question: How do individuals experience climate change not only in their bodies and livelihoods but also in their mind, heart and in their communities? What do cultural practices, social networks and systems of governance contribute to resilience and adaptation? And above all, what can we think about well-being in a way that respects both its quantifiable aspects and its qualitative, subjective aspects?

The chapters have been structured into six parts that deal with a different lens through which the narrative of climate change and well-being is presented. The book starts with the psychological and emotional consequences, the climate anxiety to the plight of migrant communities. Then, the subject of health, with global trends and local inequalities, enters the discussion. The economic and livelihood part is based on how scarcity, uncertainty, and reliance on natural resources change the lives of people, whereas the social and cultural section identifies how the traditions, knowledge and inequalities interact with the effects of climate change.

Later on in the book, the subject matter shifts to resilience and governance. In this case, community practice, adaptive strategies, and legal or policy intervention case studies can give us an idea of what works, what does not, and where new thinking is desperately required. When put together, these chapters are not merely documenting these impacts, but indicate ways to develop systems that safeguard the planet and the health of individuals.

The book has a broad audience; it can be used by students and researchers who want to find an interdisciplinary view of a topic, policymakers who want to find evidence-based information, and practitioners interested in climate change questions, as well as by those who just want to get a clearer picture of how climate change reaches the human experience in all its complexity. This book is intended to catalyze new discussions, contemplations, and practices in the direction of more resilient future by traversing the subjective and objective worldviews.

Kharagpur, West Bengal, India
18 May 2026

Pritha Datta
Pulak Mishra
Bhagirath Behera

Acknowledgments

This book represents the work of many minds and many hands. We owe a debt of gratitude to all of the authors who took their time and wrote their chapters with diligence and care, and the reviewers and others who made a wise comment that enriched the volume. We also owe the communities and individuals whose experiences make us understand why it is critical to bridge the gap between climate change research and human well-being. We are thankful to the team of publishers at Springer who assisted us professionally and guided on the path of manuscript to book. Most importantly, we owe it to our families, friends, and well-wishers who have been supportive and patient during this process.

Kharagpur, West Bengal, India

Pritha Datta
Pulak Mishra
Bhagirath Behera

Contents

Editors and Contributors

About the Editors

Dr. Pritha Datta is an Assistant Professor at the Rekhi Centre of Excellence for the Science of Happiness, Indian Institute of Technology Kharagpur, West Bengal, India. Prior to this, she served as an Assistant Professor in the Department of Policy and Management Studies at the TERI School of Advanced Studies (TERI SAS), New Delhi, India. In her doctoral work, she explored the intersections of climate change, food systems, and ecological sustainability. Over the years, her work has focused on issues that lie at the heart of today's climate crisis; how climate change shapes food security, biodiversity, and human–wildlife interactions, and what can be done to help vulnerable communities adapt.

Her research combines science, policy, and sustainability with the lived experiences of local people, aiming to design adaptation strategies that are practical, equitable, and resilient. She has published her work in reputed international journals and regularly presents at conferences in India and abroad. She is also actively involved in collaborative projects that seek to bring solutions to the ground, especially in regions most at risk from climate impacts. Outside of her academic role, Dr. Datta contributes to regional and global efforts to advance climate resilience. She serves on the Steering Committee of the Indo-Pacific Climate Hub (INPACC Hub), where she works with partners across the region to strengthen cooperation and drive sustainable, climate-resilient development.

Prof. Pulak Mishra currently works as a Professor at the Department of Humanities and Social Sciences of the Indian Institute of Technology Kharagpur. He is also a Joint Faculty Member of the Centre of Excellence on Safety Engineering and Analytics of this Institute.

He has teaching and research experience of more than 20 years. He holds an MPhil in Applied Economics from Jawaharlal Nehru University, New Delhi (through the Centre for Development Studies, Thiruvananthapuram), and a Ph.D. in Economics from Vidyasagar University, West Bengal. He has research interests in the fields of Industrial Economics, Public Economics and Policy, and Economics of Rural Development. He has been engaged in carrying out research on issues relating to corporate response to economic reforms (particularly, on implications of mergers and acquisitions, technology strategies, non-price competition, etc., in respect of market structure, corporate performance, and public policies and regulations). His research works also include issues relating to market–policy–institution linkages for inclusive and sustainable development, socio-economic and ecological implications of rural infrastructure development, value chain development in the agriculture sector, cleaner energy transition, etc.

He has been associated with research projects funded by the Competition Commission of India, the National Foundation for Corporate Governance, the Ministry of Education (Government of India), the Ministry of Rural Development (Government of India), the Department of Science and Technology (Government of India), the Ministry of Environment and Forests (Government of India), the Indian Council of Social Science Research (ICSSR), the National Bank for Agriculture and Rural Development (NABARD), the Australian Centre for International Agricultural Research (ACIAR), the Social Science and Humanities Research Council (SSHRC, Canada), the University of Edinburgh (UK), the University of Osnabruck (Germany), etc. He has also carried out consultancy studies for the National Bank for Agriculture and Rural Development (NABARD), South Eastern Coalfields Limited, ACC Limited, etc.

He has published his research outputs in leading national and international journals such as *Innovation and Development* (Routledge), *Journal of Industry, Competition and Trade* (Springer), *South Asia Economic Journal* (SAGE), *Margin: The Journal of Applied Economic Research* (SAGE), *Journal of Cleaner Production* (Elsevier), *Energy* (Elsevier), *Land Use Policy* (Elsevier), *Renewable and Sustainable Energy Reviews* (Elsevier), *Journal of Urban Management* (Elsevier), *The Extractive Industries and Society* (Elsevier), *Agricultural Systems* (Elsevier), *Development Policy Review* (Wiley), *Review of Development and Change* (SAGE), *Economic and Political Weekly*, *Eurasian Journal of Business and Economics*, etc.

Prof. Bhagirath Behera holds a Ph.D. in Economics from the University of Bonn and has been serving in the Department of Humanities and Social Sciences at the Indian Institute of Technology Kharagpur since 2007. He has made significant contributions to environmental and natural resource economics. His research focuses on the sustainable use and management of natural resources, exploring their economic and livelihood impacts, particularly in India and surrounding regions. His expertise spans critical areas such as water, energy, biodiversity, and climate change, and he has over 20 years of research experience in these fields. His work often addresses development economics, with a focus on environmental policy, resource management, and community-based conservation strategies. He has published numerous research articles in international journals and book chapters, contributing to both theoretical and policy-oriented studies. In addition, he has undertaken 23 research and consultancy projects sponsored by Indian and international funding agencies. He is also deeply interested in broader areas of research, including human behavior, inner development, and the role of economics in achieving sustainable production and consumption for a livable planet.

Contributors

Irshad Ahmad Department of Economics, Aligarh Muslim University, Aligarh, Uttar Pradesh, India

Dastgir Alam Department of Economics, Aligarh Muslim University, Aligarh, Uttar Pradesh, India

Ananya Department of Policy and Management Studies, TERI School of Advanced Studies, New Delhi, India

Rakhi Baidya Dustha Shasthya Kendra (DSK), Dacope, Khulna, Bangladesh

Subhanil Banerjee Gedu College of Business Studies, Royal University of Bhutan, Gedu, Chhukha, Bhutan

Bhagirath Behera Department of Humanities and Social Sciences, Indian Institute of Technology Kharagpur, Kharagpur, West Bengal, India

Deepen Chettri Department of Humanities and Social Sciences, Indian Institute of Technology Kharagpur, Kharagpur, India

Sukanya Das Department of Policy and Management Studies, TERI School of Advanced Studies, New Delhi, India

Uttam Das Department of Humanities and Social Sciences, Indian Institute of Technology Kharagpur, Kharagpur, West Bengal, India

Pritha Datta Rekhi Centre of Excellence for the Science of Happines, Indian Institute of Technology, Kharagpur, West Bengal, India

Mohammed Majeed Hameed Upper Euphrates Centre for Sustainable Development Research, University of Anbar, Ramadi, Iraq

Md Lokman Hossain Department of Geography and Resource Management, The Chinese University of Hong Kong, Shatin, New Territories, Hong Kong, China; Department of Environment Protection, German University Bangladesh, Gazipur, Bangladesh

Abid Hussain Department of Economics, Aligarh Muslim University, Aligarh, Uttar Pradesh, India

Md. Humayain Kabir Wegener Center for Climate and Global Change, University of Graz, Graz, Austria;
Department for Knowledge and Communication Management, Danube University Krems, Krems an der Donau, Austria

Yu Kojima The Climate Panel, State College, Pennsylvania, USA

Derrick Y. F. Lai Department of Geography and Resource Management, The Chinese University of Hong Kong, Shatin, New Territories, Hong Kong, China

Jianfeng Li Department of Geography and Resource Management, The Chinese University of Hong Kong, Shatin, New Territories, Hong Kong, China;
Centre for Environmental Policy and Resource Management, The Chinese University of Hong Kong, Shatin, New Territories, Hong Kong, China

Anjali Majumdar School of Psychological Sciences, Christ University, Bengaluru, India

Maansi Malik Department of Geography, Delhi School of Economics, University of Delhi, New Delhi, India

Adil Masood Department of Natural and Applied Sciences, TERI School of Advanced Studies, New Delhi, India

Abhimita Paul Mazumder Department of Policy and Management Studies, TERI School of Advanced Studies, New Delhi, India

Pulak Mishra Department of Humanities and Social Sciences, Indian Institute of Technology Kharagpur, Kharagpur, West Bengal, India;
Rekhi Centre of Excellence for the Science of Happiness, Indian Institute of Technology Kharagpur, Kharagpur, West Bengal, India

Prasanta Moharaj School of Law, Dayananda Sagar University, Bangalore, India

Chandan Naik Department of Humanities and Social Sciences, Indian Institute of Technology Kharagpur, Kharagpur, India

Deepak Panda Indian Institute of Technology, Kharagpur, India

Rizwan Qasim Department of Economics, Aligarh Muslim University, Aligarh, Uttar Pradesh, India

S. C. Rai Department of Geography, Delhi School of Economics, University of Delhi, New Delhi, India

Santhosh Kareepadath Rajan School of Psychological Sciences, Christ University, Bengaluru, India

Anuska Roy Department of Natural and Applied Sciences, TERI School of Advanced Studies, New Delhi, India

Ashik Rubaiyat Centre for International Studies, Ohio University, Athens, OH, USA

Dibakar Sahoo Department of Economics, Faculty of Social Sciences, Banaras Hindu University, Varanasi, Uttar Pradesh, India

Swosti Subhrajyoti Sahoo Department of Policy and Management Studies, TERI School of Advanced Studies, New Delhi, India

Anwesha Sarma Department of Policy and Management Studies, TERI School of Advanced Studies, New Delhi, India

Khusnur Jahan Shapna Department of Geography and Resource Management, The Chinese University of Hong Kong, Shatin, New Territories, Hong Kong, China; Centre for Environmental Policy and Resource Management, The Chinese University of Hong Kong, Shatin, New Territories, Hong Kong, China

Divyansh Sharma Department of Sustainable Engineering, TERI School of Advanced Studies, New Delhi, India

Ramandeep Singh Sidhu Bathinda College of Law, Punjab, India

Sneha Sindhuja School of Psychological Sciences, Christ University, Bengaluru, India

Neelam Singh Department of Economics, Faculty of Social Sciences, Banaras Hindu University, Varanasi, Uttar Pradesh, India

Chiranjeevi Tallapragada Livelihoods and Natural Resources Management Institute, Hyderabad, India

Shikha Vasishta Research Scholar (Political Science), School of Liberal Arts, Bennett University, Greater Noida, Uttar Pradesh, India

Anurag Krishna Vippala Livelihoods and Natural Resources Management Institute, Hyderabad, India

Part I
Psychological, Emotional, and Behavioral Impacts

Chapter 1
Exploring the Nexus Between Climate Change and Well-Being: Subjective and Objective Dimensions

Pritha Datta, Pulak Mishra, and Bhagirath Behera

Abstract Climate change is not simply an environmental problem; it is a phenomenon that is impacting social, economic, and psychological aspects of people significantly. The present paper discusses the effects of climate change on both objective well-being including health, income, and living conditions and subjective well-being including happiness, life satisfaction, and emotional resilience. Through the prism of hedonic and eudaimonic well-being, we demonstrate that people are affected by climate change not only by their livelihoods, infrastructure, and access to resources but also by stress, anxiety, and uncertainty about the future. Globally, the evidence is pointing at increased mental health issues, eco-anxiety, and ecological grief, economic losses, decreased agricultural output, and destruction of homes and basic services. Though some communities show resilience by adapting together and social support, a number of them have difficulties with the growing vulnerability and escalating inequalities. This paper highlights the necessity of policies that extend beyond economic growth and are concerned with the entire human effects of climate change by considering both the material and emotional aspects of well-being.

Keywords Climate resilience · Health · Income · Living conditions · Life satisfaction · Happiness

---

P. Datta (✉)
Rekhi Centre of Excellence for the Science of Happines, Indian Institute of Technology, Kharagpur, West Bengal, India
e-mail: dattapritha3@gmail.com

P. Mishra · B. Behera
Department of Humanities and Social Sciences, Indian Institute of Technology Kharagpur, Kharagpur, West Bengal, India

P. Mishra
Rekhi Centre of Excellence for the Science of Happiness, Indian Institute of Technology Kharagpur, Kharagpur, West Bengal, India

P. Datta et al. (eds.), *Climate Change and Human Well-being*, Advances in Global Change Research 83, https://doi.org/10.1007/978-3-032-25479-5_1

1.1 Introduction

The effects of climate change are no longer an environmental issue; they have social, economical and psychological consequences that usually affect the society negatively. The number of climate-related calamities around the world has been increasing, especially floods and severe weather patterns, and a significant trend in this regard is the increase in frequency since the 1980s (Fig. 1.1). Although the mortality rates caused by these disasters have been reducing as a result of better early warning as well as disaster preparedness (Fig. 1.2), the financial burden is ever increasing with damages induced by the climate taking a considerable portion of the total economic costs globally (Fig. 1.3). Such occurrences do not only disorient livelihoods and infrastructure but also result in poor well-being.

Although traditional well-being measures, including GDP per capita, levels of income, and lifespan have remained popular in determining the progress of a society, they usually do not reflect all the implications of climate change on human life (Tov & Diener, 2013). The increase of the GDP, the most common metric of economic growth, does not always correlate with increased happiness or higher quality of life, a phenomenon that is commonly known as the Easterlin Paradox (Easterlin, 1974a, 1974b). According to this contradiction, although income and economic growth are associated with well-being to a particular extent, there is a point after which material

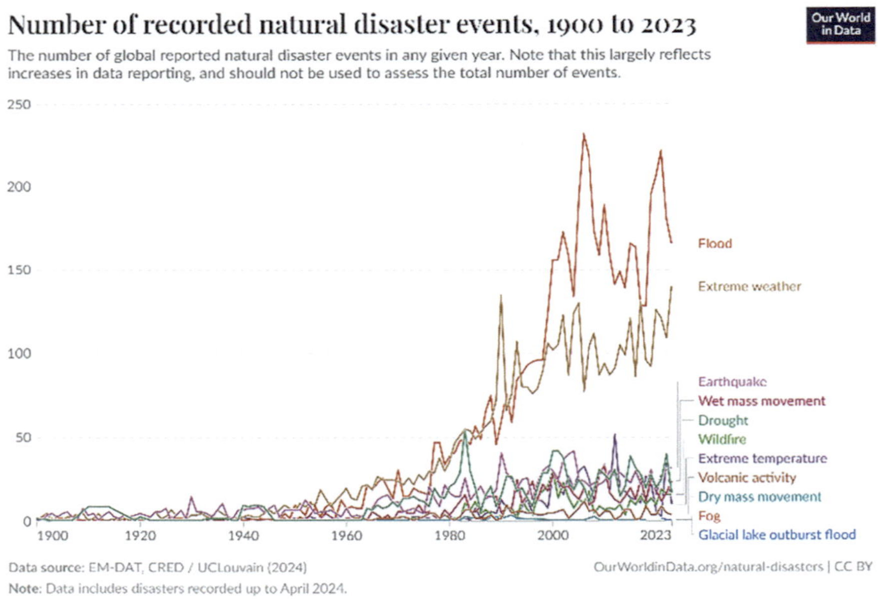

Fig. 1.1 Number of recorded natural disaster events in the world (1900–2023) [*Source* https://ourworldindata.org/]

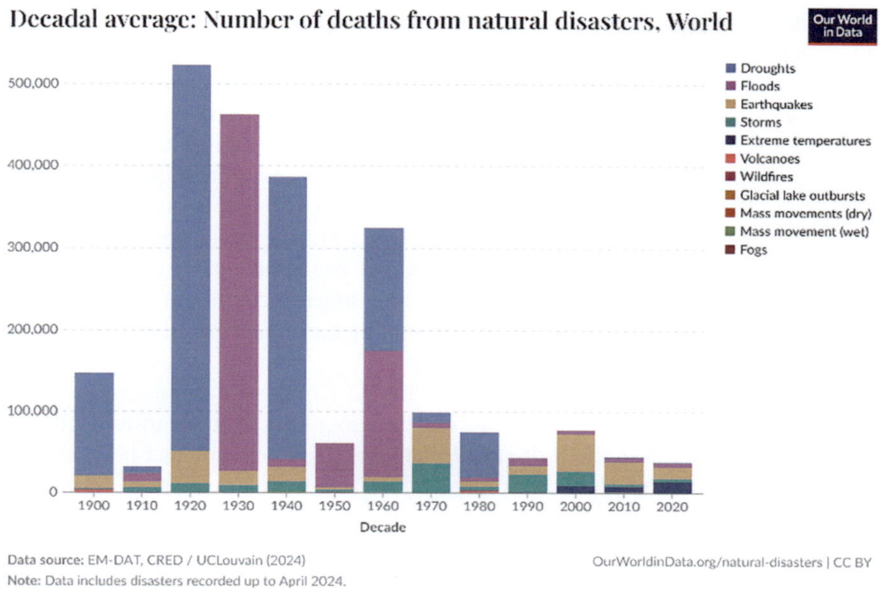

Fig. 1.2 World decadal average of the number of deaths from natural disasters [*Source* https://ourworldindata.org/]

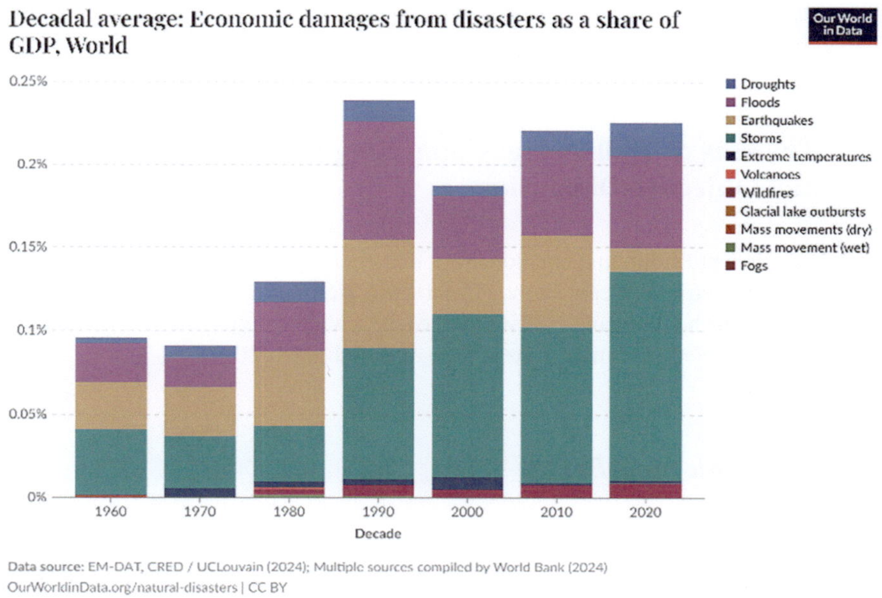

Fig. 1.3 World decadal average of economic damages incurred from disasters as a share of GDP [*Source* https://ourworldindata.org/]

wealth does not positively correlate with the level of life satisfaction (Easterlin & O'Connor, 2022).

The Easterlin Paradox has certain implications, which are especially important in the case of climate change. Even economically successful areas might lose their well-being due to climate-based disasters (e.g., cyclones, extreme rainfall, heatwaves, cold waves, droughts, floods, etc.), depletion of resources, and environmental degradation (Dasgupta, 2024; IPCC, 2022). The increase in the GDP can be accompanied by the deterioration of the quality of the air, severe climate conditions, and mental discomfort associated with the uncertainty about the environment (Hickel, 2020). The communities that are rapidly growing economically could still be affected by the decline in environmental conditions, prompting the enhancement of health risks, biodiversity destruction, and climate anxiety.

As an example, the Yale Program on Climate Change Communication (Leiserowitz et al., 2018) has been monitoring the emotional attitude toward climate change over time, and in 2018, it stated that 69% of Americans had been at least slightly concerned about global warming, and 29% were very concerned. Moreover, almost half (49%) thought that they would be affected personally, and 56% were worried about their family, and 57% care about other people in their locality. Thus, subjective aspects of well-being, including emotional stability, perceived security, and community cohesion, become fundamentally significant to understanding the implications of climate change on human existence including conventional economic indicators. The next section conceptualizes well-being in both subjective and objective terms and discusses how climate changes and their resulting disruptions influence human experiences and policy interventions.

1.2 Theories of Well-Being: Connecting Subjective and Objective Dimensions

Well-being is a multidimensional construct that encompasses both subjective experiences and objective life conditions. Theories of well-being typically fall into two broad categories: hedonic and eudaimonic, each of which aligns differently with the subjective and objective dimensions (Lamb & Steinberger, 2017).

1.2.1 Hedonic Well-Being

The foundation of hedonic well-being lies in the work of Epicurus and classical liberalists like Mill and Bentham (Lamb & Steinberger, 2017). It dwells on pleasure seeking and pain avoiding (Kahneman et al., 1999). It is in parallel with subjective aspect of well-being where the focus is on the happiness, the life satisfaction, and the emotional states of individuals (e.g., positive/negative mood) self-reported by them (Lamb & Steinberger, 2017).

This stream of thought, usually implemented in the form of the Satisfaction with Life Scale (Diener et al., 1985), perceives the well-being as a state of mind dependent on the personal perceptions, preferences, and temporary states of affect. Questions commonly posed to test it include (Diener et al., 1985): "In most aspects, my life is ideal", "The circumstances of my life are ideal", and "I am pleased with my life". These objects are designed to reflect the cognitive appraisal of a life of an individual.

Subjective well-being is important since it represents the way individuals perceive their lives, without considering their external conditions. It provides valuable information about mental health, emotional resilience, and life satisfaction not always present in material or objective measures.

1.2.2 Eudaimonic Well-Being

Unlike hedonic, eudaimonic well being which is based on Aristotelian philosophy focuses on living according to his or her real self, realizing his or her potential, and living a good life (Ryan & Deci, 2001). According to this perspective, it is not only how we feel that is relevant but also our lives and decisions and the extent to which we fulfill our potential (Lamb & Steinberger, 2017; Ryan et al., 2008). This is more related to the objective aspect of well-being that comprises capabilities, achievements, and access to resources and opportunities (Lamb & Steinberger, 2017; Nussbaum, 2011; Sen, 1999).

This idea is grounded in many theoretical frameworks (Lamb & Steinberger, 2017), including: (i) the capability approach by Sen (1990) and Nussbaum (2003), which focuses on ensuring people have the freedom to live lives they value; (ii) the Multidimensional Poverty Index, which measures poverty through indicators beyond just income; and (iii) theories of basic human needs by Max-Neef (1991) and Doyal & Gough, which list universal needs like affection, understanding, and participation (Lamb & Steinberger, 2017).

Climate change affects both hedonic and eudaimonic well-being, which means it influences both subjective and objective aspects of human lives. Hedonic well being, the one that concerns personal happiness, enjoyment, and emotional conditions, is influenced since the environmental stressors interfere with psychological stability. Eudaimonic well-being that focuses on living a good life and living to its fullest potential is undermined once climate change constrains the ability of people and access to resources that support good living.

1.3 Climate Change and Subjective Well-Being

The following sections discuss the climate change influence on SWB through psychological health, social relationships, and perceptions of control over climate risks and the ability to adapt.

1.3.1 Psychological Distress

Stressors such as extreme weather patterns, loss of biodiversity, and environmental degradation, which are the result of climate change, are becoming more and more linked to the increase in the level of mental disorders, such as anxiety, depression, post-traumatic stress disorder (PTSD), and even domestic violence (Morganstein & Ursano, 2020). To give an example, Obradovich et al. (2018) compared the patterns of daily weather and mental health in the United States and discovered that extreme heat conditions, hurricanes, and more sustained droughts had a significant effect on the prevalence of self-reported days of poor mental health. Heightened CO_2 emissions are linked to lower happiness and life contentment in Southeast Asians, which underscores the negativity behind climate change on subjective well-being (Priyanto et al., 2024). Conversely, Rahman et al. (2023), concluded that climate catastrophes do not have any significant impact on happiness but adversely influence life satisfaction in Indonesia.

These effects are particularly great in vulnerable groups. Studies in Australia and India indicate that extended droughts are a factor in the increasing rates of suicide in farmers, many of whom are under economic pressure and the breakdown of traditional livelihoods (Hanigan et al., 2012; Mishra et al., 2021). Climate disasters have a negative impact on life satisfaction, especially among rural inhabitants in Indonesia, who are more susceptible as compared to their urban counterparts (Rahman et al., 2022a, b). In China, a recent study discovered that daily high temperatures (more than 30 °C) had a significant negative impact on the life satisfaction of older adults (50 years old or older) (Li et al., 2024). Meanwhile, in the arctic, the indigenous communities are facing cultural dislocation and identity crises as the permafrost melts and the traditional hunting and fish methods are threatened, leading to mental health problems (Cunsolo & Ellis, 2018).

The effects of extreme climatic events are experienced in mental health as well. According to the reports of survivors of Hurricane Katrina in the U.S. and new Australian wildfires, the experience of a psychological distress, especially in children and marginalized groups, lasted over time (Berry et al., 2018; Gibbs et al., 2019). Outside the immediate impacts of climate events, increasing uncertainty about the future, including fear of food security, shelter, and financial stability, has led to eco-anxiety, chronic fear of environmental doom (Kurth & Pihkala, 2022). This is particularly common among young generations, which are becoming increasingly concerned about the world they are living in as they age (Clayton, 2020; Léger-Goodes et al., 2022). This profound worry is often expressed in despair and a negative attitude towards life.

Alongside eco-anxiety, there is ecological grief, which is a deep feeling of loss in connection with the loss of ecosystems, species, and culturally valued landscapes (Cunsolo & Ellis, 2018). It is especially bitter because those whose identity and livelihood depended on the natural environment are suffering the most. A similar phenomenon is solastalgia which has been defined as the emotional or existential suffering that accompanies the degradation of familiar landscapes with changes in the environment (Galway et al. 2019). This sensation reflects the experienced reality of the world that has been changing as communities experience the changes to the visual environment as a result of climate change (Tupou et al., 2023).

1.3.2 Social Cohesion and Resilience

Climate change is not only affecting the individual well-being, it is also destabilizing the social fabric of communities undermining social cohesion, trust, and collective resilience (Datta, 2023). When environmental stressors become more pronounced, they have a potential to undermine traditional support structures, to uproot populations, to put pressure on local institutions, and to increase social tensions, particularly in those areas where socio-economic vulnerabilities exist in the first place (Datta et al., 2024).

In some areas such as the Sahel in Africa, frequent droughts and land degradation have aggravated tensions between pastoralists, farmers, and other communities, which has led to intra and inter-community conflict (Buhaug, 2015; Raleigh et al., 2015). Such strains are usually augmented by poor governance, poverty, and ethnic conflicts, which is why climate change is a threat multiplier (UNDP, 2020). Another issue to social cohesion in South Asia is the migration brought about by changes in climate. Some Bangladesh-based studies reveal that rural-to-urban migration due to sea-level rise, salinity intrusion and cyclones causes immense strain on infrastructure and services in urban areas and results into informal settlements and heightened social tension between migrant and host groups (Islam & Shamsuddoha, 2017a, 2017b; McLeman, 2018). Influx of climate migrants intensifies the feeling of lack of resources, marginalization as well as exclusion especially in situations where policy responses are insufficient or discriminatory.

Nevertheless, under certain circumstances, extreme climatic events may enhance social cohesion because communities come up with common adaptation and disaster management strategies. As an example, a study of Hurricane Katrina in the United States found that certain neighborhoods experienced great post-disaster solidarity and they organized local support networks to recover their lives (Aldrich, 2012). Moreover, the current social networks can provide emotional support and material resources that reduce the mental health effects of both long-term and abrupt climate-related disasters (Torres & Casey, 2017). Indicatively, Cannings et al. (2024) noted that agricultural communities that are affected by climate in Ghana tend to report low degrees of objective well-being, which can probably be attributed to the vulnerability of livelihoods and a lack of access to services. Nevertheless, they indicate high subjective well-being because they have strong social relationships and reduced relative happiness expectations.

1.3.3 Perceived Control and Adaptation

The sense of control over climate risks and capacity to adapt comprise a key factor affecting the subjective well-being. The people who possess access to adaptive resources, including education, financial stability, early warning systems, and government assistance, will report higher levels of life satisfaction even after being affected by climate threats (2022b; Rahman et al., 2022a). To illustrate, the study of farmers in Vietnam and Nepal implies that farmers with access to climate adaptation training and resources to diversify their crops face less stress and are more satisfied with their lives than those who do not have such opportunities (Thapa et al., 2021). In a similar manner, Adger et al. (2013) observed that UK communities living along the coasts that were engaged in local climate adaptation planning were more psychologically resilient and reported experiencing less climate-related distress. In addition, a study conducted by Rahman et al (2023) in Indonesia revealed that adaptation of the small fishermen to climate change increases their happiness and life satisfaction to a significant level compared to the non-adopters.

A study conducted in the Netherlands and Denmark illustrates that the people who believe that government institutions can handle climate risks record a lower anxiety over climate change than people in nations with less powerful governance systems (Brugger et al., 2015). This brings out the importance of institutional capacity in influencing how the society views climate resilience. In addition, the relationship between climate change and subjective well-being is also mediated by personal values and cultural worldviews. Studies indicate that people who engage in pro-environmental actions, including lowering their carbon footprint or participating in conservation efforts, tend to feel a sense of purpose and increased subjective well-being, which supports the notion that active participation could alleviate climate-related distress (Suárez-Varela et al., 2016; Prinzing et al., 2024).

1.4 Climate Change and Objective Well-Being

The following sections highlight the impact of climate change on objective well-being through physical health, economic conditions, infrastructure, and living conditions.

1.4.1 Physical Health Impacts

Climate change directly and indirectly affects the physical well-being of human beings in terms of increased temperatures, disease patterns, and environmental degradation. Greater temperatures have increased the cases of heat-related health conditions, including heatstroke and dehydration, especially in old age and in workers in the outdoor environment and low-income groups that lack cooling systems (Ebi et al., 2021; Haines & Ebi, 2019). To offer an example, heat-related excess mortality will increase significantly under high-emission conditions (RCP8.5) with the increase in mortality rates of 1.9 to 2.4% in the 2010s and 2030s, respectively and up to 5.5% by the 2090s in China (Yang et al., 2021).

Warmer temperatures and changed precipitation patterns, in turn, breed mosquitoes, so the spread of such diseases as malaria and dengue to new areas occurs (Dhimal et al., 2021; Rocklöv & Dubrow, 2020). It has been observed that the Anopheles and Aedes mosquitoes are changing to higher altitudes and hitherto temperate regions resulting in increased disease burdens (Reiter, 2008).

Climate change also causes respiratory illnesses because the air becomes more polluted by wildfires, industrial emissions, and urban heat island. Particulate matter (PM2.5) and ground-level ozone are known to cause more asthma, chronic obstructive pulmonary disease (COPD), and cardiovascular diseases (Xu et al., 2020). Additionally, dust storms are also caused by prolonged droughts and desertification, which also worsen the quality of air and respiratory health (Middleton & Kang, 2017).

Health outcomes are also influenced by food insecurity that is caused by the negative effects of climate in the agricultural sector (Chettri et al., 2024). Extreme weather events destabilize food production, reduce access to healthy foods, and raise food commodity prices, raising the rates of malnutrition and stunting, especially in low-income countries (Datta et al., 2024; FAO, 2021).

1.4.2 Economic Vulnerability

The effects of climate change on the economy are profound as they can destroy lives, make the economy less productive, and lead to financial instability. These economic implications are especially pronounced in those areas where agricultural activities are predominant and crop failures, water shortage, and weather extremities have direct effects on income and food security (Datta et al., 2024).

The shifting rainfall patterns, rising temperature, and soil erosion are lowering agricultural productivity (Datta & Behera, 2022). Research indicates that the production of staple crops (wheat, rice and maize) has been on the decrease in most areas, specifically South Asia and sub-Saharan Africa (Lobell et al., 2011). In addition to diminishing food security, low yields decrease the income of farmers and push rural people into poverty (Nelson et al., 2014).

Another economic threat caused by climate change is the water shortage. Long droughts and decreased groundwater flows lower water supplies used in irrigation, drinking, and industries, which impacts the economic productivity of various industries (Vörösmarty et al., 2010). Competition over the water resources is growing in countries such as India and China, which creates both social and economic tensions (Gleick, 2014).

Rise in sea level and extreme weather pose a threat to the economy of the coasts as they destroy fisheries, tourism and infrastructure. Frequent flooding and saltwater intrusion into freshwater sources are causing small island countries and coastal urban areas to lose more and more of their economies (Nicholls & Cazenave, 2010). According to estimates by the World Bank, more than 100 million people may fall into extreme poverty because of losses and displacement by the year 2050 because of climate change (Hallegatte et al., 2016).

Extreme heat is also lowering labor productivity. Research indicates that elevated temperatures lower mental processing and physical ability to work, which impacts the economic performance of sectors, including construction, agriculture, and manufacturing (Zhao et al., 2016). The loss in economy due to lowered productivity of labor in hot areas may cost more than billions of dollars per year (Kjellstrom et al., 2016).

1.4.3 Infrastructure and Living Conditions

Climate change has detrimental impacts on infrastructure and living conditions due to the rise in frequency and intensity of floods, cyclones, droughts, and wildfires (Datta et al., 2024). Such extreme weather will lead to destruction of homes, roads, energy systems, water supply systems, deteriorating the quality of life and raising the financial burden on governments and individuals. One of the most devastating effects of climate change is flooding, especially in areas with low altitude and cities located near the coastline. Megacities like Mumbai, Jakarta, and New York have lost billions of dollars in economic values and displaced millions of residents by urban flooding that has caused a health crisis among people (Hallegatte et al., 2013). Poor drainage systems and informal settlements in developing nations further increase the risks of damage of infrastructure due to flooding and outbreaks of diseases (Douglas et al., 2008).

The increase in sea surface temperatures has led to stronger cyclones and hurricanes, which result in general destruction. Superstorm Sandy (2012) and Cyclone Amphan (2020) have shown that extreme weather can paralyze transportation systems, affect the electricity grid, and leave a large number of people homeless (Knutson et al., 2020).

In the arid areas, water security and infrastructure are at risk of drought and desertification. Drying up of major rivers, including the Colorado River and the Indus River, has decreased the production of hydroelectric power, irrigation capabilities, and drinking water and has caused extreme impacts on the socio-economic life (Bates et al., 2008). The growing deserts in the Sahel region of Africa are causing people to move out in search of water and land to grow foodstuffs, which is creating a likelihood of conflict over limited resources (Theisen et al., 2011).

The number of wildfires and their intensity is also on the rise especially in places such as California, Australia, and the Mediterranean. The damage caused by such fires includes the destruction of homes, forests, and farmland with billions of dollars in losses and ecological degradation in the long term (Abatzoglou & Williams, 2016).

The rising expenses of disaster recovery put a heavy cost strain on governments and individuals. The cost of insurance is increasing in climatically vulnerable regions, and the low-income groups cannot afford to rebuild their lives following climate-related disasters (Surminski et al. 2016). Most governments are suffering to free up enough resources to adapt to climate and to have infrastructure resilience, and inequalities in the living conditions are further increasing (OECD, 2013).

1.5 Conclusion

Climate change is transforming the human well-being beyond the environmental or economic changes. It has direct influence on objective factors like health, earnings, and infrastructure, and also on the way individuals feel; level of stress, anxiety, and happiness, and life satisfaction generally. These effects can be complicated, interdependent, and lived differently by communities, and this is why interventions need to focus on tangible and personal aspects of well-being.

These issues demand multi-disciplinary efforts to manage them. Policymakers, economists, social researchers, psychologists, and climate scientists must collaborate in order to come up with measures that protect the physical and emotional aspects of well being. With these two approaches combined, societies will be able to achieve resilience, ensure livelihoods, as well as improve the quality of life amidst the current climatic changes.

Future studies must attempt to comprehend these associations further with the view that a particular region might be more vulnerable or adaptable. Long-term research may demonstrate the time-dependent changes in climate stressors and how these changes can affect the resilience of an individual and community. Attention to the correlations between material conditions and subjective experiences can lead to the creation of interventions that actually make the life of people better.

References

Abatzoglou, J. T., & Williams, A. P. (2016). Impact of anthropogenic climate change on wildfire across western US forests. *Proceedings of the National Academy of Sciences, 113*(42), 11770–11775.

Adger, W. N., Barnett, J., Brown, K., Marshall, N., & O'Brien, K. (2013). Cultural dimensions of climate change impacts and adaptation. *Nature Climate Change, 3*(2), 112–117.

Aldrich, D. P. (2012). *Building resilience: Social capital in post-disaster recovery*. University of Chicago Press.

Bates, B. C., Kundzewicz, Z. W., Wu, S., & Palutikof, J. P. (Eds.). (2008). *Climate change and water*. IPCC.

Berry, H. L., Waite, T. D., Dear, K. B., Capon, A. G., & Murray, V. (2018). The impact of climate change on mental health: A conceptual framework. *Current Opinion in Environmental Sustainability, 39*, 51–57.

Brugger, A., Morton, T. A., & Dessai, S. (2015). Proactive and reactive adaptation to climate change: The role of individual responses. *Global Environmental Change, 35*, 227–238.

Buhaug, H. (2015). Climate–conflict research: Some reflections on the way forward. *Wiley Interdisciplinary Reviews: Climate Change, 6*(3), 269–275.

Cannings, L., Hutton, C. W., Sorichetta, A., & Nilsen, K. (2024). Opposing objective and subjective wellbeing outcomes within an environmentally vulnerable delta: A case study of Volta Delta, Ghana. *Frontiers in Psychology, 15*, 1401480.

Chettri, D., Datta, P., & Behera, B. (2024). Climate change and household food security in the Himalayas: A systematic review of the challenges and household adaptative measures. *Environmental Development*, 101019.

Clayton, S. (2020). Climate anxiety: Psychological responses to climate change. *Journal of Anxiety Disorders, 74*, Article 102263.

Cunsolo, A., & Ellis, N. R. (2018). Ecological grief as a mental health response to climate change-related loss. *Nature Climate Change, 8*(4), 275–281.

Dasgupta, P. (2024). *The economics of biodiversity*. Cambridge university Press.

Datta, P., & Behera, B. (2022). Climate change and Indian agriculture: A systematic review of farmers' perception, adaptation, and transformation. *Environmental Challenges, 8*, 100543.

Datta, P., Behera, B., & Rahut, D. B. (2024). Climate change and water-related threats in the Indian Sundarbans: Food security and management implications. *International Journal of Water Resources Development, 40*(3), 323–344.

Datta, P. (2023). *Climate change and agriculture: An assessment of farmers' adaptation in the foothills of West Bengal, India* (Doctoral dissertation, IIT Kharagpur).

Dhimal, M., Kramer, I. M., Phuyal, P., Budhathoki, S. S., Hartke, J., Ahrens, B., Kuch, U., Groneberg, D. A., Nepal, S., Liu, Q. Y., & Huang, C. R. (2021). Climate change and its association with the expansion of vectors and vector-borne diseases in the Hindu Kush Himalayan region: A systematic synthesis of the literature. *Advances in Climate Change Research, 12*(3), 421–429.

Diener, E., Emmons, R. A., Larsen, R. J., & Griffin, S. (1985). The satisfaction with life scale. *Journal of Personality Assessment, 49*(1), 71–75.

Douglas, I., Alam, K., Maghenda, M., Mcdonnell, Y., McLean, L., & Campbell, J. (2008). Unjust waters: Climate change, flooding and the urban poor in Africa. *Environment and Urbanization, 20*(1), 187–205. https://doi.org/10.1177/0956247808089156

Easterlin, R. A. (1974a). Does economic growth improve the human lot? Some empirical evidence. In *Nations and households in economic growth*. Academic Press.

Easterlin, R. A. (1974b). Does economic growth improve the human lot? Some empirical evidence. In P. A. David & M. W. Reder (Eds.), *Nations and households in economic growth* (pp. 89–125). Academic Press.

Easterlin, R. A., & O'Connor, K. J. (2022). The easterlin paradox. *Handbook of labor, human resources and population economics* (pp. 1–25). Springer International Publishing.

Ebi, K. L., Capon, A., Berry, P., Broderick, C., de Dear, R., Havenith, G., Honda, Y., Kovats, R. S., Ma, W., Malik, A., & Morris, N. B. (2021). Hot weather and heat extremes: Health risks. *The Lancet, 398*(10301), 698–708.

FAO. (2021). *The state of food security and nutrition in the world 2021*. FAO.

Galway, L. P., Beery, T., Jones-Casey, K., & Tasala, K. (2019). Mapping the solastalgia literature: A scoping review study. *International Journal of Environmental Research and Public Health, 16*(15), 2662.
Gibbs, L., Nursey, J., Cook, J., Ireton, G., Alkemade, N., Roberts, M., Gallagher, H. C., Bryant, R., Block, K., Molyneaux, R. & Forbes, D. (2019). Delayed disaster impacts on academic performance of primary school children. *Child Development, 90*(4), 1402–1412.
Gleick, P. H. (2014). Water, drought, climate change, and conflict in Syria. *Weather, Climate, and Society, 6*(3), 331–340. https://doi.org/10.1175/WCAS-D-13-00059.1
Haines, A., & Ebi, K. L. (2019). The imperative for climate action to protect health. *New England Journal of Medicine, 380*(3), 263–273.
Hallegatte, S., Bangalore, M., Bonzanigo, L., Fay, M., Kane, T., Narloch, U., & Vogt-Schilb, A. (2016). *Shock waves: Managing the impacts of climate change on poverty*. World Bank Publications.
Hallegatte, S., Green, C., Nicholls, R. J., & Corfee-Morlot, J. (2013). Future flood losses in major coastal cities. *Nature Climate Change, 3(*9), 802–806. https://doi.org/10.1038/nclimate1979
Hanigan, I. C., Butler, C. D., Kokic, P. N., & Hutchinson, M. F. (2012). Suicide and drought in New South Wales, Australia, 1970–2007. *Proceedings of the National Academy of Sciences, 109*(35), 13950–13955.
Hickel, J. (2020). *Less is more: How degrowth will save the world*. Random House.
IPCC. (2022). *Climate change 2022: Impacts, adaptation and vulnerability*. Contribution of working group II to the sixth assessment report. Cambridge University Press.
Islam, M. M., & Shamsuddoha, M. (2017a). Socioeconomic consequences of climate-induced human displacement and migration in Bangladesh. *International Sociology, 32*(3), 277–298.
Islam, M. M., & Shamsuddoha, M. (2017b). Socioeconomic consequences of climate-induced migration in Bangladesh. *International Sociology, 32*(3), 277–298.
Kahneman, D., Diener, E., & Schwarz, N. (1999). *Well-being: The foundations of hedonic psychology*. Russell Sage Foundation.
Knutson, T., Camargo, S. J., Chan, J. C. L., Emanuel, K., Ho, C.-H., Kossin, J., Mohapatra, M., Satoh, M., Sugi, M., Walsh, K., & Wu, L. (2020). Tropical cyclones and climate change assessment: Part II. Projected response to anthropogenic warming. *Bulletin of the American Meteorological Society, 101*(3), E303–E322. https://doi.org/10.1175/BAMS-D-18-0194.1
Kjellstrom, T., Briggs, D., Freyberg, C., Lemke, B., Otto, M., & Hyatt, O. (2016). Heat, human performance, and occupational health: A key issue for the assessment of global climate change impacts. *Annual Review of Public Health, 37*, 97–112. https://doi.org/10.1146/annurev-publhealth-032315-021740
Kurth, C., & Pihkala, P. (2022). Eco-anxiety: What it is and why it matters. *Frontiers in Psychology, 13*, Article 981814.
Lamb, W. F., & Steinberger, J. K. (2017). Human well-being and climate change mitigation. *Wiley Interdisciplinary Reviews: Climate Change, 8*(6), Article e485.
Léger-Goodes, T., Malboeuf-Hurtubise, C., Mastine, T., Généreux, M., Paradis, P. O., & Camden, C. (2022). Eco-anxiety in children: A scoping review of the mental health impacts of the awareness of climate change. *Frontiers in Psychology, 13*, Article 872544.
Leiserowitz, A., Maibach, E., Rosenthal, S., Kotcher, J., Ballew, M., Goldberg, M., & Gustafson, A. (2018). Climate change in the American mind: December 2018. *Yale University and George Mason University. New Haven, CT: Yale Program on Climate Change Communication, 48.*
Li, H., Chen, Y., & Ma, M. (2024). Temperature and life satisfaction: Evidence from Chinese older adults. *Ecological Economics, 225*, Article 108342.
Lobell, D. B., Schlenker, W., & Costa-Roberts, J. (2011). Climate trends and global crop production since 1980. *Science, 333*(6042), 616–620. https://doi.org/10.1126/science.1204531
Max-Neef, M. (1991). *Human-scale development: Conception, application and further reflection*. Apex Press.
McLeman, R. A. (2018). Migration and displacement risks due to mean sea-level rise. *Bulletin of the Atomic Scientists, 74*(3), 148–154.

Middleton, N., & Kang, U. (2017). Sand and dust storms: Impact mitigation. *Sustainability, 9*(6), 1053.

Mishra, A., Malik, A., Singh, R. P., & Sarkar, A. (2021). Climate change and farmer suicides in India: A study of 14 states. *Environmental Research, 196*, Article 110937.

Morganstein, J. C., & Ursano, R. J. (2020). Ecological disasters and mental health: Causes, consequences, and interventions. *Frontiers in Psychiatry, 11*, 1.

Nelson, G. C., Valin, H., Sands, R. D., Havlík, P., Ahammad, H., Deryng, D., Elliott, J., Fujimori, S., Hasegawa, T., Heyhoe, E., Kyle, P., von Lampe, M., Lotze-Campen, H., Mason-D'Croz, D., van Meijl, H., van der Mensbrugghe, D., Müller, C., Popp, A., Robertson, R., Robinson, S., Schmid, E., Schmitz, C., Tabeau, A., & Willenbockel, D. (2014). Climate change effects on agriculture: Economic responses to biophysical shocks. *Proceedings of the National Academy of Sciences, 111*(9), 3274–3279. https://doi.org/10.1073/pnas.1222465110

Nicholls, R. J., & Cazenave, A. (2010). Sea-level rise and its impact on coastal zones. *Science, 328*(5985), 1517–1520. https://doi.org/10.1126/science.1185782

Nussbaum, M. (2011). Creating capabilities: The human development approach. Harvard University Press.

Nussbaum, M. (2003). Capabilities as fundamental entitlements: Sen and social justice. *Feminist Economics, 9*, 33–59.

Obradovich, N., Migliorini, R., Paulus, M. P., & Rahwan, I. (2018). Empirical evidence of mental health risks posed by climate change. *Proceedings of the National Academy of Sciences, 115*(43), 10953–10958.

OECD. (2013). *How's life? 2013: Measuring well-being*. OECD Publishing. https://doi.org/10.1787/9789264201392-en

Prinzing, M., Lades, L. K., Weber, T. O., Fredrickson, B., & Laffan, K. (2024). Pro-environmental behaviors and well-being in everyday life. *Journal of Environmental Psychology, 98*, Article 102394.

Priyanto, M. W., Wijayanti, D. E., & Hilmi, Y. S. (2024). CO_2 emissions and subjective well-being: Empirical evidence from The Southeast Asian population. *Jurnal Ekonomi Pembangunan: Kajian Masalah Ekonomi Dan Pembangunan, 25*(1), 32–47.

Rahman, M. A., Alam, K., & Rolfe, J. (2022a). Climate change, adaptation, and subjective well-being. *Ecological Economics, 191*, Article 107235.

Rahman, M. S., Andriatmoko, N. D., Saeri, M., Subagio, H., Malik, A., Triastono, J., Oelviani, R., Kilmanun, J. C., da Silva, H., Pesireron, M., & Senewe, R. E. (2022b). Climate disasters and subjective well-being among urban and rural residents in Indonesia. *Sustainability, 14*(6), 3383.

Rahman, M. S., Huang, W. C., Toiba, H., Putritamara, J. A., Nugroho, T. W., & Saeri, M. (2023). Climate change adaptation and fishers subjective well-being in Indonesia: Is there a link? *Regional Studies in Marine Science, 63*, Article 103030.

Raleigh, C., Jordan, L., & Salehyan, I. (2015). Assessing the impact of climate change on migration and conflict. *Social Dimensions of Climate Change*, 37–51.

Reiter, P. (2008). Climate change and mosquito-borne disease: Knowing the horse before hitching the cart. *Revue Scientifique et Technique de l OIE, 27*(2), 383–398.

Rocklöv, J., & Dubrow, R. (2020). Climate change: An enduring challenge for vector-borne disease prevention and control. *Nature Immunology, 21*, 479–483.

Ryan, R. M., & Deci, E. L. (2001). On happiness and human potentials: A review of research on hedonic and eudaimonic well-being. *Annual Review of Psychology, 52*, 141–166.

Ryan, R. M., Huta, V., & Deci, E. L. (2008). Living well: A self-determination theory perspective on eudaimonia. *Journal of Happiness Studies, 9*, 139–170.

Sen, A. (1990). Development as capability expansion. In K. Griffin & J. Knight (Eds.), *Human development and the international development strategy for the 1990s* (pp. 41–58). Macmillan.

Sen, A. (1999). *Development as freedom*. Oxford University Press.

Suárez-Varela, M., Guardiola, J., & González-Gómez, F. (2016). Do pro-environmental behaviors and awareness contribute to improve subjective well-being? *Applied Research in Quality of Life, 11*, 429–444.

Thapa, B., Nepal, S. K., & Rahman, A. (2021). Climate adaptation and well-being in rural Nepal. *Climate and Development, 13*(3), 285–296.

Theisen, O. M., Holtermann, H., & Buhaug, H. (2011). Climate wars? Assessing the claim that drought breeds conflict. *International Security, 36*(3), 79–106. https://doi.org/10.1162/ISEC_a_00065

Torres, J. M., & Casey, J. A. (2017). The centrality of social ties to climate migration and mental health. *BMC Public Health, 17*(1), 600.

Tov, W., & Diener, E. (2013). *Subjective well-being.*

Tupou, T., Tiatia-Siau, J., Newport, C., Langridge, F., & Tiatia, S. (2023). Is the concept of solastalgia meaningful to pacific communities experiencing mental health distress due to climate change? An initial exploration. *International Journal of Environmental Research and Public Health, 20*(22), 7041.

United Nations Development Programme (UNDP). (2020). *Human development report 2020: The next Frontier—human development and the Anthropocene.* New York: UNDP. https://hdr.undp.org/content/human-development-report-2020

Vörösmarty, C. J., McIntyre, P. B., Gessner, M. O., Dudgeon, D., Prusevich, A., Green, P., Glidden, S., Bunn, S. E., Sullivan, C. A., Reidy Liermann, C., & Davies, P. M. (2010). Global threats to human water security and river biodiversity. *Nature, 467*(7315), 555–561. https://doi.org/10.1038/nature09440

Xu, M., Sbihi, H., Pan, X. & Brauer, M. (2020). Modifiers of the effect of short-term variation in PM2. 5 on mortality in Beijing, China. *Environmental Research, 183*, 109066.

Yang, J., Zhou, M., Ren, Z., Li, M., Wang, B., Liu, D. L., Ou, C. Q., Yin, P., Sun, J., Tong, S., & Wang, H. (2021). Projecting heat-related excess mortality under climate change scenarios in China. *Nature Communications, 12*(1), 1039.

Zhao, Y., Dijst, M., de Jong, M., & Stead, D. (2016). The influence of the physical environment on mental health: A review of the evidence. *Cities, 55*, 131–143.

Chapter 2
Addressing Climate Anxiety in India: Role of Governmental and Non-governmental Organizations

Sneha Sindhuja, Anjali Majumdar, and Santhosh Kareepadath Rajan

Abstract The increasing environmental challenges have posed significant risks to mental health. Climate anxiety is an emerging mental health concern in India, fuelled by the increasing frequency and severity of climate-related events such as floods, droughts, air pollution, etc. This chapter seeks to explore the role of governmental and non-governmental organizations in mitigating the mental health consequences of climate anxiety in India, focusing on understanding their approaches, effectiveness, and suggesting areas for improvement. The primary objective is to evaluate the initiatives introduced by these organizations to address and alleviate the psychological impact of climate change and analyze their policies, programs, and interventions aimed at raising awareness, providing emotional support, and building resilience in affected communities. The chapter adopts an Ecopsychological perspective, and incorporates the Power Threat Meaning Framework (PTMF). An Interpretive Policy Analysis (IPA) is conducted, relying on secondary data, followed by a comparative analysis to examine the strategies of key governmental agencies and prominent non-governmental organizations. The findings reveal significant gaps and challenges in policy, practice, and coordination; both sectors face challenges such as limited funding, political constraints, and insufficient integration of mental health support within climate policies. Implications of the study include recommendations for policies for better mental health support systems and culturally sensitive interventions to reduce climate anxiety, which is critical for India's long-term resilience to climate change.

Keywords Climate anxiety · Climate change · Governmental and non-governmental organizations · India

S. Sindhuja (✉) · A. Majumdar · S. K. Rajan
School of Psychological Sciences, Christ University, Bengaluru, India
e-mail: sneha.sindhuja@res.christuniversity.in

A. Majumdar
e-mail: anjali.majumdar@christuniversity.in

S. K. Rajan
e-mail: santhosh.kr@christuniversity.in

P. Datta et al. (eds.), *Climate Change and Human Well-being*, Advances in Global Change Research 83, https://doi.org/10.1007/978-3-032-25479-5_2

2.1 Introduction

The Earth's climate and overall health have been disrupted and declining rapidly over the past five decades, and scientists have systematically assessed human activities as the primary force behind this change. US National Climate Assessment implied that the "Earth's climate is now changing faster than at any point in the history of modern civilization, primarily due to human activities" (Jay et al., 2018, p. 34). Climate change refers to the long-term variations in the Earth's temperature and weather patterns. These changes are attributed to either natural shifts in the sun's activity and major volcanic eruptions or to human activities such as burning fossil fuels like coal, gases, and oil, which cause increased emissions of harmful greenhouse gases. The significant consequences of climate change include calamities like floods and droughts, melting ice caps and rising sea levels, wildfires, and biodiversity loss.

Climate change can affect our health, food security, housing and safety, and employment stability. These increasing challenges have posed significant risks to our mental health. Climate anxiety is a chronic fear or emotional distress caused by the growing awareness of deteriorating environmental health and the increasing consequences of climate change. This is particularly prevalent among the youth, environmental and climate scientists, those living in areas impacted by climate change, and children (Clayton, 2020). Women showed statistically significant elevated scores regarding pathological worry and pro-ecological views than men (Verplanken et al., 2020). A survey by Hickman et al. (2021) of 10,000 young people aged 16–25 from 10 countries showed that 59% were strongly worried about climate change, with more than 45% experiencing daily challenges because of climate change concerns. A large segment of 75% declared that the future appeared frightening, while 56% suggested that humanity was destined for doom.

The American Psychological Association (APA) studied the rising psychological impact of climate change, which causes stress, anxiety, and depression primarily in people who face direct exposure to climate disasters (Clayton et al., 2017). Climate-related distress continues to rise globally, according to the Lancet Countdown Report on Health and Climate Change, since it emerges from direct experiences like extreme weather events and indirect effects of future uncertainty (Watts et al., 2021). Clinical Ecopsychology attempts to conduct a systematic study of the direct and indirect consequences to mental health because of the ongoing changes in the climate, pollution, decay of the environment, deterioration of the air, soil, and water bodies, as well as the ecosystems (Thoma et al., 2021). Yet, climate anxiety is not considered pathological. The mental health community regards climate anxiety as a significant public health concern despite its lack of diagnosis as a clinical disorder, because it does affect people at substantial levels. Many countries face major shortcomings between mental health requirements of the communities and the availability and accessibility of service offerings. The World Health Organization (WHO) developed a policy brief which illustrates connections between mental health and climate change while providing essential policies to tackle mental health effects of the environmental crisis. According to the WHO (2024), authorities advocate linking climate and mental health policies because vulnerable areas such as South Asia require immediate attention and action, as many social, environmental, and economic risk factors pose problems in mental health and psychosocial wellbeing.

Addressing and mitigating climate anxiety requires significant efforts and investments by both governmental and non-governmental organizations. As the climate crisis heightens, collective action to build resilience and resolve the physical and mental health impacts of climate change is crucial (Clayton et al., 2017). This chapter investigates the initiatives introduced by these organizations to address and alleviate the psychological impact of climate change. It also examines their policies, awareness programs, support measures, and community resilience strategies. Adopting qualitative research methods through an extensive literature review in combination with governmental reports and NGO case studies, the chapter also examines the strategies of key agencies, such as the Ministry of Environment, Forests and Climate Change (MoEFCC) and the Ministry of Health and Family Welfare (MoHFW), alongside major NGOs like Greenpeace India and TERI.

2.2 Understanding Climate Anxiety

The term eco-anxiety was coined by Glenn Albrecht, an Australian environmentalist and notable Professor of sustainability. Climate anxiety is used interchangeably with eco-anxiety, which is defined as "the chronic fear of environmental doom" (Clayton et al., 2017, p. 68). Albrecht (2011) explores mental health challenges linked to environmental issues, describing them as 'psychoterratic' effects that stem from negatively experienced and perceived changes in the environment. The term "psychoterratic" combines "psyche" (mind or soul) with "terra" (earth) and includes symptoms such as anxiety, grief, sadness, despair, solastalgia and place-based distress, depression, panic attacks, and sleep disturbances. He explains these conditions to help better understand and address the psychological consequences of ongoing environmental and climate change. According to Barnwell et al. (2020), various symptoms develop through multiple sociocultural influences, together with power relations and justice considerations. The growing number of social crises during this decade could potentially result in climate anxiety getting tangled with other forms of anxiety (Pihkala, 2020). Literature also suggests a correlation between climate anxiety and existential anxiety. Clayton (2020) stated that climate anxiety can lead to eco-paralysis, holding individuals back from taking necessary climate action.

The phenomenon of climate anxiety and the lack of satisfaction regarding governmental climate responses affect the daily lives of children and young people throughout different nations across the world. According to Hickman et al. (2021), when governments demonstrate inadequate action against the climate crisis, it results in higher levels of anxiety among the people. There is an urgent need to investigate the emotional effects of climate change among children and teenagers, along with a call for action for governments to immediately respond to the climate crisis.

Engaging in problem-focused coping leads to "pro-environmental behaviour" because those who pursue it allocate significant time to study climate change and seek personal solutions to address the issue (Schmidt, 2023). Although climate change exceeds individual control, people may become distressed because problem-focused coping reaches its personal limitations, thus preventing them from overcoming negative emotions. Ojala (2012) focuses on how young adults approach global environmental threats through their thought processes and emotional responses. Young people use emotion-focused coping, which entails denial and avoidance of climate change to escape emotions of anxiety about it. Effective treatment involves a multilayered approach that integrates psychological and emotional support, community engagement, and structural acknowledgment of the anthropogenic climate crisis.

2.3 Indian Climate Landscape: Stressors and Vulnerabilities

India is among the many South Asian countries that are highly vulnerable to the anticipated consequences of climate change. Tam et al. (2023) stated that climate anxiety is found to be higher in China and India than in Japan and America. Among these four countries, India demonstrated not only the highest vulnerability to climate change but also lacked readiness for climate action. The country faced around 1058 climate-related disasters, such as erratic rainfalls and floods, extreme heat and droughts, and cold waves, between 1995 and 2020, which has been linked to the higher rates of suicides among the affected farmers, i.e., around 59,000 suicides in the last three decades (Madge, 2024). Climate change directly affects farmers and hampers the functioning of the Indian agriculture sector (Datta et al., 2022). This leads to the large-scale migration of farmers as an adaptation strategy to help ease the climate shocks (Jha et al., 2018). In rural areas, the agrarian distress is worsening due to erratic rainfall and rapidly changing monsoon patterns, resulting in with devastating floods in states like Kerala and Assam. Although both urban and rural populations tend to be affected by the changing environmental scenario, Menon (2024) notes that rural areas tend to hide the issue behind bigger issues that mainly involve economic problems and social caste and class groups. Godsil (1991) defines environmental racism as systemic and intentional or unintentional policies, practices, and decisions that disproportionately expose marginalized communities, especially those based on race, caste, ethnicity, or socioeconomic status, to environmental hazards. For instance, in India, the Adivasi communities are frequently displaced and forced to migrate due to dam and mining projects. Similarly, the Dalit communities of lower socioeconomic status are more likely to be settled near waste dumps or polluted water bodies. This oppression can have a significant impact on the physical, mental, economic, and environmental well-being.

Climate change also aggravates rising sea levels and coastal vulnerabilities in India (Sundaresan et al., 2013). It affects the country's coastlines and low-lying island territories. Mumbai, Chennai, and several cities of Orissa have become highly vulnerable to cyclones and tidal flooding, which leads to the saline intrusion into the freshwater systems. India's fragile island ecosystems, such as Lakshadweep and the Andaman and Nicobar Islands, are at high risk of coastal erosion, land submergence, and loss of biodiversity. Climate change impacts the mangrove ecosystems, coral reefs, marine fisheries, livestock, and agriculture (Srivastava, 2012). These islands also face the risk of long-term displacement if the coastal ecosystems collapse, affecting the livelihoods of the locals and the vast fishing communities living along the shores. Place attachment refers to the psychological, emotional, and cultural bond that people form with their surroundings (Lewicka, 2010). The threats of displacement from the ancestral lands can cause climate anxiety among these vulnerable communities whose identities are closely tied to the land and the sea.

A UNICEF report shows that Indian children stand in position 26 among all children worldwide who face high risk from climate change effects. Climate-induced rise in temperature along with irregular monsoon patterns and flooding and polluted air result in the poor health-related outcomes in children, such as respiratory illnesses, heat-related stress, water-borne diseases, and malnutrition. Climate change can cause long-term household food security challenges and nutritional deficiencies in young children (Muttarak & Dimitrova, 2019). Patra (2016), conducted a study to see if climate shocks affect childhood malnourishment in the India. Climate shocks have a direct impact on malnutrition, especially in the climate vulnerable regions like the Sundarbans. Children are disproportionately affected, not only physically, but also mentally. Experiences of climate-related disasters can lead to anxiety, trauma, and a sense of helplessness. Despites these enormous challenges, children remain largely absent from climate and mental health policy frameworks in India. Councils, such as the 28th Conference of Parties under the United Nations Framework Convention on Climate Change and the World Health Organization, have established various forums to discuss the psychological effects of environmental issues, which include grief and stress alongside trauma (Venkataraman & Aggarwal, 2024). The safety of Indian children depends on climate policy, which should remain sensitive toward children by providing protective measures, continuous education, and health services during climate adaptation and disaster risk reduction.

2.4 Theoretical Framework

This chapter adopts an Ecopsychological perspective in understanding the impact of climate change on mental welfare in the Indian context. "Ecopsychology explores humans' psychological interdependence with the rest of nature and the implications for identity, health, and well-being. Topics within ecopsychology include emotional responses to nature; the impacts of environmental issues such as natural disasters and global climate change; and the transpersonal dimensions of environmental identity and concern" (Roszak et al., 1996, *Ecopsychology* section, para. 10). Most societies in India, especially Indigenous groups, rural farmers, and the coastal population, obtain not only their livelihoods but also their cultural identity and spiritual value out of the natural environment (Mishra et al., 2009). As floods, droughts, or deforestation in the ancestral lands force people to leave their origin, it is not merely the material that is lost.

Ecopsychology also challenges the traditional climate policies, as it states that healthy psychological well-being and healthy environmental conditions are two components that are inextricably linked. By emphasizing physical infrastructure or economic recovery, policies may overlook the psyche and culture around the issue of climate impact. In the case of governmental and non-governmental organizations, this implies that the resilience programmes need to incorporate ecological restoration and psychosocial support. These might include post-disaster rescue and recovery efforts such as mangrove replantation, also enhancing mental health with the community through workshops to learn the background of the climate change and the way to deal with it; school programs teaching both the way of climate and how to cope with it.

The chapter also draws from the Power Threat Meaning Framework (PTMF), developed collectively by a group of psychologists, psychiatric survivors, and activists. According to Read and Harper (2020), the PTMF was a reaction to the persistent dominance of a reductionist biomedical approach to psychological distress and distressing conduct. This framework focuses on how climate anxiety is framed as a socio-environmental response to the perceived existential threat, mediated through the structures of power, inequality, and a sense of helplessness. It emphasizes the role of power in significantly shaping one's interpersonal, social, economic, as well as ideological experiences. Therefore, this lens enables the interpretation of personal and group emotional reactions to be considered in a wider sociopolitical framework and provides implications as to how institutions can aid resilience.

Barnwell et al. (2020) attempted to gain insight into the PTMF in the context of climate crises through a case study in Rustenburg, South Africa. The authors argue that this framework offers a more ethical and socially informed view of the issues related to climate-related distress. Therefore, PTMF can recast climate anxiety in larger social and political terms, rather than as an individual psychopathology. Addressing power, structural injustice, and communal processes of meaning-making as central, the framework provides a de-pathologizing framework of power, context, and meaning-making that becomes especially applicable to the Global South.

2.5 Methodological Framework

This chapter employs interpretive policy analysis through qualitative examinations of governmental and non-governmental organizations deal with climate anxiety in India. Interpretive Policy Analysis (IPA) is a research methodology that emphasizes the subjective meanings and experiences individuals and groups attach to policies and their processes. The approach examines policies beyond the rational-positivist thinking and integrates language aspects together with emotional and historical dimensions. IPA proves to be effective for analyzing complicated matters like climate anxiety, since it properly recognizes the significant role emotions, context, and ethics. IPA investigates the barriers to resolving various problems faced by the communities.

The analysis relies on secondary data from 2010 to 2024, consisting of policy documents, reports, case studies, and academic literature on climate change, mental health, policy and decision-making, and environmental psychology in the Indian context to understand the psychological and emotional effects of response strategies of key stakeholders. The selection of documents was made through their direct relevance to the study of climate change and mental health in India, as well as the preference towards those that describe possible concrete interventions, mechanisms of the involved institutions, or policy priorities, in particular.

A comparative matrix was created to provide a comparison of the approaches of governmental bodies and NGOs in a systematic way. This was a structured tool that was used to organise information available in different sources in a customised way. It was organized as follows into four major parameters, such as;

(a) Inclusivity, i.e., the degree to which programs attract marginalized and climate-stressed populations like farmers, Indigenous groups, women, and young people.
(b) Responsiveness, i.e., the pace and effectiveness of the response by institutions to each climate-related mental health need, including during and after events of extreme weather.
(c) Policy Depth, i.e., the thoroughness of policy documents concerning climate anxiety, and the extent to which mental health has been included in policy responses to problems of climate anxiety in general.
(d) Psychosocial Support, i.e., availability of systematic interventions (such as counselling programs, community-based emotional support, and culturally appropriate treatment practices).

2.6 Addressing Climate Anxiety: Roles of Key Organizations

Both the Governmental and Non-Governmental Organisations have addressed the psychological distress caused by the growing awareness of the anthropogenic climate change. They have complemented various efforts in order to mitigate climate anxiety in their own distinctive ways.

2.6.1 Role of Governmental Organizations

India has launched various important measures to fight climate change. The National Action Plan for Climate Change & Human Health (NAPCCHH) outlines the steps that India took to address climate change. First, MoEFCC was created as the lead agency, and then the 2006 National Environmental Policy document was created while forming the Prime Minister's Council on Climate Change to coordinate climate-related matters. The MoEFCC launched the National Action Plan on Climate Change (NAPCC) with eight initial missions until four additional ones were introduced. New health-related and four other missions were added to the original eight missions in the National Action Plan on Climate Change. The Ministry of Health and Family Welfare (MoHFW) formulated the National Expert Group on Climate Change and Health (NEGCCH) chaired by Dr. Vishwa Mohan Katoch, since he served as former Secretary of Health Research and Director General of the Indian Council of Medical Research (ICMR). This team received instructions to establish action plans and present advice about monitoring indicators and developing prevention strategies emphasizing building capabilities. However, these initiatives are implemented to mitigate climate-sensitive illnesses such as vector-borne diseases, food and water-borne diseases, cardio-respiratory illnesses, zoonotic diseases, renal diseases, and those caused by nutritional deficiencies. NAPCCHH loosely talks about non-communicable diseases (NCDs) and mental health disorders, which have been linked to climate variability, environmental pollutants, and occupational hazards, and just suggests the implementation of national health programs to control and reduce NCDs and mental illnesses.

The Ministry of Environment and Forest (MoEF) was established in the year 1980, and it was only in May 2014 that the government conceived the need to address and act in response to the Anthropocene climate change. Therefore, MoEFCC focuses on sustainable development and improving well-being, and it also works towards attaining environmental protection and conservation of forests and wildlife. This ministry is particularly responsible for planning, promoting, coordinating, and overseeing the implementation of environmental and forestry policies and programs. It also coordinates India's strategies on climate change, both domestically and internationally, at forums such as the United Nations Framework Convention on Climate Change (UNFCCC) and the Paris Agreement. The National Green Corps (NGC) is a national program initiated in 2001 by the MoEFCC to sensitize school children about the environment, its problems, and conservation. The aim of NGC is to provide environmental education, create awareness of local environmental issues, and encourage environmental decision-making to take relevant actions. Though MoEFCC represents the country in climate negotiations and articulates the country's stand on SDGs and climate finance, it gives the least focus on the mental health impacts of climate change.

2.6.2 Role of Non-governmental Organizations

Several prominent organizations in India have participated in addressing climate change and its psychosocial consequences and are actively taking action to mitigate it. These organizations bridge the gap between policy and practice, thereby offering emotionally responsive and culturally relevant interventions at the grassroots levels. They promote environmental education and awareness, support mental health and advocacy for those affected by climate-related disasters, and initiate community mobilization.

Indian Youth Climate Network (IYCN) was formed in 2008 and aims to raise awareness of climate change and its impact. It also advocates for policy changes, supports the SDGs, and encourages the reduction of the emission of harmful greenhouse gases. IYCN also engages in youth-based campaigns like Greenpeace India and integrates climate awareness and emotional resilience. They work directly with local communities to address their needs and empower them to take action on environmental issues for a sustainable future. The Energy and Resources Institute (TERI) operates with a multidisciplinary lens, integrating scientific research with education and outreach. TERI supports climate adaptation strategies, but its focus is policy, tech, and institutional readiness, not psychological impacts like climate anxiety. However, TERI initiated the GREEN Olympiad in 1999 to sensitize school children to environmental sustainability and the development of green skills. The Centre for Science and Environment (CSE), founded in 1980, supports sustainable agriculture and water management and promotes practices to improve resilience in climate-vulnerable communities. CSE, though influential in informing policy and educating the public, does not address climate-related mental health impacts or emotional responses to climate change.

These organizations are intensely involved in climate initiatives that provide awareness and adaptation strategies for climate issues throughout India. However, they do not actively address the issue of climate anxiety and mental health effects, despite these problems representing substantial gaps in the country's climate strategy.

2.7 Comparing Governmental Versus Non-governmental Approaches

Addressing climate anxiety in India requires a comprehensive, multi-sectoral approach involving governmental and non-governmental organizations, even though they both function under different conceptual systems. A comparative analysis helps contrast the different dimensions of their strategies in relation to their structural orientation, engagement methodology, and collaborative dynamics.

2.7.1 Policy Versus Practice Orientation

All government programs base their operations on policy-based systems. The Government of India implements two national programs, including NAPCC and NMHP, which unite environmental sustainability with public health objectives

(Government of India, 2008; Ministry of Health and Family Welfare, 2018). The policies are designed to establish primary objectives and use sector-wide resources to fulfil their goals. Climate change-related mental health effects, alongside anxiety regarding climate change, are frequently left out of consideration when examining these frameworks. The Indian climate policy fails to recognize mental health as a fundamental part of its structure, instead of treating it as an insignificant secondary matter (Singh et al., 2020).

NGOs are usually defined by their field-based practical approaches. Through dedicated interventions, Greenpeace India and The Energy and Resources Institute (TERI) work with affected communities in the field. TERI, together with Greenpeace India, implements community-sharing spaces and educational initiatives that promote environmental education, which can possibly help with emotional skills. Moreover, NGOs often pioneer innovative psychosocial methodologies better than governmental organizations due to their close participation at the grassroots level.

2.7.2 Top-Down Versus Bottom-Up Engagement

A top-down approach refers to initiatives and policies designed and implemented by central authorities such as governments, national agencies, or international organizations, where the decision-making flows from higher levels to local or community levels. Whereas, a bottom-up approach emerges from the local level, where communities, grassroots organizations, or individuals actively participate in identifying problems and implementing solutions. Most governmental responses follow vertical structures. The State Action Plans on Climate Change (SAPCCs) move downward from central ministries through various levels of government to support uniformity across wide areas. The standardized policymaking process protects policy consistency but fails to address the various situations that differ across India's communities. Groups at the lower end of adaptive capacities, including tribal populations and agricultural and urban slum labourers, feel their specific psychological needs are not properly dealt with (Jha et al., 2017).

The engagement strategies of NGOs involve working from the ground level up. Their approach to implementing active community involvement helps them develop programs that align with local needs. Through these efforts, communities enhance their psychological resilience, which remains crucial for communities living on the frontlines of climate change.

2.7.3 Areas of Collaboration and Conflict

Numerous experts now understand that effective collaboration between different sectors represents a key requirement to resolve the effects of climate change on mental health. Through partnerships between government departments and NGOs that develop joint awareness campaigns for mental health outreach in schools, the combined effect of their collaboration results in increased impact (UNICEF India, 2021). States such as Maharashtra and Kerala have implemented successful pilot projects that demonstrate the potential to unite government resources with NGO creative approaches for strengthening local mental health resilience (Varughese & Purushothaman, 2021). Public–Private Partnerships (PPPs) are a collaborative and effective strategy in minimising the consequences of climate-led disasters (Auzzir et al., 2014). In Kerala, collaborative flood relief involved mental health camps that were supported by NGOs as well as state health services.

Such collaborative efforts between organizations encounter multiple obstacles that impede their success. NGO efforts to critique governmental economic growth initiatives meeting environmental justice standards will produce ideological differences, especially when they criticize infrastructure projects that disrupt ecosystems (Fisher, 2014). The autonomy of NGOs is threatened when they resist partnerships involving bureaucratic controls or policy goals that do not match ground realities (Table 2.1).

Although the government has an institutional and national policy framework, the integration of mental health in its climate action is poor. Conversely, NGOs are also closer to the affected communities; however, they lack the funding and resources for support. The inter-departmental coordination within the government (ranging across the MoHFW to MoEFCC) is poor, and the collaboration (typically formal and funded poorly or stricken with conflicts) with NGOs is weak. There is a lack of a formal inter-agency platform to merge psychosocial support in climate strategies.

Table 2.1 Compares NGOs and the government across key dimensions

Dimension	Government agencies	NGOs
Main focus	Policy, disaster relief, and physical health	Education, awareness, and youth empowerment
Mental health action	Indirect via NAPCCHH (not climate-specific)	Not specifically focused on climate anxiety
Approach	Top down: centralized decisions	Bottom up: local engagement
Collaboration	Weak cross-ministry links (MoEFCC, MoHFW)	Work with schools, grassroots groups
Challenges	Bureaucratic delays, lack of funding	Funding issues, dependency on donations

2.8 Structural and Systemic Barriers to Addressing Climate Anxiety in India

Addressing climate anxiety in India is marked by psychological, institutional, cultural, and socioeconomic challenges, and both governmental and non-governmental organizations face distinct and overlapping challenges.

2.8.1 Lack of Institutional Recognition and Policy Integration

India increasingly recognizes climate anxiety, especially among youth and vulnerable communities. However, it lacks nationwide response and is not a mainstream policy priority. India's climate policy primarily focuses on the mitigation and adaptation strategies and often overlooks the mental health impacts of climate change. It emphasises improving the physical infrastructure, energy transitions, and disaster management and resilience. Despite the empirical evidence, the integration of mental health policies remains unimportant. Despite the pronouncement of major policies such as the National Mental Health Policy, 2014, and the Mental Healthcare Act, 2017, India allocates less than 1% of its total health budget to mental health, leading to a significant treatment gap (Shastri & Mahashur, 2023). Fragmented governance and a lack of proper categorization hamper the recommendation of mental health strategies. Mental health policies come under the Ministry of Health and Family Welfare (MoHFW), whereas the climate-related policies come under the Ministry of Environment, Forests, and Climate Change (MoEFCC). This lack of inter-ministerial coordination depicts that the mental health challenges are neither systematically mapped nor strategically addressed, and thus, require an integrated policy approach. To meaningfully address climate anxiety in the country, the existing policy frameworks must transcend beyond the conventional boundaries of environment and health.

2.8.2 Socioeconomic Inequality and Priority Gap

Socioeconomic inequalities limit the willingness of the communities to recognise, recount, and respond to the climate crisis. Climate anxiety is often dismissed as an overreaction, overthinking, and is often overshadowed by the concerns of poverty and other basic survival needs (Mondal et al., 2024). In India, large sections of the population have limited access to basic services such as healthcare, education, clean water, nutritious food, and housing. These inequalities tend to create a significant "priority gap" in policy-making, giving importance to the immediate social and economic concerns rather than focusing on the long-term environmental strategies. The effects of climate disasters on communities from a lower socioeconomic background are often framed as an economic or physical vulnerability (Barbier & Hochard, 2017).

The availability of mental health care services in India is both unbalanced and predominantly focused on urban locations. The population of rural areas, together with economically disadvantaged groups, misses out on professional mental health care services as well as community-based mental health programs. The lack of accessible treatment would worsen the treatment gap while blocking the creation of specific interventions that address both psychological and economic factors related to climate risks. At the policy level, the adaptation strategies and disaster response plans fail to incorporate climate-related mental health concerns in their frameworks. They often prioritize physical infrastructure and economic recovery and pay little or no attention to the psychological and emotional impacts of climate-related events. NGOs, especially grassroots ones, often struggle with limited funding, which hinders their ability to initiate and sustain their programs. Addressing the intricate challenges of climate change requires a holistic approach that includes socioeconomic vulnerabilities and promotes social equity. It lacks interdisciplinary encouragement and collaboration between various stakeholders outside policymaking. Local communities, those affected and prone to the effects of climate change, should be included in the decision-making process.

2.8.3 Cultural Stigma and Challenges in Communication

Mental health remains highly stigmatized in India, and climate anxiety is a relatively new term that is poorly understood and is often labelled as a "privilege-driven concern" that stems from the advantages, along with distinctive rights, which individuals obtain because of their social position or financial resources. Both GOs and NGOs find it difficult to navigate the strong social stigma related to mental help-seeking behaviours. One of the cultural barriers is the religious or fatalistic worldview that is significantly prevalent in India. A notable segment of the population believes environmental disasters to be the manifestations of the divine will or the karmic force (Levy et al., 2009). This perception of the effects of climate change can make people feel powerless as they view the consequences as events predetermined and beyond their control.

India is a highly multilingual country with over 121 languages and 270 mother tongues, and is home to more than 2000 ethnic communities. Many of these languages lack specific terms to express modern psychological concepts like climate anxiety, and this lack of vocabulary makes it difficult for people to understand or articulate their emotional distress related to climate change. In the absence of culturally appropriate terms and climate literacy, it becomes challenging to communicate and raise awareness on the current landscape of anthropogenic climate change and its anticipated psychological consequences. Effective climate communication is often hindered by cultural and linguistic barriers. Climate communication is a field within science communication that focuses on educating the causes and consequences of anthropogenic climate change and suggests significant actions to mitigate the effects (Cox, 2010). In India, climate communication strategies overlook the need to fuse local languages and continue using a universal language without linguistic and cultural adaptation. Hence, it fails to engage the target audience and perpetuate meaningful action.

2.8.4 Climate Denialism and Political Barriers

Hoofnagle and Hoofnagle (2007) defined denialism as "the employment of rhetorical tactics to give the appearance of argument or legitimate debate, when in actuality there is none". Climate denialism refers to the rejection of scientific facts and the refusal to acknowledge anthropogenic climate change. This dismissal can significantly hinder the efforts taken to mitigate the consequences of climate change. The denial of climate science implies there are no climate-related issues at hand. Members of the public, along with certain media outlets, choose to accept denialist analyses, which propels them into becoming skeptics of climate change (Capstick & Pidgeon, 2014), and hence, denialism becomes an arm against the climate policies. Unlike the western countries where climate denial is often manifested through the rejection of scientific facts, in India, it operates on a more subtle level, or say "soft denial", through minimization and deflection. Political and industrial stakeholders in India often acknowledge climate change but ignore the urgency to act or just redirect the blame towards global inequities (Mishra, 2024). Therefore, this subtle form of denialism masks the increasing risks and consequences, and limits public discussions on the emotional and psychological dimensions of the climate crisis.

While India officially recognizes climate change and participates in major international pacts like the Paris Agreement, the domestic political narratives often downplay the severity of the environmental threats and simplify the problem to developmental and energy priorities (Lakshmanan et al., 2017). Political leaders and policy officials usually select pressing economic growth needs instead of sustainability goals for the environment, thus creating conflicts that ignore climate action and dismissing mental health impacts as a result of ecological disruptions. Environmental concerns in India frequently become controversial when politicians, economic groups, and ideological factions align their views according to regional, economic, and ideological standings. Certain narratives present environmental limitations as barriers that prevent national advancement and poverty elimination by construing climate activism as either an upper-class movement or international board control (G et al., 2023). Hence, the politicization process makes it harder for the government to agree on developing a socially supportive climate and mental health policies together.

2.9 Toward a Holistic Framework for Addressing Climate Anxiety in India

The Multiple Needs Framework was originally developed by Russell Cropanzano and his colleagues, but was later adopted by Bingley et al. (2022) to understand the effectiveness of climate anxiety interventions, and therefore, applicable to several stakeholders, such as educators, community leaders, and environmental activists, who might be interested in alleviating climate change anxiety, as well as addressing other significant outcomes. The Multiple Needs Framework suggests that interventions to reduce climate change anxiety differ in their impacts on individual, social, and environmental outcomes (Bingley et al., 2022). It helps conceptualize the effectiveness of climate change anxiety intervention not only in terms of the effects on individual well-being, but also in a more thorough representation of the effects of this global issue. The framework targets individual needs, social needs, and environmental needs.

The individual needs focus on minimizing the emotional suffering (anxiety, grief, or solastalgia) produced by climate change through enhancing individual psychological well-being. It emphasizes on interventions that include managing people's emotions and being problem-oriented when coping with their emotions in the wake of incidents due to climate change. The social needs focus on moving beyond individual gains, such as building group cohesiveness, trust, and connectedness. It prioritizes the strengthening of social connections and participation in community or group activities, e.g., community-building initiatives. The environmental needs lay emphasis on systemic and structural change in coping with the climate crisis and motivating a pro-environmental conduct, focusing on interventions that not only help with coping with the effects of climate change but also encourage advocacy and action. The framework acknowledges that individual-level and community-strength actions are necessary, but alone are not enough to necessitate a change in character on a larger scale of policy.

A holistic approach to tackling climate anxiety remains essential because climate change causes challenges across both natural environments and the mental states of affected people. Mental health integration into climate policies remains essential because climate stress and anxiety, together with grief, develop substantial psychological effects that risk being disregarded by communities unless proper measures are implemented. Therefore, such a framework builds mental resilience through community-based coping while creating an environmentally and emotionally inclusive response that increases resilience to face climate change.

2.9.1 *Empirical Research and Policy Integration*

India needs more extensive scientific evidence regarding the psychological effects of climate change because ongoing global acknowledgment fails to support understanding of Indian-specific environmental and cultural factors. A significant lack of data exists to comprehend the frequency rates, as well as origins and stress management techniques of environmental anxiety among Indians (Shukla, 2016). Current investigations of climate anxiety stem from Western points of view, which fail to guide policy development in the Global South because environmental risks combine with poverty barriers, caste system divisions, and healthcare access problems. The wide-scale survey by Hickman et al. (2021) shows Indian adolescents express impressive levels of climate unease and distrust regarding government actions, but the study does not analyze these findings by geographic, economic, or environmental vulnerability dimensions, which distinguish India's various territories. Longitudinal interdisciplinary studies about the psychological impacts of climate change on different regions and social groups across India must become a top priority due to immediate research requirements. Government agencies, along with academic institutions, must use their funds for collecting data that would guide them to create appropriate interventions specific to regional cultures.

Policy integration is a critical component in addressing anthropogenic climate change and its consequences. The government needs to develop a formal collaboration between the Ministry of Health and Family Welfare (MoHFW) and the Ministry of Environment, Forests, and Climate Change (MoEFCC). This will serve to merge mental health together with climate adaptation and disaster risk reduction plans while developing cohesive policies that address eco-anxiety and linked mental health problems. Climate Anxiety needs to be added as a priority in India's countrywide mental healthcare initiatives. India must integrate climate stress modules in all National Mental Health Programme (NMHP) sessions. Mental health services should be integrated into state programs that deal with climate action. The State Action Plans on Climate Change (SAPCCs) need updates through the addition of psychosocial adaptive elements, including emergency mental health services that support residents before and after climate disasters. The developed plans need to treat mental health platforms as resources that defend against harm but also construct resilience.

The number of programs that merge local leadership with mental well-being practices alongside traditional ecological knowledge should expand across Indian society. Various programs, including climate clubs, art sessions, counseling groups, and rural and urban nature healing activities, reinforce group resistance while developing emotional endurance. Health authorities should acknowledge and approve climate anxiety as a matter of public health concern. When the Indian public health system recognizes climate anxiety, it will create a path to include it in medical diagnoses, public health messaging, and treatment plans. The recognition of climate anxiety would simultaneously lower prejudice from society while encouraging people to seek support.

2.9.2 Public Campaigns and Climate Education

The rising impact of climate change on Indian ecosystems and communities requires public campaigns and climate education for sustainable awareness, resilience development, and climate anxiety reduction. Through public education systems, people gain better environmental knowledge and develop psychological readiness through learning how to understand and emotionally handle environmental threats, according to Clayton (2020). The Indian situation with socio-economic inequalities coupled with geographic differences alongside substantial youth numbers makes integrated educational measures and public outreach about climate issues highly impactful. Both governmental and non-governmental organizations in India successfully continue to conduct public campaigns that raise environmental awareness. Reaching maximum effectiveness requires Indian climate education initiatives and public campaigns to develop strategies that respect cultural values and engage emotions while including community participation. The current marketing approach needs to be developed by adding dialogues about ecological grief together with Indigenous storytelling and professional mental healthcare.

Climate education must be added to the curriculum to address the emotional well-being in times of climate crisis. Educational boards like NCERT and state boards need to create teaching materials that equip students with emotional coping skills, climate literacy, and resilience-building practices. The School Health and Wellness Programme operated by the government stands to gain effectiveness through the implementation of climate emotional literacy.

2.9.3 Climate Cafes and Art-Based Interventions

As the climate crisis becomes more widespread, it is important to uncover better and creative strategies to act and adapt. Some of the innovative trends include the emergence of a grassroots movement called the climate cafes and art-based interventions to address the mental health impacts of climate change. Climate café is a relatively new concept, an open space which serves as a non-professional hospitable environment where individuals can share their psychological experiences with environmental destruction with others in the community (Calabria & Marks, 2024). Climate Cafés are open in many countries, including India, Africa, Nepal, and the USA, making the movement an active global force for local change (Consalvo, 2024). However, this concept exists in its initial stages in India but attracts growing interest among young city dwellers, students from universities, as well as activists. Climate Cafés act as beneficial venues to promote communication about these concerns. People can express climate-related fears as well as grief and hopes within informal group discussions called climate cafés, which stem from the 'Death Café' model and was developed by the Climate Psychology Alliance (Calabria & Marks, 2024). The emotional response to the climate crisis lets people form strategies together to create real-world solutions for their communities. Such grassroots practices with cultural grounding function as an encouraging additional approach to standard therapeutic methods in India because the country faces restricted access to formal mental healthcare, coupled with significant stigma (Majra & Gur, 2009). Grassroots climate pioneers express their views about support spaces, which serve as venues for connecting people to express their rising feelings of helplessness toward climate change and government inaction.

Art-based interventions have been gaining popularity in addressing the emotional distress caused due to ecological destruction. These have emerged as a meaningful approach to help people deal with climate anxiety, grief, and ecological loss. Art functions differently from traditional therapy since it enables people to communicate advanced emotions without words by utilizing symbolic visual expressions. Artistic expression through climate change contexts helps people express their fear, together with helplessness, as well as hope to build both social participation and emotional strength (Van Lith, 2024). The art-based interventions in India serve to promote environmental awareness and psychological well-being. Gond painting practitioners operating as artists and therapists demonstrate how Indigenous cultural art functions as a healing force that restores natural connections. The research by Dua et al., (2024 demonstrates that Gond art therapy integration enhances ecological consciousness through the preservation of therapeutic value and traditional environmental values. Such initiatives create a space for environmental storytelling and cultural expression for the affected tribal communities, and this would foster a sense of connectedness and security during a climate-related crisis. These interventions have great worth in settings where mental health resources remain scarce or culturally stigmatized. Collective programs that integrate art-based approaches may function as essential components for achieving both climate adaptation and psychological resilience in public health, education, and community activities, and therefore, the government must integrate these into its post-disaster care policies (Table 2.2).

2.10 Conclusion

Anthropogenic climate change is a pressing concern, the consequences of which have been under-addressed, especially in a country like India, where ecological vulnerability meets socioeconomic inequality, cultural diversity, and limited mental health infrastructure. This chapter examined the roles of governmental and non-governmental organizations in dealing with climate anxiety and reveals significant gaps in policy, practice, and coordination. India's environmental and climate policies, such as the National Action Plan on Climate Change, focus mainly on mitigation and adaptation but do not explicitly address mental health impacts. Despite rising risks, India's mental health infrastructure remains underdeveloped, especially in rural and disaster-prone regions. The lack of official recognition for climate anxiety, together with insufficient funding for mental health support, leaves multiple communities

Table 2.2 Building community resilience and coping with climate anxiety in India

Area of action	What to do	Who should do it
Policy integration	Include climate anxiety in the existing national mental health policies	MoHFW, MoEFCC
School education	Add emotional resilience training to the curriculum of schools in the climate-related disaster-prone areas	NCERT, state boards
Public campaigns	Use storytelling techniques in regional languages to reduce stigma and relate to the communities	NGOs, Doordarshan, social media
Community interventions	Set up climate cafés and art-based healing programs for the affected communities	NGOs, community colleges, local bodies
Health services	Train local healthcare workers in climate-informed counselling	MoHFW, state health departments, mental health institutions

without sufficient emotional care, particularly affecting youth groups, Indian farmers, indigenous communities, and coastal populations.

Future predictions indicate that climate anxiety will worsen for young people and the vulnerable population groups due to extreme heat worsened by changes in monsoons and increasing sea levels. Preliminary investigations show that a lack of proper psychosocial support and climate-specific mental health services will expose more than 50% of Indian youth to clinical levels of eco-anxiety by 2030. Addressing climate anxiety in India requires an integrated action, infusing mental health into policies regarding climate change adaptation as well as disaster response efforts and educational and public health structures. The approach should focus on improved interaction between ministries, together with backing local initiatives, while conducting studies that target specific geographical areas. Acknowledging climate anxiety goes beyond just addressing emotional distress. It involves building resilience, promoting emotional well-being, and ensuring that the country's climate governance is inclusive. A holistic strategy will equip the country in times of climate crisis and its anticipated psychological consequences.

2.11 Disclosure Statement

The authors declare that there is no conflict of interest.

2.12 Declaration

The authors declare that the content is not AI-generated.

References

Albrecht, G. (2011). Chronic environmental change: Emerging psychoterratic syndromes. In: Weissbecker, I. (eds) *Climate change and human well-being. International and cultural psychology.* Springer. https://doi.org/10.1007/978-1-4419-9742-5_3

Auzzir, Z. A., Haigh, R. P., & Amaratunga, D. (2014). Public-private partnerships (PPP) in disaster management in developing countries: A conceptual framework. *Procedia Economics and Finance, 18*, 807–814. https://doi.org/10.1016/s2212-5671(14)01006-5

Barbier, E. B., & Hochard, J. P. (2017). The impacts of climate change on the poor in disadvantaged regions. *Review of Environmental Economics and Policy, 12*(1), 26–47. https://doi.org/10.1093/reep/rex023

Barnwell, G., Stroud, L., & Watson, M. (2020). Critical reflections from South Africa: Using the power threat meaning framework to place climate-related distress in its socio-political context. *Clinical Psychology Forum, 1*(332), 7–15. https://doi.org/10.53841/bpscpf.2020.1.332.7

Bingley, W. J., Tran, A., Boyd, C. P., Gibson, K., Kalokerinos, E. K., Koval, P., Kashima, Y., McDonald, D., & Greenaway, K. H. (2022). A multiple needs framework for climate change anxiety interventions. *American Psychologist, 77*(7), 812–821. https://doi.org/10.1037/amp0001012

Calabria, L., & Marks, E. (2024). What happens in Climate Cafés? Exploring responses to the psychological burdens of climate change in seven UK women. *The Cognitive Behaviour Therapist, 17*. https://doi.org/10.1017/s1754470x24000096

Capstick, S. B., & Pidgeon, N. F. (2014). What is climate change scepticism? Examination of the concept using a mixed methods study of the UK public. *Global Environmental Change, 24*, 389–401. https://doi.org/10.1016/j.gloenvcha.2013.08.012

Clayton, S. (2020). Climate anxiety: Psychological responses to climate change. *Journal of Anxiety Disorders, 74*, Article 102263. https://doi.org/10.1016/j.janxdis.2020.102263

Clayton, S., Manning, C. M., Krygsman, K., & Speiser, M. (2017). *Mental health and our changing climate: Impacts, implications, and guidance.* American Psychological Association, and ecoAmerica.

Consalvo, C. (2024). *How climate cafés can have a ripple effect on the wider community and spark action.* Icarus Complex Magazine. https://icaruscomplexmagazine.com/how-climate-cafes-can-have-a-ripple-effect-on-the-wider-community-and-spark-action/

Cox, J. R. (2010). Beyond frames: Recovering the strategic in climate communication. *Environmental Communication, 4*(1), 122–133. https://doi.org/10.1080/17524030903516555

Datta, P., Behera, B., & Rahut, D. B. (2022). Climate change and Indian agriculture: A systematic review of farmers perception, adaptation, and transformation. *Environmental Challenges, 8*, Article 100543. https://doi.org/10.1016/j.envc.2022.100543

Dhanapal, G., Gupta, D., & Prakash, A. (2023). Barriers and opportunities in achieving climate and sustainable development goals in India: a multilevel analysis. *Journal of Integrative Environmental Sciences, 20*(1), 1–16. https://doi.org/10.1080/1943815x.2022.2163665

Dua, R., Malhotra, B., & St John Tager, P. A. (2024). Indigenous gond art in art therapy: A nature-based creative process. *Art Therapy, 41*(3), 135–144. https://doi.org/10.1080/07421656.2024.2356969

Fisher, S. (2014). The emerging geographies of climate justice. *Geographical Journal, 181*(1), 73–82. https://doi.org/10.1111/geoj.12078

Godsil, R. D. (1991). Remedying environmental racism. *Michigan Law Review, 90*(2), 394. https://doi.org/10.2307/1289559

Hickman, C., Marks, E., Pihkala, P., Clayton, S., Lewandowski, R. E., Mayall, E. E., Wray, B., Mellor, C., & Van Susteren, L. (2021). Climate anxiety in children and young people and their beliefs about government responses to climate change: A global survey. *The Lancet Planetary Health, 5*(12), e863–e873. https://doi.org/10.1016/s2542-5196(21)00278-3

Hoofnagle, M., & Hoofnagle, C. J. (2007). What is denialism? *SSRN Electronic Journal.* https://doi.org/10.2139/ssrn.4002823

Jay, A., Reidmiller, D. R., Avery, C. W., Barrie, D., DeAngelo, B., et al. (2018). Overview. In D. R. Reidmiller, C. W. Avery, D. R. Easterling, K. E. Kunkel, & K. L. M. Lewis (Eds.), *Impacts, risks, and adaptation in the United States: Fourth national climate assessment* (vol. 2, pp. 33–71). U.S. Global Change Research Program.

Jha, C. K., Gupta, V., Chattopadhyay, U., & Amarayil Sreeraman, B. (2018). Migration as adaptation strategy to cope with climate change: A study of farmers migration in rural India. *International Journal of Climate Change Strategies and Management, 10*(1), 121–141. https://doi.org/10.1108/IJCCSM-03-2017-0059

Jha, S. K., Mishra, S., Sinha, B., Alatalo, J. M., & Pandey, R. (2017). Rural development program in tribal region: A protocol for adaptation and addressing climate change vulnerability. *Journal of Rural Studies, 51*, 151–157. https://doi.org/10.1016/j.jrurstud.2017.02.013

Lakshmanan, P. K., Singh, S., & Lakshmi, S. A. (2017). The Paris agreement on climate change and India. *Journal of Climate Change, 3*(1), 1–10. https://doi.org/10.3233/jcc-170001

Levy, B. R., Slade, M. D., & Ranasinghe, P. (2009). Causal thinking after a tsunami wave: Karma beliefs, pessimistic explanatory style and health among Sri Lankan survivors. *Journal of Religion and Health, 48*, 38–45. https://doi.org/10.1007/s10943-008-9162-5

Lewicka, M. (2010). Place attachment: How far have we come in the last 40 years? *Journal of Environmental Psychology, 31*(3), 207–230. https://doi.org/10.1016/j.jenvp.2010.10.001

Madge, V. (2024). The hidden toll of climate change: India's struggle with eco-anxiety and mental health. *The Daily Pioneer*. https://www.dailypioneer.com/2024/columnists/the-hidden-toll-of-climate-change--india---s-struggle-with-eco-anxiety-and-mental-health.html

Majra, J., & Gur, A. (2009). Climate change and health: Why should India be concerned? *Indian Journal of Occupational and Environmental Medicine, 13*(1), 11. https://doi.org/10.4103/0019-5278.50717

Menon, A. (2024). *Eco-anxiety soars as planet health plummets.* Mongabay-India. https://india.mongabay.com/2024/05/eco-anxiety-soars-as-planet-health-plummets/

Ministry of Health & Family Welfare. (2018). *National action plan for climate change and human health.* Government of India. https://ncdc.mohfw.gov.in/wp-content/uploads/2024/04/27505481411548674558.pdf

Mishra, A. (2024). Comforting lies: Authoritarianism, anti-environmentalism and climate change denial. In: Sarkar, A., Bandyopadhyay, N., Singh, S., & Sachan, R. (eds) Risk, uncertainty and maladaptation to climate change. Disaster Risk Reduction. Springer. https://doi.org/10.1007/978-981-99-9474-8_2

Mishra, S., Mazumdar, S., & Suar, D. (2009). Place attachment and flood preparedness. *Journal of Environmental Psychology, 30*(2), 187–197. https://doi.org/10.1016/j.jenvp.2009.11.005

Mondal, S., Kar, J., Bagchi, E., & Mukhopadhyay, U. (2024). Eco-anxiety in adolescents: A meta-analysis of the impact of climate change on mental health. In *Climate change management* (pp. 347–368). https://doi.org/10.1007/978-3-031-58261-5_15

Muttarak, R., & Dimitrova, A. (2019). Climate change and seasonal floods: Potential long-term nutritional consequences for children in Kerala, India. *BMJ Global Health, 4*(2), Article e001215. https://doi.org/10.1136/bmjgh-2018-001215

Ojala, M. (2012). How do children cope with global climate change? Coping strategies, engagement, and well-being. *Journal of Environmental Psychology, 32*(3), 225–233. https://doi.org/10.1016/j.jenvp.2012.02.004

Patra, N. (2016). Climatic shock, community capability and childhood malnutrition: Case study from the Indian Sundarbans. *International Journal of Child and Adolescent Health, 9*(3).

Pihkala, P. (2020). Anxiety and the ecological crisis: An analysis of eco-anxiety and climate anxiety. *Sustainability, 12*(19), 7836. https://doi.org/10.3390/su12197836

Read, J., & Harper, D. J. (2020). The power threat meaning framework: Addressing adversity, challenging prejudice and stigma, and transforming services. *Journal of Constructivist Psychology, 35*(1), 54–67. https://doi.org/10.1080/10720537.2020.1773356

Roszak, T., Gomes, M. E., & Kanner, A. D. (1996). Ecopsychology: Restoring the earth, healing the mind. *Choice Reviews Online, 33*(07), 33–4213. https://doi.org/10.5860/choice.33-4213

Schmidt, C. (2023). Climate anxiety. *Harvard Medicine Magazine.* Retrieved from https://magazine.hms.harvard.edu/articles/climate-anxiety

Shastri, M., & Mahashur, S. (2023). *Mind matters: India's mental health budget crisis.* Speaking of Medicine and Health. PLOS. https://speakingofmedicine.plos.org/2023/05/25/mind-matters-indias-mental-health-budget-crisis/

Shukla, J. (2016). Extreme weather: mental health challenges and community response strategies. In *Extreme weather and society* (pp. 245–270). https://doi.org/10.1007/978-3-319-30626-1_11

Singh, N., Singh, S., & Mall, R. (2020). Urban ecology and human health: Implications of urban heat island, air pollution and climate change nexus. In *Urban Ecology* (pp. 317–334). https://doi.org/10.1016/b978-0-12-820730-7.00017-3

Srivastava, R. C. (2012). Ecological threats to an Islands ecosystem due to climate change: The Andaman experience. *Proceedings of the National Academy of Sciences, India, Section B Biological Science, 82*(Suppl 2), 335–340. https://doi.org/10.1007/s40011-012-0099-9

Sundaresan, J., Sreekesh, S., Ramanathan, A., Sonnenschein, L., & Boojh, R. (2013). Climate change and island and coastal vulnerability. In *Springer eBooks.* https://doi.org/10.1007/978-94-007-6016-5

Tam, K., Chan, H., & Clayton, S. (2023). Climate change anxiety in China, India, Japan, and the United States. *Journal of Environmental Psychology, 87*, Article 101991. https://doi.org/10.1016/j.jenvp.2023.101991

Thoma, M. V., Rohleder, N., & Rohner, S. L. (2021). Clinical ecopsychology: The mental health impacts and underlying pathways of the climate and environmental crisis. *Frontiers in Psychiatry, 12.* https://doi.org/10.3389/fpsyt.2021.675936

Van Lith, T. (2024). Introduction to the special issue: Art therapy and climate action. *Art Therapy, 41*(3), 116–117. https://doi.org/10.1080/07421656.2024.2394367

Varughese, A., & Purushothaman, C. (2021). Climate change and public health in India: The 2018 Kerala floods. *World Medical and Health Policy, 13*(1), 16–35. https://doi.org/10.1002/wmh3.429

Venkataraman, M., & Aggarwal, R. (2024). *Rising prevalence of eco-anxiety and climate anxiety is a growing concern.* Down To Earth. https://www.downtoearth.org.in/health/rising-prevalence-of-eco-anxiety-and-climate-anxiety-is-a-growing-concern-94686

Verplanken, B., Marks, E., & Dobromir, A. I. (2020). On the nature of eco-anxiety: How constructive or unconstructive is habitual worry about global warming? *Journal of Environmental Psychology, 72*, Article 101528. https://doi.org/10.1016/j.jenvp.2020.101528

Watts, N., Amann, M., Arnell, N., Ayeb-Karlsson, S., Belesova, K., Boykoff, M., et al. (2021). The 2021 report of the Lancet Countdown on health and climate change: Code red for a healthy future. *The Lancet, 398*(10311), 1619–1662. https://doi.org/10.1016/S0140-6736(21)01787-6

World Health Organization. (2024). *Factor in climate change impact to strengthen mental health services: WHO.* https://www.who.int/southeastasia/news/detail/19-08-2024-factor-in-climate-change-impact-to-strengthen--mental-health-services–who

Chapter 3
Climate Migration and Sexual Violence: Vulnerabilities of Women Migrant Workers from Bihar

Shikha Vasishta

Abstract With an increased pace of climate change in India, women from rural villages in flood- and drought-stricken Bihar are being pushed into migration for employment, leaving behind destroyed homes and livelihoods. Their migration is not just a survival imperative but also a trek with increased vulnerability to sexual violence, labour abuse, and disregarded for reproductive health. In Delhi, Mumbai, and Punjab's urban slums, garment factories, and agricultural camps, these women are constantly at risk and exposed to unsafe situations, but their plight is invisible in the policy and legal sphere. This chapter explores how climate migration puts Bihari women at intersecting risks, analysing the implications for reproductive rights and how effective the current safeguards are. Based on ethnographic field research, interviews, and NGO case studies, it enhances the voices of migrant women's lived experience and dissects legal proceedings in which justice has proved elusive. Employing ecofeminist and structural violence theories, the chapter exposes systemic failures, ranging from abusive landlords to poor healthcare and unaffordable justice that ensure continued harm. It concludes with policy implications of labour reforms, mobile reproductive healthcare, and increased implementation of gender-sensitive legislation, calling for a new vision of climate governance that puts the dignity and rights of migrant women at the forefront.

Keywords Climate migration · Sexual violence · Bihari women · Reproductive rights · Ecofeminism · Structural violence · Gendered labour · Migrant workers

---

S. Vasishta (✉)
Research Scholar (Political Science), School of Liberal Arts, Bennett University, Greater Noida, Uttar Pradesh, India
e-mail: Shikhavasishta61@gmail.com

P. Datta et al. (eds.), *Climate Change and Human Well-being*, Advances in Global Change Research 83, https://doi.org/10.1007/978-3-032-25479-5_3

3.1 Introduction

In a span of recent decades, climate change has become one of the most urgent survival threats facing India, characterized by soaring temperatures, unpredictable rainfalls, and more frequent instances of extreme climatic phenomena. One of the most climate-exposed regions within the nation is that of Bihar, an eastern Indian state that has experienced frequent floods, droughts, and erosion of riverbanks. These environmental shocks have not only decimated rural livelihoods but have also caused large-scale displacement, with whole communities having to migrate for economic sustenance (Pattenden, 2016). Yet climate-driven migration in Bihar possesses a strongly gendered nature. While male household members migrate to cities or leave rural areas in the face of declining agrarian economies, women are increasingly becoming the primary migrants. This feminization of movement from areas of natural disaster represents a decisive break from classical gender roles and poses significant questions concerning safety, agency, and rights in cases of forced mobility.

In spite of increasing numbers of women in migrant labour markets in India, their lives are still invisible to policy discourses as well as the scholarly literature. Bihar's Dalit, OBC, and Muslim women migrants, in particular, are situated at the nexus of intersecting marginalities. After being displaced, they find their way into urban employment markets like domestic service in Delhi, garment factories in Mumbai, or farm work in Punjab; employment sectors that are informal, badly regulated, and exploitative. They experience not just economic vulnerability but also systemic violence in the form of sexual harassment, abuse, and reproductive neglect. Their vulnerability is heightened by the fact that they are migrants, belong to a lower caste, and have precarious employment. However, their lives are usually hidden behind statistical abstractions and policy summations that do not reveal the richness of their lived experiences (Santha et al., 2015).

A progressively expansive research literature has sought to examine the link between climate change and migration in India. Studies have given attention to the manner in which environmental degradation fuels rural poverty and hastens displacement (Deshingkar & Akter, 2009). Researchers have also focused on the dynamics of internal migration and the novel challenges it brings to urban, labour, and social protection governance (Tumbe, 2018; Hashim, 2024). Feminist researchers have also begun to problematize the gendered nature of displacement, pointing out that women face specific challenges in transit points and destinations (Agnihotri & Mazumdar, 2009). However, there is a huge gap in the literature regarding the reproductive rights of climate migrants, especially in the Indian context. The intersections of gender violence, exploitation at work, and reproductive health are also not adequately researched, especially among women migrants from climate-stricken areas like Bihar (Anjum & Aziz, 2025).

This chapter aims to fill this gap by putting the voices and experiences of Bihari women migrants at its focal point. It inquires: (1) How does climate migration expose Bihari women to increased risks of sexual violence and exploitation of labour? (2) What are the implications for their reproductive health and rights of such violence? (3)

How strong are existing legal and policy instruments in addressing their protection? These are not abstract questions. They resonate with critical questions of justice, dignity, and survival in an increasingly ecologically crisis-dominated world. By asking these questions, the chapter seeks to place centre stage the structural conditions that determine migrant women's lives and to contribute insights towards more gender-inclusive, rights-based responses to climate migration.

The objectives of this chapter are four. First, it proposes to record the particular vulnerabilities that women migrant workers from Bihar confront, with special reference to sexual violence and reproductive abandonment. Second, it tries to deconstruct the structural and institutional failures—legal, social, and economic, behind these vulnerabilities. Third, it provides a theoretical window to examine these dynamics through the lenses of ecofeminism, structural violence, and gendered migration. Lastly, the chapter suggests a range of policy recommendations that go beyond symbolic gender incorporation to meaningful protection and empowerment of migrant women.

Ecofeminism offers a theoretical basis for the analysis. Emerging from the writings of researchers such as Agarwal (1992), ecofeminism suggests that women, and especially those from poor communities, disproportionately suffer from environmental degradation because of their socially constructed closeness to nature and reliance on natural resources. This model serves to shed light on how environmental disruption in Bihar by way of floods, droughts, or soil erosion gets converted into more labour-intensive burdens, displacement, and exploitation for women. Most significantly, it refutes anthropocentric and androcentric concepts of development by connecting environmental justice with gender justice.

To this is added Galtung's (1969) theory of structural violence, which is concerned with how social hierarchies and institutional arrangements create and sustain harm in normalized and invisible ways. Structural violence in the context of this research takes the form of insufficient labour protection, unaffordable healthcare, absence of legal recourse, and societal stigmatization conditions that subject women migrants to sexual violence and yet make their suffering institutionally invisible.

Finally, the chapter engages with gendered migration literature, notably the work of de Haas et al. (2019) to highlight that migration is not a gender-neutral phenomenon. Women's migration paths are influenced by distinct push and pull factors, and their labour market experiences are conditioned by gender norms, cultural taboos, and institutional discrimination. The combination of these lenses permits analysis in multidimensional terms that documents harm as well as probes its structural origins.

Drawing on primary ethnographic fieldwork and case studies, this chapter is organized in the following manner. The subsequent section, Data and Methodology, explains the qualitative research methodology, fieldwork locations, data, and ethical issues underpinning the research. This is followed by the Results section, which reports important findings from interviews conducted among migrant women in Seemapuri (Delhi), Dharavi (Mumbai), and Bathinda (Punjab). It analyzes issues like sexual coercion within the workplace, transactional housing arrangements, and reproductive health neglect. The Discussion section provides a critical reading of these findings with the ecofeminist, structural violence, and gendered migration theory frameworks. It analyzes the inadequacies of existing legal tools like the Inter-State Migrant Workmen Act, Protection of Women from Sexual Harassment at Workplace Act (POSH), and reproductive health programs like the Pradhan Mantri Matru Vandana Yojana (PMMVY). The last part, Conclusion and Policy Implications, integrates the research findings and provides tangible suggestions with the aim to reform labour legislation, improve access to reproductive healthcare, and adopt a gender perspective within climate governance.

Furthermore, by intertwining climatic realities, migratory distress, and gendered violence, this chapter aims to trigger a reconsideration of policy paradigms. It argues that responses to climate migration which do not meaningfully engage with women's sexual and reproductive rights are not only inadequate but also fundamentally inequitable. In a world of global flows in which mobility has so frequently been a prerogative and not an option, the protection of the rights of the weakest of the weak must be the starting point for adaptation to climate and for urban policy. The narratives of women migrant workers from Bihar are not ancillary to this agenda; they are integral to any serious vision of inclusive and equitable development.

3.2 Data and Methodology

This chapter utilizes a qualitative, feminist, and exploratory research design to question the interconnections of climate-induced migration, labour abuse, sexual violence, and reproductive health marginalization among Bihari women migrant workers. With the scarcity of disaggregated data on women migrants particularly those made invisible by informal labour markets and systemic state neglect; this research takes an ethnographic approach anchored in the epistemological assumption that lived experience is not just data but theory (Smith, 1987). The intention is not just to document suffering, but to place women's agency in the foreground, to critique structural injustice, and to reimagine the politics of care and rights in the face of accelerating climate crises.

The research is based on women migrants from Bihar, in particular from the flood-affected districts of Darbhanga, Madhubani, Sitamarhi, and Supaul, which are situated in the Kosi river basin area of North Bihar. The districts are among the most climate-exposed regions of eastern India with a combined population of more than 15 million, of whom approximately 22–25% are poor (Census of India, 2011; Bihar SDMA, 2022). Livestock and agri-based livelihood constitute the area, but frequent floods, soil erosion and waterlogging frequently result in displacement and starvation, with large-scale seasonal and permanent migration. With the average household size from 5.5 to 6.2 family members, as revealed in Bihar's district-level population profiles, the school burden on facilities is higher, particularly when considered in the context of the state's own overall literacy rate at 61.8%, which is way below the national rate of 74%; further compounded by a yawning gender gap, where male literacy stands at 71.2% and that of females is 51.5% (Census of India, 2011), showing deep structural cleavages in accessing education. Religious and caste groupings delineate a substantial proportion of Scheduled Castes and Muslims, erstwhile groups living at the margins who are further disproportionately affected by climate and livelihood shocks. These socioeconomic realities form a critical background against which the push factors causing migration and enhanced vulnerability among women that occur with displacement are evaluated.

The use of ethnography and qualitative interviews is based on feminist methodological principles that prioritize subjectivity, reflexivity, and knowledge co-production (Harding, 1987). Such methods enable rich, narrative-heavy accounts that statistical data cannot provide. Heeding women in their own voices; frequently in dialects imbued with the traces of cultural and environmental loss, reveals the way ecological displacement is felt not abstractly, but sensually: in broken routines, frayed kinship, body violation, and the uncertain navigation of everyday survival.

Fieldwork was conducted for nine months, July 2023 to March 2024, across three large urban and peri-urban locations: Seemapuri in Delhi, Dharavi in Mumbai, and Bathinda in Punjab. They were purposively chosen because of their high density of migrant workers from Bihar, and specifically for representing the gendered architecture of India's informal labour markets. Seemapuri, that congested resettlement colony on the fringes of East Delhi, is an important site for women domestic workers, several of whom have been driven out of Bihar's flood-affected districts. Dharavi, a name most commonly deployed as evidence of the entrepreneurial mettle of Mumbai, is at the same time a crucible of informal garment labour, in which women's work is both vital and disposable. Bathinda, located in the legacy of the green revolution, relies upon seasonal migrant women for farm labour but provides them with extraordinarily little in terms of security or status. These locations together demonstrate how spatial, occupational, and environmental forces interact to create and entrench gendered precarity.

A total of 30 in-depth, semi-structured interviews were carried out with 18–45-year-old women migrants from a representative sample drawn from a pooled demographically diverse range of caste, religion, and marital status. The sample consisted of 12 women working as domestic helpers, 10 working in garment factories, and 8 as agricultural workers. They all came from climate-stressed districts of Bihar, i.e., Darbhanga, Madhubani, Sitamarhi, and Supaul, which have been traditionally affected by repeated flooding, extreme land degradation, and long-term agrarian decline. Most of them were from Dalit, OBC, and Muslim communities, which captured the intersectionality of environmental risk, exclusion on the grounds of caste, and gendered displacement. Snowball sampling was made possible due to partnerships with grassroots movements: Jagori (Delhi), Ghar Bachao Ghar Banao Andolan (Mumbai), and Lok Bhalai Trust (Punjab), who had long-standing roots in these communities and therefore facilitated trust and access.

As a complement to the main interview data, this research conducted a close textual and legal reading of five anonymised case files, all involving women migrants from Bihar who had filed legal cases against sexual violence, work-place harassment, or exploitative labour relations. Made available through networking with legal aid collectives and rights-based NGOs, these case files provided an institutional perspective to situate personal narratives against larger juridical-administrative processes. In contradistinction to analysing such legal documents as fixed bureaucratic texts, they were studied as narrative documents; highly fractured, disputed, and incomplete, yet informative about the structural limitations that make migrant women's access to justice possible. Through this cross-referencing of juridical documents with oral testimonies, a multilayered evidentiary triangulation was facilitated, highlighting the systemic tensions between legal recognition and lived experience.

Fieldwork also consisted of extended participant observation in urban slum clusters, informal work sites, and neighbourhood meetings, where migrant life's embodied rhythms; unpaid work, waiting, care work, and haggling with landlords or middlemen might be seen in action. Extended field notes captured not just verbal content but also affective atmospheres, silences, and emotional registers that evanesce from formal interviews. Specific focus was placed on non-verbal modes of communication; gesture, tone, hesitation, and pause that within the environments of trauma populations, serve as key data points. This ethnographic sensitivity was important to the comprehension of the relational ecologies of exposure in which these women live.

The spatial movement of participants is illustrated below in Fig. 3.1, which shows the primary migration corridors between climate-stressed districts in Bihar and informal labour destinations across Delhi, Mumbai, and Punjab.

The ethical framework of the study was derived from feminist theories of situated knowledge production and relational accountability. Recruits were solicited with the help of trust-based mediators in their local settings and informed orally and in writing

Fig. 3.1 Migration corridors from flood-affected districts of Bihar (Darbhanga, Madhubani, Sitamarhi, Supaul) to informal labour destinations: Seemapuri (Delhi), Dharavi (Mumbai), and Bathinda (Punjab). *Note* Map created by the author using Google My Maps

about the objectives of the study, the voluntary nature of the response, and steps taken to maintain confidentiality. Informed consent was elicited in Hindi, Maithili, or Bhojpuri, as subjects preferred. Where conversations engaged potentially retraumatizing ground, e.g., history of coercion, pregnancy loss, or institutional betrayal, interviews were held in the presence or advance consultation of trauma-sensitive counsellors who were affiliated with collaborating NGOs.

Pseudonyms and contextual anonymization practices were used across the transcription and analysis stages to protect against markers of identifiability, for example, workplace contexts or kinship affiliations, being left unmasked. In a feminist epistemological approach, the study took an active role in removing the researcher's authority from centre stage by seeing the migrant women as neither blank slates to be filled with research data nor mere information providers but as knowledgeable actors and co-theorizers of their own worlds.

The analysis was carried out with thematic analysis based on the six-phase model described by Braun and Clarke (2006), and iterative coding was performed manually as well as using NVivo software. The emergent codes were sorted into higher-level thematic categories that represented both the manifest and latent content of the participants' accounts. They were: "sexual coercion in casual economies," "infrastructural abandonment of reproductive health", "coercive landlordism and housing insecurity" and "moving through juridical obscurity." By cross-analysing these across sites and occupational fields, the study mapped structural patterns of dispossession, and situated localized practices of adaptation and resistance. Thematic saturation was not attained through repetition of content alone, but through the research deepening of patterns, contradictions, and silences across interviews, legal files, and observation data.

Data triangulation was used as a methodological precaution in the study to increase credibility and reflexive reliability. Interview data were contrasted with the legal case file content and NGO reports, as well as direct field observations of working and living conditions. This triangulation not only served to verify findings, but also to reveal the discrepancies between rights "on paper" and rights in practice. For instance, although numerous women were legally entitled to maternity benefits under national labour systems, none of the domestic workers in our interview sample had any knowledge of or access to such provisions; a discrepancy only revealed by multi-source corroboration.

As with all qualitative research based on changing field realities, this study was influenced by a number of methodological and logistical constraints. To begin with, the comparatively modest and purposively sampled group; albeit rich in narrative richness, restricts the generalizability of the findings to larger migrant groups. In addition, linguistic flexibility and translation raised interpretative issues, most noticeably in Bathinda, where respondents tended to blend Punjabi terms with Bhojpuri or Maithili idioms. While attempts were made to preserve linguistic texture in translation, semantic textures may nevertheless have been lost or shifted imperceptibly in the course of doing so. Third, conditions of spatial access restriction barred access to a node in the migration trajectory; temporary lodging near railway terminals or transit points where emergent forms of precarity may also be generated. Finally, while the study foregrounds gender as a central analytic category, it acknowledges its own limitations in engaging substantively with the experiences of queer, trans, or non-binary migrant individuals, whose realities often fall outside the purview of mainstream data collection frameworks.

The theoretical scaffolding of this work is constructed around an interdisciplinary theory that draws on ecofeminism, structural violence, and gendered migration studies. Agarwal's (1992) ecofeminist framework offers the foundation for knowledge on how environmental degradation is not just a material loss but also a gendered crisis that disproportionately displaces rural women whose social roles are intertwined with ecological labour; gathering water, gathering firewood, working in fields. These disruptions shatter their socio-economic autonomy and drive them into informal regimes of labour in urban spaces where exploitation is institutionalized.

The structural violence lens, pioneered by Galtung (1969), explains how institutional structures; legal, bureaucratic, and social, inflict harm not in the form of direct aggression but through patterned disregard, denial of rights, and policy lethargy. The public policy invisibility of migrant women, the denial of reproductive healthcare by the state, and the police indifference to sexual harassment charges, each is a structural violence that undermines dignity and multiplies vulnerability.

Supporting these viewpoints is the theory of gendered migration de Haas et al. (2019), which is a critical approach that challenges masculinist assumptions of migrancy studies on the whole that have historically focused on male labour migration while neglecting the unique trajectories, burdens, and violence faced by women. Migration for the women in this research is not a straightforward movement from starting point to endpoint but a continuum of precariousness that is punctuated by sexual coercion, reproductive disregard, and uncounted labour. Their mobility is both a gesture of economic survival and a space of gendered dispossession.

Together, these frameworks prepare this study to go beyond descriptive accounts of suffering and toward a critical examination of the systems that make such suffering possible, permissible, and lasting. The aim is not to sentimentalize the vulnerability of Bihari migrant women, but to insist on the epistemic, legal, and political imperative of centering their voices in climate and migration governance in India.

3.3 Results

The narratives that unfolded in Seemapuri in Delhi, Dharavi in Mumbai, and Bathinda in Punjab; all highly concentrated with informal economy migrant workers from Bihar; betray deeply ingrained dynamics of gendered exploitation, erasure, and institutional abandonment. While the women had also come from flood and drought-affected districts like Darbhanga, Madhubani, Sitamarhi, and Supaul, it was here, in these receiving areas, that their marginalization was aggravated: no longer defined just by environmental dispossession, but by the everyday economies of coercion, silence, and survival.

The female interviewees for this research were between the ages of 18 and 47, with the majority falling in the 25–35 years category. Most had arrived in the past five to seven years and were working in precarious jobs like domestic work, garment production, and seasonal agricultural work. A large proportion were main income earners in their families, many supporting dependents remaining behind in Bihar. Educational levels were low, about two-thirds had not finished primary school, yet most expressed a high level of resilience, community information, and informal knowledge in negotiating city systems.

Most women recounted their migration as a series of displacements and forced movement in the wake of failed agriculture, debt spirals, or male migration. A woman in her thirties, now a domestic worker in Seemapuri, remembered: "Year by year the water kept rising. First up to the road, then into our houses. The land disappeared. My husband went away to Punjab. I came a year later with my son. There was no work, no food, and no land remaining in the village." Her words resonate with so many other women migrants who came not by aspiration or opportunity, but by necessity, driven by the loss of land and livelihood, and often left as caregivers after male family members had already migrated in search of employment. These travels were characterized by dislocation but also by a deep reconfiguring of gender roles; women entering economic mobility but being structurally excluded from its protections.

During transport and at destination, almost all women interviewed had experienced some type of gendered abuse or sexual assault. Vulnerabilities started right from the journey. A few women described traveling in crowded private buses or lorries, with no access to safety whatsoever. A young worker in the garment industry in Dharavi narrated: "We had no money for the proper training.". In the private bus, men sat too close, feigning to fall on you. When I complained, the conductor smirked and said, 'Then don't travel.'" For women traveling alone or with young children, these experiences created an ambient sense of danger that lasted long after their destination.

This vulnerability was further deepened in the workplace. In both Bathinda's farms and Mumbai's garment factories, women workers reported being bullied into silence upon facing overtures from contractors or supervisors. One woman who worked in a small workshop sewing clothes said simply, "He said, 'Say yes or go home.' And if I said no, I was off the list the next day. There are ten more waiting for my place." Her statement makes it evident that consent, in such exploitative labour conditions, turns into an empty concept. Economic vulnerability, lack of complaint channels, and threats of reprisal effectively criminalize defiance. Another woman, seasonally employed in the cotton fields of Bathinda, remembered being approached by *thekedaars* (contractors) and asked to "stay back and assist in the storage room", a ruse, she later learned, for tolerating sexual overtures. Those who refused usually found themselves out of work the next day.

This coercion logic carried over into housing. In all three locations, housing was mostly informal; crowded rooms, unregistered tenancies, and one-room settlements with unrelated men. The lack of tenure or documentation provided another axis of exploitation. In Seemapuri, there was a domestic worker who explained: "He [the landlord] said, 'You can stay, but you need to go to my room.' I didn't have any other place to go." Most women reported the same thing happening, with rent becoming commoditized, and the threat of eviction used as leverage to have them be sexually compliant. Having children present with them under these conditions made the psychological cost even higher. A Bathinda mother of two explained closing the door on her kids when she left for work: "I don't trust the men in the neighbourhood. They lecher even at my daughter. So I lock the door, feed her, and go." Her account highlights not only the threat of physical harm but also the emotional weight of parenting in a climate of all-around surveillance and terror.

While the contours of exploitation differed somewhat at each field site, the underlying frameworks were frighteningly similar. In Seemapuri, transactional housing dynamics were supported by the unofficial landlord-tenant economy, and in Bathinda, employer-managed housing camps internalized these dynamics into the labour contract. Likewise, while harassment in the garment industry of Mumbai was frequently spatial and verbal, farm contractors in Punjab used women's reliance on day wages to extract physical submission. These differences highlight the flexibility, much more than the exceptionalism of patriarchal dominance within India's informality-based labour geography.

Health risks came through as a recurring and cumulative issue. At the three field locations, researchers interviewed no women who had access to formal reproductive health care. Many of them had contracted sexually transmitted diseases, unplanned pregnancies, or unsafe abortions but not many had looked for or received medical help. One woman in Dharavi recounted a terrifying experience with a recent abortion: "I bled for ten days. There was no doctor in the basti. The dai [local midwife] offered me some leaves to boil and asked me to lie down. I was going to die." This kind of silent suffering with lack of access, cultural taboo, and economic helplessness was prevalent. Even when women did want to go for care, they did not have the language, trust, or time. In most interviews, women explained symptoms using words such as "naramipan," "kamzori," or "garam peth," both limited biomedical vocabulary and the cultural awkwardness of talking about reproductive issues.

During pregnancy, women said they continued working until their last month, getting no maternity leave or even days off. One Delhi housemaid explained: "If I quit, they will get another maid. That's how the city operates. So I swept the floor with one hand and supported my belly with the other." This unseen reproductive labour exploited and unsupported at the same time is evidence of the systemic breakdown in labour protections in the informal economy of India.

Legal redress was largely missing from the women's narratives. Just three of the thirty complainants had gone to the police or any institution at all, and all three had a degrading and useless experience. One of the farm workers from Bathinda reported that when she went to file a complaint, the officer asked her why she was late in the evening and what kind of work she did. "They treated me as though I had committed something wrong," she reported. Few women who had reached NGOs or legal aid spoke of the assistance as significant but fleeting. "They assisted me in exiting the factory, but thereafter, I was on my own. Where are we going next?" a woman asked.

These constraints were reaffirmed by actors operating in the supporting infrastructure. One legal aid officer with an NGO based in Delhi reported, "We witness the same pattern over and over again: displacement, casual work, exploitation, silence, and intimidation. The system destroys them in ways that are statistically unseen." These accounts by intermediary actors reaffirmed the larger trends represented in women's narratives, a failure to implement gender-sensitive policy frameworks at the ground level.

The NGO landscape, although critical, frequently did not have the resources or coverage to meet long-term needs particularly for women who were highly mobile and working in fragmented areas such as domestic labour. Quantitatively, the field data revealed astounding trends: over 80% of the women interviewed had experienced some kind of harassment, be it in transit, at work, or in residence. Only three had ever reported to the police or an NGO. None had ever received any kind of reproductive healthcare in their workplace jurisdiction, and none had maternity leave or legal protections of labour. More than 70% resided in insecure or informal housing setups.

These results are reflective of broader structural patterns which have always been found within comparable socio-economic environments. These field-based observations are supported by national-level data as well. NFHS-5 (2019–21) only reports 32.4% of presently wedded women belonging to urban poor households to have had access to public health centres providing reproductive services. Additionally, the ILO (2018) has reported that more than 90% of female informal workers in India work without any written contract, social security, or sexual harassment protection. The congruence of such macro-level data with the experiences recorded in this study attests to the critical need for systemic reform.

These are not exceptional cases; they are symptoms of an underlying ecosystem of gendered structural violence manifesting across India's informal labour markets. What is revealed by this empirical terrain is a highly intermeshed web of risk: the breakdown of rural ecosystems, forced mobility, hazardous transit, sexual coercion in work and in housing, reproductive neglect, and near-total juridical invisibility. And yet, in this desolate landscape, women still manage to survive not as passive victims, but as active agents navigating deeply unequal frameworks with meagre resources. Their voices, recorded in this research, call not merely for recognition, but for radical change to the policies and systems of governance that have long made them invisible (Chattoraj, 2024).

In this thick landscape of exploitation, women quietly, sometimes in ways that were subversive, claimed authority over their own lives. Some roomed only with relatives or same-sex groups to evade male oversight. Others joined informal savings associations to acquire some measure of financial independence, or quietly obtained contraceptives through respected female intermediaries. These actions while seldom evident in official documents were habitual expressions of resistance, resilience, and moral defiance against a regime that attempted to make them invisible and expendable.

Together, these empirical trends shed light on how migration, gender, work, and state inaction come together to form multilayered sites of vulnerability. But they also unveil a crucial fact: the women at the centre of these narratives are not simply victims of suffering; they are survivors moving through broken systems, frequently with more foresight than those who make their rules. It is to this intersection between structural violence and survival that the next section addresses.

3.4 Discussion

The empirical evidence presented in this research puts centre stage a profoundly disturbing, yet structurally integrated pattern of gendered precarity within India's informal labour market; one that is deeply interrelated with climate-induced displacement, institutional abandonment, and deeply embedded patriarchal norms. These weaknesses, although experientially specific to the women spoken to, are analytically connected to wider processes of eco-social displacement, in which environmental degradation of Bihar's agrarian heartlands meets exploitative urban economies to generate what can be thought of as a feminised geography of exclusion.

The narratives emerging from Delhi's Seemapuri, Mumbai's Dharavi, and Bathinda in Punjab reflect a recurring arc: from the collapse of rural livelihoods due to recurrent flooding, soil erosion, and crop failure, to the informalisation and invisibilisation of female migrant labour in urban centres. As previously theorized by Agarwal (1992), women's involvement in ecological systems is not symbolically abstracted but materially grounded based on their everyday responsibilities of food production, water gathering, and childcare. Their destruction thus does not just "impact" women but displaces them, economically, geographically, and reproductively. And it does so disparately. Women's displacement is not a reflection of men's outmigration it is constructed by reduced bargaining capacity, asset-lessness, and increased care responsibility, leading to increased dependency and increased structural vulnerability in the course of migration and at its urban destination.

In this context, climate migration is no longer a gender-neutral economic category but becomes a deeply gendered phenomenon. Current policy literature, like India's National Action Plan on Climate Change, is still blind to these gender-specific migration outcomes. The prevailing account still renders migrants as economically mobile, male subjects, thus effacing the uniquely reproductive and embodied trajectories of women displaced not only as workers, but as caregivers, mothers, daughters, and wives faring in unfamiliar landscapes without institutional protections (Serraglio & Thornton, 2024).

This lack of recognition directly supports what Galtung (1969) theorises as structural violence; slow, cumulative, and embedded in bureaucratic, legal, and social structures that systematically deprive people of their fundamental needs and protections. The women in this research do not tell their narratives through the vocabulary of rights, because for many the law is not an available or credible interlocutor. Rather, they convey meaning through silence, metaphor, and evasion, indicating a fundamental distrust of institutions. Landlords' sexual coercion, supervisors' verbal abuse, and transactional arrangements for housing are not necessarily reported since these experiences are normalized in the informal economy. In fact, according to one respondent, "We learn to ignore or we cannot work. The city does not allow complaints."

In this regard, the collapse of protection labour regimes most notably the failure to implement the Sexual Harassment of Women at Workplace Act (2013) or the Inter-State Migrant Workmen Act (1979) is not merely an issue of implementation but one of epistemic exclusion. These legislations were not created with regard to female migrant domestic workers, garment workers, or seasonal farmhands. Their lives cannot be easily articulated in the terms of 'workplace' or 'employer,' nor does their violence always take the shape the law can imagine as 'violence.' The failure of the legal system to capture these living complexities amounts to a de facto legal void; one that makes migrant women both legally invisible and politically irrelevant (Social and Political Research Foundation (SPRF), 2021; Serraglio & Thornton, 2024).

This is where gendered studies of migration, especially as theorized by de Haas et al. (2019), become so vital. Their writings show how not just the motivations and patterns of migration are gendered, but also the available labour types, the forms of vulnerability one faces, and the (in)visibility of these experiences in overarching policy discourses. The conclusions of this research confirm that female migration is seldom a linear process of economic advancement. Instead, it is nested migration; a series of dislocations usually triggered by environmental crisis, economic hardship, or family dissolution, and characterized by an ongoing negotiation of risk, survival, and silence.

Of particular importance, the reproductive body is also a focal point of negotiation within this risky landscape. Migrant women do not just work in the fields or factories, they also work to control fertility, care for children, avoid pregnancy, and heal from sexual violence without the support of institutions. In so doing, reproductive injustice is both a cause and result of their precarity. None of the women had access to maternity leave, institutional health care, or formalized paths of reproductive care. The idea of reproductive autonomy, taken for granted in rights-based frameworks, is made meaningless when there are situations in which women are denied access to safe contraception, are forced to attend untrained midwives, or are required to trade bodily compliance for employment or housing security. This resonates with Ross and Solinger's (2017) definition of reproductive justice as being inherently tied up with economic, racial, and spatial justice, such that the right to be a parent (or not), and the terms on which one becomes one, are structurally limited.

And still, the existence of resistance in everyday life is undeniable. Migrant women, enmeshed as they are within adverse systems, are not passive agents. Their agency appears in hidden, frequently unofficial ways; selecting safer co-tenants, collectively denying wage theft, learning local idioms in order to move through urban environments, or quietly obtaining contraceptives via solidarity networks. These actions do not fit within conventional activism, but they make up substantive survival and assertion techniques within hazardous environments. These strategies resonate with Das's (2007) concept of the "everyday state," in which subaltern subjects continually navigate, subvert, and reinterpret sites of power through intimate and embodied awareness.

In parallel, one should avoid romanticizing resilience. These strategies do not stem from empowerment but from necessity; a response to the state's inability to provide meaningful protections. The restricted domain of NGOs, while formidable, is also a feature of the wider outsourcing of welfare roles to under-resourced civil society organizations operating at the periphery of legal accommodation. One legal counsellor in Delhi aptly observed, "We're always plugging holes in a dam that keeps breaking." The state's abdication of responsibility not only places undue burden on NGOs, but creates a tiered system of rights access where justice becomes a matter of geography, funding, and luck.

From a more sweeping analytical view, what this research discovers is not simply the gendered expense of climate migration, but also the inability of existing scholarly and policy language to understand the multi-scalar violence it involves. The women in this research hold what political theorist Fraser (2009) would call "subaltern counterpublics", their words in parallel to dominant publics, and infrequently intersecting them. Their lives disrupt the tidy divisions between the environmental, the economic, and the legal. In their worlds, the village is here in the city, the climate is here in the body, and the law is everywhere and nowhere.

Furthermore, this research calls for a rethink of the way in which we think about both vulnerability and migration. The narratives of migrant women are not residual but foundational to the operation of India's informal economy. They are not statistical outliers but indicative of a systemic state in which patriarchy, poverty, and precarity are sustained through silence and institutional neglect. Their narratives call not for sympathy but for structural change. The following section presents policy directions informed by this imperative.

3.5 Conclusion and Policy Implications

This research illustrates how climate displacement, especially from Bihar, envelops migrant women in structurally violent city economies that capture work while withholding legitimate and reproductive rights. These are not accidental exclusions but created through the interplay of ecological disintegration, unorganized regimes of labour, and policy unconsciousness. To properly respond to this, we need to shift the understanding of climate migration from a spatial move to a gendered disruption that eliminates women from formal protections; whether labour, housing, health, or justice.

Policy responses need to start by formally acknowledging climate-displaced women as rights-bearing workers. Kerala's Domestic Workers Act of 2011, for example, provides a useful template on formal registration, minimum wage guarantee, and grievance redressal systems that can be incorporated into a gender-sensitive national migration register. Registers must record female migrants as soon as they arrive to ensure that they have access to maternity benefits, provident fund entitlement, and access to legal aid. More generally, mainstreaming climate migrants into the upcoming National Climate Adaptation Fund would place urban, gender-specific distribution first to ensure mobility driven by environmental degradation receives organized protection.

To broaden protections into informal segments, one needs to formalize contractual minimums in domestic homes, garment workshops, and seasonal agricultural agreements. This might be built on Brazil's Bolsa Família programme that makes cash transfers contingent on health check-ups and school attendance, showing that support tied to rights is more inclusive protection (Hellmann, 2015; Baroni et al., 2021). India can follow suit by making labour registration contingent on entitlements like maternity leave, social security cover, and employer registration under the Inter-State Migrant Workmen Act (1979). Routine, gender-specific labour inspections incorporated in municipal urban health and fire safety inspections would assist in implementing these protections across informal work locations.

Reproductive health needs to be at the centre of climate migration policy, not on the periphery. From Latin America's Women's Health and Rights Initiative, the model of Tamil Nadu's mobile health van can be replicated in urban migrant camps with staff trained in maternal care, contraception, and safe abortion referral. Services have to be free of cost, anonymous, and offered in Bhojpuri, Maithili, Hindi, and Punjabi to ensure that language is no barrier. A national directive may be instituted under the NHM–Urban Health Mission to reach unregistered urban migrants, combining reproductive justice with climate justice.

Housing policy has to be redefined as a climate–labour imperative. Safety and security in housing with defined tenancy rights like Pune's municipal support of women's collectives can radically rebuild informal rental markets. Women's right-based hostels, inspired by Gujarat's urban working women's hostels, combined with rent-free transition shelters and rent-to-own schemes offering access to sanitation, cooking facilities, and legal protection against eviction or sexual coercion, should be promoted by municipal authorities.

Justice and security become the last frontier. Community Legal Clinics for Labourers in Tamil Nadu indicate locally funded legal assistance centres integrated into the labour settlements. These clinics may be scaled up throughout migrant-dense urban wards, with paralegals and counsellors having local language capabilities and close coordination with police. Frontline police officers must be trained in gender-sensitive, migration-conscious procedures, with performance measures tied to responsiveness to violence complaints by migrants. The complaint mechanisms must be made responsive without discrimination, that is, instead of "why were you here," ask "what happened." This is crucial to changing institutional behaviour.

The insights drawn from these lessons indicate that piecemeal reforms will not suffice. Policy efficacy needs multi-scalar governance, ranging from village resettlement plans and gender-targeted relief during disasters to state-level migrant enrolment and national reproductive health entitlements. A federated "Safe Migration and Women's Agency Act" can have enforceability, which will bind central and state governments to institute grievance redressal mechanisms, safe shelter, and provision of services at the nexus of climate, labour, and gender.

Moreover, this chapter exhorts researchers and policy-makers to change diagnosis as well as response. No longer can climate migration be isolated into "environmental," "urban," or "labour" policy, the lives of women from Bihar demand a more integrated, justice-based governance response. Guaranteeing displaced women's safety from coercion, their status as workers, support in their reproductive work, and access to legal recourse is not merely a policy imperative; it is a moral, constitutional, and developmental necessity. The moment has arrived for policy to catch up with the lived wisdom of women who, despite being rendered invisible, continue to refuse and with that refusal, map new possibilities toward more just futures.

Conflict of Interest Statement The author declares that there is no conflict of interest with respect to the research, authorship, or publication of this chapter.

AI-Use Declaration The content of this chapter is entirely human-authored. No generative AI tools were used in the drafting, writing, or analysis presented in this manuscript.

References

Agarwal, B. (1992). The gender and environment debate: lessons from India. *Feminist Studies, 18*(1), 119–158. https://doi.org/10.2307/3178217

Anjum, G., & Aziz, M. (2025). Climate change and gendered vulnerability: A systematic review of women's health. *Women's Health, 21*. https://doi.org/10.1177/17455057251323645

Baroni, L., Alves, R. F. S., Boccolini, C. S., Salles, R., Gritz, R., Paixão, B., & Boccolini, P. M. M. (2021). Database on the coverage of the Bolsa-Família conditioning cash-transfer program: Brazil, 2005 to 2021. *BMC Research Notes, 14*(1), 435. https://doi.org/10.1186/s13104-021-05849-5

Bihar State Disaster Management Authority. (2022). *Annual flood and disaster risk management reports*. Government of Bihar. Retrieved from BSDMA Publications & Reports https://bsdma.oustadji.com/en/publication-reports

Braun, V., & Clarke, V. (2006). Using thematic analysis in psychology. *Qualitative Research in Psychology, 3*(2), 77–101. https://doi.org/10.1191/1478088706qp063oa

Census of India. (2011). *District census handbook: Bihar (Part A & B)*. Ministry of Home Affairs, Government of India. Retrieved from Bihar NIC district portal https://patna.nic.in/document/district-census-handbook-2011-part-b/

Chattoraj, D. (2024). Breaking barriers: Migration, and fight against sexual harassment in India. In A. A. Ullah (Ed.), *Handbook of migration, international relations and security in Asia*. Springer. https://doi.org/10.1007/978-981-99-8001-7_57-1

Das, V. (2007). *Life and words: Violence and the descent into the ordinary* (1st ed.). University of California Press. http://www.jstor.org/stable/10.1525/j.ctt1pprtk

Deshingkar, P., & Akter, S. (2009). Migration and human development in India.

de Haas, H., Castles, S., & Miller, M. J. (2019). *The age of migration: International population movements in the modern world*. Bloomsbury Publishing.

Fraser, N. (2009). *Scales of justice: Reimagining political space in a globalizing world*. Columbia University Press. http://www.jstor.org/stable/10.7312/fras14680

Galtung, J. (1969). Violence, peace, and peace research. *Journal of Peace Research, 6*(3), 167–191. https://doi.org/10.1177/002234336900600301

Hashim, S. R. (2024). Challenges to sustainable urbanization in India—Migration, poverty and urban crimes. In A. Alakshendra, A. Datta, & B. Reddy (Eds.), *Development, inclusion and sustainability* (pp. 193–208). Springer. https://doi.org/10.1007/978-981-97-6863-9_11

Harding, S. G. (Ed.). (1987). *Feminism and methodology: Social science issues.* Indiana University Press.

Hellmann, A. G. (2015). *How does Bolsa Familia work?: Best practices in the implementation of conditional cash transfer programs in Latin America and the Caribbean.* https://doi.org/10.18235/0000141

Pattenden, J. (2016). Labour, state and society in rural India: A class-relational approach. Manchester University Press. https://doi.org/10.7228/manchester/9780719089145.001.0001

Ross, L. J., & Solinger, R. (2017). *Reproductive justice: An introduction* (1st ed.). University of California Press. http://www.jstor.org/stable/10.1525/j.ctv1wxsth

Santha, S. D., Jaswal, S., Sasidevan, D., Datta, K., Khan, A., & Kuruvilla, A. (2015). Migration, vulnerability and urban livelihoods. In Climate change, livelihoods and health inequities: The vulnerability of migrant workers in Indian cities (pp. 11–29). International Institute for Environment and Development. http://www.jstor.org/stable/resrep01310.10

Serraglio, D. A., & Thornton, F. (2024). Women on the move? Mainstreaming gender in policies and legal frameworks addressing climate-induced migration. *Comparative Migration Studies, 12*, Article 53. https://doi.org/10.1186/s40878-024-00412-0

Smith, D. E. (1987). *The everyday world as problematic: A feminist sociology.* Northeastern University Press. https://archive.org/details/everydayworldasp00smit

Social and Political Research Foundation (SPRF). (2021). *Domestic workers in India: An invisible workforce.* https://sprf.in/wp-content/https://sprf.in/wp-content/uploads/2021/02/Domestic-Workers-in-India-An-Invisible-Workforce.pdfuploads/2021/02/Domestic-Workers-in-India-An-Invisible-Workforce.pdf?

Tumbe, C. (2018). *India moving: A history of migration.* Penguin Random House India Private Limited.

Part II
Health Impacts of Climate Change

Chapter 4
Carbon Emission and Physical Wellbeing: A Global Exploration into the Infant Mortality Rate

Subhanil Banerjee and Deepak Panda

Abstract Carbon emissions continue to pose a significant obstacle to the attainment of infant mortality reduction targets outlined in SDG-3. The present study investigated the association between key developmental indicators and IMR across 168 countries, focusing on the influence of per capita CO_2 emissions, GDP per capita, population density, growth rate, unemployment rate, and the distribution between rural and urban populations. The analysis reveals a nonlinear, U-shaped association between per capita CO_2 emissions and IMR, suggesting that while moderate industrial emissions are related to lower infant mortality, excessive pollution leads to adverse health outcomes. Moreover, the study finds that higher per capita GDP and greater population density are consistently associated with reduced IMR across all quantiles, reflecting the beneficial effects of economic growth and infrastructure concentration on child mortality. In contrast, population growth and a higher rural-to-urban population ratio are positively correlated with IMR, highlighting challenges related to resource allocation, service delivery, and disparities in healthcare access in rapidly growing or predominantly rural regions. Interestingly, the unemployment rate does not demonstrate a statistically significant impact on IMR. These findings collectively emphasise the intricate interaction between industrialisation and health, advocating for threshold-sensitive environmental regulation and strategic healthcare investments to mitigate the adverse effects of development on vulnerable populations.

Keywords Infant mortality rate · CO_2 emissions · Per-capita GDP · Population density · Population growth · Quantile regression

---

S. Banerjee (✉)
Gedu College of Business Studies, Royal University of Bhutan, Gedu, Chhukha, Bhutan
e-mail: subhanil11@gmail.com

D. Panda
Indian Institute of Technology, Kharagpur, India
e-mail: deepak.panda.d1@gmail.com

P. Datta et al. (eds.), *Climate Change and Human Well-being*, Advances in Global Change Research 83, https://doi.org/10.1007/978-3-032-25479-5_4

4.1 Introduction

The increase in carbon dioxide (CO_2) emissions is one of the main causes of environmental degradation and global warming. The increased temperature subsequently results in increased sea level rises and the number of natural disasters like drought and flood, exacerbating the rising greenhouse gas concentrations. The increase in carbon dioxide emissions, along with other challenges of climate change, is also one of the major reasons for numerous health problems that further directly affect the wellbeing of a nation. A number of studies have highlighted the negative health consequences of carbon dioxide emissions, particularly on children (Guo et al., 2024), resulting in mortality among infants. The lower ability to metabolise to remove pollutants, under-developed respiratory system and a weak immune system make children more susceptible to the harmful effects of the air pollutants. Currie et al. (2009) noted that one unit reduction in neonatal average levels of carbon monoxide in the early fourteen days of life can save about 18 lives per one-lakh births. Further, Arceo-Gomez et al. (2012) observed substantial effects of particulate matter and carbon monoxide on neonatal and infant mortality throughout Mexico City. Latifi et al. (2024) highlighted that prenatal exposure to air pollutants, like poly-cyclic hydrocarbons, emitted through crude oil and charcoal burning, has been associated with worse birth outcomes, reduced birth weight and lesser head circumference.

Infant mortality, defined as the death of children under one year of age, is a widely recognised indicator of a country's physical health status, an important parameter of wellbeing. Despite significant progress in reducing infant mortality rates globally, disparities persist, particularly in South Asian (SA) and Sub-Saharan African (SSA) countries (Nwani & Ujah, 2024). Concerning the larger implication of children's morbidity and mortality, the United Nations has dedicated Goal 3, particularly Sect. 3.2, for reducing the mortality rate among children. SDG report 2025 claims that the under-5 child mortality rates (U5CMR) dropped to 37 and the neonatal child mortality rate fell to 17, in 2023, down by 16 and 12 percent since 2015, respectively. Emission of carbon compounds is the most critical impairment to attain SDG 3 targets in SA and SSA countries. The multifaceted impacts further highlight the urgency of addressing CO_2 emissions as part of a comprehensive strategy to improve children's health and well-being.

Building upon this critical backdrop, the present study endeavours to rigorously assess the multifaceted effects of a range of development parameters, such as CO_2 emission, GDP, population growth, population density, unemployment, rural & urban population on infant mortality rates, thereby contributing to the understanding of their interconnectedness.

4.2 Review of Literature

4.2.1 Carbon Emissions and Infant Mortality

Among the numerous determinants of infant mortality, environmental factors, especially those linked to anthropogenic activities, have gained increasing attention in recent decades. CO_2 emissions, as both a greenhouse gas and a proxy for broader air pollution, play a central role in shaping environmental health outcomes, including those affecting the infant populations. Tavassoli et al. (2021) demonstrated that IMR in the US spiked during the converging period of the nineteenth and twentieth centuries, resulting from increased industrial activities and concomitant pollution. Recent research has indicated the negative effects of CO_2 on health, such as IMR. Adeleye et al. (2022) examined the association between income, carbon emissions and mortality rates in SSA, depicting positive correlation between CO_2 emissions and the mortality rates. The result was further established by Adeleye et al. (2023) in the Asia–Pacific. In a similar manner, Sinha (2014) carried out causal analysis in India and established a bi-directional causal relationship between IMR change and CO_2 emissions growth. Guo et al. (2024) also supported the findings among SAARC nations. Carbon emissions have also been highlighted to mediate the link between energy consumption and mortality among children. Nwani and Ujah (2024) further built upon this to discuss the SDG 3 targets, highlighting the pivotal role of carbon emissions, in addressing IMR in SA and SSA countries.

A growing body of empirical research underscores the direct and indirect pathways through which CO_2 emissions influence infant mortality. CO_2 emissions are a major contributor to air pollution, which in turn is associated with a range of hostile health conditions. Recent studies have demonstrated a robust positive association between CO_2 emissions and IMR. Oyedele (2022) investigated the health impacts associated with environmental quality resulting from CO_2 emissions in Nigeria, covering the years 1980 to 2016. The findings and a sensitivity analysis indicated that total carbon dioxide emissions significantly accounted for both IMR and U5CMR. Javanshirova (2024) analysed the relationship in Azerbaijan spanning four decades found a strong positive correlation coefficient of 0.78 between CO_2 emissions and IMR, with econometric modelling found that 1% increase in CO_2 emissions leads to a 1.69% rise in IMR. In Nepal, the association between the indicators is lower, where 1% CO_2 emissions resulted in a 0.11% rise in child mortality, emphasising the adverse consequences of environmental dilapidation (Khatri et al., 2024). Cross-country analysis has revealed that the association between the amount of CO_2 and IMR is both statistically significant and persistent. Studies employing panel data approaches across developing countries have found that a one percent increase in carbon emissions causes a 0.4–0.5% rise in IMR. Associations among countries and across all geographical locations are not uniform. Heft-Neal et al. (2020) estimated that a one percent surge in air pollution impacts infant mortality by a staggering 24% in African economies. These effects are particularly pronounced in regions where externalities related to carbon emissions are not adequately mitigated, and where infants and young children are especially susceptible due to their underdeveloped immune systems.

The association between higher carbon emissions and IMR is, however, not uniform across countries or regions. Evidence from Mexico indicates that 1 part per billion increases in carbon monoxide leads to 0.0032 additional infant deaths per 1,00,000 births (Arceo-Gomez et al., 2012). Notably, the magnitude and nature of these effects may differ between the global north and south, reflecting differences in emission levels, population vulnerability, and the capacity for avoidance behaviours. Erdoğan et al. (2019) examined the effects of CO_2 on health indicators during 1971–2016 in Turkey and presented that higher CO_2 emissions raise IMR. Additionally, there is a long-run co-integration between CO_2 emissions and health indicators. Naeem et al. (2021) analysed the impact of CO_2 emissions on infants' health, considering the data between 1975 and 2013, in Pakistan and found no significant evidence to establish the association in Pakistan. Additionally, Husnain (2016) conducted an econometric estimation into the association between IMR and CO_2 emission in four south Asian nations, namely, Pakistan, India, Sri Lanka and Bangladesh, taking into account data for a period of 1978–2010 and considering controlling variables such as socio-economic factors, rural population growth, trade, remittance and inflation etc. that may have a potential influence on the IMR, indicated that excessive CO_2 results in the increase in IMR but the relationship is rather indirect. Finally, Sial (2022) evaluated the influence of fossil fuel energy utilisation on IMR in fifteen Asian countries for the years 1996 to 2019 and concluded that the association between the predictor and outcome variable follows a U-shaped relationship.

4.2.2 Economic Parameters and Infant Mortality

Economic determinants, specifically GDP per capita, strongly influence child health outcomes. O'Hare et al. (2013) performed a meta-analysis and showed strong negative correlation between rising income levels in the developing nations and child mortality, proving that increased income leads to better healthcare facilities, resulting in reduced mortality among children. This is complemented by Maruthappu et al. (2017), who identified that economic recessions negatively affect the mortality rate among children all over the world, with recessions especially raising child mortality in developing nations. Onanuga and Onanuga (2014) examined the effect of CO_2 emission on IMR, holding constant other covariates of human well-being in SSA, by employing a panel FGLS and GMM for 1990–2012 showed GDP per capita is negatively correlated with infant mortality rate. Besides these observations, Trondillo (2016) emphasised that the changes in child and maternal health are primarily responsible for growth in GDP per capita, highlighting the benefit of investing in the health sector. With a focus on child health outcomes including IMR and U5CMR as explained variables, Salahuddin et al. (2020) examined the degree to which economic growth, FDI and the use of the internet affect the explained variables in South Africa, covering the period of 1985–2016. Using ARDL for the study, the results indicated that foreign investments and economic development have effects on reducing U5CMR and IMR. In addition, Testoni et al. (2021) identified strong correlation between the clusters of neonatal mortality and per capita GDP, indicating direct association between higher economic status and neonatal health. Adeleye et al. (2022) examined an unbalanced panel sample of 47 SSA nations between 2005 and 2019 and found that real per capita earning is linked to declining mortality. Shrestha and Jung (2023) established that health programs in Nepal resulted in substantial declines in gender-specific IMR, highlighting the significance of health intervention policies in reducing IMR and improving children health.

4.2.3 Population Distribution and Infant Mortality

Sims et al. (2007) examined the association between urban poverty and IMR following 1990–1994 data in the US and found that high poverty significantly predicts IMRs in urban areas, with disparities varying across race and ethnicity. Maternal health and regional factors largely explain the variance in IMR among different population groups. Sideeg and Mardi (2011) identified IMR in urban and rural areas of Sudan is 95.2 and 25.4 respectively and concluded a better health outcome for urban infants resulting in a lower IMR compared to rural areas. Kumar and Singh (2014) measured the effects of economic inequality on IMR in both urban and rural areas of India following three successive rounds of the NFHS data and concluded that IMR has significantly dropped for the poor category in rural areas and the rich category in urban areas. Ortigoza et al. (2021) investigated the correlations between urban features and IMR, conducted from 2014 to 2016, utilising data from 286 cities in eight countries and concluded that higher population growth in urban settings is associated with lower IMR. Enhancing living conditions, better provision of services, and easier public transit positively influences IMR in Latin American cities. Naeem et al. (2021) highlighted that the urbanisation rate has positive effects in improving health in Pakistan, over a long period. Interestingly, Macharia et al. (2023) found that neonatal and perinatal mortality rates were highest in central urban areas in Tanzania, suggesting a reversal of the urban advantage, with urban populations experiencing poorer birth outcomes compared to rural areas.

4.2.4 Research Gaps

Researchers aiming to contribute to the literature have predominantly examined the connection between CO_2 emissions and health indicators, particularly IMR and U5CMR, based on the data of a single or some regional countries. Identification of the connection between CO_2 emissions and IMR across the world is paramount from the perspective of finding a linkage across the world that further gives suggestions regarding the adverse impacts of global environmental degradation. Further, existing researches has drawn non-uniform, rather opposite, associations between various predictors and IMR that create adversities for policy development in both health and environmental domains.

4.2.5 *Objectives*

Previous research suggests that IMR is influenced by an interaction of a number of environmental, economic, and social determinants. Carbon emissions, population density, economic capacity, and rate of urbanisation together determine the health outcomes, especially IMR. The study's primary objective is to identify the association of CO_2 emissions, GDP per capita, population density, population growth, unemployment, rural & urban population on the health variable. IMR, often employed in cross-country studies as an indicator of health and wellbeing, is employed as the dependent variable and other development variables as predictors. The current study intended to make an overall global estimate of the relationship of the predictors with IMR, but the current study is restricted to 168 countries, considering the availability of secondary data for estimation.

4.3 Methodology

The methodology section comprises of three parts: data under consideration and their sources, the econometric model and the pertinent approximation methods.

4.3.1 *Sources of Data*

The information on infant mortality rate (IMR), per capita CO_2 emission measured in metric tons (PCO), per capita gross domestic product measured in constant USD 2015 (PGDP), population density (PD), annual population growth rate (PG), rate of unemployment (UEP), rural and urban population in numbers have been obtained from World Development Indicator (2024) for 168 countries, for a period of 21 years (2000–2020).

4.3.2 *Model Specification*

The IMR_{ij} is considered as the regressand. The PCO, square of PCO (PCO2), PGDP, PD, PG, UEP and rural–urban population ratio (RUR) have been considered as the regressors.

In functional form the relation might be represented as

$$IMR_{ij} = f\left(PCO_{ij}, PCO2_{ij}, PGDP_{ij}, PD_{ij}, PG_{ij}, UEP_{ij}, RUR_{ij}\right) \quad (4.1)$$

The regression equation following the aforementioned functional form might be represented as

$$IMR_{ij} = a + b_1 \times PCO_{ij} + b_2 \times PCO2_{ij} + b_3 \times PGDP_{ij} + \ldots + b_7 \times RUR_{ij} + \varepsilon \quad (4.2)$$

where i = 1, 2, … 168 and j = 2000 … 2020. ε_{ij} is a stochastic random disturbance. a refers to the intercept term, and b_1……b_7, are the slopes of the adjacent variables.

4.3.3 Estimation Techniques

The estimation technique starts with the creation of novel digitalization index through principal component analysis and followed by descriptive statistics. Then, considering the data is a panel data and resorts to multiple regression, it has been tested for multicollinearity (Variance inflation factor), autocorrelation (Wooldridge, 2010), heteroskedasticity (Greene, 2000), CD (Fan et al., 2015; Juodis & Reese, 2021; Pesaran, 2004, 2015, 2021; Pesaran & Xie, 2021); and endogeneity problem (Semykina & Wooldridge, 2010; Wooldridge, 2010). Then, the stationarity test was adopted, followed by the cointegration tests before resorting to the primary regression technique for the present article, the MMQR. Following the analysis, the Granger causality test was applied, and finally, the robustness of the model was proven and discussed, feeding the same data to twelve kinds of panel data models.

4.3.3.1 Contemporaneous Correlation and Heterogeneous Slope Tests

Ignoring CD if it exists can lead to spurious regression, specifically in short panel analysis (Cameron & Trivedi, 2005; Chudik et al., 2011). The logic behind providing different CD tests is that they portray successive improvements over the previous tests. Starting with Pesaran (2015) general diagnostic test of CD in panels that mainly focused on average correlation CD and revisited to test weak CD (2021). Fan et al. (2015) who focused on power-enhanced weak CD in panels and Juodis and Reese (2021) who reflected on the weak CD among variables from the perspective of the instrumental parameters problem. Finally, the Pesaran and Xie (2021) CD test addresses the biases involved in the previous CD tests. It presents a bias-corrected CD test that also provides direct corrections on the asymptotic biases of the existing CD methods. Further, to test the slope level heterogeneity, Pesaran and Yamagata (2008) slope level heterogeneity test has been applied. Slope level heterogeneity test becomes specifically important in the presence of CD as it can be used to compare with the parameter values.

$$CD = \sqrt{\frac{2T}{n(n-1)}}\left(\sum_{i=1}^{n-1}\sum_{j=i+1}^{n} pj, T\right) \tag{4.3a}$$

$$CDw = \sqrt{\frac{2}{Tn(n-1)}}\sum_{t-1}^{T}\sum_{i=2}^{n}\sum_{j-1}^{t-1}(w_t\varepsilon_{it})(w_t\varepsilon_{it}) \tag{4.3b}$$

$$CD_{w+} = CD_w + \Delta_{nT} \tag{4.3c}$$

$$CD^* = \frac{CD + \sqrt{\frac{T}{2}}\theta_n}{1-\theta_n} \tag{4.3d}$$

$$\Delta HT = (N)\frac{1}{2}(2k) - \frac{1}{2}\left(\frac{1}{N}S - k\right) \tag{4.4a}$$

$$\Delta AHT = (N)\frac{1}{2}\left(\frac{2k-(T-k-1)}{T+1}\right)\frac{1}{2}\left(\frac{1}{N}S - k\right) \tag{4.4b}$$

Considering the above equations, the (4.3a) represents the Pesaran (2015) CD test. Similarly, Eq. (4.3b) represents the newly invented Juodis and Rees (2021) CD test. Equation (4.3c) represents Fan et al.'s weak CD test (2015). Furthermore, Pesaran and Xie (2021) invented a bias-corrected CD test that is superior to other CD tests and immune to asymptotic biases in prevailing CD tests.

Curiously, $\theta_n = 1 - \frac{1}{n}\sum_{i=1}^{n} \text{Ø}_{i,n}^2$ settles the presence of CD and instrumental dependency in functional form in the mathematical construct as represented by Eq. (4.3d). Moreover, Eqs. (4.4a) and (4.4b) are the Pesaran & Yamagata Slope Heterogeneity test in functional forms. Again, ΔHT stands for the nominal tilde and ΔAHT represents the adjusted delta tilde of the across-nation slopes.

4.3.3.2 Panel Stationarity Test Techniques

The traditional unit-root tests are susceptible to the presence of CD (Uche et al., 2024). In the presence of CD, the CD-adjusted Dickey-Fuller (CADF) Test (Pesaran, 2007) and Cross-sectionally adjusted lm-Pesaran-Shin (CIPS) tests (lm et al., 2003) have been followed. Both are 2nd generation unit-root tests and work perfectly in detecting an incidence of unit-root even under CD. Both of these unit root test has been joined by Karavias and Tzavalis (2014) a third-generation unit root test, which is not only works perfectly under CD but also capable of putting structural breaks where needed and can be adjusted for slope level heterogeneity.

$$CIPS = \frac{1}{N}\sum_{i=1}^{N} \partial, (N, T) \quad (4.5)$$

$$\Delta_{it} = \sigma_{it} + \beta_{it-1} + \theta_1 T + \sum_{j-1}^{N} \partial_{ij} \Delta \times_{it-1} + \mu_{it} \quad (4.6)$$

$$\varphi' = \left(\sum_{i=1}^{n} \theta'_{t-1}\theta_m^{(\lambda)}\theta_{i-1}\right)^{-1}\left(\sum_{i=1}^{n} \theta'_{t-1}\theta_m^{(\lambda)}\theta_{i-1}\right), m(M1, M2) \quad (4.7)$$

Equations (4.5), (4.6) and (4.7) are respective depictions of CIPS, CADF, and KT tests. Again, considering the Karavias and Tzavalis panel stationarity test technique, the H_0 states that some panel series possess a unit root. On the other hand, the H_a states that some or all the panel time series are bereft of unit root. Similarly, in Eq. (4.7), $\theta_m^{(\lambda)}$ is the conversion matrix of the pertinent longitudinal datasets.

4.3.3.3 Panel Cointegration Check Methods

The first-generation panel cointegration tests often provide spurious results in the presence of cross-sectional dependency and are subject to common factor restrictions (Uche et al., 2024). The Westerlund (2007) test with bootstrapping acts precisely in the presence of contemporaneous correlation and is also free from common factor boundaries (Persyn & Westerlund, 2008).

4.3.3.4 Primary Empirical Approach: A Detailed Description

The data has been verified for normality through the Jarque Bera test and supported by Shapiro-Francia, Shapiro–Wilk and the Chen-Shapiro test (Adefisoye et al., 2016) of normality. These steps are necessary to decide between Least Squares and quantile regression. In case errors are normally distributed, and error structures are known. Least square regression is still preferred over quantile regression. Furthermore, the previously mentioned unit root test results have been considered as the Novel MMQR method can only be applied if all the variables are stationary at the level or first difference or a combination of them.

Justification of Choosing MMQR Over Ordinary Least Squares

Consequently, the MMQR panel model is an asymmetric panel analysis similar to Machado and Silva (2019). It is robust to heteroskedasticity and outliers. As per Machado and Silva (2019), the MMQR validates extra expansively, using the methods of the moment, the stages and likely impacts of the variations in the regressors on the regressand along the distributions. The method also unearths heterogeneity within conditional distributions of the regressand (Guan et al., 2023; Machado & Silva, 2019).

MMQR might be used on other occasions but acts best while modelling the entire conditional distribution. Whereas the accuracy of the least squares is mean-centric and assumes constant variance, MMQR is free from these parametric assumptions. This particular attitude makes its findings more robust than least squares. Cook and Manning (2013) favoured quantile regression over other forms of regression as it offers greater flexibility. Quantile regression identifies the different impacts of the different parts of the distribution on the regressand by the regressors. However, as mentioned before, the Least Square Estimate should still be preferred in case of known error distribution. In such cases, error distribution is normally distributed. However, quantile regression is the best option if the error distribution is unknown. Cleophas et al. (2021) have shown that even in the presence of autocorrelation, the quantile regression provides accurate and precise estimation. As per Rodriguez and Yao (2017), quantile regression also provides unbiased inference in the presence of heteroskedasticity. Moreover, though OLS with Driscoll and Kraay (1998) standard error correction can adjust for heteroskedasticity, autocorrelation and contemporaneous correlation, "Nevertheless, the QR was the chosen statistical tool based on its superior to the OLS for different reasons. The standard circulation, as well as the zero, mean approval of the OLS error concept, is rather unrealistic, since there may be multiple distribution models for socioeconomic measures. The QR reinforces this deficit. The methodology (QR) does not presume the function of the period. In the case of outlier forecasts remain robust" (Gyamfi et al., 2021, p. 5). Again, the novel MMQR also takes care of the CD (Bakhsh et al., 2024; Damrah et al., 2022; Machado & Silva, 2019).

In this context, if the previously mentioned error tests favour the presence of mentioned errors, the data distribution is not normal then quantile regression is preferred over any other forms of least square estimates. However, the present article moves one step ahead of the quantile regression, as illustrated below.

Novel MMQR: A Step Ahead

In a recent article, Jahanger et al. (2023) criticized the traditional quantile regression methods presented by Koenker and Bassett (1982) due to their inability to address the ill effect of unobserved heterogeneity across the panel cross-section. However, the Novel MMQR unearths a fountain of benefits. It provides unbiased inferences and is not subject to outliers, heterogeneity, autocorrelation, contemporaneous correlation, and endogeneity (Machado & Silva, 2019). The mentioned method takes care of even the non-linearity and non-normality matters (Jahanger et al., 2023). The concerned regression is robust to country-specific fixed effects that counterattack incidental parameter challenges across distribution (Bista & Khan, 2023). Moreover, the method is sensitive to variables subject to non-normal distributions. Finally, apart from predicting the interaction of variables at location and scale level, the mentioned method also includes the interaction of the variables at the different lower, middle and upper quantiles, enabling it in short, intermediate and long-run impact analysis of the regressors on the regressand.

Location Coefficients: These coefficients capture the central tendency (like the mean) of the regressand variable at a specific quantile, given the regressors. They are estimated in the first step, often using Ordinary Least Squares (OLS) on a transformed data version. The transformation accounts for the targeted quantile (e.g., median, 10th percentile).

Scale Coefficients: These coefficients represent the regressand variable's variability (like standard deviation) around the estimated location at a specific quantile. They capture the phenomenon of heteroskedasticity, where the variance of the error term changes with the regressors. Scale coefficients are estimated in a second step using an auxiliary regression that leverages the residuals from the first step.

Here is a key point: Novel MMQR combines the estimation of location and scale to provide a more comprehensive picture of the quantile regression compared to standard quantile regression, which only estimates location coefficients.

$$Q_{\gamma}(\tau|X_{it}) = (\alpha_i + \theta_i\delta_i) + X_{it}'\beta + K_{it}'\gamma\rho(\tau) \tag{4.8}$$

In Eq. (4.8), $Q_{\gamma}(\tau|X_{it})$ represents the conditional quantile distributions of the regressand. X_{it} stands for the vector of the regressors and constant variables. $\alpha_i(\tau) = \alpha_i + \theta_i\delta_i(\tau)$ portrays the scalar coefficient of the τ-*th* quantile of the individual (i) specific fixed effects. K_{it}' is the K-vector, while $\rho(\tau)$ stands for the τ-th quantile formulated through an optimization technique represented in Eq. (4.9).

$$\min \theta \sum_i \sum_t \sigma_\tau\left(F_{it} - \left(\delta + K_{it}'\gamma\right)\theta\right) \tag{4.9}$$

In Eq. (4.9), $\rho_\tau(A) = (\tau - 1)\, AI\, \{A \leq 0\} + \tau AI(\{A > 0\})$ shows the quantile conditional check function. Thus, this exclusive time-varying panel estimator would efficiently complement any possible deviance from normal distributions, structural breaks, and time variations.

4.3.3.5 Panel Level Granger Casualty Test

A recent improved Granger non-causality test has been applied (Xiao et al., 2023).

$$Y_{i,t} = \emptyset_{0,i} + \sum_{p=1}^{p} \emptyset_{p,i} Y_{i,t-p} + \sum_{p=1}^{p} \beta_{p,i} X_{i,t-p} + \varepsilon_{i,t} \tag{4.10}$$

- $\beta_{p,i}$ are the heterogeneous feedback coefficients or Granger causality parameters.

The hypothesis of Granger non-causality is:$H_0 : \beta_{p,i} = 0$ *for all i and p*.

The alternative hypothesis is:$H_1 : \beta_{p,i} \neq 0$ *for all i and p.*
The JKS test: key estimation ideas.

- Idea 1: Under the null hypothesis: $\beta_{p,i} = 0$ for all i and p, and thus we can use a more efficient $\left(\sqrt{NT}\right)$ pooled estimator

$$\hat{\beta} = \left(\sum_{i=1}^{N} X_i' M_{Z_i} X_i\right)^{-1} \left(\sum_{i=1}^{N} X_i' M_{Z_i Y_i}\right) \tag{4.11}$$

Distributed as.

$$\sqrt{NT}\left(\hat{\beta} - \beta_0\right) \rightarrow J^{-1} N(-kB, V), \tag{4.12}$$

- Idea 2: Remove the bias using the half-panel jackknife estimator (Dhaene & Jochmans, 2015):

$$\tilde{\beta} = \hat{\beta} + \left(\hat{\beta} - \frac{1}{2}\left(\hat{\beta}_{1/2} + \hat{\beta}_{2/1}\right)\right) = \hat{\beta} + T^{-1}\hat{B} \tag{4.13}$$

- As N, T $\rightarrow \infty$ with$\frac{N}{T} \rightarrow k^2 \in (0, \infty)$, we have:

$$\hat{W}_{HPJ} = NT\tilde{\beta}'\left(\hat{J}^{-1}\hat{V}\hat{J}^{-1}\right)^{-1}\tilde{\beta} \rightarrow x^2(P), \tag{4.14}$$

where

$$\hat{J} = (NT)^{-1} \sum_{i=1}^{N} X_i' M_{Z_i} X_i \tag{4.15}$$

4.4 Econometric Analysis

The econometric analysis starts with descriptive statistics as depicted in Table 4.1.

The huge difference between the estimated mean and median barring PG, UEP and RUR points to the probability of outliers in the data under consideration. However, as mentioned earlier that Novel MMQR is robust to outliers. The standard deviations barring PG and RUR are quite high, which implies high variability over time in the data. The Kurtosis test out IMR, PCO, PCO2, PG, PGDP, PD, PG, UEP, and RUR analysis depicts that all the variables have relatively heavier tails as the kurtosis values for these three variables are way over the threshold limit 3.

Table 4.1 Descriptive statistics

Values	IMR	PCO	PCO2	PG	PGDP	PD	UEP	RUR
Mean	28.3621	4.7880	65.7396	1.4998	12,465.12	181.2171	7.8148	1.2721
Median	18.0500	2.3739	5.6355	1.3551	4430.896	72.6603	6.0955	0.7755
Maximum	234.9	53.5984	2872.787	21.7003	112,417.9	7965.878	37.32	11.1271
Minimum	1.8	0.0242	0.0006	−11.3567	233.0324	1.5434	0.1	0
Standard Deviation	26.3622	6.5442	217.0745	1.5920	17,892.8	580.8125	5.8267	1.4055
Skewness	1.2754	2.9651	8.0366	1.8178	2.2323	10.4749	1.4393	2.1887
Kurtosis	4.9238	15.8777	84.4156	20.3185	8.4918	126.3379	5.2998	9.2918
Jarque–Bera test of normality	< 0.001	< 0.001	< 0.001	< 0.001	< 0.001	< 0.001	< 0.001	< 0.001
Shapiro-Francia test of normality	< 0.001	< 0.001	< 0.001	< 0.001	< 0.001	< 0.001	< 0.001	< 0.001
Sum	100,061.5	16,892.1	231,929.3	5291.3	43,976,943.4	639,333.9	27,570.6	4488.0
Sum Sq deviation	2,451,644.0	151,094.36	166,244,123	8941.7	112896X106	1,190,146,457	119,775.6	6969.2

Source Computed

Table 4.2 Test results for multicollinearity, Heteroskedasticity and autocorrelation

IMR_{ij} (Regressand)	Regressors: PCO_{ij}, $PCO2_{ij}$, $PGDP_{ij}$, PG_{ij}, PD_{ij}, UEP_{ij}, RUR_{ij}	
Group (unit) effect		
	Stat	Prob
Multicollinearity	$2 < 2.96 < 5$	NA
Autocorrelation	2.038	0.1553
Heteroskedasticity	1.5e + 07	0.0000
Endogeneity	0.26	0.7930

Source Computed

Considering the skewness value, all the variables have a rightward tail. Again, the variables under consideration are non-symmetrical, considering their standard deviation. The Jarque–Bera statistics also reject that the data are normally distributed. The result has been supported by Shapiro-Francia test.

Now, the analysis moves to the test of multicollinearity, autocorrelation, heteroskedasticity and endogeneity as illustrated in the introductory part of Sect. 3.3.

It is apparent from Table 4.2 that there is no multicollinearity, endogeneity and autocorrelation problem, but the error structure has heteroskedasticity. Table 4.3 represents the mentioned CD and slope coefficient heterogeneity tests as depicted in Sect. 3.3.1.

It is apparent from Table 4.3 that the variables are cross-sectionally dependent and subject to slope coefficient heterogeneity. In this background, as reasoned in

Table 4.3 Results of different CD tests and slope Heterogeneity test

Variables	CD (Pesaran, 2015, 2021)	CD (Judios & Rees, 2021)	CD (Fan et al., 2015)	CD (Pesaran & Xie, 2021)
IMR	447.56***	−2.92***	56,256.82***	−0.27
PCO	37.06***	−0.82	34,997.47***	9.98***
PCO2	32.27***	−0.42	34,313.91***	10.74***
PGDP	287.15***	−0.35	45,219.14***	−0.26
PG	22.13***	2.47**	23,505.01***	13.92***
PD	297.36***	8.55***	58,447.54***	−1.00
UEP	33.46***	0.88	23,348.33***	17.87***
RUR	317.16	14.75	54,289.17***	−1.27
Slope heterogeneity test (Pesaran & Yamagata, 2008)				
	Delta		p-value	
	5.161		0.0000	
adj	9.305		0.0000	

Source Computed (*, ** and *** imply 90%, 95% and 99% confidence level respectively)

Sect. 3.3.2, CADF and CIPS, along with Karavias and Tzavalis (2014) test of unit root, have been opted and findings are depicted in Table 4.4.

As depicted in Table 4.4, all the variables are stationary at the level I(0) or the first difference I(1) at 95% to 99% statistical significance level. Now, the mentioned test of cointegration as portrayed in Sect. 3.3.3 has been followed, and the findings are depicted in Table 4.5.

In Table 4.5, Gt and Ga represent the cross-sectional statistics, and Pt and Pa represent the panel statistics. One statistically significant robust p-value (since bootstrap has been applied, hence robust p-value has been chosen) is adequate to discard the null hypothesis of zero cointegration. However, there are two at 99% confidence level so the variables are cointegrated.

As discussed in Sect. 3.3.4, Novel MMQR handles the mentioned errors without further adjustments, so the method has been applied and results are depicted in Table 4.6.

It is evident from Table 4.6 that PCO has a negative and statistically significant impact on IMR over the location, scale and all the quantiles. However, it's square value though remains positive but insignificant in quantile 5 but exerts a positive impact on IMR in a statistically significant way over the location scales and quantile

Table 4.4 Unit root test statistics

Variables	CADF test statistics and probability		CIPS Test Statistics and Probability		Karavias and Tzavalis	
	I (0)	I (1)	I (0)	I (1)	Statistics	Break Date
IMR	−2.226***	−0.118	−2.302***	−2.419***	3.7144***	2001
PCO	12.214	−2.33***	−1.027	−3.883***	−0.8194	NA
PCO2	11.300***	−1.703**	−1.122	−3.899**	−8.3326	NA
PGDP	2.890	−2.542***	−1.457	−2.586***	0.9180***	2019
PG	5.936	−15.470***	−1.776	−3.004***	−10.9762	NA
PD	−0.489	−3.067***	−1.788	−1.992	4.2586***	2006
UEP	5.623	−5.222***	−1.339	−2.991***	−0.3792	NA
RUR	3.394	−2.064**	−2.059**	−2.009*	0.0012***	2001

Source Computed (*, ** and *** imply 90%, 95% and 99% confidence level respectively)

Table 4.5 Westerlund bootstrap test for cointegration

Statistics	Value	Z value	p-value	Robust p-value
Gt	−1.378	10.448	1.000	0.000***
Ga	−0.176	19.492	1.000	0.000***
Pt	−57.053	−26.729	0.000	1.000
Pa	−2.529	8.968	1.000	1.000

Source Computed (*** imply 99% confidence level)

Table 4.6 MMQR estimates (regressand, IMR)

Series	Location	Scale	Quantile grids				
			q_tile 5	q_tile 25	q_tile 50	q_tile 75	q_tile 95
PCO	−2.1348***	−0.5878***	−1.2241***	−1.5990***	−1.9527***	−2.5563***	−3.5825***
PCO2	0.03389***	0.0167***	0.0080	0.0186***	0.0287***	0.0459***	0.0751***
PGDP	−0.0003***	−0.00004***	−0.0002***	−0.0002***	−0.0003***	−0.0003***	−0.0004***
PD	−0.0023***	0.0008	−0.0035***	−0.0030***	−0.0025***	−0.0018***	−0.0004***
PG	5.3726***	1.5802***	2.9242***	3.9321***	4.8829***	6.5057***	9.2647***
UEP	−0.0271	0.0385	−0.0868	−0.0622	−0.0391	0.0005	0.0676
RUR	3.9148***	1.3566***	1.8129***	2.6782***	3.4944***	4.8875***	7.2561***
Constant	27.3316***	10.8857***	10.4654***	17.4087***	23.9587***	35.1375***	54.1437***

Source Computed (** and *** imply 95% and 99% confidence level, respectively)

25 onwards. This strange finding proposes that in the initial stages as PCO increases the IMR will decline. This may come from the proliferation of the pharmaceutical industries that are polluting in nature but important in reducing the IMR. However, the unprecedented growth of polluting industries will eventually higher the IMR. The PGDP, and PD reduces IMR in a statistically significant way over location, scale and all the quantiles. Signifying that with the betterment of livelihood that accompanies rise in PGDP, IMR will decline. Similarly rise in PD might refer to rise in infrastructure and that will eventually decline the IMR. On the other hand, growth of population and rural–urban population ratio increases the IMR. Growth of population might refer to the less attention to the existing infants, quick pregnancy leading to weak progenies and thus an ambience of uncertainty around the survival of the infants. The rural places are usually less equipped with health facilities than the urban places so rise in rural–urban population ratio is detrimental to infant survival. The unemployment though remains insignificant in its impact on IMR over location, scale and all the quantiles.

Now, as mentioned in Sect. 3.4.5 the analysis resorts to the mentioned Granger Causality Test. The result is given in Table 4.7.

The Granger non-causality test results (Table 4.7) demonstrate that the selected covariates have significant predictive power over carbon emission, as evidenced by the strong rejection of the null hypothesis at conventional significance levels. This finding aligns with a growing body of literature that emphasises the complex interconnections between economic, policy, and technological variables and carbon emissions dynamics. The use of an optimal lag structure, as determined by the BIC, strengthens the robustness of the result, ensuring that the temporal dependencies in the data are appropriately captured.

Table 4.7 Xiao et al. (2023) Granger non-causality test

Number of units	168	Observation per unit	19
Number of lags	2	BIC	1547.382
H0: selected co-variates do not granger cause coe			
H1: H0 is violated			
HPJ Wald test	41.3136		
p-value	0.0002		
BIC selection			
Lags = 1	BIC	6419.6546	
Lags = 2	BIC	1547.3819*	
Lags = 3	BIC	6593.7044	
Lags = 4	BIC	5947.9398	

Source Computed

4.5 Conclusions and Policy Implications

The present study tries to assess the relation between PCO and IMR across the world and found that they have a nonlinear, complex relationship, with initial industrialisation having a dramatic reduction in IMR in all quantiles, apparently due to enhanced availability of pharmaceuticals and healthcare facilities with industrial development, but showing a U-shaped reversal impact from the median quantile onwards, indicated by the statistically significant positive coefficient of PCO's square term, that indicates too much pollution ultimately overrules these advantages and raises IMR, thereby affecting the wellbeing aspects. This contributes to the research gap, which present a diverse claim about the relationship between the variables. Concurrently, per capita GDP and population density exhibit statistically significant negative impacts on IMR across all quantiles, indicating how urbanisation and economic development enhance infant survival through improved healthcare accessibility, nutrition, and infrastructure, whereas population growth as well as increased rural–urban population ratios substantively add to IMR, possibly as a result of overpopulation, decreased per-capita healthcare attention, and urban–rural disparity in quality of medical services, with unemployment having no statistically significant impact, suggesting its effects are either being negated by other socioeconomic determinants or are taking place by means of indirect mechanisms.

These results call for a multi-faceted policy response to balance industrialisation with infant health protections: (1) Pollution-Controlled Industrialisation- governments need to enforce stringent emission standards on pharmaceutical and other polluting industries and encourage green technologies through tax credits and subsidies to avoid the U-shaped pollution-IMR tipping point, supplemented by real-time air/water quality monitoring around industrial agglomerations and health centres to anticipatively neutralise threats and benefit the wellbeing of the upcoming generations; (2) Rural Healthcare Rejuvenation-considering the detrimental impact of rural–urban gaps, policy should focus on constructing neonatal intensive care facilities in rural areas, educating community health workers in maternal/infant care, and promoting mobile health units to eliminate accessibility deficiencies, funded through earmarked shares in GDP growth revenues; (3) Urban Planning for Health Equality-cities need to have infant-health-sensitive zoning regulations that concentrate low-cost housing around healthcare clusters, while improving slum sanitation and nutrition initiatives to negate IMR risks caused by over-saturation. By contributing to these factors, governments will create a better liveable social environment that contributes to the physical wellbeing of the residents; (4) Demographic-Responsive Health Budgeting—to counteract population growth's pressure on infant survival, governments need to implement dynamic health budgeting models that automatically adjust allocations between regions according to birth rate information, along with pregnancy spacing campaigns and enhanced postnatal care networks; (5) Economic Growth-Health Nexus Maximisation- since Per capita GDP decelerates IMR, development policies need to make explicit the correlation of industrial/urban investments with health improvements, such as requiring corporate social responsibility funding for paediatric hospitals in high-growth areas and anchoring infrastructure projects to maternal health clinics; and (6) Data-Driven Early Warning Systems-drawing on the quantile-specific results, policymakers need to implement AI-powered IMR surveillance systems and spatial clustering approach that identify high-risk areas based on real-time economic/pollution data so that targeted interventions can be made before mortality rates accelerate. Most importantly, such actions should be mainstreamed into national climate adaptation plans, since industrial pollution's lagged IMR impacts reflect climate health threats that necessitate intersectoral coordination among health, industry, and environmental policymakers to coordinate infant survival objectives with sustainable development goals, to secure that economic advancement does not compromise long-term public health. By implementing this multi-scale, evidence-based strategy, societies are able to tap into climate change's adverse affecting potential and, systematically, minimise its harms, and navigate towards a sustainable growth path promoting all aspects of wellbeing.

4.6 Limitations and Scope for Further Research

The current research highlighted the adverse effects of environmental externalities, in form of per capita carbon emission, on physical health aspects across 168 countries by taking data into account over a span of 21 years. The research utilised a Novel MMQR method and established a U-shaped reversal impact after the median quantile that established that the effect of PCO on IMR, changes over due course. However, the present study limits in addressing the effect from regional perspectives. Further researches can address the limitation by exploring the relationship between the regressor and regressand by taking regional groups of countries into account. Further researches also can explore the relationship on a larger time frame that might bring deeper novel insights between the parameters.

References

Adefisoye, J., Golam Kibria, B., & George, F. (2016). Performances of several univariate tests of normality: An empirical study. *Journal of Biometrics and Biostatistics, 7*(4), 1–8.

Adeleye, B. N., Azam, M., & Bekun, F. V. (2023). Infant mortality rate and non-renewable energy consumption in Asia and the Pacific: The mediating role of carbon emissions. *Air Quality, Atmosphere and Health, 16*(7), 1333–1344. https://doi.org/10.1007/s11869-023-01347-8

Adeleye, B. N., Olohunlana, A. O., Ibukun, C. O., Soremi, T., & Suleiman, B. (2022). Mortality rate, carbon emissions, renewable energy and per capita income nexus in Sub-Saharan Africa. *PLoS ONE, 17*(9), Article e0274447. https://doi.org/10.1371/journal.pone.0274447

Arceo-Gomez, E., Hanna, R., & Oliva, P. (2012). Does the effect of pollution on infant mortality differ between developing and developed countries? Evidence from Mexico City. HKS Working Paper No. RWP12-050. https://doi.org/10.2139/ssrn.2137022

Bakhsh, S., Zhang, W., Ali, K., & Oláh, J. (2024). Strategy towards sustainable energy transition: The effect of environmental governance, economic complexity and geopolitics. *Energy Strategy Reviews, 52*, Article 101330.

Bista, R., & Khan, M. I. (2023). Financial development and the distribution of trade flows. *Applied Economics Letters, 30*(17), 2370–2376.

Cameron, A. C., & Trivedi, P. K. (2005). *Microeconometrics: Methods and applications.* Cambridge University Press.

Chudik, A., Pesaran, M. H., & Tosetti, E. (2011). Weak and strong cross-section dependence and estimation of large panels.

Cleophas, T. J., Zwinderman, A. H., Cleophas, T. J., & Zwinderman, A. H. (2021). More on nonlinear regressions: Assessment of nonlinear effects in clinical research. *Regression Analysis in Medical Research: for Starters and 2nd Levelers*, 291–312.

Currie, J., Neidell, M., & Schmieder, J. F. (2009). Air pollution and infant health: Lessons from New Jersey. *Journal of Health Economics, 28*(3), 688–703. https://doi.org/10.1016/j.jhealeco.2009.02.001

Damrah, S., Satrovic, E., Atyeh, M., & Shawtari, F. A. (2022). Employing the panel quantile regression approach to examine the role of natural resources in achieving environmental sustainability: Does globalization create some difference? *Mathematics, 10*(24), 4795.

Dhaene, G., & Jochmans, K. (2015). Split-panel jackknife estimation of fixed-effect models. *The Review of Economic Studies, 82*(3(292)), 991–1030. http://www.jstor.org/stable/43551728

Driscoll, J. C., & Kraay, A. C. (1998). Consistent covariance matrix estimation with spatially dependent panel data. *Review of Economics and Statistics, 80*(4), 549–560.

Erdoğan, S., Yıldırım, D. Ç., & Gedikli, A. (2019). The relationship between CO_2 emissions and health indicators: The case of Turkey. *Econometrics Letters, 6*(1), 28–39.

Fan, J., Liao, Y., & Yao, J. (2015). Power enhancement in high-dimensional cross-sectional tests. *Econometrica, 83*(4), 1497–1541.

Greene, W. H. (2000). *Econometric analysis* (4th ed., pp. 201–215). Prentice Hall.

Guan, Y., Yan, J., Shan, Y., Zhou, Y., Hang, Y., Li, R., et al. (2023). Burden of the global energy price crisis on households. *Nature Energy, 8*(3), 304–316.

Guo, A., Ullah, O., Zeb, A., Din, N. U., Hussain, S. (2024). Unveiling health dynamics: Exploring the impact of CO_2 emissions, Urbanization, and renewable energy on life expectancy and infant mortality in SAARC Countries (1990–2022), pp. 1–28. Natural Resources Forum.

Gyamfi, B. A., Adedoyin, F. F., Bein, M. A., Bekun, F. V., & Agozie, D. Q. (2021). The anthropogenic consequences of energy consumption in E7 economies: Juxtaposing roles of renewable, coal, nuclear, oil and gas energy: Evidence from panel quantile method. *Journal of Cleaner Production, 295*, Article 126373.

Heft-Neal, S., Burney, J., Bendavid, E., Voss, K. K., & Burke, M. (2020). Dust pollution from the Sahara and African infant mortality. *Nature Sustainability, 3*(10), 863–871. https://doi.org/10.1038/s41893-020-0562-1

Husnain, M. I. U., Haider, A., Salman, A., Zahid, H. M., Khan, M., & Shaheen, F. (2016). An econometric analysis of the statistical relationship between carbon dioxide emissions and infant mortality in South Asia. *Journal of Scientific Research and Reports, 9*(2), 1–8.

Im, K. S., Pesaran, M. H., & Shin, Y. (2003). Testing for unit roots in heterogeneous panels. *Journal of Econometrics, 115*(1), 53–74.

Jahanger, A., Zaman, U., Hossain, M. R., & Awan, A. (2023). Articulating CO_2 emissions limiting roles of nuclear energy and ICT under the EKC hypothesis: An application of non-parametric MMQR approach. *Geoscience Frontiers, 14*(5), Article 101589.

Javanshirova, Z. (2024). The impact of carbon emissions on infant mortality rate in Azerbaijan. *Journal of Sustainable Development Issues*, 2(2), 104–114. https://doi.org/10.62433/josdi.v2i2.34

Juodis, A., & Reese, S. (2021). The incidental parameters problem in testing for remaining cross-section correlation. *Journal of Business & Economic Statistics, 40*(3), 1191–1203. https://doi.org/10.1080/07350015.2021.1906687

Karavias, Y., & Tzavalis, E. (2014). Testing for unit roots in short panels allowing for a structural break. *Computational Statistics and Data Analysis, 76*, 391–407.

Khatri, B. B., Poudel, O., Timsina, T. R., & Sapkota, K. N. (2024). Effects of carbon dioxide emissions on child mortality in Nepal. *International Journal of Energy Economics and Policy, 15*(1), 566–577. https://doi.org/10.32479/ijeep.17804

Koenker, R., & Bassett Jr, G. (1982). Robust tests for heteroscedasticity based on regression quantiles. *Econometrica: Journal of the Econometric Society*, 43–61.

Kumar, A., & Singh, A. (2014). Is economic inequality in infant mortality higher in urban than in rural India? *Maternal and Child Health Journal, 18*(9), 2061–2070. https://doi.org/10.1007/s10995-014-1452-9

Latifi, A. M., Abdi, F., Miri, M., Ashtari, S., Ghalandarpoor-attar, S. N., Mohamadzadeh, M., Imani Fooladi, A. A., Uddin, S., & Vahedian-azimi, A. (2024). Association between maternal exposure to polycyclic aromatic hydrocarbons and birth anthropometric outcomes: A systematic review and meta-analysis. In *Ecotoxicology and environmental safety* (vol. 287). Academic Press. https://doi.org/10.1016/j.ecoenv.2024.117290

Le Cook, B., & Manning, W. G. (2013). Thinking beyond the mean: A practical guide for using quantile regression methods for health services research. *Shanghai Archives of Psychiatry, 25*(1), 55.

Machado, J. A., & Silva, J. S. (2019). Quantiles via moments. *Journal of Econometrics, 213*(1), 145–173.

Macharia, P. M., Beňová, L., Pinchoff, J., Semaan, A., Pembe, A. B., Christou, A., & Hanson, C. (2023). Neonatal and perinatal mortality in the urban continuum: A geospatial analysis of the household survey, satellite imagery and travel time data in Tanzania. *BMJ Global Health, 8*, Article e011253. https://doi.org/10.1136/bmjgh-2022-011253

Maruthappu, M., Watson, R. A., Watkins, J., Zeltner, T., Raine, R., & Atun, R. (2017). Effects of economic downturns on child mortality: A global economic analysis, 1981–2010. *BMJ Global Health, 2*(2), Article e000157.

Naeem, M., Ejaz, A., Arshad, S., Hayat, M., Birau, R., Popescu, J., & Spulbar, C. (2021). Investigating the impact of CO_2 emission and economic factors on infants health: A case study for Pakistan. *Industria Textilă, 72*, 39–49. https://doi.org/10.35530/IT.072.01.1784.

Nwani, S. E., & Ujah, J. C. (2024), Achieving infant mortality SDG 3 target in South Asia and Sub-Saharan Africa: Does carbon emission matter?. *Green and Low-Carbon Economy, 2*(4), 299–309. https://doi.org/10.47852/bonviewGLCE3202929

O'Hare, B., Makuta, I., Chiwaula, L., & Bar-Zeev, N. (2013). Income and child mortality in developing countries: A systematic review and meta-analysis. *Journal of the Royal Society of Medicine, 106*(10), 408–414.

Onanuga, O., & Onanuga, A. (2014). Economics of the environment and infant mortality in Sub-Saharan Africa. MPRA Paper No. 83323, posted 20 Dec 2017 16:34 UTC. Online at https://mpra.ub.uni-muenchen.de/83323/

Ortigoza, A. F., Tapia Granados, J. A., Miranda, J. J., et al. (2021). *Journal of Epidemiol Community Health, 75*, 264–270. https://doi.org/10.1136/jech-2020-215137

Oyedele, O. (2022). Carbon dioxide emission and health outcomes: Is there really a nexus for the Nigerian case? *Environmental Science and Pollution Research, 29*(37), 56309–56322.

Persyn, D., & Westerlund, J. (2008). Error-correction–based cointegration tests for panel data. *The STATA Journal, 8*(2), 232–241.

Pesaran, M. H. (2004). General diagnostic tests for cross section dependence in panels. Cambridge working papers. *Economics, 1240*(1), 1.

Pesaran, M. H. (2007). A simple panel unit root test in the presence of cross section dependence. *Forthcoming in Journal of Applied Econometrics.*

Pesaran, M. H., & Xie, Y. (2021). A bias-corrected CD test for error cross-sectional dependence in panel data models with latent factors. arXiv preprint arXiv:2109.00408.

Pesaran, M. H. (2015). Testing weak cross-sectional dependence in large panels. *Econometric Reviews, 34*(6–10), 1089–1117.

Pesaran, M. H. (2021). General diagnostic tests for cross-sectional dependence in panels. *Empirical Economics, 60*(1), 13–50.

Pesaran, M. H., & Yamagata, T. (2008). Testing slope homogeneity in large panels. *Journal of Econometrics, 142*(1), 50–93.

Rodriguez, R. N., & Yao, Y. (2017). Five things you should know about quantile regression. In *Proceedings of the SAS global forum 2017 conference, Orlando, FL, USA* (pp. 2–5).

Salahuddin, M., Vink, N., Ralph, N., & Gow, J. (2020). Effects of economic growth, foreign direct investment and internet use on child health outcomes: Empirical evidence from South Africa. *Development Studies Research, 7*(1), 1–17. https://doi.org/10.1080/21665095.2020.1717362

Semykina, A., & Wooldridge, J. M. (2010). Estimating panel data models in the presence of endogeneity and selection. *Journal of Econometrics, 157*(2), 375–380.

Shrestha, V., & Jung, J. (2023). Healthcare reform and gender specific infant mortality in rural Nepal. *World Development, 161*, Article 106018.

Sial, M. H., Arshed, N., Amjad, M. A., & Khan, Y. A. (2022). Nexus between fossil fuel consumption and infant mortality rate: A non-linear analysis. *Environmental Science and Pollution Research, 29*(38), 58378–58387.

Sideeg, T., & Mardi, G. (2011). Infant mortality rate among urban and rural population in Abu-Haraz administrative area, shikan locality, North Kordofan State, Sudan. *Journal of Science and Technology, 12*(1), 79–83. https://search.emarefa.net/detail/BIM-298798

Sims, M., Sims, T. L., & Bruce, M. A. (2007). Urban poverty and infant mortality rate disparities. *Journal of the National Medical Association., 99*(4), 349–356.

Sinha, A. (2014). Carbon emissions and mortality rates: A causal analysis for India (1971–2010). *International Journal of Economic Practices and Theories, 4*(4), 486–492.

Tavassoli, N., Noghanibehambari, H., Noghani, F., & Toranji, M. (2021). Upswing in industrial activity and infant mortality during late 19th century US. arXiv preprint arXiv:2101.02590. https://doi.org/10.48550/arXiv.2101.02590

Testoni, C. D., Kawakami, M. D., Areco, K. C. N., Sanudo, A., Balda, R. C. X., Marinonio, A. S. S., & Kiffer, C. R. V. (2021). Clusters of cause specific neonatal mortality and its association with per capita gross domestic product: A structured spatial analytical approach. *PLoS ONE, 16*(8), Article e0255882.

Trondillo, J. (2016). Estimating the effect of maternal and child health outcomes to GDP per Capita. *British Journal of Economics, Management and Trade, 12*(1), 1–12.

Uche, E., Ngepah, N., Das, N., & Dey, L. (2024). Dynamic interactions of green innovations, green transitions and ecological load capacity factor in BRICS. *Renewable Energy, 231*, Article 120905.

Westerlund, J. (2007). Testing for error correction in panel data. *Oxford Bulletin of Economics and Statistics, 69*(6), 709–748.

Wooldridge, J. M. (2010). *Econometric analysis of cross section and panel data*. MIT press.

Xiao, J., Karavias, Y., Juodis, A., Sarafidis, V., & Ditzen, J. (2023). Improved tests for Granger noncausality in panel data. *The Stata Journal, 23*(1), 230–242.

Chapter 5
Climate-Induced Mortality in India: Regional Disparities and Emerging Hotspots

Uttam Das

Abstract Extreme climatic events have appeared significantly over the years in the face of rapidly changing climate parameters, which sometimes appear to be a reason for accidental human deaths. In the above context, the present study explores the trend and patterns of accidental deaths associated with extreme climatic events in India. The data on deaths were gathered from the data on the number of accidental deaths by natural causes provided by the Government of India from 2001 to 2012. The study applies different descriptive statistics, and the death rate was computed. Further, quartile division is estimated to identify hotspot region regarding climate-induced deaths in the country. The data shows that a significant number of people died due to extreme climatic phenomena such as cold and exposure, cyclone/tornado, floods, heat stroke, lightning, and torrential rains, with a major proportion due to lightning (41.83%). However, the spatial pattern of accidental deaths for climate events is not homogenously distributed; the concern regarding such significant deaths throughout the country is undeniable. Although the highest number of deaths was found in Uttar Pradesh, Maharashtra, and Bihar, a high severity of death rate was found in Madhya Pradesh, Odisha, Chhattisgarh, and Punjab. Indeed, the burden of climate event-induced deaths was highest in the central-eastern part of the country. Extensive and proper strategies are highly inevitable in fighting against rapid climate change and minimising the irreversible loss of human deaths due to such emerging extreme climatic events.

Keywords Deaths · Climate · Lightning · Human health · India

U. Das (✉)
Department of Humanities and Social Sciences, Indian Institute of Technology Kharagpur, Kharagpur, West Bengal, India
e-mail: uttamdastfgm@gmail.com

P. Datta et al. (eds.), *Climate Change and Human Well-being*, Advances in Global Change Research 83, https://doi.org/10.1007/978-3-032-25479-5_5

5.1 Introduction

Global climate change is an evident phenomenon that has manifold impacts on the Earth (Bellard et al., 2012; Bunyavanich et al., 2003; Hitz & Smith, 2004; Hoegh-Guldberg & Bruno, 2010; Mall et al., 2006). As per the Intergovernmental Panel on Climate Change (IPCC), climate change refers to any changes in climate over time, whether it is because of natural variability or the consequence of anthropogenic activities (Change, 2007). The energy balance of climate conditions is altered due to changes in aerosols and greenhouse gases, solar radiation, and land surface properties (Change, 2007). Global warming has resulted in various effects on humans, making climate change a disquieting issue (Benjamin et al., 2017; Change, 2001). Since the 1860s, the average temperature has fluctuated in response to natural processes, including variations of solar radiation output, in atmospheric and oceanic currents (Change, 2001).

Climate change is likely to have a large impact on human health, which is mostly adverse in nature and results in a significant loss of life. However, apart from the direct impact of climate change on human deaths, it introduces chronic and acute morbidity through the creation of extreme environmental conditions across the globe (Bunyavanich et al., 2003; Chen et al., 2023; Haines et al., 2006; Kim et al., 2014; Patz et al., 2005). The direct prime effect of global warming is reflected in excessively hot weather. Worldwide, several countries have witnessed deaths due to heat and cold every year in these current weather conditions (Hajat et al., 2014; Vicedo-Cabrera et al., 2021). As a result of excessive heat, the health of all, in general and older people in particular, could be aggravated. For instance, the heat wave in 1995 killed hundreds of people in Chicago, several thousand people in Uttar Pradesh of India and parts of central China. Further, the summer of 1998 generated unprecedented severe heat, which struck North America, Europe, China, and India, causing forest fires, property loss and deaths (Change, 2001). During the summer of 2003, the excessively high temperature caused more than seventy thousand deaths in Europe (Mitchell et al., 2016). In addition to the adverse effects of hot weather, climate change has complex effects on human societies and the ecosystem, disrupting human lives and well-being (Change, 2001).

In India, the changing pattern of climate is also evident[1] (Dash et al., 2007; Revadekar et al., 2012), resulting in the acceleration of extreme climatic events. Important climatic events influencing India are drought, flood, cyclone, heat waves, and cold waves (Dash et al., 2007). A significant loss or damage has occurred due to the extreme climatic events in various spheres, either directly or indirectly. The impact of climate change imprints on manifold aspects ranging from forest ecosystem, agriculture, water resources, to food security (Aggarwal, 2003; Ahmad et al., 2011; Gosain et al., 2011; Pathak, 2023; Ravindranath et al., 2006). In addition, extreme climatic events have severe consequences on human life. Every year, many people lose their lives due to such climatic events in the country. In fact, the socio-economic

---

[1] Study reported surface temperature has risen by around 1 and 1.1 °c in India during winter and post monsoon months respectively (Dash et al., 2007).

impacts of these extreme climatic events have been increasing compared to the past (De et al., 2005; Singh & Patwardhan, 2012). Moreover, the IPCC also predicted that extreme weather events would increase in India (Tolani et al., 2024). Despite the fact, most of the recent research studies focused on heat-induced mortality in the country (Azhar et al., 2014; de Bont et al., 2024; Guin et al., 2025; Kumar & Singh, 2021; Nori-Sarma et al., 2019; Ray et al., 2021), while other climatic events such as cyclone/tornado (Ray et al., 2021), floods (Ray et al., 2021; Singh & Kumar, 2013), lightning (Mishra et al., 2022; Ray et al., 2021; Yadava et al., 2020), and torrential rains were largely overlooked. The study by Ray et al. (2021) studied several extreme climatic events-induced deaths in the country, but was limited to only a few selected states at the state level. Hence, a lack of systematic studies on the spatial and temporal variation of mortality due to extreme climatic events in India prepares the fertile ground for studying this issue, especially in the face of rapidly changing climate conditions. Owing to the above drawbacks, the present study throws light on the spatial and temporal variation of human deaths due to extreme climatic events in the country.

The findings of the study will be supportive for the adaptation of climate resilience strategies and overall better disaster management, more specifically, the prevention of significant mortality in the country by providing vital information on the trend of extreme climatic events-induced deaths and their spatial distribution across the country.

5.2 Data Sources and Methods

While detailed mortality data, by age and gender, were publicly available at the national level for the 19-y period (2001–2019), such granular data at the state level were only available for the 14-year period (2001–2014). Data on mortality due to heat stroke and exposure to cold from the National Crime Records Bureau (NCRB) was used in the previous study (Guin et al., 2025). In the present study, cold and exposure, cyclone/tornado, flood, heat stroke, lightning, and Torrential rains have been considered for analysing climate-induced deaths in the country. The accidental deaths data were obtained from the data on the number of accidental deaths by natural causes provided by the Government of India as part of the National Data Sharing and Accessibility Policy (NDSAP) from 2001 to 2012.

The data analysis has been done at two administrative levels, specifically the national level and the state level. In order to show spatial variation across the states, thematic maps were prepared using the number of deaths in each state. However, to make the comparable statistics, the death rate was estimated using the 2001 and 2011 census population of the concerned states. It is important to mention that due to the unavailability of population data for other years apart from 2001 and 2011, the death rate of the state due to various climatic events was calculated for those two years. The following formula has been adopted to compute the death rate due to climatic events.

$$DR_c = \frac{ND_{ic}}{TP_i} \times 100,000 \tag{5.1}$$

where, DR_c is the death rate due to climatic events; ND_{ic} indicates the number of deaths due to climatic events in the state in the ith year; TP_i indicates the total population in the state in the ith year;

Further, the average death per year was estimated using the following formula:

$$AD_y = \frac{TD}{NY} \tag{5.2}$$

where, AD_y is the average deaths per year; TD is the total number of deaths; and NY is the number of years.

Further, the quartile range of normalised death rate (death per 10,000,000 population) was done for thematic mapping to identify the states that belong to different quartiles, such as Quartile 1 (Q1), Quartile 2 (Q2), Quartile 3 (Q3), and Quartile 4 (Q4).

The two correlation techniques, namely Pearson and Spearman Rank correlation, have been employed to reveal any correlation between the total population of the state and total deaths due to climatic events in the state.

5.3 Results and Discussions

5.3.1 National-Level Analysis of Human Deaths Due to Climatic Events

5.3.1.1 Deaths Due to Cold and Exposure

A significant number of deaths have taken place due to cold and exposure in the country. Overall, a total of 9074 deaths were recorded in the country from 2001 to 2012. The annual deaths show a less fluctuating trend in the death number over the years. The deaths due to cold and exposure ranged from 525 (2002) to 997 (2012), with an average of around 756 per year.

5.3.1.2 Deaths Due to Cyclone and Tornado

Similar to cold and exposure, the number of deaths caused by cyclones and tornadoes is substantial. The data indicate that 3936 persons died from cyclones and tornadoes from 2001 to 2012 in India. However, most importantly, 2004 and 2005 alone were responsible for 2950 deaths out of 3936. Except for these two years, other years had comparatively low deaths, such as 47 deaths in 2012. The average number of deaths per year was about 328 due to cyclones and tornadoes in the country.

5.3.1.3 Deaths Due to Floods

Flood becomes a very copious apprehension of huge human deaths every year. It is evident that every year, many people lose their lives due to devastating floods. As per the data, a total of 8877 deaths were reported in India from 2001 to 2012, making it one of the major concerns of climate-induced disasters.

5.3.1.4 Deaths Due to Heat Stroke

The number of heat stroke cases has been increasing in the country over the years. A significant number of people lose their lives to heat stroke every year. As per the data recorded, a total of 10,550 deaths took place due to heat stroke from 2001 to 2012 in the country. Worryingly, the intensity of deaths due to heat stroke seems to be rising over the years. On average, 879 people die each year because of heat stroke in the country.

5.3.1.5 Deaths Due to Lightning

Lightning is also intensely accountable for human deaths in the country. In comparison to cold and exposure, cyclones and tornadoes, floods, and heat stroke, lightning strikes are significantly more deadly in human deaths. The data suggest that a total of 25,866 deaths occurred because of lightning from 2001 to 2012 in the country. Every year, a substantial number of deaths occur from lightning, which range from 1383 (2002) to 2790 (2007). Lightning causes 2156 people's deaths per year, on average.

5.3.1.6 Deaths Due to Torrential Rains

In India, extreme precipitation is a significant concern given the substantial number of fatalities and damages. Torrential rain is another phenomenon that takes many human lives every year. As per the data, a total of 3492 deaths were documented from 2001 to 2012 in the country. Interestingly, 2002 experienced a sharp surge in human deaths (1296 deaths) due to torrential rains, followed by 2005 (557 deaths). On average, 291 people die every year due to torrential rains in the country.

However, the intensity of each climatic event varies over the temporal scale. The number of cold and exposure-related deaths was on an increasing trend, with the highest number in 2012. The cyclone/tornado-related deaths were highest in 2004 and 2005, and afterwards sharply declined. The number of deaths due to floods showed an increasing trend up to 2007 and afterwards a declining trend. The number of deaths due to heat stroke has fluctuated, with the highest number in 2010, and is mostly on an increasing trend. Lightning had a sharp trend of increasing up to 2007, and then fluctuated. The number of deaths due to torrential rains is mostly around steady state, except for the year 2002.

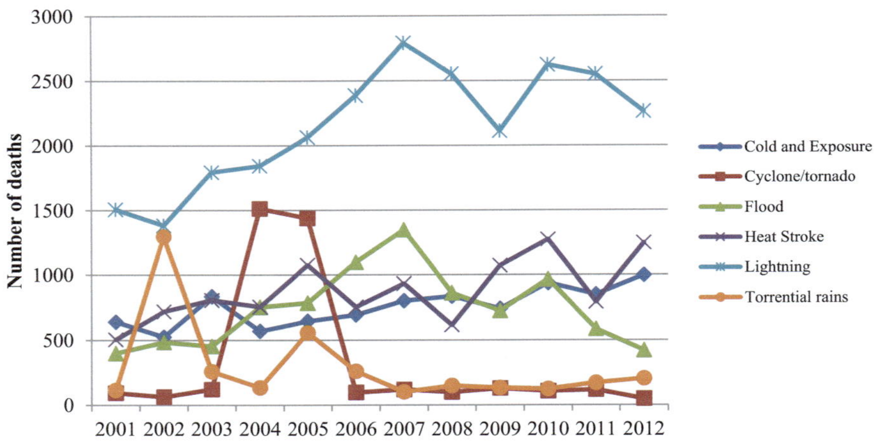

Fig. 5.1 Number of deaths due to various climatic events in India from 2001 to 2012 (Data source: NCRB)

However, Fig. 5.2 shows that the total number of deaths due to extreme climatic events has been increasing over the years in the country. The maximum number of deaths was found in 2004–2005 due to the excessive deaths caused by a cyclone/tornado in the Andaman and Nicobar Islands. On average, from 2001 to 2012, a total of about 5149 people died from climatic events such as cold and exposure, cyclone/tornado, floods, heat stroke, lightning, and torrential rains in the country.

Further, the data shows that lightning has the maximum share in the total deaths (41.83%), followed by heat stroke (17.06%), cold and exposure (14.73%), flood (14.35%), cyclone/tornado (6.36%), and torrential rains (5.35%) (See Fig. 5.3). Every year from 2001 to 2012, lightning consistently contributed the maximum share in total deaths (See Fig. 5.1), indicating the harshness of it.

5.3.2 Gender-Specific Deaths to Climatic Events

As per the gender-wise distribution, Fig. 5.2 clearly depicts that males are more disposed to deaths due to climatic events than females. In the case of cold and exposure, more than 80% of deaths were associated with males. However, females are relatively found to be highly vulnerable to cyclone/tornado-induced deaths, with a share of more than 40% compared to other climatic events. More than 70% deaths due to flood were male. Similarly, around 80% deaths in the case of heat stroke were male. In the case of lightning, female deaths were less than 30%. Also, males are more susceptible to death due to torrential rains, accounting for more than 70% deaths (Figs. 5.2, 5.3 5.4).

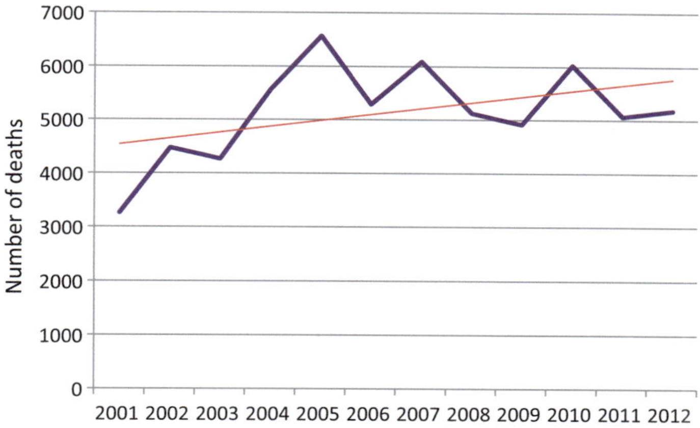

Fig. 5.2 Trend of deaths due to climatic events in India from 2001 to 2012 (Data source: NCRB)

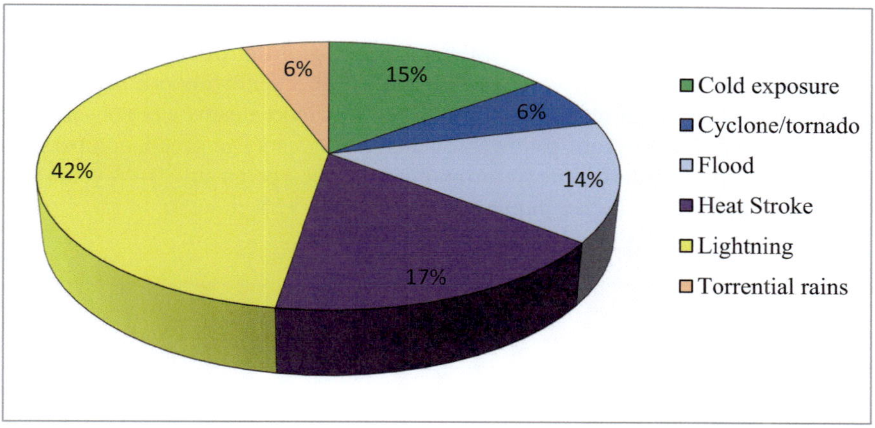

Fig. 5.3 Share of each climatic event in total deaths from 2001 to 2012 in India (Data source: NCRB)

5.3.3 Analysis of Age-Specific Deaths

Figure 5.3 represents the human deaths across age groups due to various climatic events. The data shows that age groups 45–59 years and more than 60 years were likely to have deaths due to cold and exposure. However, in the case of the more than 60 years age group, female had a higher death propensity in comparison to their counterparts. On the contrary, cyclones/tornado affect the age group 30–44 years more, and had similar results for both male and female groups. Further, the 30–44 years age group was also more affected due to the flood, followed by the 15–29 years age group. Females less than 14 years of age were relatively more exposed

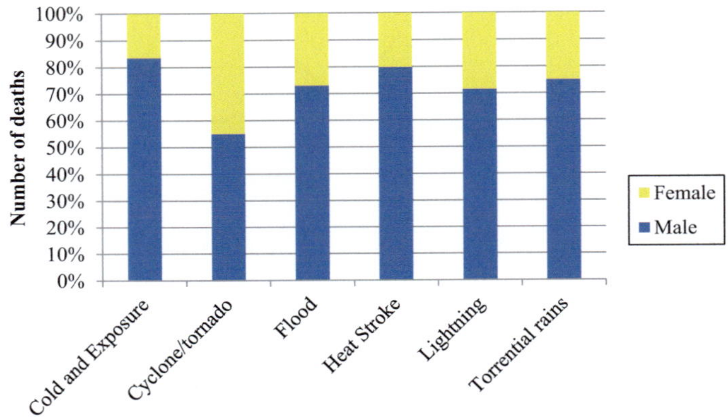

Fig. 5.4 Gender-specific deaths due to various climatic events in India from 2001 to 2012 (Data source: NCRB)

to death due to flood than males of the same age. Heat stroke was more prominent among the age group of 45–59 years, followed by the age group of more than 60 years. Interestingly, females aged 60 years or older had a higher death proportion compared to the same age group of males. In case of lightning, more deaths were observed in the age group of 30–44 years, followed by the 15–29 years age group, compared to other age groups. Similarly, due to torrential rains, the age group 30–44 years was more affected, having a higher number of deaths (Fig. 5.5).

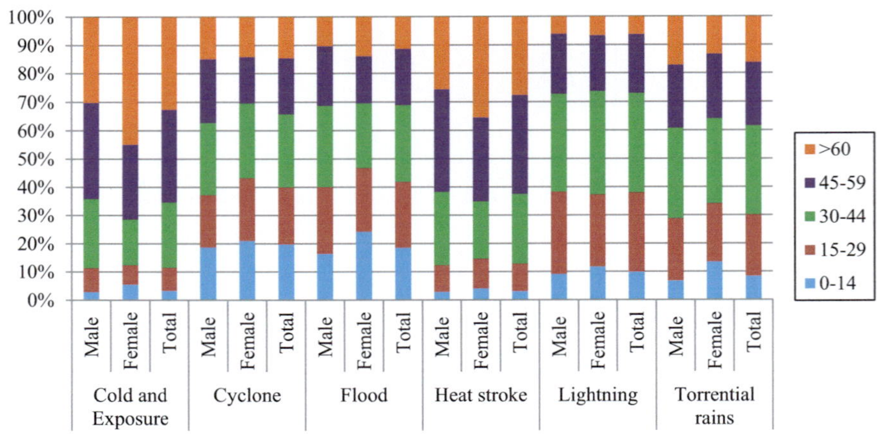

Fig. 5.5 Age-sex specific deaths due to various climatic events in India from 2001 to 2012 (Data source: NCRB)

5.3.4 *State-Level Analysis of Human Deaths Due to Climatic Events*

The number of human deaths is not similar across the states of the country. It is evident that some states are more likely to experience high human deaths due to extreme climatic events, while some states are more likely to be in a low-risk zone. However, as per total count, Uttar Pradesh, the state with the largest population in the country, recorded the highest number of deaths (7635 deaths), followed by Maharastra (6981 deaths), and Bihar (5098 deaths) due to climatic events from 2001 to 2012, while the states like Nagaland (8 deaths), Manipur (11 deaths), Goa (12 deaths), and Mizoram (17 deaths) recorded low number of deaths (Fig. 5.6).

Over the temporal scale, it has been found that the maximum number of deaths was found in Maharashtra, followed by Madhya Pradesh and Uttar Pradesh in 2001. In 2002, Uttar Pradesh witnessed excessive deaths, around two times compared to the next highest death in Andhra Pradesh. In 2003, Uttar Pradesh again made it to the top of the list, followed by Madhya Pradesh. Andaman and Nicobar Islands experienced the highest number of deaths in 2004 and 2005. Maharashtra experienced the highest number of deaths in 2006 and 2007. In 2008, the highest number of deaths was found in Uttar Pradesh, followed by Bihar. In 2009, Uttar Pradesh, followed by Andhra Pradesh, witnessed the highest number of deaths. Similarly, in 2010, Uttar Pradesh, followed by Maharashtra, experienced the highest deaths. Interestingly, the highest death rate was recorded in Bihar, followed by Uttar Pradesh in 2011. In 2012, the highest death rate was observed in Uttar Pradesh, followed by Bihar (Table 5.1).

As per each climatic event, the number of human deaths also varies from state to state. Cold and exposure are more prominent in states like Bihar, Punjab and Uttar Pradesh. These three states remained at the top of the hierarchy in terms of cold and exposure-related deaths up to 2019 (Guin et al., 2025). Mainly Uttar Pradesh,

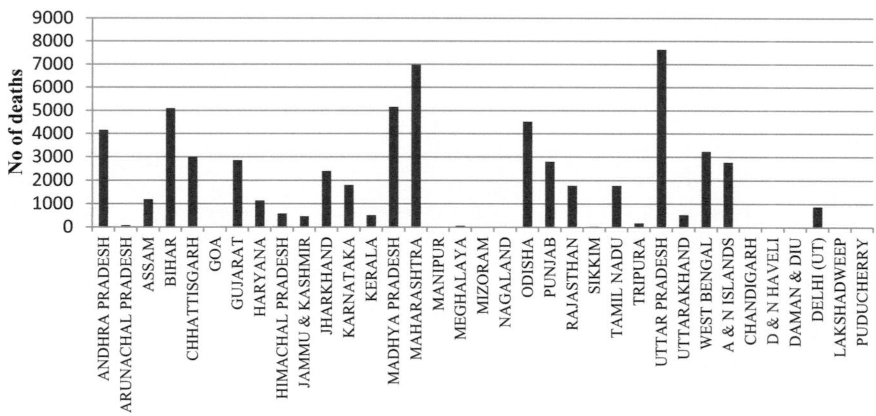

Fig. 5.6 Number of deaths due to climatic events in India from 2001 to 2012 in different states and union territories (*Data source* NCRB)

the coastal state Andhra Pradesh and the Andaman and Nicobar islands experienced the most fatalities due to cyclones and tornadoes. Andaman and Nicobar Islands experienced a devastating number of deaths during 2004 and 2005. In the previous study, Ray et al. (2025) highlighted that coastal states are more susceptible to cyclones and floods. Floods appear to be distressing, leading to a huge number of deaths in Bihar, Maharashtra, and Gujarat. The flood in Bihar is inescapable being located in the sub-Himalaya plain (Das et al., 2021). During a massive flood, large areas of the state were inundated, exposing people to flood-related disaster (Senapati, 2022). In case of heat stroke-related death, Andhra Pradesh, Punjab, Uttar Pradesh and West Bengal are more prominent. Further, Ray et al. (2025) highlighted that extreme temperature are more prevalent in inland states. Lightning struck in the states like Chhattisgarh, Madhya Pradesh, Maharashtra and Odisha, with a significant number of deaths. The number of deaths due to torrential rains is highest in Uttar Pradesh, followed by Andhra Pradesh (Table 5.2).

As per the quartile range based on the death rate per 10 million people, states that fall under different severity levels are presented in Table 5.3. Madhya Pradesh, Odisha, Chhattisgarh, and Punjab were in the high quartile in both years 2001 and 2011. On the contrary, A & N Islands, D & N Haveli, Daman & Diu, and Lakshadweep were in the very marginal severity zone for both years 2001 and 2011 (Fig. 5.7).

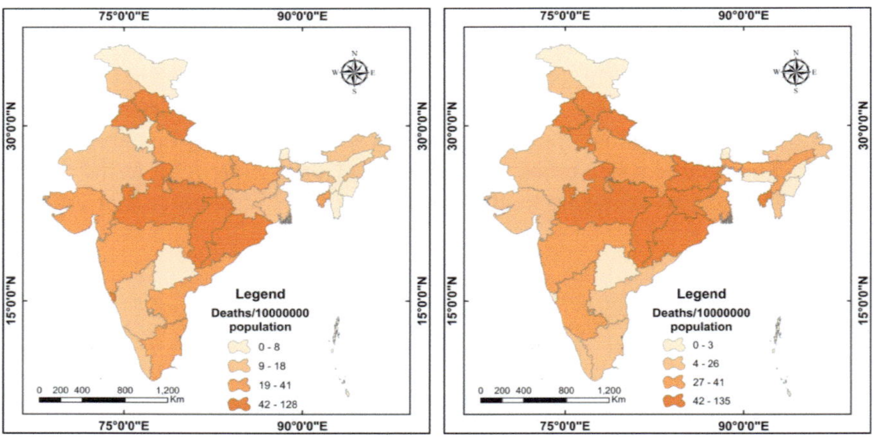

Fig. 5.7 Human deaths due to climatic events in the states and union territories of India (*Data source* NCRB & Census of India)

Table 5.1 Number of deaths in different states and union territories of India from 2001 to 2012 due to extreme climatic events

State name	2001	2002	2003	2004	2005	2006	2007	2008	2009	2010	2011	2012
ANDHRA PRADESH	245	593	268	225	322	407	397	324	463	380	188	344
ARUNACHAL PRADESH	2	10	2	6	6	1	12	16	14	2	3	9
ASSAM	19	52	54	316	34	71	170	40	89	123	102	116
BIHAR	164	234	377	360	178	245	706	632	330	536	693	643
CHHATTISGARH	266	198	326	265	246	237	301	267	229	232	236	197
GOA	7	1	0	0	1	0	0	0	3	0	0	0
GUJARAT	179	208	207	257	287	387	386	182	145	252	158	200
HARYANA	16	51	35	31	59	48	171	132	106	145	138	205
HIMACHAL PRADESH	40	35	94	24	36	45	49	67	59	21	39	70
JAMMU & KASHMIR	10	5	18	15	21	41	20	19	6	284	14	15
JHARKHAND	31	45	53	86	264	198	260	262	219	268	329	375
KARNATAKA	83	106	117	131	160	129	167	147	279	195	175	109
KERALA	41	48	37	36	53	34	51	36	48	60	33	29
MADHYA PRADESH	336	244	503	382	451	651	458	453	306	457	484	423
MAHARASHTRA	399	468	423	513	1053	866	888	464	486	664	419	338
MANIPUR	0	2	0	0	1	0	2	4	0	1	0	1
MEGHALAYA	3	4	3	6	11	4	2	8	8	7	1	3
MIZORAM	0	0	1	0	0	0	2	0	2	2	1	9
NAGALAND	3	0	1	0	0	2	0	0	0	2	0	0
ODISHA	309	297	345	380	350	319	425	446	379	509	385	379
PUNJAB	142	102	127	143	238	161	286	242	283	382	374	313
RAJASTHAN	87	113	127	101	140	260	161	155	159	168	150	158
SIKKIM	0	15	1	6	1	4	3	0	1	0	0	5
TAMIL NADU	122	130	191	116	157	191	110	134	210	214	108	96
TRIPURA	7	1	2	18	27	16	18	19	19	25	22	7
UTTAR PRADESH	335	1186	665	407	659	526	578	655	548	736	650	690
UTTARAKHAND	66	13	30	31	27	37	48	27	45	61	46	100
WEST BENGAL	141	167	172	221	379	353	317	301	364	218	282	315
A & N ISLANDS	0	0	1	1430	1344	0	0	3	0	0	0	0
CHANDIGARH	4	1	0	0	0	0	0	1	0	0	3	0
D & N HAVELI	0	0	0	3	3	4	0	1	0	0	0	0
DAMAN & DIU	0	0	0	0	0	1	0	0	0	0	0	0
DELHI (UT)	200	141	84	58	56	49	102	84	112	83	21	28
LAKSHADWEEP	0	0	0	0	0	0	0	0	0	0	0	0
PUDUCHERRY	2	0	0	0	1	0	0	0	0	0	10	0

Data source NCRB

5.3.5 *Correlation Between Deaths Due to Climatic Events and Population*

The two correlations, namely Pearson and Spear Rank correlation, reveal that there is a strong positive correlation. Pearson correlation value (0.783 and 0.826), which is statistically significant at 1% level, suggests that with the increase of population, the number of deaths due to climatic events also increases. Further, the Spear Rank correlation value (0.907 and 0.903), which is also statistically significant at a 1% level, indicates that the number of deaths due to climatic events rises with the increase in the total population of the states.

Type of correlation	Year	Correlation value	Sig. (2-tailed)
Pearson correlation	2001	0.783**	0.000
	2011	0.826**	0.000
Spearman's rho	2001	0.907**	0.000
	2011	0.903**	0.000

** Significance at 0.01 level

Table 5.2 Number of deaths due to different climatic events in different states and union territories of India from 2001 to 2012

State name	Cold and exposure	Cyclone/tornado	Flood	Heat Stroke	Lightning	Torrential rains
Andhra Pradesh	87	199	477	1531	1299	596
Arunachal Pradesh	19	1	37	17	5	5
Assam	10	51	749	67	297	12
Bihar	1237	121	1790	770	1023	157
Chhattisgarh	61	7	171	125	2573	71
Goa	0	0	3	6	3	0
Gujarat	345	100	1113	254	728	308
Haryana	658	8	0	381	87	3
Himachal Pradesh	289	1	166	18	44	61
Jammu & Kashmir	53	8	307	17	47	36
Jharkhand	645	5	26	420	1240	54
Karnataka	15	7	249	89	1311	127
Kerala	0	0	24	9	450	23
Madhya Pradesh	253	49	377	483	3894	92
Maharashtra	214	73	1790	750	3655	499
Manipur	0	1	4	0	2	0
Meghalaya	9	7	3	1	40	0
Mizoram	0	4	4	5	1	3
Nagaland	0	0	2	1	5	0
Odisha	108	32	69	986	3290	38
Punjab	1427	11	85	1175	92	3
Rajasthan	289	27	151	493	760	59
Sikkim	7	0	6	0	21	2
Tamil Nadu	16	55	403	228	1025	52
Tripura	2	7	2	63	102	5
Uttar Pradesh	2234	260	513	1376	2184	1068
Uttarakhand	288	16	89	6	48	84
West Bengal	197	70	219	997	1614	133
A & N islands	0	2773	0	0	5	0
Chandigarh	5	0	0	1	2	1
D & N Haveli	0	0	4	0	7	0
Daman & Diu	0	0	0	0	1	0
Delhi (UT)	639	33	44	281	9	12
Lakshadweep	0	0	0	0	0	0
Puducherry	0	10	0	0	3	0

Data source NCRB

However, it is worth mentioning that the non-existence of 'safe limit' of climate change regarding health impact, since current climate variability in some regions is not adequately coped with by health systems (Kovats et al., 2005). As per the estimation, heat-related deaths are likely to increase, but cold-related deaths will decline in the UK in the absence of any adaptation strategies (Hajat et al., 2014). In another study, Vicedo-Cabrera et al. (2021) observed that a significant portion of warm-season heat-related deaths could be attributed to anthropogenic climate change. They also

Table 5.3 Severity of deaths due to climatic events in different states and union territories of India

Interquartile range	Severity of death	Year	Range (Deaths/ 1,000,0000 population)	States
Quartile 1 (Q1)	Very marginal	2001	< 8	Manipur, Mizoram, Sikkim, A & N Islands, D & N Haveli, Daman & Diu, Lakshadweep, Assam, Haryana
		2011	< 3	A & N Islands, D & N Haveli, Daman & Diu, Goa, Lakshadweep, Manipur, Nagaland, Sikkim, Meghalaya
Quartile 2 (Q2)	Low	2001	9–18	Jammu & Kashmir, Jharkhand Kerala, Meghalaya, Nagaland, Rajasthan, Karnataka, West Bengal, Arunachal Pradesh
		2011	4–26	Mizoram, Kerala, Jammu & Kashmir, Delhi (UT), Tamil Nadu, Arunachal Pradesh, Rajasthan, Andhra Pradesh, Gujarat
Quartile 3 (Q3)	Moderate	2001	19–44	Tamil Nadu, Bihar, Uttar Pradesh, Puducherry, Tripura, Andhra Pradesh, Gujarat, Maharashtra
		2011	27–57	Chandigarh, Karnataka, West Bengal, Uttar Pradesh, Assam Maharashtra, Uttarakhand, Haryana, Himachal Pradesh

(continued)

Table 5.3 (continued)

Interquartile range	Severity of death	Year	Range (Deaths/ 1,000,0000 population)	States
Quartile 4 (Q4)	High	2001	> 45	Chandigarh, Goa, Madhya Pradesh, Punjab, Himachal Pradesh, Uttarakhand, Odisha, Chhattisgarh, Delhi (UT),
		2011	> 58	Tripura, Bihar, Madhya PradeshPuducherry, Odisha, Chhattisgarh, Jharkhand, Punjab

found that increased mortality is evident on each continent (Vicedo-Cabrera et al., 2021).

In the previous study, based on 50 years' data (1970–2019) of the India Meteorological Department, floods accounted for the maximum mortality followed by tropical cyclones. However, the intensity of lightning on mortality was on an increasing trend (Ray et al., 2021).

5.4 Conclusion

The present study explores the climate events' induced human deaths in India from 2001 to 2012, specifically focusing on climatic events such as cold and exposure, cyclone/tornado, floods, heat stroke, lightning, and torrential rains. It is evident that these climatic events have led to human deaths in the country. However, the intensity of deaths is found to be disproportionate and varies over the spatial–temporal scale. Some states of the country exhibited a severe risk for human deaths due to such climatic events, while some states came under low risk for human deaths. Among these climatic events, lightning has strikingly recorded the maximum number of human deaths from 2001 to 2012. Further, it also revealed that male are more susceptible to death because of harsh climatic events in comparison to their counterparts. The death rate is also heterogeneous across the different age groups. Statistically, the number of deaths is positively associated with the population at the state level.

The rising concern of human deaths due to unforgiving climatic events in the country necessitates proper management and mitigation strategies to adopt and implement, applying appropriate management approaches. Owing to the increasing trend of climate-induced disasters, the National Disaster Management Authority (NDMA) has developed action plans to tackle extreme climatic events in general, for specific events in particular. It would be better to adopt action plans that focus on the current nature and pattern of each climatic event and its geographical distribution. In this direction, the findings of the study could be beneficial for formulating and adopting suitable action plans through region-specific execution at different levels based on their severity.

Future research could be navigated towards the analysis of each climate-induced death across different physiographic and climatic zones of the country. Further, future studies could emphasise seasonal variations in deaths. Also, further micro-level studies, such as district-level studies, could be crucial for identifying vulnerable districts and assisting in the implementation of appropriate action plans.

Acknowledgements I would like to thank Dr. Pritha Datta for assisting in extracting the data used in the study, and I am grateful to the editors for valuable feedback and suggestions on the earlier version of the manuscript, which helped to improve the manuscript considerably.

References

Aggarwal, P. K. (2003). Impact of climate change on Indian agriculture. *Journal of Plant Biology-New Delhi, 30*(2), 189–198.

Ahmad, J., Alam, D., & Haseen, M. S. (2011). Impact of climate change on agriculture and food security in India. *International Journal of Agriculture, Environment and Biotechnology, 4*(2), 129–137.

Azhar, G. S., Mavalankar, D., Nori-Sarma, A., Rajiva, A., Dutta, P., Jaiswal, A., & Hess, J. J. (2014). Heat-related Mortality in India: Excess all-cause Mortality associated with the 2010 Ahmedabad heat wave. *PLoS ONE, 9*(3), Article e91831.

Bellard, C., Bertelsmeier, C., Leadley, P., Thuiller, W., & Courchamp, F. (2012). Impacts of climate change on the future of biodiversity. *Ecology Letters, 15*(4), 365–377.

Benjamin, D., Por, H. H., & Budescu, D. (2017). Climate change versus global warming: Who is susceptible to the framing of climate change? *Environment and Behavior, 49*(7), 745–770.

Bunyavanich, S., Landrigan, C. P., McMichael, A. J., & Epstein, P. R. (2003). The impact of climate change on child health. *Ambulatory Pediatrics, 3*(1), 44–52.

Change, C. (2001). Climate change. *Synthesis report.*

Change, I. P. O. C. (2007). Climate change 2007: The physical science basis. *Agenda, 6*(07), 333.

Chen, X. M., Sharma, A., & Liu, H. (2023). The impact of climate change on environmental sustainability and human Mortality. *Environments, 10*(10), 165.

Das, A., Santra, P. K., & Bandyopadhyay, S. (2021). The 2016 flood of Bihar, India: An analysis of its causes. *Natural Hazards, 107*(1), 751–769.

Dash, S. K., Jenamani, R. K., Kalsi, S. R., & Panda, S. K. (2007). Some evidence of climate change in twentieth-century India. *Climatic Change, 85*(3), 299–321.

de Bont, J., Nori-Sarma, A., Stafoggia, M., Banerjee, T., Ingole, V., Jaganathan, S., Ljungman, P., et al. (2024). Impact of heatwaves on all-cause Mortality in India: A comprehensive multi-city study. *Environment International, 184*, Article 108461.

De, U. S., Dube, R. K., & Rao, G. P. (2005). Extreme weather events over India in the last 100 years. *The Journal of Indian Geophysical Union, 9*(3), 173–187.

Gosain, A. K., Rao, S., & Arora, A. (2011). Climate change impact assessment of water resources of India. *Current Science*, 356–371.

Guin, P., Bhan, N., & Sethi, K. (2025). Mortality due to heatstroke and exposure to cold: Evidence from India. *Temperature, 12*(2), 179–199.

Haines, A., Kovats, R. S., Campbell-Lendrum, D., & Corvalán, C. (2006). Climate change and human health: Impacts, vulnerability and public health. *Public Health, 120*(7), 585–596.
Hajat, S., Vardoulakis, S., Heaviside, C., & Eggen, B. (2014). Climate change effects on human health: Projections of temperature-related Mortality for the UK during the 2020s, 2050s and 2080s. *Journal of Epidemiology and Community Health, 68*(7), 641–648.
Hitz, S., & Smith, J. (2004). Estimating global impacts from climate change. *Global Environmental Change, 14*(3), 201–218.
Hoegh-Guldberg, O., & Bruno, J. F. (2010). The impact of climate change on the world's marine ecosystems. *Science, 328*(5985), 1523–1528.
Kim, K. H., Kabir, E., & Ara Jahan, S. (2014). A review of the consequences of global climate change on human health. *Journal of Environmental Science and Health, Part C, 32*(3), 299–318.
Kovats, R. S., Campbell-Lendrum, D., & Matthies, F. (2005). Climate change and human health: Estimating avoidable deaths and disease. *Risk Analysis: An International Journal, 25*(6), 1409–1418.
Kumar, A., & Singh, D. P. (2021). Heat stroke-related deaths in India: An analysis of natural causes of deaths, associated with the regional heatwave. *Journal of Thermal Biology, 95*, Article 102792.
Mall, R. K., Singh, R., Gupta, A., Srinivasan, G., & Rathore, L. S. (2006). Impact of climate change on Indian agriculture: A review. *Climatic Change, 78*(2), 445–478.
Mishra, M., Acharyya, T., Santos, C. A. G., Da Silva, R. M., Chand, P., Bhattacharyya, D., Singh, O., et al. (2022). Mapping main risk areas of lightning fatalities between 2000 and 2020 over Odisha state (India): A diagnostic approach to reduce lightning fatalities using statistical and spatiotemporal analyses. *International Journal of Disaster Risk Reduction, 79*, Article 103145.
Mitchell, D., Heaviside, C., Vardoulakis, S., Huntingford, C., Masato, G., Guillod, B. P., Allen, M., et al. (2016). Attributing human Mortality during extreme heat waves to anthropogenic climate change. *Environmental Research Letters, 11*(7), Article 074006.
Nori-Sarma, A., Anderson, G. B., Rajiva, A., ShahAzhar, G., Gupta, P., Pednekar, M. S., Bell, M. L., et al. (2019). The impact of heat waves on Mortality in Northwest India. *Environmental Research, 176*, Article 108546.
Pathak, H. (2023). Impact, adaptation, and mitigation of climate change in Indian agriculture. *Environmental Monitoring and Assessment, 195*(1), 52.
Patz, J. A., Campbell-Lendrum, D., Holloway, T., & Foley, J. A. (2005). Impact of regional climate change on human health. *Nature, 438*(7066), 310–317.
Ravindranath, N. H., Joshi, N. V., Sukumar, R., & Saxena, A. (2006). Impact of climate change on forests in India. *Current science*, 354–361.
Ray, K., Giri, R. K., Ray, S. S., Dimri, A. P., & Rajeevan, M. (2021). An assessment of long-term changes in mortalities due to extreme weather events in India: A study of 50 years' data, 1970–2019. *Weather and Climate Extremes, 32*, Article 100315.
Revadekar, J. V., Kothawale, D. R., Patwardhan, S. K., Pant, G. B., & Rupa Kumar, K. (2012). About the observed and future changes in temperature extremes over India. *Natural Hazards, 60*(3), 1133–1155.
Senapati, S. (2022). Vulnerability and risk in the context of flood-related disasters: A district-level study of Bihar, India. *International Journal of Disaster Risk Reduction, 82*, Article 103368.
Singh, O., & Kumar, M. (2013). Flood events, fatalities and damages in India from 1978 to 2006. *Natural Hazards, 69*, 1815–1834.
Vicedo-Cabrera, A. M., Scovronick, N., Sera, F., Royé, D., Schneider, R., Tobias, A., Gasparrini, A., et al. (2021). The burden of heat-related Mortality attributable to recent human-induced climate change. *Nature Climate Change, 11*(6), 492–500.
Yadava, P. K., Soni, M., Verma, S., Kumar, H., Sharma, A., & Payra, S. (2020). The major lightning regions and associated casualties over India. *Natural Hazards, 101*, 217–229.
Tolani, H., Neogi, S. B., Gupta, S. D., Mishra, S. S., & Samtani, R. (2024). Analyzing dynamics of extreme weather events (EWE) in India: Unfolding trends through statistical assessment of 50 years data (1970–2019). *BMC Environmental Science, 1*(1), 12.

Singh, A., & Patwardhan, A. (2012). Spatio-temporal distribution of extreme weather events in India. *APCBEE Procedia, 1*, 258–262.

Chapter 6
Exploring the Nexus Between Climate Change, Air Pollution, and Human Health: Status and Future Directions

Divyansh Sharma, Anuska Roy, Adil Masood, and Mohammed Majeed Hameed

Abstract A complex interplay between climate change, air pollution, and human health has emerged as a global issue of high priority. Pollutants such as particulate matter, ozone, sulfur dioxide, nitrogen dioxide, and volatile organic compounds-key precursors of photochemical smog- are primarily generated by transportation and industry as well as biomass burning. These pollutant species pose significant local health risks and contribute to global climate change by altering atmospheric chemistry and intensifying the greenhouse effect. Against the backdrop of shifting temperature and precipitation patterns, air pollution may further exacerbate health risks, leading to increased incidences of respiratory and cardiovascular diseases, infectious illnesses, and premature mortality. This chapter presents a comprehensive overview of the interconnections between climate dynamics and air quality, and how their interaction impacts public health. The review will serve as a baseline study of recent research methods and results on the health outcomes of temperature variations as a result of climate change and their effects on air quality.

Keywords Air pollution · Climate change · Human health · Particulate matter · Temperature

---

D. Sharma (✉)
Department of Sustainable Engineering, TERI School of Advanced Studies, New Delhi, India
e-mail: divyansh.sharma@terisas.ac.in

A. Roy · A. Masood
Department of Natural and Applied Sciences, TERI School of Advanced Studies, New Delhi, India
e-mail: anuska.roy@terisas.ac.in

A. Masood
e-mail: adil.masood@terisas.ac.in

M. M. Hameed
Upper Euphrates Centre for Sustainable Development Research, University of Anbar, Ramadi, Iraq
e-mail: mohammed.hameed@uoanbar.edu.iq

P. Datta et al. (eds.), *Climate Change and Human Well-being*, Advances in Global Change Research 83, https://doi.org/10.1007/978-3-032-25479-5_6

6.1 Introduction

Climate has a vital role in air quality as it affects the main processes, ranging from emission, transport, dispersion, chemical transformation, and deposition of air pollutants. Such processes are carefully controlled by weather variables such as temperature, humidity, see-saw wind forces, and vertical mixing in the air (Jury & Buthelezi, 2022). One likely impact of climate change is that it will increase the extent of air pollution in most populated regions due to changes in atmospheric circulation, precipitation, and removal processes, as well as the interaction of the atmosphere. Poor air quality jeopardizes the health of the population directly, and also has the possibility of affecting the food chain, ecosystem functions, with consequences to the population's health and feedback processes related to climate change (Ogwu et al., 2024). This issue of air pollution is deeply tied to the choice that society makes in relation to the development of the urban environment, energy, and urban consumption patterns. Although conventionally linked to the process of industrialization and afflictions of the urban jungle, air pollution has now become a global health issue, which poses a threat to human health much further than the industrial hubs (Ma et al., 2021). According to the World Health Organisation, over 90% of the world's population breathes air that doesn't meet safety standards (Donzelli et al., 2021). This causes millions of deaths every year. The harmful pollutants, such as particulate matter (≤ 2.5), oxides of nitrogen (NOx) and sulphur (SOx), and black carbon (BC), not only affect people's health but also contribute to global warming.

In recent years, vehicular emissions have played a significant role in worsening air pollution; however, industrial activities continue to be the primary source of air pollution in most industrialized countries (Herath Bandara & Thilakarathne, 2025; Masood & Ahmad, 2023). Recent studies have also suggested a possible correlation between meteorological factors, particularly temperature and air pollution levels (Chen et al., 2023; Grigorieva & Lukyanets, 2021; Sharma et al., 2025). Climate change may influence the dispersion of primary pollutants, especially particulate matter, and may also enhance the formation of secondary pollutants such as ozone (O_3), whose production is affected by temperature (Baidrulhisham et al., 2022). Empirical investigations have reported a positive linear correlation between ambient temperature and pollutant concentrations, including PM_{10}, $PM_{2.5}$, and NO_2. With rising summer temperatures, it is projected that both the minimum and maximum hourly ozone levels will increase, along with the number of days during which ozone concentrations exceed regulatory standards.

The Fifth Assessment Report (AR5) of the Intergovernmental Panel on Climate Change (IPCC) showed that the average global temperature on land and ocean increased by a total of about 0.85 °C between 1880 and 2012 (Shen et al., 2022). Moreover, the average temperature in the world over the last ten years, from 2003 to 2012, was approximately 0.78 °C above the period of 1850 to 1900. With a scenario of no other climate policy, it is estimated that the average surface temperature in the entire world may increase to between 3.7 and 4.8 °C in the next hundred years

(Shivanna, 2022). At the same time, there has been a significant rise in the atmospheric levels of greenhouse gases. Furthermore, the increasingly intensive industrial activity over the last one hundred years has also played a big role in deteriorating the quality of air through high levels of pollution emissions.

Many studies revealed that there is a strong correlation between the mortality rates and temperature variations (Heutel et al., 2021; Wu et al., 2022a, 2022b). At the same time, the effects of exposure to air pollution on human health are well known to be extremely deleterious (Masood & Ahmad, 2017). Several studies have connected ambient air pollution relatively in urban, suburban, or rural environments to premature deaths (Wang et al., 2024; Wu et al., 2022a, 2022b). Evidence indicates that the death rate attributed to outdoor air pollution is about 1.4 percent of all deaths across the globe. Furthermore, there is also a possibility of synergy between temperature and air pollution as a result of changing climatic conditions, which compounds the effects of the two factors on human health. Therefore, it is essential to comprehend such linked challenges of environmental health to formulate effective measures of public health. This chapter provides a short reading of the possible effects of climate change and air pollution on population health and discusses the most important issues and future needs of scientific studies and policy development.

6.2 Human Health Impacts Associated with Climate Change

Climate change and rising extreme weather events (EWEs) impact our environment surroundings which provides us with necessities, including potable water, clean air, safe food, and shelter (Khan et al., 2023). Climate change, coupled with other natural and anthropogenic health stressors, endangers human health and well-being through various direct and indirect pathways (Vafeiadou et al., 2023). Figure 6.1 highlights the interconnected pathways of the impact of climate change on human health. Moreover, the impacts of climate change are disproportionally distributed, affecting regions and populations with socioeconomic disadvantage (Paavola, 2017). Rapid urbanization and rising global population place additional stress on limited resources, driving the vulnerability of the human population to health risks associated with climate change (Chaudhry, 2023). It is expected that 60% of the global population will reside in cities by 2030, and most of the urbanization is expected to occur in developing countries (Harlan & Ruddell, 2011). Thus, it is crucial to continue to investigate and determine the impact of climate change on human health.

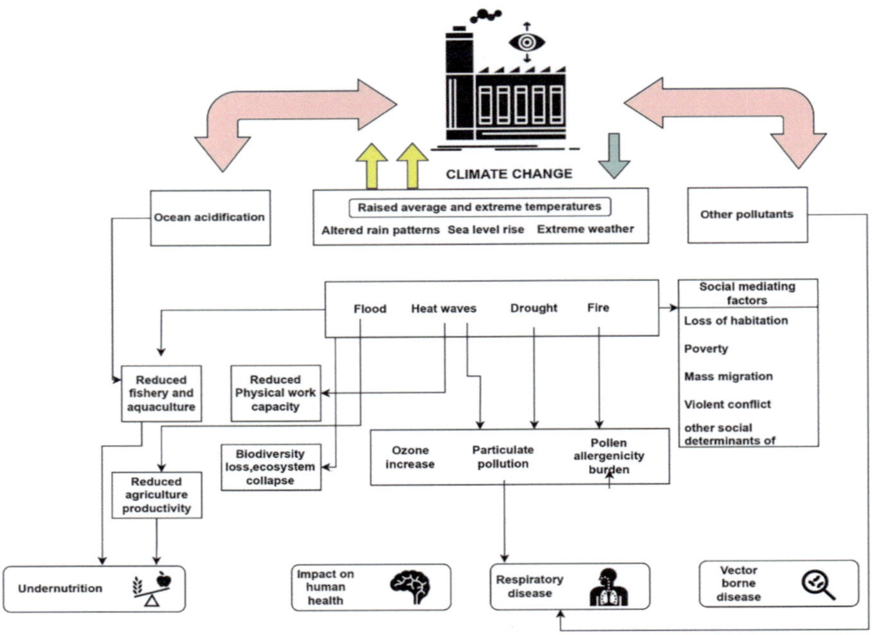

Fig. 6.1 Pathways between climate change and health outcomes

6.2.1 Direct Impacts

The health implications of climate change are far-reaching, affecting physical and mental health, and are influenced by socioeconomic, demographic, and other factors (Sorensen et al., 2023; Zhao et al., 2022). In its assessment report, the Intergovernmental Panel on Climate Change (IPCC) stated that extremes of climate change, such as altered precipitation and temperature patterns, may have a substantial effect on human health by raising mortality and morbidity rates. They may also cause ecological changes that can result in the spread of various infectious and contagious diseases, as well as shortages of food and water (IPCC, 2007). An increase in the frequency of extreme weather events has resulted in significant displacement of populations, injuries, and fatalities, with cascading effects on mental health and well-being. (Lina et al., 2025) reported that between 1993 - 2022, these extreme events have resulted in the loss of about 0.8 million lives and USD 4.3 trillion in economic losses.

Climate projections suggest an increase of 1.4–5.8 °C by the end of this century (Adamo et al., 2021; Hassan, 2024; Nunes, 2023). It is understood that such significant changes can have adverse effects, which could be difficult to predict and harder to deal with. This increase in baseline temperatures, especially in urban areas, compounded by extreme heat events (EHEs), is projected to increase rates of cardiovascular and respiratory diseases, particularly in vulnerable populations such as the elderly and those with pre-existing health conditions (Fagliano & Roux, 2018). Rise in uncompensable heat stress events in which heat-humidity levels overwhelm the human body's cooling ability, are expected to increase in tropical regions (Hess et al., 2018). A study conducted by Justine et al. (2023) has reported that 105 events of wet bulb temperature exceeding the physiological threshold for human thermoregulation, resulting in a cumulative exposure of no less than 300 h to uncompensable heat stress for the study period 1995–2020 in South Asia. This leads to heat stroke, dehydration, and exacerbation of chronic diseases.

Recent events, like the deadly heat wave of Karachi 2015, where more than 1200 deaths were attributed to high heat-high humidity events (Salim et al., 2015), highlight how unprepared urban areas can be to protect vulnerable populations from extreme heat. Excessive heat is one of the dominant causes of weather-related deaths in North America (Schatz & Kucharik, 2015; Vaidyanathan et al., 2020). According to studies, a number of conditions and illnesses linked to higher mortality during EHEs, including acute renal failure, cardiovascular disease, and heat-related illnesses like heat stroke, heat exhaustion, and hyperthermia, are correlated with more hospital/ED visits during heat waves (Kim et al., 2020; Knowlton et al., 2008; Schmeltz et al., 2016). Moreover, mental health is also compromised during heat waves, possibly due to potential impairment of physiological and behavioural thermoregulation. The rise in suicides and hospitalizations for mental illnesses has been observed during the hot season (Ebi et al., 2021; Rony & Alamgir, 2023). It has been reported that an increase by 2 °C in global mean temperatures above the pre-industrial levels is expected to impact vulnerable populations in 35% of global land area, whereas 4–5 °C warming would drastically increase this area (Matthews et al., 2025). Moreover, the elevated temperatures driven by climate change will impair productivity and working capacity and directly raise worker health hazards (Dasgupta et al., 2021).

Intensified storms, rainfalls, and subsequent flooding are becoming more frequent and severe, causing widespread damage to property and infrastructure, and loss of lives, while also increasing the risk of accidental and non-accidental mortality and morbidity (Bell et al., 2018; Zhao et al., 2022). Ban et al. (2023) reported an association between exposure to floods and risk of all-cause, non-accidental, and accidental mortality, particularly among older adults. According to (Keim, 2020), the most frequent extreme weather events globally from 1969 to 2018 were floods (47%) and storms (30%). These events, which show a rising trend, collectively accounted for about 55% of all mortality. Beyond direct mortality, floods contribute to injuries and the risk of infections, including respiratory and gastrointestinal diseases (Rocque et al., 2021). Furthermore, the devastating impact of floods extends to mental health, with prolonged exposure often leading to post-traumatic stress disorder, depression, and anxiety among affected populations (Martínez et al., 2015).

Furthermore, climate-driven sea level rise and changes in coastal weather events can result in the inundation of coastal communities, salinization of freshwater sources, and destruction of critical infrastructure, further exacerbating health vulnerabilities (Luber & Prudent, 2009). Increase in the frequency and intensity of extreme weather events, such as hurricanes and typhoons, also poses significant threats to human health through direct injury, displacement, and disruption of essential services (Berrang-Ford et al., 2021). These events disproportionately impact marginalized communities, who often lack the resources for effective preparation and recovery, amplifying existing health disparities (Butler, 2024).

Several studies have applied quantitative methods to predict health outcomes under various climate change scenarios and to determine the types and rates of weather-related mortality and morbidity (Bell et al., 2018; Daalen et al., 2020; Li et al., 2021; Piscitelli & Miani, 2024; Portier et al., 2010). These studies highlight that the global improvement in health achieved over the past several decades could be hindered or even reversed by climate change, and that inequalities in health between the rich and poor may be widened (Khan et al., 2023). It has become a key research challenge to detect and attribute health impacts to climate change.

6.2.2 Indirect Impacts

Rising frequency and intensity of extreme weather events—including floods, storms, droughts, and wildfires—pose direct threats to human health. These events result in immediate outcomes such as injuries, fatalities, and population displacement, while also generating cascading effects on mental health and well-being. A variety of environmental changes mediate how climate change affects human health (Vafeiadou et al., 2023). However, these impacts are mediated by factors such as geography, socioeconomic status, infrastructure, and access to healthcare (Chakraborty et al., 2019; Howe et al., 2019; Ke et al., 2024; Lee & Brown, 2022). The economic repercussions of water-related disasters, specifically floods and droughts, have been substantial, with global losses reaching approximately USD 600 billion between 2001 and 2018 (Lee et al., 2020).

Events like changing precipitation patterns, floods, droughts, and wildfires can affect food security, cause injuries, and contribute to the spread of infectious diseases (Arshad et al., 2020). These pathways expose different people to different risk intensities, creating varied impacts and vulnerabilities. Even though they are frequently evaluated separately, exposure to several climate change concerns can happen at the same time and have compounding or cascading effects on one's health (Paavola, 2017). For instance, changing weather patterns and extreme events, including increased precipitation or rising temperatures, have affected the waterborne disease (WBDs) and vector-borne disease (VBDs) transmission, and spread, and the consequences are likely to worsen (Cissé, 2019; Rocklöv & Dubrow, 2020). VBDs are transmitted to humans through hematophagous arthropods (such as mosquitoes, ticks, and sandflies), with tropical and subtropical regions bearing a disproportionate impact on public health (Ayinde et al., 2024).

Climate change, through altered temperature and precipitation patterns, modifies vector survival and reproduction rates, which in turn affect pathogen incubation periods and biting rates. A study conducted by (Romanello et al., 2022), have observed that between 1951–60 and 2012–21, the number of months that were conducive to the spread of malaria climbed by 31·3% in the highlands of the Americas and 13·8% in the highlands of Africa, while the probability of dengue transmission increased by 12% during the same time frame. The risk of vector-borne diseases is also increasing as a result of deforestation, which is altering natural habitats and increasing contact between people and disease vectors (Afrane et al., 2008; Ayinde et al., 2024; Dejene et al., 2020). Similarly, WBDs, which are linked to the ingestion of pathogens through contaminated food and water, are projected to increase, increasing the global burden of disease (Jung et al., 2023; Nichols et al., 2018).

Direct effects such as floods and sea-level rise can lead to contamination of water with fecal–oral pathogens (Cissé, 2019). These could cause a rise in cases of cholera and other diarrheal diseases (Cissé, 2019; Husman, 2013). Increase in frequency and severity of droughts, and heavy rainfall events, as well as the number of warm days and nights, affect the abundance, persistence, and geographic range of waterborne pathogens (Ahmed et al., 2020; Nichols et al., 2018; Piscitelli & Miani, 2024). Drought conditions can harbour pathogens in limited water supplies, thus increasing their concentrations (Fadhil et al., 2021). Severe rainfall events can mobilize pathogens and jeopardize water and sanitation infrastructure. Moreover, high temperatures can change pathogen survival, replication, and transmission rates, as well as the host susceptibility to infection. Changing weather patterns can interact with other factors, such as human behavior and the effectiveness of health systems in reducing risks (Ebi & Hess, 2020). Climate change is projected to continue to contribute to the spread of Lyme disease and tick-borne encephalitis, particularly in North America and Europe (Semenza et al., 2022). Therefore, it becomes crucial to reduce greenhouse gas emissions in order to lower the risk of infectious diseases (Mojahed et al., 2022).

6.3 Human Health Impacts Associated with Air Pollution

Air pollution has emerged as a critical global health threat, intricately linked with both climate change and public health outcomes. Rapid urbanization and extensive growth in developing countries have resulted in significant environmental degradation at both local and regional levels, exhausting natural resources and disturbing ecological processes (Kaginalkar et al., 2022; Waly et al., 2020). This urban expansion has led to dramatic alterations in land use and land cover, reducing limited green spaces and natural habitats. These changes contribute to problems including increased Land Surface Temperature (LST), intensification of the Urban Heat Island (UHI) effect, soil degradation, vulnerability to flooding, and exacerbated levels of pollution (Shi et al., 2020).

Among the myriad global environmental concerns, ambient (outdoor) air pollution stands out as a prominent issue exacerbated by intensive and uncontrolled urban development. Poor ambient air quality is primarily caused due to elevated levels of pollutants such as particulate matter (PM10, PM2.5), gaseous pollutants like oxides of nitrogen (NOx) and sulfur (SOx), carbon monoxide (CO), and ozone (O3) (Devi et al., 2025; Erin & David, 2012). These are largely emitted by anthropogenically generated emissions, including vehicular, industrial, and construction activities, as well as domestic burning, while certain emissions originate from natural sources (Baklanov et al., 2015; Ganguly et al., 2019; Shafie et al., 2022). It is well established that the concentration of these pollutants can be influenced and adjusted by synoptic patterns and meteorological parameters (Zhang et al., 2015).

The growing demand for public infrastructure and transportation systems in urban areas has led to elevated vehicular and industrial emissions. These emissions significantly contribute to the rising mortality and morbidity rates, with their impact varying across regions due to differences in regulatory frameworks and public health infrastructure (Baca-López et al., 2021; Ganguly et al., 2021). It is estimated that 9 out of 10 people worldwide breathe air containing high levels of pollutants, resulting in approximately 7 million deaths annually, including 660,000 children (Wu et al., 2021). The adverse health effects of air pollution, a consequence of these emissions, have been studied extensively around the globe. Epidemiological studies have conclusively established a causal relationship between exposure to air pollutants and health issues. These include cardio-pulmonary and respiratory disorders, diseases affecting the central nervous system, and complications leading to premature delivery in pregnant women (Manisalidis et al., 2020; Sharma et al., 2017).

Sun and Zhu (2019) have categorized the health outcomes associated with modern-day outdoor air pollution. These include respiratory diseases such as asthma and lung infections, cardiovascular diseases including hypertension, various blood disorders, heart diseases, and stroke, as well as several types of cancer. Additionally, they have noted adverse effects on pregnancy and children, including issues related to birth weight and infant mortality. Particulates of differing sizes and diameters have been positively correlated with elevated rates of overall morbidity and mortality (Jiang et al., 2019). When inhaled, these particles penetrate our respiratory system and impair the functioning of both the respiratory and cardiovascular systems, while also damaging our reproductive and central nervous systems (Kim et al., 2014; Lim & Kim, 2024; Nik Hassan, 2023).

Samoli et al. (2011), in a study conducted in Athens, Greece, reported a 2.54% increase in pediatric asthma cases corresponding to a 10 $\mu g/m^3$ increase in PM10 concentrations. Additionally, Chen et al. (2013), in their research in northern Chinese cities, highlighted that exposure to particulates could reduce life expectancy due to increased mortality from cardiorespiratory issues. A study conducted by Khanh et al. (2024) in Barcelona, Spain, examined the impact of air pollution on children's health and well-being, including its effects on metabolic health and weight. The findings suggested that adhering to World Health Organization (WHO) air quality guidelines could help mitigate childhood obesity to some extent. Building on these studies, it is clear that exposure to poor air quality poses a significant risk to human health. Additionally, the health consequences of air pollution led to increased personal and public healthcare expenditures, straining the existing public health infrastructure. Ultimately, an unhealthy workforce and escalated medical costs due to air pollution can impede the economic growth of a country, as highlighted by Patankar and Trivedi (2011) (Table 6.1).

Moreover, the health impacts of exposure to air pollution are not uniformly distributed across the globe, as the burden varies significantly between regions and

Table 6.1 Documented health impacts of air pollution

Health outcome/system	Specific effects	References
Respiratory system	Asthma, lung infections, impaired lung function	Manisalidis et al. (2020), Sharma et al., (2017), Samoli et al. (2011)
Cardiovascular system	Hypertension, heart disease, stroke; reduced life expectancy from cardiorespiratory mortality	Manisalidis et al. (2020), Sun and Zhu (2019), Chen et al. (2013)
Central nervous system	Neurological disorders, cognitive impairment	Kim et al. (2014), Lim and Kim (2024); Nik Hassan (2023)
Reproductive and pregnancy outcomes	Premature delivery, reduced birth weight, and infant mortality	Manisalidis et al. (2020), Sun and Zhu (2019)
Child health and development	Paediatric asthma, metabolic health issues, and obesity risk linked to PM2.5	Samoli et al. (2011); Khanh et al. (2024)
Cancer	Various cancers are linked to long-term pollutant exposure	Sun and Zhu (2019)
General morbidity and mortality	Elevated morbidity & mortality rates associated with PM10 and PM2.5; ~ 7 million deaths annually (660,000 children)	Jiang et al. (2019), Wu et al. (2021)

countries. According to the (IQAir, 2025) report, the top 100 most polluted cities are concentrated in fewer than 10 countries, underscoring the disproportionate nature of pollution exposure. A study by (Sang et al., 2022) assessed the spatio-temporal patterns of disease burdens attributed to ambient PM2.5 and found that regions such as North Africa, the Middle East, South Asia, and East Asia bear significantly higher burdens compared to others. Similarly, (Yu et al., 2024) highlighted widespread inequality in PM2.5 exposure, with substantial regional variation. Their analysis revealed that while air pollution levels have declined in many high-income countries, exposure inequality has increased, particularly in urban areas. In contrast, rapidly industrializing regions in East and South Asia experience severe pollution and growing exposure disparities, especially in the suburban areas (Apte et al., 2021; Chen et al., 2020). These findings emphasize the persistence of high exposure inequality over decades and the need for tailored mitigation strategies to address these disparities and ensure equitable access to cleaner air globally. Further complicating the issue, research indicates that lower socioeconomic status communities and populations often experience higher concentrations of ambient air pollutants (Hekmatpour & Leslie, 2022).

Therefore, the multifaceted evidence from these studies underscores the urgent need for comprehensive and sustained efforts to mitigate air pollution and safeguard public health, especially in vulnerable populations and high-risk regions. Air pollution contributes significantly to the global burden of disease, a burden that has escalated over the past 25 years due to demographic and epidemiological shifts, as well as increasing pollution levels in low and middle-income countries. The unequal spatial distribution of particulate and gaseous pollutants, clearly documented across and within nations, reveals that the benefits of cleaner air remain inequitably accessible, thereby perpetuating environmental injustice and undermining broader sustainable-development goals.

6.4 Nexus Between Urbanization, Climate Change, Air Pollution, and Human Health

Climate change and air pollution are intertwined global environmental challenges that pose a significant threat to human health and well-being (Shaddick et al., 2020). Rapid urbanization, characterized by the increasing concentration of populations in urban centers, is a major driver of both climate change and air pollution (Chen et al., 2020). Urban areas are hotspots for greenhouse gas emissions, primarily from transportation, energy consumption, and industrial activities. These emissions contribute to global warming and climate change, resulting in rising temperatures, altered precipitation patterns, and increased frequency of extreme weather events (Campbell-Lendrum & Corvalán, 2007). As discussed earlier, cities emit various pollutants like particulate matter, sulfur dioxide, nitrogen oxides, ozone (O3), and volatile organic compounds, which degrade air quality. Air pollution not only exacerbates climate change but is also worsened by it, creating a feedback loop with profound implications for environmental sustainability and public health (Dawson, 2014; Fiore et al., 2015; Kumar, 2021). The intricate interactions among urbanization, climate change, air pollution, and human health are of paramount importance, especially as urbanization rates continue to rise and global climate patterns undergo unprecedented shifts (Das et al., 2024) (Fig. 6.2).

Meteorological variables, such as temperature, precipitation, humidity, and wind speed, are critical in influencing air pollution concentrations and dispersion. They play a critical role in understanding the patterns of air quality over multiple spatio-temporal scales. The fate of air pollutants, from source to receptor, is sensitive to prevailing meteorological conditions that determine atmospheric stability, mixing, transport, and deposition processes (Kumar, 2021; Logothetis et al., 2025). For instance, the formation of secondary pollutants like ozone is highly dependent on temperature and sunlight (Jin et al., 2020). Control strategies to improve air quality often consider the existing meteorological circumstances of a particular area and the intricate relationships between these factors and air quality (Thambiran & Diab, 2010). However, with expected climate change, the understanding of the impacts of changing the mean and extreme meteorological conditions on air quality is an important consideration in developing effective control strategies to safeguard public health. It is increasingly acknowledged that climate change can either exacerbate or

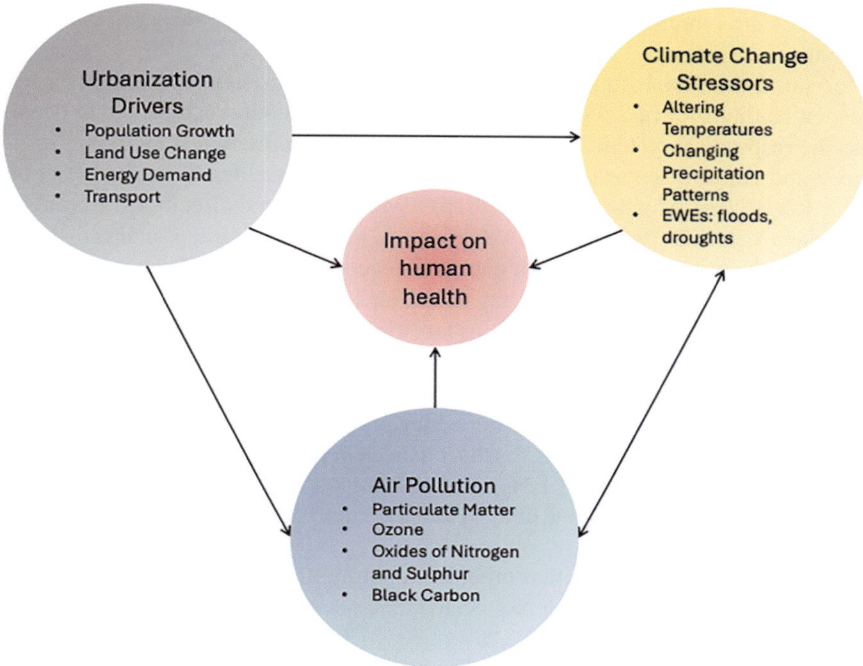

Fig. 6.2 Nexus between urbanization, climate change, air pollution, and human health

improve air quality, depending on the region, pollutant, and specific climate scenario (Fifth National Climate Assessment, 2023; Harlan & Ruddell, 2011; Vilcins et al., 2024). Thus, evaluating prospective climate conditions and their impact on achieving air quality goals should be a part of developing the air quality management plans to limit future levels of important health-relevant pollutants like ozone and fine particulates (Kinney, 2018).

Ozone and fine particulate matter are the two major air pollutants which pose significant threats to human health and cause climate change (Jacob & Winner, 2008). Ozone is created in the atmosphere by chemical reactions between nitrogen oxides and volatile organic compounds in the presence of sunlight (Fiore et al., 2015). While ozone in the stratosphere protects the Earth from harmful ultraviolet radiation, ground-level ozone is a harmful air pollutant. Higher temperatures and increased sunlight intensity can accelerate the formation of ozone, leading to higher concentrations during the warmer months. Moreover, emissions of ozone precursors, mainly oxides of nitrogen (NOx) and volatile organic compounds (VOCs), could also rise due to extreme heat conditions (Lefohn et al., 2018). These are emitted mainly by anthropogenic activities, including the burning of fossil fuels, industrial processes, and vehicles. They can also be released from evaporation from vegetation in the form of biogenic emissions. Increased ozone levels can cause respiratory problems, such as asthma, reduced lung function, and increased susceptibility to respiratory infections.

In contrast to Ozone, fine particulate matter ($PM_{2.5}$) consists of tiny airborne particles with a diameter of 2.5 μm or less, which upon inhalation can penetrate deeply into the respiratory system and cause a variety of health problems, including respiratory and cardiovascular diseases, cancer, and premature death (Anderson et al., 2011; Köse, 2020; Morakinyo et al., 2016). PM2.5 is emitted directly from combustion sources, such as power plants, vehicles, and industrial facilities, and can also be formed in the atmosphere through chemical reactions (Singh & Thajudeen, 2023). Small size of PM2.5 allows it to stay in the air for longer periods and travel long distances, affecting air quality in regions far from the source. Therefore, high temperatures can lead to increased evaporation of VOCs that can react in the atmosphere to form secondary organic aerosols, a major component of PM2.5. Moreover, climate change is projected to increase the frequency and severity of wildfires, which are a major source of $PM_{2.5}$ emissions. Lastly, changes in precipitation patterns can affect the deposition and removal of $PM_{2.5}$ from the atmosphere (Fifth National Climate Assessment, 2023; Jindamanee et al., 2025; Montero-Montoya et al., 2018; Sahoo et al., 2023).

These examples highlight that urbanization amplifies emissions, how air pollution and climate change are interconnected, and have synergistic effects on human health. (Adamkiewicz et al., 2020; Porwal, 2015). Moreover, increased urbanization contributes significantly to both air pollution and climate change through increased energy demand, industrial activity, and transportation emissions (Hassan et al., 2024). Many researchers believe that air pollution and climate change is closely related to urban land use patterns and socio-economic activities (Zlatev et al., 2024). Urban landscapes, through the built environment and green spaces, directly or indirectly impact air quality and climate change via the spatial distribution of land use, land cover, and transportation systems (Sharma et al., 2024). For instance, compact, mixed-use developments can reduce vehicle miles traveled and associated emissions, while green infrastructure can help to mitigate the urban heat island effect and improve air quality (Schneidemesser et al., 2015). Another study, conducted by Gao and Bukovsky (2023), found that efficiently and smartly planned urban land use patterns can reduce the exposure of the population to air pollution and the impacts of climate change, promoting more sustainable and resilient cities.

Therefore, addressing air pollution and mitigating climate change requires an integrated approach that recognizes the interconnectedness of these challenges and their impact on public health (Africa et al., 2019; Cohen et al., 2017; Xie et al., 2018).

6.5 Future Research Directions

The association between air quality and climate-related health risks is well-positioned to be a critical study focus in the persistence of the global burden of disease from air pollution and the increase in the impact of climate change on environmental exposures

(Akasha et al., 2023). An important avenue of future research effort is to unravel the complex and interactive processes through which climate change can increase or alter the health effects of anthropogenic air pollution. This will involve exploring the compound effects of various climate vulnerabilities, including heatwaves, droughts, wildfires, and flooding- all of which can directly or indirectly exacerbate poor air quality and compound health vulnerabilities. As an example, in the future, episodes of extreme heat can only increase ground-level ozone and particulate matter but may be synergistically combined with wildfire smoke exposure and healthcare infrastructure stress, resulting in consequent effects on health. Studies of such cascading, non-linear effects will require coupled modelling systems bringing together atmospheric science, epidemiology, climate projections, and health systems analysis. In addition, it will be essential to develop the potential to perform high-resolution empirical and modeling studies at several spatial and time scales, particularly at the local decision-making and planning of the population. To deal with these challenges, multidisciplinary interactions will be required. Future studies will need to draw in new data sources, including remote sensing and live exposure measurements, and use forward-looking data analysis techniques in order to further characterize risk factors in different populations. Conclusively, further knowledge regarding the climate-air pollution-health nexus will prove instrumental in the modeling of adaptive strategies and policies to safeguard the health of the citizens in the face of the rapidly changing environment.

6.6 Conclusion

The chapter has presented an overview of recent research methods and results on the health consequences of temperature variations as a result of climate change and effects on air quality. It highlighted the interaction of climate change and air pollution in health outcomes, and specifically in mortality. Recent studies show that anticipated climate change would probably increase near-surface ozone and particulate matter levels, which would lead to poor health outcomes with significant uncertainties. Since there are not too many choices for minimizing human exposure to air pollution, in the warming atmosphere, the protection of the population will need and demand serious regulatory actions, and this means a great reduction in the emission of greenhouse gases. Measures that induce policy changes with respect to emission standards of air pollutants should be implemented in sectors including transport, industry, urban planning, and energy. Most of these policy-related initiatives will most likely be effective in reducing greenhouse gas emissions in the long term. However, further studies are required to estimate the levels and patterns of future health hazards, including morbidity and mortality, which would ultimately help in crafting even better and informed environmental policies.

References

Adamkiewicz, G., Liddie, J., & Gaffin, J. M. (2020). The respiratory risks of ambient/outdoor air pollution [Review of the respiratory risks of ambient/outdoor air pollution]. *Clinics in Chest Medicine, 41(4), 809.* https://doi.org/10.1016/j.ccm.2020.08.013

Adamo, N., Al-Ansari, N., & Sissakian, V. K. (2021). Review of climate change impacts on human environment: past, present and future Projections. *Engineering, 13(11), 605.*https://doi.org/10.4236/eng.2021.1311044

Afrane, Y. A., Little, T., Lawson, B. W., Githeko, A. K., & Yan, G. (2008). Deforestationand vectorial capacity of anopheles gambiae giles mosquitoes in malaria transmission, Kenya. https://wwwnc.cdc.gov/eid/article/14/10/pdfs/07-0781.pdf

Africa, A. of S. of S., Sciences, B. A. of, Leopoldina, G. N. A. of S., Medicine, U. S. N. A.of, & Sciences, U. S. N. A. of. (2019). Air Pollution and health—A science-policyInitiative. *Annals of Global Health, 85*(1). https://doi.org/10.5334/aogh.2656

Ahmed, M., Rizvi, D. R., Masood, S., Masood, A., & Ahmad, F. (2020). Machine learning approach for predicting the quality of water. *International Journal of Advanced Science and Technology 29*(5s), 275–282.

Akasha, H., Ghaffarpasand, O., & Pope, F. D. (2023). Climate change, air pollution and the associated burden of disease in the Arabian peninsula and neighbouring regions: A critical review of the literature. *Sustainability, 15*(4), 3766. https://doi.org/10.3390/su15043766

Anderson, J. O., Thundiyil, J. G., & Stolbach, A. (2011). Clearing the air: A review of theeffects of particulate matter air pollution on human health [Review of Clearing theAir: A Review of the Effects of Particulate Matter Air Pollution on Human Health]. *Journal of Medical Toxicology, 8*(2), *166. Springer Science+Business Media.* https://doi.org/10.1007/s13181-011-0203-1

Apte, J. S., Seraj, S., Chambliss, S., Hammer, M. S., Southerland, V., Anenberg, S. C.,Donkelaar, A. van, Bräuer, M., & Martin, R. D. (2021). Air inequality: global divergence in urban fine particulate matter trends. https://doi.org/10.26434/chemrxiv.14671908

Arshad, A., Ashraf, M., Sundari, R. S., Qamar, H., Wajid, M., & Hasan, M. (2020). Vulnerability assessment of urban expansion and modelling green spaces to build heatwaves risk resiliency in Karachi. *International Journal of Disaster Risk Reduction, 46*, Article 101468. https://doi.org/10.1016/j.ijdrr.2019.101468

Ayinde, A. A., Adebisi, Y. A., Mustapha, A. A., Taiwo, O., Faith, O. O., Favour, O. A., &Abdulmumin, D. I. (2024). Climate change, vector-borne diseases, and conflict: Intersecting challenges in vulnerable states. *Journal of Infectious Diseases and Epidemiology, 10*(8). https://doi.org/10.23937/2474-3658/1510326

Baca-López, K., Fresno, C., Espinal-Enríquez, J., Martínez-García, M., Camacho-López, M.A., Flores-Merino, M. V., & Hernández-Lemus, E. (2021). Spatio-temporalrepresentativeness of air quality monitoring stations in Mexico city: Implicationsfor public health. *Frontiers in Public Health, 8*. https://doi.org/10.3389/fpubh.2020.536174

Baidrulhisham, S. E., Noor, N. M., Hassan, Z., Sandu, A. V., Vizureanu, P., Ul-Saufie, A. Z., Zainol, M. R. R. M. A., Kadir, A. A., & Deák, G. (2022). Effects of weather and anthropogenic precursors on ground-level ozone concentrations in Malaysian cities. *Atmosphere, 13*(11), 1780. https://doi.org/10.3390/atmos13111780

Baklanov, A., Molina, L. T., & Gauss, M. (2015). Megacities, air quality and climate. *Atmospheric Environment, 126*, 235. https://doi.org/10.1016/j.atmosenv.2015.11.059

Ban, J., Sutton, C., Ma, Y., Lin, C., & Chen, K. (2023). Association of flooding exposure with cause-specific mortality in North Carolina United States. *Nature Water, 1*(12), 1027. https://doi.org/10.1038/s44221-023-00167-5

Bell, J. E., Brown, C., Conlon, K. C., Herring, S. C., Kunkel, K. E., Lawrimore, J. H., Luber, G., Schreck, C. J., Smith, A., & Uejio, C. K. (2018). Changes in extreme events and the potential impacts on human health. *Journal of the Air & Waste Management Association 68*(4), 265. https://doi.org/10.1080/10962247.2017.1401017

Berrang-Ford, L., Sietsma, A. J., Callaghan, M., Minx, J. C., Scheelbeek, P., Haddaway, N.,Haines, A., & Dangour, A. D. (2021). Systematic mapping of global research onclimate and health: a machine learning review [Review of Systematic mapping ofglobal research on climate and health: a machine learning review]. *The LancetPlanetary Health, 5*(8). https://doi.org/10.1016/s2542-5196(21)00179-0

Butler, C. D. (2024). Climate change and human health: Primary, secondary, and tertiaryeffects. In *Elsevier eBooks* (p. 213). Elsevier BV. https://doi.org/10.1016/b978-0-44318515-1.00007-1

Campbell-Lendrum, D., & Corvalán, C. (2007). Climate change and developing-country cities: implications for environmental health and equity. *Journal of Urban Health, 84*, 109. https://doi.org/10.1007/s11524-007-9170-x

Chakraborty, T., Hsu, A., Manya, D., & Sheriff, G. (2019). Disproportionately higherexposure to urban heat in lower-income neighborhoods: A multi-city perspective. *Environmental Research Letters, 14*(10), 105003. https://doi.org/10.1088/17489326/ab3b99

Chaudhry, D. (2023). Climate change and health of the urban poor: The role ofenvironmental justice. *The Journal of Climate Change and Health, 15*, 100277. https://doi.org/10.1016/j.joclim.2023.100277

Chen, J., Liu, Z., Yin, Z., Liu, X., Li, X., Yin, L., & Zheng, W. (2023). Predict the effect ofmeteorological factors on haze using BP neural network. *Urban Climate, 51*, 101630. https://doi.org/10.1016/J.UCLIM.2023.101630

Chen, Y., Ebenstein, A., Greenstone, M., & Li, H. (2013). Evidence on the impact ofsustained exposure to air pollution on life expectancy from China's Huai River policy. *Sustainability Science, 110*(32), 12936. https://doi.org/10.1073/pnas.1300018110

Chen, Y., Senthilkumar, N., Shen, H., & Shen, G. (2020). Environmental inequality deepened during the COVID-19 in the developing world. *Environmental Science and Technology, 55*(1), 7. https://doi.org/10.1021/acs.est.0c06193

Cissé, G. (2019). Food-borne and water-borne diseases under climate change in low- andmiddle-income countries: Further efforts needed for reducing environmental healthexposure risks. *Acta Tropica, 194*, 181. https://doi.org/10.1016/j.actatropica.2019.03.012

Cohen, A. J., Bräuer, M., Burnett, R., Anderson, H. R., Frostad, J., Estep, K., Balakrishnan,K., Brunekreef, B., Dandona, L., Dandona, R., Feigin, V. L., Freedman, G., Hubbell,B., Jobling, A., Kan, H., Knibbs, L. D., Liu, Y., Martin, R. V., Morawska, L., Forouzanfar, M. H., et al. (2017). Estimates and 25-year trends of the global burden ofdisease attributable to ambient air pollution: an analysis of data from the GlobalBurden of Diseases Study 2015. *The Lancet, 389*(10082), 1907. https://doi.org/10.1016/s0140-6736(17)30505-6

Daalen, K. R. van, Jung, L., Dhatt, R., & Phelan, A. (2020, February 1). Climate change andgender-based health disparities. In *The lancet planetary health* (Vol. 4, No. 2). Elsevier BV. https://doi.org/10.1016/s2542-5196(20)30001-2

Das, S., Choudhury, M. R., Chatterjee, B., Das, P., Bagri, S. C., Paul, D., Bera, M., & Dutta,S. (2024). Unraveling the urban climate crisis: Exploring the nexus of urbanization,climate change, and their impacts on the environment and human well-being—Aglobal perspective. *AIMS Public Health, 11*(3), 963. https://doi.org/10.3934/publichealth.2024050

Dasgupta, S., Maanen, N. van, Gosling, S. N., Piontek, F., Otto, C., & Schleußner, C.-F.(2021). Effects of climate change on combined labour productivity and supply: anempirical, multi-model study. *The Lancet Planetary Health, 5*(7). https://doi.org/10.1016/s2542-5196(21)00170-4

Dawson, J. P. (2014). Understanding the meteorological drivers of U.S. particulate matterconcentrations in a changing climate. https://journals.ametsoc.org/view/journals/bams/95/4/bams-d-12-00181.1.xml

Dejene, S. W., Tassie, N., & Boer, W. F. de. (2020). Present and future climatic suitability fordengue fever in Africa. *Infection Ecology and Epidemiology, 10*(1). https://doi.org/10.1080/20008686.2020.1782042

Devi, L. P., Chandana, R., & Bandhu, D. (2025). Assessment and forecasting of particulatematter emissions and structural health monitoring of buildings in Bangalore. *Scientific Reports, 15*(1). https://doi.org/10.1038/s41598-025-00814-9

Donzelli, G., Cioni, L., Cancellieri, M., Llopis-Morales, A., Morales-Suárez-Varela, M., & Bustin, S. (2021). Air quality during covid-19 lockdown. *Encyclopedia, 1*(3), 519–526. https://doi.org/10.3390/ENCYCLOPEDIA1030043

Ebi, K. L., Capon, A., Berry, P., Broderick, C., Dear, R. de, Havenith, G., Honda, Y., Kovats,S., Ma, W., Malik, A., Morris, N. B., Nybo, L., Seneviratne, S. I., Vanos, J., & Jay, O.(2021). Hot weather and heat extremes: Health risks. *The Lancet, 398(10301), 698.* https://doi.org/10.1016/s0140-6736(21)01208-3

Ebi, K. L., & Hess, J. (2020). Health risks due to climate change: Inequity in causes and consequences. *Health Affairs, 39*(12), 2056. https://doi.org/10.1377/hlthaff.2020.01125

Erin, M., & David, C. (2012). Understanding human illness and death following exposure to particulate matter air pollution. In *Tech eBooks*. https://doi.org/10.5772/30264

Fadhil, M. Y., Ruswan, Ismail, M., Helyani, & Tosepu, R. (2021). An impact of climaticchange on water-borne diseases: A review. *IOP Conference Series Earth andEnvironmental Science, 755*(1), 12081. https://doi.org/10.1088/17551315/755/1/012081

Fagliano, J., & Roux, A. V. D. (2018). Climate change, urban health, and the promotion ofhealth equity. *PLoS Medicine, 15*(7). https://doi.org/10.1371/journal.pmed.1002621

Fifth National Climate Assessment. (2023). https://doi.org/10.7930/nca5.2023

Fiore, A. M., Naïk, V., & Leibensperger, E. M. (2015). Air Quality and climate connections. *Journal of the Air & Waste Management Association, 65*(6), 645. https://doi.org/10.1080/10962247.2015.1040526

Ganguly, R., Sharma, D., & Kumar, P. (2019). Trend analysis of observational PM10concentrations in Shimla city, India. *Sustainable Cities and Society, 51*, 101719. https://doi.org/10.1016/j.scs.2019.101719

Ganguly, R., Sharma, D., & Kumar, P. (2021). Short-term impacts of air pollutants in three megacities of India during COVID-19 lockdown. *Environment Development and Sustainability, 23*(12), 18204. https://doi.org/10.1007/s10668-021-01434-9

Gao, J., & Bukovsky, M. (2023). Urban land patterns can moderate population exposures toclimate extremes over the 21st century. *Nature Communications, 14*(1). https://doi.org/10.1038/s41467-023-42084-x

Grigorieva, E., & Lukyanets, A. (2021). Combined effect of hot weather and outdoor airpollution on respiratory health: Literature review. *Atmosphere, 12*, 790. https://doi.org/10.3390/ATMOS12060790

Harlan, S. L., & Ruddell, D. (2011). Climate change and health in cities: impacts of heat andair pollution and potential co-benefits from mitigation and adaptation. *Current Opinion in Environmental Sustainability, 3*(3), 126. https://doi.org/10.1016/j.cosust.2011.01.001

Hassan, A. K. A., Saraya, M. S., Ali, H., & Abdelsalam, M. M. (2024). Low-cost IoT air quality monitoring station using cloud platform and blockchain technology. *Applied Sciences, 14*(13), 5774. https://doi.org/10.3390/app14135774

Hassan, N. E. (2024). Global warming: Causes, impacts and urgent strategies for asustainable future: A review. *GSC Advanced Research and Reviews, 20(3), 73.* https://doi.org/10.30574/gscarr.2024.20.3.0338

Hekmatpour, P., & Leslie, C. M. (2022). Ecologically unequal exchange and disparate deathrates attributable to air pollution: A comparative study of 169 countries from 1991 to 2017. *Environmental Research, 212,* 113161. https://doi.org/10.1016/j.envres.2022.113161

Herath Bandara, S. J., & Thilakarathne, N. (2025). Economic and public health impacts of transportation-driven air pollution in South Asia. *Sustainability, 17*(5), 2306. https://doi.org/10.3390/SU17052306

Hess, J., LM, S., Knowlton, K., Saha, S., Dutta, P., Ganguly, P., Tiwari, A., Jaiswal, A.,Sheffield, P. E., Sarkar, J., Bhan, S. C., Begda, A., Shah, T., Solanki, B., & Mavalankar, D. (2018). Building resilience to climate change: Pilot evaluation ofthe impact of India's First heat action plan on all-cause mortality. *Journal ofEnvironmental and Public Health, 1.* https://doi.org/10.1155/2018/7973519

Heutel, G., Miller, N. H., & Molitor, D. (2021). Adaptation and the mortality effects oftemperature across U.S. Climate regions. *The Review of Economics and Statistics,103*(4), 740–753. https://doi.org/10.1162/REST_A_00936

Howe, P. D., Marlon, J. R., Wang, X., & Leiserowitz, A. (2019). Public perceptions of thehealth risks of extreme heat across US states, counties, and neighborhoods. *Proceedings of the National Academy of Sciences, 116*(14), 6743. https://doi.org/10.1073/pnas.1813145116

Husman, M. A. B. A. M. de R. (2013). Dimensions of effects of climate change on watertransmitted infectious diseases. *Air & Water Borne Diseases, 2*(1).https://doi.org/10.4172/2167-7719.1000109

IQAir. (2025). World's most polluted cities—The most polluted cities according to data aggregated from over 80,000 data points.

IPCC. (2007). Climate Change 2007: Impacts, Adaptation and Vulnerability. Contribution ofWorking Group II to the Fourth Assessment Report of the Intergovernmental Panel onClimate Change. In: M. L. Parry, O. F. Canziani, J. P. Palutikof, P. J. van der Linden & C. E. Hanson, (Eds.), Cambridge University Press, Cambridge, UK, 976pp. https://www.ipcc.ch/site/assets/uploads/2018/03/ar4_wg2_full_report.pdf

Jacob, D. J., & Winner, D. A. (2008). Effect of climate change on air quality. *Atmospheric Environment, 43*(1), 51. https://doi.org/10.1016/j.atmosenv.2008.09.051

Jiang, Y., Chen, R., & Kan, H. (2019). The interaction of ambient temperature and airpollution in China. *Springer eBooks*, 105.https://doi.org/10.1007/978-981-13-2583-0_7

Jin, L., Schubert, S., Salim, M., & Schneider, C. (2020). Impact of air conditioning systemson the outdoor thermal environment during summer in Berlin, Germany. *International Journal of Environmental Research and Public Health, 17*(13), 4645. https://doi.org/10.3390/ijerph17134645

Jindamanee, K., Keawboonchu, J., Pinthong, N., Meeyai, A., Inchai, P., & Thepanondh, S.(2025). Environmental impacts and emission profiles of volatile organic compoundsfrom petroleum refineries. *Scientific Reports, 15*(1). https://doi.org/10.1038/s41598-025-99932-7

Jung, Y., Khant, N. A., Kim, H., & Namkoong, S. (2023). Impact of climate change on water-borne diseases: Directions towards sustainability. *Water, 15*(7), 1298. https://doi.org/10.3390/w15071298

Justine, J., Monteiro, J. M., Shah, H., & Rao, N. (2023). The diurnal variation of wet bulbtemperatures and exceedance of physiological thresholds relevant to human health inSouth Asia. *Communications Earth & Environment, 4(1).* https://doi.org/10.1038/s43247-023-00897-0

Jury, M. R., & Buthelezi, M. S. (2022). Air pollution dispersion over Durban. *South Africa. Atmosphere, 13*(5), 811. https://doi.org/10.3390/ATMOS13050811/S1

Kaginalkar, A., Kumar, S., Gargava, P., Kharkar, N., & Niyogi, D. (2022). Smart AirQ: a bigdata governance framework for urban air quality management in smart cities. *Frontiers in Environmental Science, 10.* https://doi.org/10.3389/fenvs.2022.785129

Ke, H., Chen, J.-H., Kao, S.-Y., Tsao, C., Kuo, C.-W., Wu, C.-C., & Shih, C. (2024). Heatstress–induced platelet dysfunction is associated with loss of fibrinogen and isimproved by fibrinogen supplementation. *Thrombosis Research, 241,* 109091.https://doi.org/10.1016/j.thromres.2024.109091

Keim, M. E. (2020). The epidemiology of extreme weather event disasters (1969–2018).*Prehospital and Disaster Medicine, 35*(3), 267. https://doi.org/10.1017/s1049023x20000461

Khan, S. U., Sheikh, M. R., Dimén, L., Batool, H., Abbas, M. A., & Nuţă, A. C. (2023). Nexus between health poverty and climatic variability in Pakistan: a geospatialanalysis. *Frontiers in Environmental Science, 11.* https://doi.org/10.3389/fenvs.2023.1180556

Khanh, H. N. T., Rigau-Sabadell, M., Khomenko, S., Barboza, E. P., Cirach, M., DuarteSalles, T., Nieuwenhuijsen, M., Vrijheid, M., Mueller, N., & de Bont, J. (2024). Ambient air pollution, urban green space and childhood overweight and obesity: A health impact assessment for

Barcelona Spain. *Environmental Research, 264*, Article 120306. https://doi.org/10.1016/j.envres.2024.120306
Kim, K., Kabir, E., & Kabir, S. (2014). A review on the human health impact of airborneparticulate matter. *Environment International, 74*, 136.https://doi.org/10.1016/j.envint.2014.10.005
Kim, S., Kim, S., Oh, J., Chae, Y., Park, J., Kim, D., & Kim, Y. (2020). Effects of the 2018heat wave on health in the elderly: implications for adaptation strategies to climatechange. *Environmental Analysis Health and Toxicology, 35*(4).https://doi.org/10.5620/eaht.2020024
Kinney, P. L. (2018). Interactions of Climate Change, Air Pollution, and Human Health. *Current Environmental Health Reports, 5*(1), 179. https://doi.org/10.1007/s40572-018-0188-x
Knowlton, K., Rotkin-Ellman, M., King, G., Margolis, H. G., Smıth, D. J., Solomon, G., Trent, R. B., & English, P. (2008). The 2006 California heat wave: Impacts on hospitalizations and emergency department visits. *Environmental Health Perspectives, 117*(1), 61. https://doi.org/10.1289/ehp.11594
Köse, S. (2020). Measurement and modelling of particulate matter emissions from harbor activities at a port area: A case study of trabzon Turkey. *Journal of ETA Maritime Science, 8*(4), 286. https://doi.org/10.5505/jems.2020.49389
Kumar, P. (2021). Climate change and cities: Challenges ahead. *Frontiers in SustainableCities, 3.* https://doi.org/10.3389/frsc.2021.645613
Lee, J., Perera, D., Glickman, T., & Taing, L. (2020). Water-related disasters and their healthimpacts: A global review. *Progress in Disaster Science, 8,* 100123. https://doi.org/10.1016/j.pdisas.2020.100123
Lee, K., & Brown, R. D. (2022). Effects of Urban Landscape and SociodemographicCharacteristics on Heat-Related Health Using Emergency Medical Service Incidents.*International Journal of Environmental Research and Public Health, 19*(3), 1287. https://doi.org/10.3390/ijerph19031287
Lefohn, A. S., Malley, C. S., Smith, L., Wells, B. B., Hazucha, M. J., Simon, H., Naïk, V.,Mills, G., Schultz, M. G., Paoletti, E., Marco, A. D., Xu, X., Zhang, L., Wang, T.,Neufeld, H. S., Musselman, R. C., Tarasick, D. W., Bräuer, M., Feng, Z., Gerosa,G. et al. (2018). Tropospheric ozone assessment report: Global ozone metrics for climatechange, human health, and crop/ecosystem research. *Elementa Science of theAnthropocene, 6*. https://doi.org/10.1525/elementa.279uns
Li, L., Jiang, C., Murtugudde, R., Liang, X., & Sapkota, A. (2021). Global population exposed to extreme events in the 150 most populated cities of the World: Implications for public health. *International Journal of Environmental Research and Public Health, 18*(3), 1293. https://doi.org/10.3390/ijerph18031293
Lim, E. Y., & Kim, G.-D. (2024). Particulate matter-induced emerging health effectsassociated with oxidative stress and inflammation. *Antioxidants, 13*(10), 1256. https://doi.org/10.3390/antiox13101256
Lina, A., Eckstein, D., Künzel, V., & Schäfer, L. (2025). Climate Risk Index 2025: Whosuffers most from extreme events? *Germanwatch.*
Logothetis, I., Mitsotakis, A., & Grammelis, P. (2025). Air quality health index anddiscomfort conditions in a heatwave episode during July 2024 in Rhodes Island, 59. https://doi.org/10.3390/engproc2025087059
Luber, G., & Prudent, N. (2009). Climate change and human health. *PubMed.* https://pubmed.ncbi.nlm.nih.gov/19768168
Manisalidis, I., Stavropoulou, E., Stavropoulos, A., & Bezirtzoglou, E. (2020). Environmental and health impacts of air pollution: A revie. *Frontiers in PublicHealth, 8*. https://doi.org/10.3389/fpubh.2020.00014
Martínez, G. S., Williams, E., & Yu, S. (2015). The Economics of health damage andadaptation to climate change in Europe: A review of the conventional and greyliterature. *Climate, 3*(3), 522. https://doi.org/10.3390/cli3030522
Masood, A., & Ahmad, K. (2023). Particulate matter forecasting using artificial neural network and support vector machine based models. In ICIDSSD 2022: Proceedings of the 3rd International

Conference on ICT for Digital, Smart, and Sustainable Development, ICIDSSD 2022, 24–25 March 2022, New Delhi, India.

Masood, A., & Ahmad, K. (2017). Air pollution dispersion modeling for diesel generators at Jamia University Campus, New Delhi, India. *IJARI, 6*, 118–121.

Matthews, T., Raymond, C., Foster, J., Baldwin, J. W., Ivanovich, C., Kong, Q., Kinney, P. L., & Horton, R. (2025). Mortality impacts of the most extreme heat events Nature. *Reviews Earth and Environment*. https://doi.org/10.1038/s43017-024-00635-w

Mojahed, N., Mohammadkhani, M. A., & Mohamadkhani, A. (2022). Climate crises and developing vector-borne diseases: A narrative review. *Iranian Journal of Public Health. Knowledge E*. https://doi.org/10.18502/ijph.v51i12.11457

Montero-Montoya, R., López-Vargas, R., & Arellano-Aguilar, O. (2018). Volatile organic compounds in air: Sources, distribution, exposure and associated illnesses in children. *Annals of Global Health, 84*(2), 225. https://doi.org/10.29024/aogh.910

Morakinyo, O. M., Mokgobu, M. I., Mukhola, M. S., & Hunter, P. R. (2016). HealthOutcomes of Exposure to Biological and Chemical Components of Inhalable andRespirable Particulate Matter. *International Journal of Environmental Research andPublic Health, 13*(6), 592. https://doi.org/10.3390/ijerph13060592

Ma, X., Zhang, T., Ji, C., Zhai, Y., Shen, X., & Hong, J. (2021). Threats to human health andecosystem: Looking for air-pollution related damage since 1990. *Renewable andSustainable Energy Reviews, 145,* 111146. https://doi.org/10.1016/J.RSER.2021.111146

Nichols, G., Lake, I., & Heaviside, C. (2018). Climate change and water-related infectious diseases. *Atmosphere, 9*(10), 385. https://doi.org/10.3390/atmos9100385

Nik Hassan, N. M. N. (2023). Cardiovascular and respiratory health effects of fineparticulate matters (PM2.5): A review on time series studies. https://www.mdpi.com/2073-4433/14/5/856

Nunes, L. J. R. (2023). The rising threat of atmospheric CO_2: A review on the causes, impacts, and mitigation strategies. *Environments, 10*(4), 66. Multidisciplinary Digital Publishing Institute. https://doi.org/10.3390/environments10040066

Ogwu, M. C., Imarhiagbe, O., Ikhajiagbe, B., & Osawaru, M. E. (2024). Toward understanding the impacts of air pollution. *Handbook of Environmental Chemistry, 133*, 3–43. https://doi.org/10.1007/698_2024_1107

Paavola, J. (2017). Health impacts of climate change and health and social inequalities in theUK. Environmental Health, 16. BioMed Central. https://doi.org/10.1186/s12940-017-0328-z

Patankar, A., & Trivedi, P. (2011). Monetary burden of health impacts of air pollution in Mumbai, India: Implications for public health policy. *Public Health, 125*(3), 157. https://doi.org/10.1016/j.puhe.2010.11.009

Piscitelli, P., & Miani, A. (2024). Climate change and infectious diseases: Navigating theIntersection through innovation and interdisciplinary approaches. *International Journal of Environmental Research and Public Health, 21*(3), 314. https://doi.org/10.3390/ijerph21030314

Portier, C. J., Tart, K. T., Carter, S., Dilworth, C., Grambsch, A., Gohlke, J. M., Hess, J.,Howard, S., Luber, G., Lutz, J., Maslak, T., Radtke, M., Rosenthal, J. P., Rowles, T.,Sandifer, P. A., Scheraga, J. D., Strickman, D., Trtanj, J., & Whung, P. (2010). A Human health perspective on climate change: a report outlining research needson the human health effects of climate change. *Environmental Health Perspectives*. https://doi.org/10.1289/ehp.1002272

Porwal, T. (2015). Paint Pollution Harmful Effects on Environment. *International Journal of Research -Granthaalayah, 3,* 1. https://doi.org/10.29121/granthaalayah.v3.i9se.2015.3204

Rocklöv, J., & Dubrow, R. (2020). Climate change: an enduring challenge for vector-bornedisease prevention and control. *Nature Immunology, 21*(5), 479. https://doi.org/10.1038/s41590-020-0648-y

Rocque, R., Beaudoin, C., Ndjaboué, R., Cameron, L., Poirier-Bergeron, L., Poulin-Rheault,R.-A., Fallon, C., Tricco, A. C., & Witteman, H. O. (2021). Health effects of climatechange: an overview of systematic reviews. *BMJ Open, 11*(6).https://doi.org/10.1136/bmjopen-2020-046333

Romanello, M., Napoli, C. D., Drummond, P., Green, C., Kennard, H., Lampard, P., Scamman, D., Arnell, N. W., Ayeb-Karlsson, S., Berrang-Ford, L., Belesova, K., Bowen, K., Cai, W.,

Callaghan, M., Campbell-Lendrum, D., Chambers, J., van Daalen, K. R., Dalin, C., Dasandi, N., Costello, A., et al. (2022). The 2022 report of the Lancet Countdown on health and climate change: Health at the mercy of fossil fuels. *The Lancet, 400*(10363), 1619. https://doi.org/10.1016/s0140-6736(22)01540-9

Rony, M. K. K., & Alamgir, H. M. (2023). High temperatures on mental health: Recognizingthe association and the need for proactive strategies—A perspective. *Health ScienceReports, 6*(12). https://doi.org/10.1002/hsr2.1729

Sahoo, A., Madheshiya, P., Mishra, A., & Tiwari, S. (2023). Combating ozone stress throughN fertilization: A case study of Indian bean (*Dolichos lablab* L.). *Frontiers in PlantScience, 14*. https://doi.org/10.3389/fpls.2023.1125529

Salim, A., Ahmad, A., Ashraf, N., & Ashar, m. (2015). Deadly Heat Wave in Karachi, July2015: Negligence or Mismanagement? *International Journal of Occupational Medicine and Environmental Health, 6*(4).

Samoli, E., Nastos, P. T., Παλιατσός, A. Γ., Katsouyanni, K., & Priftis, K. N. (2011). Acuteeffects of air pollution on pediatric asthma exacerbation: Evidence of association andeffect modification. *Environmental Research, 111*(3), 418. https://doi.org/10.1016/j.envres.2011.01.014

Sang, S., Chu, C., Zhang, T., Chen, H., & Yang, X. (2022). The global burden of disease attributable to ambient fine particulate matter in 204 countries and territories, 1990–2019: A systematic analysis of the Global Burden of Disease Study 2019. *Ecotoxicology and Environmental Safety, 238*, 113588. https://doi.org/10.1016/j.ecoenv.2022.113588

Schatz, J., & Kucharik, C. J. (2015). Urban climate effects on extreme temperatures inMadison, Wisconsin, USA. *Environmental Research Letters, 10*(9), 94024. https://doi.org/10.1088/1748-9326/10/9/094024

Schmeltz, M. T., Marcotullio, P. J., Himmelstein, D. U., Woolhandler, S., & Sembajwe, G. (2016). Outcomes of hospitalizations for common illnesses associated with acomorbid heat-related illness in the United States, 2001–2010. *Climatic Change, 138*, 567. https://doi.org/10.1007/s10584-016-1747-5

Schneidemesser, E. von, Monks, P. S., Allan, J. D., Bruhwiler, L., Forster, P. M., Fowler, D.,Lauer, A., Morgan, W. T., Paasonen, P., Righi, M., Šindelářová, K., & Sutton, M. A.(2015). Chemistry and the linkages between air quality and climate change. *Chemical Reviews, 115*(10), 3856. https://doi.org/10.1021/acs.chemrev.5b00089

Semenza, J. C., Rocklöv, J., & Ebi, K. L. (2022). Climate Change and cascading risks frominfectious disease. *Infectious Diseases and Therapy, 11*(4), 1371. https://doi.org/10.1007/s40121-022-00647-3

Shaddick, G., Thomas, M. L., Mudu, P., Ruggeri, G., & Gumy, S. (2020). Half the world'spopulation are exposed to increasing air pollution. *Npj Climate and AtmosphericScience, 3*(1). https://doi.org/10.1038/s41612-020-0124-2

Shafie, S. H. M., Mahmud, M., Mohamad, S., Rameli, N. L. F., Abdullah, R., & Mohamed, A. F. (2022). Influence of urban air pollution on the population in the Klang Valley, Malaysia: A spatial approach. *Ecological Processes, 11*(1). https://doi.org/10.1186/s13717-021-00342-0

Sharma, A. K., Baliyan, P., & Kumar, P. (2017). Air pollution and public health: Thechallenges for Delhi, India. *Reviews on Environmental Health, 33*(1), 77. https://doi.org/10.1515/reveh-2017-0032

Sharma, D., Thapar, S., Jain, D., & Sachdeva, K. (2024). Mapping the SpatiotemporalVariability of Particulate Matter Pollution in Delhi: Insights from Land UseRegression Modelling. *Journal of the Indian Society of Remote Sensing, 52*(6), 1329. https://doi.org/10.1007/s12524-024-01879-1

Shi, Y., Bilal, M., Ho, H. C., & Omar, A. (2020). Urbanization and regional air pollutionacross South Asian developing countries—A nationwide land use regression forambient PM2.5 assessment in Pakistan. *Environmental Pollution, 266,* 115145. https://doi.org/10.1016/j.envpol.2020.115145

Singh, A., & Thajudeen, T. (2023). A hybrid particle swarm optimization-tuningalgorithm for the prediction of nanoparticle morphology from microscopic images. *Aerosol and Air Quality Research, 23*(4), 220453. https://doi.org/10.4209/aaqr.220453

Sorensen, C., Campbell, H., Depoux, A., Finkel, M. L., Gilden, R., Hadley, K., Haine, D.,Mantilla, G., McDermott-Levy, R., Potter, T., Sack, T. L., Tun, S., & Wellbery, C.(2023). Core competencies to prepare health professionals to respond to the climatecrisis. *PLOS Climate, 2*(6). https://doi.org/10.1371/journal.pclm.0000230

Sun, Z., & Zhu, D. (2019). Exposure to outdoor air pollution and its human health outcomes: A scoping review. *PLoS ONE, 14*(5). https://doi.org/10.1371/journal.pone.0216550

Sharma, D., Thapar, S., Masood, A., & Sachdeva, K. (2025). Fourier-enhanced deep learning and machine learning models for predicting multi-scale PM2. 5 dynamics in megacities: A case study of Delhi. *Earth Systems and Environment,* 1–26.

Shen, B., Song, S., Zhang, L., Wang, Z., Ren, C., & Li, Y. (2022). Temperature trends in some major countries from the 1980s to 2019. *Journal of Geographical Sciences, 32*(1), 79–100. https://doi.org/10.1007/S11442-022-1937-1/METRICS

Shivanna, K. R. (2022). Climate change and its impact on biodiversity and human welfare. *Proceedings of the Indian National Science Academy, 88*(2), 160–171. https://doi.org/10.1007/S43538-022-00073-6/METRICS

Thambiran, T., & Diab, R. D. (2010). A review of scientifc linkages and interactions betweenclimate change and air quality, with implications for air quality management in South Africa. *Academy of Science of South Africa.* https://doi.org/10.4102/sajs.v106i3/4.56

Vafeiadou, A., Banissy, M. J., Banissy, J. F. M., Higgins, J. P. T., & Howard, G. (2023). Theinfluence of climate change on mental health in populations of the western Pacifiregion: An umbrella scoping review. *Heliyon, 9*(11). https://doi.org/10.1016/j.heliyon.2023.e21457

Vaidyanathan, A., Malilay, J., Schramm, P., & Saha, S. (2020). Heat-related deaths United States, 2004–2018. *MMWR Morbidity and Mortality Weekly Report, 69*(24), 729. https://doi.org/10.15585/mmwr.mm6924a1

Vilcins, D., Christofferson, R. C., Yoon, J.-H., Nazli, S. N., Sly, P. D., Cormier, S. A., & Shen, G. (2024). Updates in air pollution: Current research and future challenges. *Annals of Global Health, 90*(1). https://doi.org/10.5334/aogh.4363

Waly, N. M., Ayad, H. M., & Saadallah, D. (2020). Assessment of spatiotemporal patterns of social vulnerability: A tool to resilient urban development Alexandria Egypt. *Ainshams Engineering Journal, 12*(1), 1059. https://doi.org/10.1016/j.asej.2020.07.025

Wu, I.-P., Liao, S., Lai, S., & Wong, K.-S. (2021). The respiratory impacts of air pollution inchildren: Global and domestic (Taiwan) situation. *Biomedical Journal, 45*(1), 88. https://doi.org/10.1016/j.bj.2021.12.004

Wang, J., Lin, J., Liu, Y., Wu, F., Ni, R., Chen, L., Ren, F., Du, M., Li, Z., Zhang, H., & Liu, Z. (2024). Direct and indirect consumption activities drive distinct urban-rural inequalities in air pollution-related mortality in China. *Science Bulletin, 69*(4), Article 544553. https://doi.org/10.1016/J.SCIB.2023.12.023

Wu, X., Vu, T. V., Harrison, R. M., Yan, J., Hu, X., Cui, Y., Shi, A., Liu, X., Shen, Y., Zhang, G., & Xue, Y. (2022a). Long-term characterization of roadside air pollutants in urban Beijing and associated public health implications. *Environmental Research, 212*, Article 113277. https://doi.org/10.1016/J.ENVRES.2022.113277

Wu, Y., Li, S., Zhao, Q., Wen, B., Gasparrini, A., Tong, S., Overcenco, A., Urban, A.,Schneider, A., Entezari, A., Vicedo-Cabrera, A. M., Zanobetti, A., Analitis, A., Zeka, A., Tobias, A., Nunes, B., Alahmad, B., Armstrong, B., Forsberg, B., Guo, Y. (2022). Global, regional, and national burden of mortality associated with short-termtemperature variability from 2000–19: A three-stage modelling study. *The Lancet Planetary Health, 6*(5), e410–e421. https://doi.org/10.1016/S2542-5196(22)00073-0c

Xie, Y., Dai, H., Xu, X., Fujimori, S., Hasegawa, T., Yi, K., Masui, T., & Kurata, G. (2018). Co-benefits of climate mitigation on air quality and human health in Asian countries. *Environment International, 119*, 309. https://doi.org/10.1016/j.envint.2018.07.008

Yu, W., Ye, T., Chen, Z., Xu, R., Song, J., Li, S., & Guo, Y. (2024). Global analysis revealsregion-specific air pollution exposure inequalities. *One Earth.* https://doi.org/10.1016/j.oneear.2024.09.017

Zhang, H., Wang, Y., Hu, J., Ying, Q., & Hu, X. (2015). Relationships between meteorological parameters and criteria air pollutants in three megacities in China. *Environmental Research, 140*, 242. https://doi.org/10.1016/j.envres.2015.04.004

Zhao, Q., Yu, P., Mahendran, R., Huang, W., Gao, Y., Yang, Z., Ye, T., Wen, B., Wu, Y., Li,S., & Guo, Y. (2022). Global climate change and human health: Pathways andpossible solution. *Eco-Environment & Health, 1*(2), 53. https://doi.org/10.1016/j.eehl.2022.04.004

Zlatev, Z., Todorov, A., Karadzhova, D., Vasilev, M., & Veleva, P. (2024). Evaluating thePerformance and Practicality of a Multi-Parameter Assessment System with Design,Comparative Analysis, and Future Directions. *Sustainability, 16*(10), 4124. https://doi.org/10.3390/su16104124

Part III
Economic and Livelihood Impacts

Chapter 7
Water Scarcity and Climate Change: Household Strategies and Coping Costs in Sub-Himalayan India

Abhimita Paul Mazumder and Pritha Datta

Abstract With growing threats from climate change, coupled with rapid urbanization, population growth, and weak water supply systems, the Eastern Himalayan foothills of India are facing acute water shortages, particularly for drinking purposes. This study examines the coping costs borne by households in Jalpaiguri district, West Bengal, and identifies the factors that shape these costs. Coping includes collecting, storing, drawing, treating, and purchasing water. The results show that income, gender, caste, and perceptions of scarcity significantly influence household coping costs. Poorer households spend disproportionately more through time and labor devoted to water collection, while higher-income households rely on water purchases to reduce this burden. Women collectors face higher costs, underscoring the gendered nature of water insecurity, and caste-based differences further shape coping patterns. Households perceiving severe scarcity incur the highest expenditures, while those reporting moderate or slight scarcity spend less. These findings highlight that water insecurity in Jalpaiguri is driven not only by supply constraints but also by social and economic inequalities. Strengthening the Jal Jeevan Mission, providing targeted support to marginalized households, and promoting community-led initiatives such as rainwater harvesting can reduce coping costs and improve resilience, with lessons applicable to other rural regions facing similar challenges.

Keywords Water scarcity · Eastern Himalayan foothills · Coping cost · SDG 6 · India

A. P. Mazumder (✉)
Department of Policy and Management Studies, TERI School of Advanced Studies, New Delhi, India
e-mail: abhimita.mazumder@terisas.ac.in

P. Datta
Rekhi Centre of Excellence for the Science of Happiness, Indian Institute of Technology Kharagpur, Kharagpur, India

P. Datta et al. (eds.), *Climate Change and Human Well-being*, Advances in Global Change Research 83, https://doi.org/10.1007/978-3-032-25479-5_7

7.1 Introduction

7.1.1 Global Water Availability

Water is crucial for sustaining life on Earth. According to the National Research Council of the National Academies of Sciences, Engineering, and Medicine, it has been noted that the minimum drinking water requirement for survival of humans is 5 L of water per person per day (Brown & Matlock, 2011). The World Health Organisation has reported that 6.3% of deaths and 9.1% of Global Disease Burden are due to unsafe drinking water, inadequate sanitation, and poor hygiene (Pal et al., 2018). Globally, it has been reported that freshwater availability is estimated to be 43,750 km^3 $year^{-1}$, with America having the highest global freshwater resource with 45%, Asia with 28%, Europe with 16% and Africa with 9%. Clean drinking water has been severely scarce, and millions of people around the globe spend hours searching for it (Chakkravarthy, 2019).

The World Bank (2019) report highlights that the United States takes out a hundred times more water (per capita) than the twenty least water-consuming nations and four times more than the average of the world. While addressing rising water scarcity, it is crucial to know where global water consumption is more concentrated. By sector-wise, globally, it has been noted that withdrawal of water from agriculture is 69%, 19% by industries and manufacturing units, and 12% which comes from the usage of municipalities (UN, 2021). These uneven distributions of the water resources by geography and depletion of freshwater through excessive consumption and wastage result in serious conflict over water in local communities and a global water crisis (Shiva, 2016).

According to the World Economic Forum Global Risk Report, it has been noted that water scarcity emerges as one of the major environmental challenges of the twenty-first century. Climate changes, which are driven by hydrological changes such as increased/decreased precipitation, temperature rise, drought, etc., have predictably given rise to water-stressed regions (Mahato et al., 2022). There has been a rapid increase in population, agricultural activities, industrial development, and improper government management, which contribute to the increasing water scarcity problems. Moreover, the problem of drinking water scarcity is further aggravated by global population increase, rapid urbanisation, and increased water demands by the agricultural and industrial sectors. According to the Sustainable Development Goals Report by the United Nations, billions of people across the globe will face extreme lack of access to safe water by 2030 (Martin, 2023).

7.1.2 India's Water Scarcity Scenario

By 2050, India is expected to be one of the world's most water-stressed nations, where it will be unable to serve its vast population with safe water (World Resources

Institute India, 2023). It has been noted that approximately 600 million people face a severe water crisis, where 3/4th of the households do not have access to clean drinking water in their houses (Gulati & Banerjee, 2016). It has been noted that the western part of the Indo-Gangetic region has shown increasing trends of precipitation, whereas eastern and central regions have shown a declining trend (Bhatla et al., 2019). This has resulted in a decline in runoff levels in most of the river basins of India. With an increase in temperature and variation in runoff, the groundwater level is also expected to decline in many regions of India, especially during the dry seasons such as winter (Mall et al., 2006; Saleth, 2011).

It has been found that the northern part of India faced a major decline in rainfall of about 8% during 1951–2021, which was accompanied by frequent droughts (Mishra et al., 2012; Roxy et al., 2015). Also, in the region, people faced a decline in groundwater storage due to an increase in demand for irrigation, recharge, and pumping (Mishra et al., 2024). The northwestern part of India, mainly states like Rajasthan, Maharashtra, Uttar Pradesh, and cities like Delhi, faces acute water shortages (Rana & Guleria, 2018). The major cities of India, including Delhi, Bengaluru, Mumbai, and Hyderabad, are reported to have already depleted groundwater resources (Ghosh, 2021). Southern and eastern India also shows a significant decline in groundwater along with regions in the northern part (Panda et al., 2020).

It has also been noted that districts with greater groundwater depletion are a major consequence of human activities rather than only seasonal precipitation (Alkon et al., 2024). According to the fourth IPCC Report (2007), the rural population of India is most vulnerable since their lives depend on natural resources and agricultural activities (Ghosh & Ghosal, 2023). The study of Green (2011) reflects on the economic outcome of declining groundwater levels, which shows a tendency for an increase in the cost of pumping. Also, this problem is aggravated by rapid population growth and economic development. Additionally, recent climatic changes have been creating hindrances to the adaptive capacities of the rural population (Giz, 2025).

Pre-independence and post-independence years have witnessed several growths in water infrastructure. According to the Government of India (2012), the National Water Policy (2012) was the first policy recognized as an advanced approach in assessing water scarcity and the need for research for water saving. However, it still remains one of the biggest challenges in addressing water scarcity. Hence, there is a need to rethink the policies and how they can be addressed to the actual problem (Katyaini & Barua, 2015).

7.1.3 Water Scarcity in the Himalayan Region of India

Mountains are considered to be 'natural water towers' for the communities residing in the arid and semi-arid lowlands (Viviroli & Weingartner, 2007). Contrastingly, it has been observed that chronic shortage of clean water has been affecting the communities' lives in the Himalayan region. These were results of population growth (4% per annum), delays in implementing and developing projects for water supply, and low-cost tariff (Pattanayak et al., 2005). The Himalayan springs, which are an important source of groundwater, have been under stress due to climatic factors, seismic activities, and human-related factors such as deforestation and urbanisation (Palanisami & Nagothu, 2024). This has further led to drying of springs, affecting the number of lives dependent on these water resources.

The Indian Himalayas are seen as a source of freshwater resources, but only a limited amount of resources is stored and consumed by the communities, which gives rise to water scarcity problems (Shah & Badiger, 2018). The Eastern Himalayan region receives an ample amount of annual rainfall, but paradoxically, the regions face acute temporal water shortages (Chhetri & Tamang, 2013; Drew & Rai, 2016; Lepcha, 2013; Rasaily, 2014). The sudden increase in population, urbanisation, increased pressure of tourism, and illegal construction have led to the problem of water scarcity in the terrain of the eastern Himalayas (Ghatani, 2021). Also, the absence of drilled wells, proper construction of pipelines and river has further aggravated the situation in the region (Bhutia, 2017). Hence, it is not unknown that the Himalayan terrain faces a water scarcity problem due to the above reasons. Additionally, a lack of safe drinking water has been creating severe health hazards (Ghosh & Ghosal, 2023).

7.1.4 *Water Scarcity in the Eastern Himalayan Foothills of India*

Shifting the focus to the Sub-Himalayan region or the Eastern Himalayan foothills, little research has highlighted the problem of water scarcity in the region. Even the foothills of the Himalayan region face the issue. Deb and Mukherjee (2021) have critically highlighted that the Sub-Himalayan region of West Bengal, also known as 'Dooars', faces a water shortage problem despite receiving high-intensity rainfall and having proximity to the rivers like Teesta, Torsha, Jaldhaka, and Raidak. The author further highlights the blocks of the Dooars region where households do not have proper access to drinking water and are hence dependent on natural or public sources of water. Also, the lack of adequate wells and taps causes such severe scarcity.

Recent studies (Datta & Behera, 2021; Datta & Das, 2019) based on the sub-Himalayan foothills of West Bengal have concluded that there is a decrease in rainfall during the monsoon and no rainfall in the winter months. Considering the marginalised section of the society, the farmers mainly, their research further concluded that they have experienced overall groundwater depletion in the dry months, which they use for both irrigation and filling up drying ponds, hence escalating the problem of water scarcity. Another study of Datta and Behera (2022) in the eastern Himalayan foothills revealed that 18.55% of households in the region face water scarcity for 2–3 months, and 20.58% experience it for more than 3 months, which is considerable. The rocky landscape has prevented drilling of wells, and hence the households depend on streams and springs, which become dry during winter (Datta et al,. 2024). Furthermore, the region is characterized by a variety of soils. The region falling under the dystric cambisol soil type has less infiltration capacity, which led to a decline in groundwater level (Roy et al., 2024).

7.1.5 *Economic Outcome of Water Scarcity*

Whittington et al (1993) in their study points out that poor households maybe unable to afford the expenses of installing electric pump or heavy-duty pump for lifting water. Therefore, these households depend on wealthier households for purchasing water which in turn leads to exploitation. Mainly for drinking purpose, chronic water scarcity makes these households cope up with various activities like collecting water, treating water before drinking and purchasing water. This leads to heavy cost and expenditures incurred by the households.

Dooars, comprising Jalpaiguri, Alipurduar, and Cooch Behar districts of West Bengal, are alluvial floodplains located in the Eastern Himalayan foothills and are facing water scarcity, hampering the well-being of local communities, especially the rural blocks (Deb & Mukherjee, 2021). Recently, it was included in the "JAL JEEVAN MISSION (WB-JJM)" scheme, which aimed at ensuring every household has access to tap water and sewerage connections. However, the current water management strategies remain ineffective in addressing the water shortage, as only 56% of the households in the given districts have access to a tap water supply. The estimated sanctioned cost of the scheme is approximately 1359 INR (in crore) (Ministry of Jal Shakti, Government of India, 2025).

In response to this water scarcity, the households tend to cope with the situation by performing various activities like collecting, storing, treating, purchasing, and drawing water, which incur them a heavy cost monthly and annually, especially for the poorer households (Pattanayak et al., 2005). Thus, to fulfil the aim of this study, the first research objective is to analyse the coping behaviours adopted and the coping costs incurred by the rural households of the region. The second objective of the research is to analyse how coping costs vary across different demographics and socio-economic factors.

7.2 Study Area

The study was conducted in six villages of the Eastern Himalayan foothills of West Bengal in India. Among the three districts of the Eastern Himalayan foothills, namely, Alipurduar, Cooch Behar and Jalpaiguri (also known as *Dooars*), Jalpaiguri district is most vulnerable to the water scarcity problem as only 52% of households have access to piped water connections among 4,38,232 total households residing in the district as of till date (Ministry of Jal Shakti, Government of India, 2025). Figure 7.1 shows the variation in water scarcity district wise.

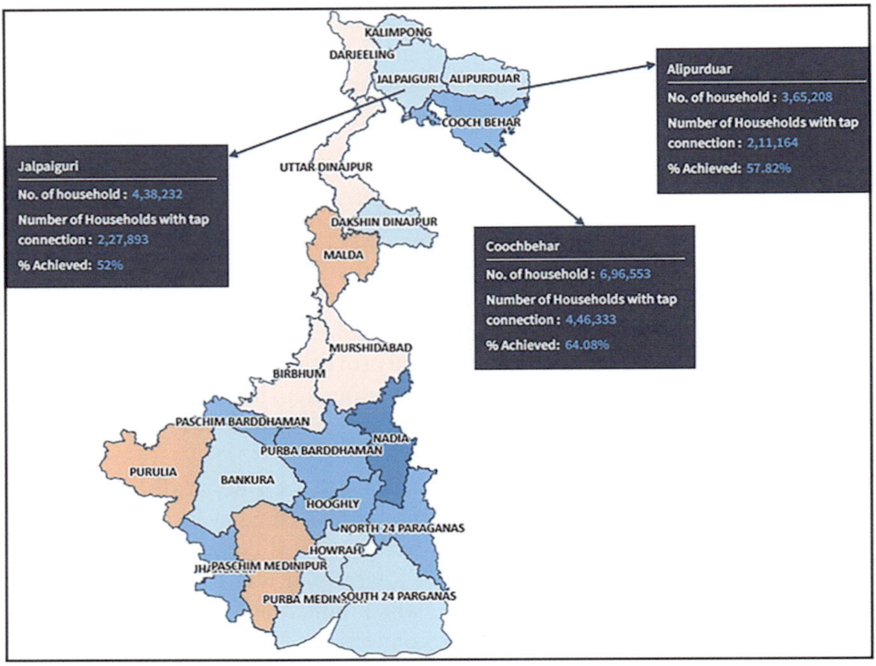

Fig. 7.1 Variation in water FHTC achievement district wise in Eastern Himalayan foothills of West Bengal. *Source* West Bengal Jal Jeevan Mission, 2025

The Jalpaiguri district consists of seven blocks, namely, Rajganj, Jalpaiguri Sadar, Mal, Dhupguri, Maynaguri, Nagrakata and Matiali. Considering the Functional Household Tap Connection (FHTC) achieved since inception under West Bengal Jal Jeevan Mission, Rajganj falls under 31–40%, Jalpaiguri Sadar and Mal at 41–50%, Dhupguri and Maynaguri at 51–60%, Nagrakata at 61–75% and Matiali at 76–99%. The study mainly conducted survey on the coping behaviour of the households belonging to villages of blocks which achieved less than 50% of FHTC. The blocks are namely, Rajganj, Jalpaiguri Sadar and Mal. Table 7.1 represents the variation in coverage of FHTC block wise for Jalpaiguri district.

Among all the three blocks, two villages from each block, i.e., total of six villages were surveyed. The selection of the villages was on the basis of 0% achievement and progress of FHTC till date. The villages are, namely, Guzrimari and Araji Belakoba II from Rajganj, Paharpur and Satkhamar from Jalpaiguri Sadar, and Rangamati Tea Garden and Tesimala from Mal. Table 7.2 represents the details of total number of households, FHTC achievement and progress by the villages selected under each block.

Table 7.1 Variation in FHTC coverage (since inception) block wise for Jalpaiguri District

Block	Household	FHTC provided	% progress
	4,38,232	**2,27,893**	**52.00**
Matiali	26,252	21,735	82.79
Nagrakata	27,468	17,318	63.05
Maynaguri	73,976	45,220	61.13
Dhupguri	99,000	55,790	56.35
Mal	49,297	23,032	46.72
Jalpaiguri	90,945	39,730	43.69
Rajganj	71,294	25,068	35.16

Source West Bengal Jal Jeevan Mission, 2025

Table 7.2 Villages, gram panchayat, number of households and FHTC achievement

Gram panchayat	Villages	Households	FHTC provided	Progress (%)
Rajganj				
Shikarpur	Guzrimari	2002	0	0.00
Panikowri	Araji Belakoba II	606	0	0.00
Jalpaiguri Sadar				
Paharpur	Paharpur	4707	1	0.02
Bahadur	Satkhamar	2700	0	0.00
Mal				
Rungamatee	Rangamati Tea Garden	1007	0	0.00
Tesimala	Tesimala	989	0	0.00

Source West Bengal Jal Jeevan Mission, 2025

7.3 Data and Methodology

7.3.1 Sampling and Primary Data Collection

The study used primary data obtained from 117 households in the selected six villages. Thirty-nine samples have been collected from each block. For sampling, the households from each village were selected at random. The household survey mainly consisted of individual interviews with the household head and focus group discussions using a semi-structured and pretested questionnaire from February to April 2025. The questionnaire emphasized collecting data on demographic and socio-economic characteristics, weather perceptions, and details about coping mechanisms adopted. Additionally, the interview was conducted by adhering to all protocols and guidelines to ensure minimal disturbance of the environment and maintain the integrity of the regions.

7.3.2 Methods

In the following sections, the description and calculation of the coping costs are discussed.

7.3.2.1 Monthly Collection Costs

The collection cost is based on the information gained from the households for the time spent collecting water from different sources (e.g., public tap, borewell, or river) in a month. The total time taken by a household to collect water (including walking to the source, waiting, collecting, and coming back) from different sources is added. Another important component of collection cost is the hourly wage rate. Generally, it is assumed that people work 8 h per day, except for the casual worker whose working duration may be less than 8 h per day. Based on the information provided by the households, the minimum hourly wage rate has been obtained.

$$\begin{aligned} \textit{Monthly Collection Cost} &= \textit{Total Time (hours) taken to collect water in a month} \\ &\quad \times \textit{Minimum hourly wage rate} \end{aligned}$$

7.3.2.2 Monthly Storage Costs

This cost is incurred by households that have invested in storage equipment such as steel pots, buckets, tanks, etc. The information about fixed costs of the storage equipment calculated over the lifetime of storage equipment (in months) and maintenance costs has been obtained.

$$\textit{Monthly Collection Cost} = \frac{\textit{Fixed Cost}}{\textit{Lifetime (in months)}} + \textit{Monthly Maintenance Cost}$$

For different storage equipment, a different expected lifetime is assumed. For steel pots, storage tanks, and jerry cans, a life of 8–10 years is assumed. Plastic buckets and bottles have an expected life of 5 years and 3 years, respectively. The cost of each storage equipment has been added to obtain the total storage cost of the households.

7.3.2.3 Monthly Treatment Costs

This cost is incurred by households that have invested in treatment devices such as water filters, chemical treatment, or boiling. For the household that boils water, the information about the monthly fuel cost has been obtained. Additionally, information about the number of times and the time taken to boil (treat) was obtained. The households mainly used fuelwood from the forest/ market and LPG for boiling water. Assuming 720 h per month, the fuel cost per hour has been obtained.

$$Monthly\ Treatment\ Cost\ (Boiling) = Fuel\ cost\ per\ hour \times Total\ Boiling\ time\ in\ a\ month\ (in\ hours)$$

For households that use a water filter, such as Reverse Osmosis (RO), for treating water, the information on the fixed cost of the water filter, calculated over the lifetime of the device (in months), and the maintenance cost has been obtained.

$$Monthly\ Treatment\ Cost\ (Water\ Filter) = \frac{Fixed\ Cost}{R(r, t)} + Monthly\ Maintenance\ Cost$$

where, $R(r, t) = \frac{1-\frac{1}{(1+r)^t}}{r}$ calculated on monthly basis. Here, r is the annual rate of interest at 10% and t is the lifetime of the asset (Chakravarty et al., 2021).

7.3.2.4 Monthly Drawing Costs

This cost is incurred by households that have invested in water drawing devices such as electric motors. The information about the motor horsepower (in units), monthly energy consumption for drawing water, number of times water is drawn, and Gramin Electricity Tariff for Jalpaiguri District has been obtained.

$$Monthly\ Drawing\ Cost = Energy\ Consumption\ Unit\ (kW) \times Total\ time\ taken\ in\ a\ month\ to\ draw\ water\ (in\ hours) \times Gramin\ Electricity\ Tariff$$

where, the Gramin Electricity Tariff for the Jalpaiguri District is 3INR up to monthly consumption in kWh from 0 to 25 units.[1]

7.3.2.5 Monthly Purchase Costs

This cost is incurred by households that have invested in purchasing water to meet their needs. The household represented the direct monthly bills of purchasing water.

[1] https://www.wbsedcl.in/irj/go/km/docs/internet/new_website/pdf/Tariff_Volumn/Tariff_chart_dated_06.03.2024_for_the_year_2024-25_12_03_1_27_03.pdf.

7.3.2.6 Monthly Coping Costs

This represents the total costs after aggregating collection cost, storage cost, drawing cost, treatment cost, and purchase cost.

$$\begin{aligned}\textit{Total monthly coping costs} = {} & \textit{Monthly collection cost} + \textit{Monthly purchase costs} \\ & + \textit{Monthly storage costs} + \textit{Monthly treatment costs} \\ & + \textit{Monthly drawing costs}\end{aligned}$$

Additionally, one standard multiple linear regression was carried out to understand the determinants of the household monthly coping cost. Table 7.3 represents the choice of explanatory/independent variables and their descriptions which also reflects the household characteristics for carrying out the standard multiple linear regression. It has been selected keeping in mind the multicollinearity issue. For the dummy variables, it shows comparisons among the categories. The base category or the reference category has been chosen according to the highest frequency in the data.

7.4 Results and Discussion

7.4.1 Household Characteristics

Table 7.4 presents the general demographic and socioeconomic characteristics of the households. It was found that 69.20% of households have a female water collector, while the remaining 30.80% have a male water collector. The proportions of households belonging to SC, ST, OBC, and General castes are 34.18%, 24.79%, 28.20%, and 12.82%, respectively. Since the study area is largely rural, a higher representation of marginalized castes was observed. In terms of religion, the majority of households identified as Hindu, accounting for 70.94% of the sample, while only a very small proportion of respondents belonged to Muslim and Christian communities. On average, household members had nearly 11 years of formal education. The average age of the water collector was about 40 years, and the average household size was approximately 4 members. Considering occupation, it has been observed that the highest proportion of households, that is, 66.67% are engaged as daily wage earners. This was mainly due to the fact that majority of the rural population engaged themselves as tea garden workers, and agricultural labourers. The remaining proportion of households had occupation as government service and business owner, with 23.07% and 10.26% respectively, which is comparatively low. The average monthly income of the households is approximately 18,555.56 INR. Besides, most of the households faced water scarcity severely (40.17%), followed by moderately (32.50%) and slightly (27.35%).

Table 7.3 Choice of explanatory variables for multiple linear regression analysis

Variables	Category	Description	Source
Gender of water collector	Male	Dummy variable; Female = 1, else 0	Mfitumukiza et al. (2024)
	Female	Base category	
Caste	General	Dummy variable; General = 1, else 0	Emile et al. (2022)
	OBC	Dummy variable; OBC = 1, else 0	
	SC	Base category	
	ST	Dummy Variable; ST = 1, else 0	
Location	Rajganj	Base category	Field Survey (2025)
	Jalpaiguri Sadar	Dummy variable; Jalpaiguri Sadar = 1, else 0	
	Mal	Dummy variable; Mal = 1, else 0	
Perception about water scarcity	Severely	Base category	Gosling and Arnell (2013)
	Moderately	Dummy variable; Moderately = 1, else 0	
	Slightly	Dummy variable; Slightly = 1, else 0	
Income of household	–	Continuous variable	Chakravarty et al. (2021)
Years of education of household	–	Continuous VARIABLE	Emile et al. (2022)
Age of water collector	–	Continuous Variable	Phiri and Velmurugan (2024)
Household size	–	Continuous variable	Arbués et al. (2010)

Source Author's Compilation; Field Survey, 2025

7.4.2 Households' Perception About Water Scarcity

7.4.2.1 Perceived Changes in Climate

From the household survey data, it has been obtained that households faced severe weather variability over the past few decades. Over 95% of households faced rising summer temperatures, shorter winters, and decrease in monsoonal precipitation. The respondents observed that summers became extremely hotter compared to past 10–15 years. Many felt that winter months have shortened and became less intense over the years. Monsoon has been reported to be irregular and delayed compared to past decades. Hence, these climatic changes caused drying up of wells, local rivers, and

Table 7.4 Demographic and socioeconomic characteristics of the sample households

Variables	Category	Count	Percentage (%)
Gender of the water collector	Male	36	30.80
	Female	81	69.20
Caste	General	15	12.82
	OBC	33	28.20
	SC	40	34.18
	ST	29	24.79
Religion	Hindu	83	70.94
	Muslim	23	19.65
	Christian	11	9.40
Occupation	Daily wager	78	66.67
	Business	12	10.26
	Govt. job	27	23.07
Perception about water scarcity	Severely	47	40.17
	Moderately	38	32.50
	Slightly	32	27.35

Variables	Mean	Minimum	Maximum
Monthly income of the household	18,555.56	4000	50,000
Years of education of household head	11.26	3	17
Age of water collector	40.12	17	67
Household size	4.26	1	10

Source Author's Compilation; Field Survey, 2025

streams in the region. Every year from February to May, water scarcity becomes severe which resulted in five coping strategies and higher coping costs. These results are similar to the study by Datta and Behera (2023) where it was reported that 91.67% of the rural population in their study area of Jalpaiguri District faced decline and irregularity in monsoonal rainfall.

7.4.2.2 Perceived Water Scarcity and Coping Cost

Due to the recent change in weather over the past decades in the region, households faced water scarcity, mainly for drinking purposes. However, the perception about water scarcity varied from household to household, which possibly resulted in variation in coping costs. This perception about the water scarcity level has been passed on through generations. About 40.17% of households that faced severe water scarcity incurred higher coping costs, 32.50% of households that faced moderate water scarcity incurred a little lower coping cost than the households incurring higher coping costs, and 27.35% of households that faced slight water scarcity incurred

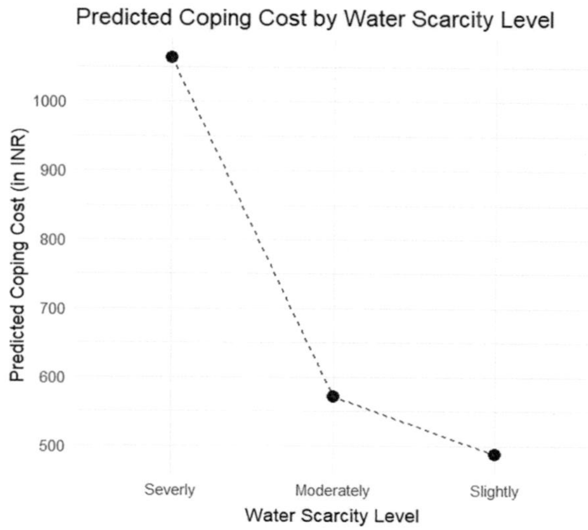

Fig. 7.2 Relation between perceived water scarcity and coping cost. *Source* Author's Compilation; Field Survey, 2025

lower coping costs (Fig. 7.2). The study by Singh et al. (2018) validates our findings. The perception is closely linked with response behaviour and willingness to adapt and cope up with the situation. Also, a similar study by Pattanayak et al. (2005) in the Himalayas supports our findings. Figure 7.2 shows that as perception about water scarcity changes from severe level to slighter level, coping cost decreases sharply.

7.4.3 Household Coping Behaviour

This following section of the study deals with the details regarding collection cost, storage cost, drawing cost, treatment and purchase cost incurred by the households in absence of proper water supply.

7.4.3.1 Water Collection

From the household survey, it has been observed that 78.63% of people collect water from various sources like public tap, public borewell, school borewell, neighbourhood tap, private borewell in search of better-quality water. These people are mostly belonging from lower- and middle-income group. Table 7.5 represents the details of various collection source, number of households engaged in collection activity, average time taken by households in collecting water and average collection cost calculated on monthly basis. For all the three blocks of Jalpaiguri district, the

Table 7.5 Details of water collection activity for all three locations of Jalpaiguri District

Location	Collection source	Number of people engaged	Average time taken for collecting water (hours in a month)	Average monthly collection cost (in INR)
Rajganj	Public tap	3	20.83	895.83
	Public pump	8	29.06	1765.62
	Neighbourhood well	5	20	1066.67
	Private borewell	11	8.37	562.67
Jalpaiguri Sadar	Public pump	26	22.19	1197.92
	Private borewell	2	9.5	625
Mal	School borewell	7	37.86	1129.46
	Public borewell	6	28.75	1197.92
	Neighbourhood tap	3	13.33	680.56
	Public pump	4	30	1531.25
	Private borewell	15	8.92	628.58

Source Author's Compilation; Field Survey, 2025

monthly collection time and collection cost are significantly higher for the households who collect water from sources like public pump, public tap/borewell, school borewell and neighbourhood well.

The monthly collection time and collection cost ranges from 20–37.86 h and 1066.67 INR–1765.63 INR, on average, respectively. The reason is that these households travel longer distances and stand in queue to collect water. Whereas the monthly collection time and collection cost are lower for the households who collect water from their personal borewell, and neighbourhood tap. Similarly, the study by Pattanayak et al. (2005) in the Himalayas validates our findings in this context. The monthly collection time and collection cost ranges from 8.38–13.33 h and 562.67 INR–680.56 INR, on average, respectively. The reason is that these household do not like spending longer time in standing in queue. Additionally, collection cost contributes the highest to coping cost calculation. From the regression model, we see that higher income group incurs higher coping cost which indirectly infers to us that higher income group has higher collection cost.

7.4.3.2 Water Storage

Like the water collection activity, 78.63% of households from all three locations store water after collecting it. They mostly use steel/silver buckets and pots, plastic bottles, buckets, and drums for storing water. The average initial cost of silver/steel

Table 7.6 Details of water storage activity for all three locations of Jalpaiguri

Storage device	Number of people engaged	Average monthly storage cost (per device) in INR
Steel/silver bucket	45	5.03
Steel/silver pot	44	6.25
Plastic jug/drum	6	1.67
Plastic bottle	22	0.83
Plastic bucket	4	12.5
Big drums	6	10.41

Source Author's Compilation; Field Survey, 2025

utensils of different sizes is around INR 500 with a lifetime of 96 months. For big drums, the average initial cost ranges from INR 250 to INR 300 with a lifetime of 24 months. For plastic bottles and jugs/drums, the average initial cost is INR 20 for 1 L and INR 40 for 2–3 L, respectively, with a lifetime of 24 months. The average initial cost of plastic buckets of different sizes is INR 300 with a lifetime of 24 months. Moreover, the monthly maintenance cost of the storage device has been reported to be nearly negligible by the households. From Table 7.6, it can be observed that people who prefer to store water in steel/silver utensils incur lower storage costs than people who store water in plastic buckets and drums. People mostly using steel/silver utensils for storage were observed to belong to the lower and middle-income groups, and people from the higher-income group preferred storage devices such as plastic bottles and buckets only. Unlike collection cost, storage cost does not make up much of the coping cost.

7.4.3.3 Drawing, Treating, and Purchasing Water

From Table 7.7, considering the other coping activities, 21.36%, 18.80% and 21.36% of households draw, treat, and purchase water, respectively. For drawing water, households mostly use a 0.5 HP motor. These are the people who have a private borewell installed in their house. People who draw water for 30 min daily incur an additional monthly drawing cost of 16,785 INR. Whereas people who draw water for 1 h daily incur an additional monthly drawing cost of 33.57INR. For treating water, households mostly boil water using fuelwood from the market and LPG.

Very few households have a water filter. It has been observed that people who boil water incur additional monthly treatment costs of INR 15.35 on average, which is much lower than the treatment cost by using a water filter (RO). The average purchase cost/initial cost of RO was reported by the households to be INR 8000 with a lifetime of 10 years at 10% rate of interest annually. These few households mainly belonged to the higher income group from Mal. For all the three locations, only the higher income households purchased water. They found it cost efficient and time saving in purchasing water. The average monthly purchase cost incurred by

Table 7.7 Average monthly coping cost of drawing, treating and purchasing activity for all three locations of Jalpaiguri District

Coping behaviour	Number of people engaged	Average monthly cost (in INR)
Drawing (0.5 HP motor)		
30 min/day	15	16.785
1 h/day	5	33.57
Treating		
Fuelwood from market (boiling)	17	16.81
LPG (boiling)	3	13.89
Water filter (RO)	2	108.48
Purchasing		
20 L water jar	25	300.8

Source Author's Compilation; Field Survey, 2025

these households is 300.8INR. Moreover, the study by Pattanayak et al. (2005) in the Himalayas validates our findings in this context. Also, a similar study by Chakravarty et al. (2021) supports our findings.

7.4.3.4 Average Monthly Coping Costs

This section of the study provides details of the average total monthly coping cost for various income groups for all three locations. The coping cost has been calculated by aggregating collection cost, storage cost, drawing cost, treatment cost, and purchase cost as described in Sect. 8.3.2. For convenience in showing how the coping cost varies with different income groups, it has been categorised into three groups. From the regression model, it has been found that income level has been one of the most significant factors influencing the coping behaviour and costs. A similar study by Chakravarty et al. (2021) supports our findings in this context. Contrastingly, the study by Pattanayak et al. (2005) in the Himalayas concludes that income level is not the sole factor determining the coping cost. In our analysis, the lower income groups have a monthly income of up to INR 10,000, the middle income group from INR 10,000–30,000, and the higher income group from above INR 30,000.

The lower- and middle-income group falls under the category of daily wager, and the higher income groups are mostly business owners or government employees. From Table 7.8, it has been observed that as income increases, coping cost tends to decrease for all three locations. The reason is that the lower-income group incurs higher collection costs due to higher collection time as described in Sect. 8.4.3. The middle-income group has lower coping costs compared to the lower-income group because they mostly have installed a private borewell in their house (lower collection cost), but they incur an additional drawing cost. Treatment cost is incurred by only a few households belonging to the lower and middle-income groups. Whereas

Table 7.8 Variation in coping cost across different income groups for all three locations of Jalpaiguri District

Location	Monthly income	Average monthly coping cost (in INR)
Rajganj	Upto INR 10,000	1490.97
	INR 10,000–30,000	636.14
	Above INR 30,000	306.53
Jalpaiguri Sadar	Upto INR 10,000	1382.46
	INR 10,000–30,000	993.22
	Above INR 30,000	312.79
Mal	Upto INR 10,000	1188.63
	INR 10,000–30,000	880.2
	Above INR 30,000	460.1

Source Author's Compilation; Field Survey, 2025

the higher income households find it cost-efficient in only purchasing water which resulted in lower coping cost as already discussed above.

7.4.4 Econometric Results

Table 7.9 represents the coefficients and its significance level of the estimated model. The findings of the study highlight how water scarcity in Jalpaiguri district, West Bengal, creates uneven burdens across households. Income plays a major role: families with higher earnings spend less on coping, while poorer households are forced to spend more on collecting, storing, or treating water. This adds to their economic stress and shows how scarcity deepens existing inequalities. Similar income-related inequalities in coping with water insecurity have been documented in rural India (Kumar et al., 2022) and across low-income countries where women and children disproportionately bear the cost of water collection (Gomez et al., 2019). Education, on the other hand, does not seem to matter much in shaping coping costs. In the study villages, where access to water is more about infrastructure and resources than knowledge, this is not surprising.

Gender differences are evident in the study villages, where households with women as the primary water collectors spend more on coping strategies. In Jalpaiguri, as in many parts of rural Bengal, women shoulder the responsibility of fetching and managing water. Long walks, waiting at collection points, and additional expenses on storage and treatment increase their household's burden. This reinforces findings from other contexts showing that water scarcity is not merely a technical issue but also a deeply gendered one, tied to unequal roles and responsibilities. For instance, Sorenson et al. (2011) highlight how women in low-income countries face disproportionate time and health costs when accessing water, while Crow and McPike

Table 7.9 Results of multiple linear regression analysis

Variables	Coefficient	Std. err.	t
Log monthly income of the household	− 0.2667**	0.1113	− 2.40
Years of education	0.0024	0.0065	0.37
Gender of water collector	0.0717**	0.0343	2.09
Caste general	0.1193**	0.0459	2.60
Caste OBC	0.0760*	0.0414	1.84
Caste ST	0.0980**	0.0464	2.11
Location Jalpaiguri Sadar	0.0425	0.0377	1.13
Location Mal	− 0.0041	0.0448	− 0.09
Household size	0.0099	0.0090	1.10
Age of water collector	− 0.0018	0.0014	− 1.34
Perceived scarcity moderately	− 0.1790***	0.0408	− 4.39
Perceived scarcity slightly	− 0.3814***	0.0687	− 5.55
Cons	4.0171***	0.4315	9.31

Significance codes: *** $p < 0.01$; ** $p < 0.05$; * $p < 0.10$
$F(12, 104) = 35.30$; $\text{Prob} > F = 0.0000$
R-squared = 0.8029; Adj R-squared = 0.7801

Source Author's Compilation; Field Survey, 2025

(2009) demonstrate that gendered water collection responsibilities can exacerbate socio-economic vulnerabilities.

Caste patterns are also important. Families from General, OBC, and ST groups tend to face higher coping costs compared to Scheduled Caste households. This does not necessarily mean that SC households have easier access, but rather that different caste groups adopt different strategies—some relying more on financial spending, others on labor-intensive collection. It reflects the complex way social identity, settlement patterns, and resource access shape coping behavior in Jalpaiguri.

Interestingly, the differences between locations—Jalpaiguri Sadar, Mal, and Rajganj—are not significant. This means coping costs are less about district boundaries and more about household circumstances. Also, household size does not come out to be significant. This analysis shows that the cost of coping with water scarcity is uniformly distributed across households, regardless of the number of members. What does stand out is the perception of scarcity. Families that feel water is severely scarce spend much more, while those who report moderate or slight scarcity spend far less. This shows how perceptions, shaped by seasonal shortages and local conditions, directly affect household decisions and expenses.

Overall, the study makes it clear that coping with water scarcity in Jalpaiguri is shaped by a mix of income, gender, caste, and perception. Despite the district's rivers and high rainfall, many households struggle with unreliable supply and poor water quality, forcing them to spend heavily. Addressing this issue will require more than just new infrastructure. It will mean paying attention to the inequalities in who bears the costs—especially poor families, women water collectors, and socially marginalized groups—and ensuring that safe, affordable water is available to all.

7.5 Conclusion and Policy Implications

The study reveals that coping with water scarcity in Jalpaiguri district involves both direct financial costs and indirect burdens such as time, effort, and lost opportunities. Households relying on private borewells often reported receiving rusty and undrinkable water, forcing them to spend more on treatment or alternative sources. While wealthier households could afford to purchase water and avoid the drudgery of collection, poorer families faced higher relative costs, particularly in terms of time and labor. The analysis also shows that gender and caste shape coping burdens in significant ways: women collectors and households from marginalized caste groups often bear the heaviest load. For many families, especially those perceiving severe scarcity, the struggle to secure safe water reduces time for education, livelihood activities, and care, thereby reinforcing existing inequalities.

These insights point to the need for sustainable and inclusive policy measures. Expanding the Jal Jeevan Mission's coverage to villages with unreliable or unsafe supply remains critical for reducing rural coping costs. Targeted support—such as subsidies for water storage systems, affordable household treatment options, or low-energy drawing equipment—can ease the burden on marginalized groups. Since women are at the center of household water management, self-help groups and community-led collectives could be strengthened to promote rainwater harvesting, water-saving practices, and small-scale infrastructure. Water literacy programs for schoolchildren can also build awareness about safe water use and long-term conservation. While this study highlights key drivers of coping costs, it may not have fully captured hidden burdens, such as workdays lost due to illness or the long-term psychological stress of water insecurity. Future research should explore these dimensions and continue to place community agency at the heart of water governance.

References

Alkon, M., Wang, Y., Harrington, M. R., Shi, C., Kennedy, R., Urpelainen, J., Kopas, J., & He, X. (2024). High resolution prediction and explanation of groundwater depletion across India. *Environmental Research Letters, 19*(4), Article 044072. https://doi.org/10.1088/1748-9326/ad34e5

Arbués, F., Villanúa, I., & Barberán, R. (2010). Household size and residential water demand: An empirical approach*. *Australian Journal of Agricultural and Resource Economics, 54*(1), 61–80. https://doi.org/10.1111/j.1467-8489.2009.00479.x

Bhatla, R., Verma, S., Pandey, R., & Tripathi, A. (2019). Evolution of extreme rainfall events over Indo-Gangetic plain in changing climate during 1901–2010. *Journal of Earth System Science, 128*(5). https://doi.org/10.1007/s12040-019-1162-1

Bhutia, S. (2017). A situational analysis of water resources in Darjeeling Municipal Town: Issues and challenges. *International Journal of Research in Geography, 3*(4). https://doi.org/10.20431/2454-8685.0304007

Brown, A., & Matlock, M. D. (2011). A review of water scarcity indices and methodologies. *White Paper, 106*(1), 19.

Chakkaravarthy, D. N. (2019). Water scarcity—Challenging the future. *International Journal of Agriculture Environment and Biotechnology, 12*(3). https://doi.org/10.30954/0974-1712.08.2019.2

Chakravarty, S., Das, S., & Das, S. (2021). *Unreliable public water supply and coping mechanisms of low-income households in Delhi (No. 448).* Institute of Economic Growth.

Chhetri, B., & Tamang, L. (2013). Population growth and associated problems: A case study of Darjeeling town. *International Journal of Humanities and Social Science Invention, 2*(5), 63–67.

Crow, B., & McPike, J. (2009). How the drudgery of getting water shapes women's lives in low-income urban communities. *Gender, Technology and Development, 13*(1), 43–68.

Datta, P., & Behera, B. (2021). What caused smallholders to change farming practices in the era of climate change? Empirical evidence from Sub-Himalayan West Bengal, India. *GeoJournal, 87*(5), 3621–3637. https://doi.org/10.1007/s10708-021-10450-1

Datta, P., & Behera, B. (2022). Assessment of adaptive capacity and adaptation to climate change in the farming households of Eastern Himalayan foothills of West Bengal, India. *Environmental Challenges, 7*, Article 100462.

Datta, P., & Behera, B. (2023). Climate change adaptation through agroforestry: Empirical evidence from Indian Eastern Himalayan Foothills. *Strategizing agricultural management for climate change mitigation and adaptation* (pp. 167–181). Springer International Publishing.

Datta, P., & Das, S. (2019). Analysis of long-term precipitation changes in West Bengal, India: An approach to detect monotonic trends influenced by autocorrelations. *Dynamics of Atmospheres and Oceans, 88*, Article 101118.

Datta, P., Rahut, D. B., Behera, B., Sonobe, T., & Naveen, N. (2024). Assessing forest villagers' livelihood vulnerability to the environmental changes in Buxa Tiger Reserve of sub-Himalayan India. *Trees Forests and People, 17*, Article 100632. https://doi.org/10.1016/j.tfp.2024.100632

Deb, P., & Mukherjee, R. (2021). Household vulnerability of tribal people to climate change in the part of Dooars region, West Bengal, India. In *Advances in geographical and environmental sciences* (pp. 385–408). https://doi.org/10.1007/978-981-16-5681-1_15

Drew, G., & Rai, R. P. (2016). Water management in post-colonial Darjeeling: The promise and limits of decentralised resource provision. *Asian Studies Review, 40*(3), 321–339. https://doi.org/10.1080/10357823.2016.1192580

Emile, R., Clammer, J. R., Jayaswal, P., et al. (2022). Addressing water scarcity in developing country contexts: A socio-cultural approach. *Humanities and Social Sciences Communications, 9*, 144. https://doi.org/10.1057/s41599-022-01140-5

Ghatani, S. (2021). Problems and challenges on urban water management in Darjeeling Hill Town. *Asian Research Journal of Arts & Social Sciences, 13*(2), 24–33. https://doi.org/10.9734/arjass/2021/v13i230209

Ghosh, P. (2021). Water stress and water crisis in large cities of India. In *Advances in geographical and environmental sciences* (pp. 131–138). https://doi.org/10.1007/978-981-15-8237-0_11

Ghosh, M., & Ghosal, S. (2023). Geographies of vulnerability to climate change: Empirical evidences from the Indian Himalayan foothills. *Asia-Pacific Journal of Regional Science, 7*(2), 405–431. https://doi.org/10.1007/s41685-022-00273-4

Giz. (2025, June 3). *National Workshop on Climate Change Adaptation in Rural Areas of India.* giz.de. https://www.giz.de/en/worldwide/27789.html

Gomez, M., Perdiguero, J., & Sanz, A. (2019). Socioeconomic factors affecting water access in rural areas of low and middle income countries. *Water, 11*(2), 202.

Gosling, S. N., & Arnell, N. W. (2013). A global assessment of the impact of climate change on water scarcity. *Climatic Change, 134*(3), 371–385. https://doi.org/10.1007/s10584-013-0853-x

Green, T. R. (2011). Beneath the surface of global change: Impacts of climate change on groundwater. *Journal of Hydrology, 405*, 532–560.

Gulati, A., & Banerjee, P. (2016). Emerging water crisis in India: Key issues and way forward. *Indian Journal of Economics, 96*(383), 681–704.

Katyaini, S., & Barua, A. (2015). Water policy at science–policy interface: Challenges and opportunities for India. *Water Policy, 18*(2), 288–303. https://doi.org/10.2166/wp.2015.086

Kumar, D., Bassi, N., & Kumar, S. (2022). *Drinking water security in rural India.* Springer Singapore.

Lepcha, N. K. C. (2013). Problems and prospects of water resource of Kurseong municipality, Darjeeling district, West Bengal. *Science Park Research Journal, 1*(19), 1–11.

Mahato, A., Upadhyay, S., & Sharma, D. (2022). Global water scarcity due to climate change and its conservation strategies with special reference to India: A review. *Plant Archives*, 64–69. https://doi.org/10.51470/plantarchives.2022.v22.no1.009

Mall, R. K., Gupta, A., Singh, R., Singh, R. S., & Rathore, L. S. (2006). Water resources and climate change: An Indian perspective. *Current Science*, 1610–1626.

Martin. (2023, October 19). Water and sanitation—United Nations Sustainable Development. United Nations Sustainable Development. https://www.un.org/sustainabledevelopment/water-and-sanitation/

Mfitumukiza, D., Twongyirwe, R., Mwesigwa, G. Y., Kebirungi, E., Nyakato, V. N., & Kabonesa, C. (2024). Typologies and determinants of coping responses to forage and water scarcity among livestock farmers in South-western Uganda: Does gender matter? *Scientific African*, e02459. https://doi.org/10.1016/j.sciaf.2024.e02459

Ministry of Jal Shakti, Government of India. (2025). *JJM block map view dashboard*. https://ejalshakti.gov.in/jjmreport/JJMBlockMapView.aspx

Mishra, V., Smoliak, B. V., Lettenmaier, D. P., & Wallace, J. M. (2012). A prominent pattern of year-to-year variability in Indian summer monsoon rainfall. *Proceedings of the National Academy of Sciences, 109*(19), 7213–7217. https://doi.org/10.1073/pnas.1119150109

Mishra, V., Dangar, S., Tiwari, V. M., Lall, U., & Wada, Y. (2024). Summer monsoon drying accelerates India's groundwater depletion under climate change. *Earth S Future, 12*(8). https://doi.org/10.1029/2024ef004516

Pal, M., Ayele, Y., Hadush, A., Panigrahi, S., & Jadhav, V. J. (2018). Public health hazards due to unsafe drinking water. *Air & Water Borne Diseases, 7*(1), 1–6. https://doi.org/10.4172/2167-7719.1000138

Palanisami, K., & Nagothu, U. S. (2024). Expanding hill water management. In *Advances in geographical and environmental sciences* (pp. 203–227). https://doi.org/10.1007/978-981-97-1785-9_10

Panda, D. K., Ambast, S. K., & Shamsudduha, M. (2020). Groundwater depletion in northern India: Impacts of the sub-regional anthropogenic land-use, socio-politics and changing climate. *Hydrological Processes, 35*(2). https://doi.org/10.1002/hyp.14003

Pattanayak, S. K., Yang, J., Whittington, D., & Kumar, K. C. B. (2005). Coping with unreliable public water supplies: Averting expenditures by households in Kathmandu, Nepal. *Water Resources Research, 41*(2). https://doi.org/10.1029/2003wr002443

Phiri, A. C., & Velmurugan, T. (2024). An exploratory study on water scarcity and coping mechanisms among women households with special reference to Chidothe Village in Zomba, Malawi. *Journal of Arts, Humanities and Social Science, 1*(1), 21–27. https://doi.org/10.69739/jahss.v1i1.11

Rana, M., & Guleria, V. (2018). Water scarcity in India: A threat to sustainable management of water. *ESSENCE Int J Env Rehab Conserv, 9*(1), 35–44.

Rasaily, D. S. (2014). *Darjeeling pahadka nagarpalika kshetra ko vikas ra khane paani ko itihas sanchipta ma, san* (pp. 1835–2012). New Delhi.

Roxy, M. K., Ritika, K., Terray, P., Murtugudde, R., Ashok, K., & Goswami, B. N. (2015). Drying of Indian subcontinent by rapid Indian Ocean warming and a weakening land–sea thermal gradient. *Nature Communications, 6*(1), 7423.

Roy, D., Barman, S., Mandal, G., Mitra, R., Sarkar, A., Hossain, G., Roy, P., Almohamad, H., Abdo, H. G., & Mandal, D. K. (2024). Extracting of prospective groundwater potential zones using remote sensing data, GIS, and multi-criteria decision-making approach in the SubHimalayan Dooars region of West Bengal, India. Applied Water Science, 14(4). https://doi.org/10.1007/s13201-024-02124-3

Saleth, R. M. (2011). Water scarcity and climatic change in India: The need for water demand and supply management. *Hydrological Sciences Journal, 56*(4), 671–686. https://doi.org/10.1080/02626667.2011.572074

Shah, R., & Badiger, S. (2018). Conundrum or paradox: Deconstructing the spurious case of water scarcity in the Himalayan Region through an institutional economics narrative. *Water Policy, 22*(S1), 146–161. https://doi.org/10.2166/wp.2018.115

Shiva, V. (2016). *Water wars: Privatization, pollution, and profit*. North Atlantic Books.

Singh, C., Osbahr, H., & Dorward, P. (2018). The implications of rural perceptions of water scarcity on differential adaptation behaviour in Rajasthan, India. *Regional Environmental Change, 18*(8), 2417–2432. https://doi.org/10.1007/s10113-018-1358-y

Sorenson, S. B., Morssink, C., & Campos, P. A. (2011). Safe access to safe water in low income countries: Water fetching in current times. *Social Science & Medicine, 72*(9), 1522–1526.

United Nations. (2021). *UN world water development report 2021*. United Nations.

Viviroli, D., & Weingartner, R. (2007). "Water towers"—A global view of the hydrological importance of mountains. In *Springer eBooks* (pp. 15–20). https://doi.org/10.1007/978-1-4020-6748-8_2

Whittington, D., Altaf, M. A., Jamal, H., & Smith, V. K. (1993). Rethinking rural water supply policy in the Punjab, Pakistan. *Water Resources Research, 29*(7), 1943–1954. https://doi.org/10.1029/92wr02848

World Bank. (2019). World development indicators. [Data File]. Retrieved from http://data.worldbank.org/indicator

World Resources Institute India. (2023). *India's first water body census: Connecting the missing links.*

Chapter 8
Climate Change Impact on the Socioeconomic Conditions and Well-Being of the South Asian Small-Scale Fishermen: A Review

Chandan Naik and Bhagirath Behera

Abstract Climate change is a major threat to the small-scale fishing communities, particularly in South Asia, a region characterized by widespread coastlines, extreme population density, and high dependence on marine resources for food, income, and employment. This review synthesizes findings from peer-reviewed articles published between 2000 and 2025 to analyse the climate change impact on socioeconomic conditions and well-being of the small-scale fishermen. The review identifies a range of climatic stressors, such as rising sea surface temperatures, ocean acidification, sea level rise, and extreme weather events, that are severely disrupting marine ecosystems and fish availability. These ecological shifts directly affect the livelihoods, income stability, and food security of fishing communities, escalating existing vulnerabilities like poverty, indebtedness, and limited occupational mobility. The study categorizes the impacts into physical, economic, and social dimensions, highlighting issues such as declining catch volumes, increased operational costs, infrastructure destruction, and disruptions in food supply and nutrition. It also examines local and regional adaptation responses, ranging from ecosystem-based solutions, such as mangrove restoration and cage aquaculture, to institutional and behavioural shifts, including migration, livelihood diversification, and early warning systems. While some adaptations enhance resilience, others pose sustainability challenges. This review highlights the pressing need for targeted policy interventions that support sustainable adaptation, enhance institutional frameworks, and prioritize vulnerable fishing communities in climate resilience planning.

Keywords Climate change · Fishermen · Adaptation · Coastal community · Socioeconomic · South Asia

C. Naik (✉) · B. Behera
Department of Humanities and Social Sciences, Indian Institute of Technology Kharagpur, Kharagpur, India
e-mail: nchandan589@gmail.com

B. Behera
e-mail: bhagirath9@gmail.com

P. Datta et al. (eds.), *Climate Change and Human Well-being*, Advances in Global Change Research 83, https://doi.org/10.1007/978-3-032-25479-5_8

8.1 Introduction

Climate Change disrupts food systems, affects livelihoods, and threatens the socioeconomic life of communities worldwide (Datta et al., 2022; Chettri et al., 2024; Chettri et al., 2025). It significantly exacerbates food insecurity by hindering agricultural and fishing productivity. South Asia, which is surrounded by three vast water bodies, i.e., the Bay of Bengal, the Arabian Sea, and the Indian Ocean, the SAARC countries are gifted with a 12,000 km long highly productive coastline, possessing highly valued coastal and marine fisheries (SAARC, 2008; Hossain & Shrestha, 2019). Millions of people in this region depend upon fisheries not only as a source of food and nutrition but also for livelihood, income, employment, and cultural identity (Islam et al., 2014). South Asia alone produced 23.83 million tonnes of fish in 2023, which is approximately 12.14% of the total global fish production (Giri, 2024). The fishery sector is a significant source of the national income and food supply of coastal South Asian countries, including India, Bangladesh, Sri Lanka, and the Maldives. The Bay of Bengal, a region recognized as one of the six most cyclone-prone areas in the world, experiences frequent and intense tropical storms that pose significant risks to the surrounding ecosystems and communities (Iwasaki et al., 2009). The high population growth, coupled with ongoing high rates of poverty and food insecurity, makes it one of the most exposed regions to the detrimental effects of climate change (Sivakumar & Stephanski, 2010).

Fisheries are an important contributor to food supply, food security, and livelihoods both at the local and global level (FAO, 2014). Seafoods have a high nutrient value and are one of the highest exported and traded products for many developing countries (FAO, 2014). Small-scale fisheries constitute approximately 90% of the global fishing activities and contribute almost 50% to the total fish production for human consumption (Islam et al., 2014; Mozumder et al., 2023). Small-scale artisanal fishing is considered a sustainable method of food production and can be used as a remedy to problems like natural resource depletion, overfishing, habitat degradation, etc. (FAO, 2022). While recognizing the importance of Fisheries in the global food production chain, the United Nations General Assembly has declared 2022 as the Year of Artisanal Fisheries and Aquaculture (FAO, 2022).

Fishing as a livelihood option is very risky because of its volatile nature, unpredictable environment of the seas, and low-lasting nature of the product (Islam et al., 2014; MRAG, 2011). Climatic changes significantly affect the fisheries sector through a rise in sea surface temperature, ocean acidification, changing salinity levels, coastal erosion, siltation, shrinking water bodies, a decrease in water level in rivers and lakes, and frequent natural disasters. Furthermore, climate-induced Sea Level Rise, flooding, and cyclones enhance the vulnerability of fishing communities by reducing fish availability, fish productivity, and loss of fishing boats and nets (Badjeck et al., 2010; Coulthard, 2008). The fisheries are also susceptible to changes in ecological conditions, such as changes in salinity and reduced pH levels, increasing weeds or invasive species in the water, which hampers the fishing activities. For South

Asia's small-scale fishermen, these ecological shifts lead to substantial socioeconomic challenges such as income disruption, loss of traditional livelihoods, reduced food security, and increased debt burdens (Ahmed et al., 2021; Badjeck et al., 2010).

Even though the ecological effects of climate change on marine and inland fisheries are well evidenced, there remains a significant gap in understanding its socioeconomic implications, especially in South Asia (Cinner et al., 2018). The literature is fragmented and often focuses on individual countries or local case studies, lacking a comprehensive regional synthesis. Moreover, while some adaptation strategies have been documented—such as changes in fishing gear, migration, or livelihood diversification- the sustainability and effectiveness of these responses remain underexplored.

This review focuses on bridging this gap by synthesizing the available literature on the socioeconomic impacts of climate change on fishing communities in South Asia. It highlights how the ecological shifts due to climate change in fish populations, transcend into socioeconomic challenges faced by the fishing-dependent communities. It specifically focuses on three key areas: (1) how climate-induced ecological changes affect the economic security of fishing households; (2) how livelihoods and income stability are influenced by climate variability; and (3) what coping and adaptation strategies are employed by fishermen in response to these changes. The review focuses primarily on small-scale fishing communities across South Asia by drawing from peer-reviewed literature published in the last two decades. The review will follow the following structure: Sect. 8.2 presents the methodology; Sect. 8.3 explores the socioeconomic impacts; Sect. 8.4 discusses adaptation strategies; and Sect. 8.5 concludes with key insights and policy recommendations.

8.2 Methodology

This paper employs a narrative literature review approach to peer-reviewed literature on the socioeconomic impact of climate change on small-scale fishermen in the South Asian region. Narrative literature review is the technique to describe and discuss a topic from a fundamental or contextual point of view. It focuses on the critical analysis of existing peer-reviewed literature. It gives the readers up-to-date knowledge about a specific topic of interest (Rother, 2007). The study used electronic databases like Google Scholar, Web of Science, Scopus, and ScienceDirect to obtain relevant peer-reviewed literature. A keyword search was conducted by using keywords like "climate change", "fishermen", "small-scale fishermen", "income", "livelihood", "food security", "South Asia", "India", "Bangladesh", "adaptation", etc. To maintain the relevance and replicability of the review, only articles and book chapters were considered, along with relevant grey literature (national and international reports). Cross-referencing and citation tracking were done to ensure no relevant studies were omitted from the review process. Given that climate change is a prolonged process, the review covers the period from 2000 to 2025 to ensure the relevance and currency of the findings for drawing well-informed conclusions. South

Table 8.1 Search String for the review

Databases	Keywords used
Google Scholar, Scopus, Web of Science, and ScienceDirect	(climate change OR climate variability OR environmental change OR global warming) AND (socio-economic OR income OR livelihood OR employment OR Poverty OR food security) AND (fishermen OR Fisherfolk OR fisher OR fishermen community OR small-scale fishermen) AND (South Asia OR India OR Bangladesh OR Pakistan OR Sri Lanka OR Maldives OR Nepal OR Bhutan)

Table 8.2 Inclusion and exclusion criteria

Parameter	Inclusion criteria	Exclusion criteria
Literature type	Article and book chapter	Review articles, conference papers, etc.
Language	English	Non-English
Title	Titles addressing the impact of CC on socioeconomic variables such as Income, Livelihood, and Food security etc.	Titles addressing the geographical distribution of the fishes, their growth under different climatic hazards, etc.
Study area	South Asia	Reporting results from the countries outside of South Asia
Population	Articles focusing on the impact of CC on the fishermen only	Articles focusing on people other than fishermen, such as farmers, forest people, etc.
Type of fishermen	Only small-scale fishermen are considered	Large-scale fishermen, Fishermen with large boats, are excluded

Asia has been selected as the study region due to its geographical location, being bordered by three major water bodies: the Bay of Bengal, the Arabian Sea, and the Indian Ocean. In this region, millions of fishermen depend upon fishing activities for income and livelihood (Islam et al., 2014). The results of the search string and final screening & selection criteria are shown in Tables 8.1 and 8.2.

8.3 Results

8.3.1 Physical and Environmental Impacts

Coastal seas, with their unique habitats and endemic species, are highly susceptible to climate change, risking biodiversity loss and ecosystem disruptions (Bindoff et al., 2019; Chefaoui et al., 2018; Rilov, 2016). Climate change affects marine ecosystems in various ways, including rising sea surface temperatures, changes in salinity levels, sea level rise, and the occurrence of frequent natural disasters. Climate change affects fish stocks through both direct and indirect pathways. Direct impacts influence their physiology and behaviour, leading to growth in reproductive potential, survival rates, and geographic distribution changes. Indirect impacts, on the other hand, modify the productivity, structure, and composition of marine ecosystems that provide essential food resources for fish (Brander, 2010). Until 1974, the global fish stocks were within biologically sustainable limits. However, this percentage declined to 65.8% by 2017 (Makwana & Patnaik, 2021).

Global warming is forcing the coastal fauna to migrate to higher latitudes, subsequently forcing the fishing-dependent people also to move poleward or modify their catch and diversity to maintain harvests (Barange et al., 2018; Townhill et al., 2019). Lam et al. (2020) stated that climate change seriously affects the fisheries sector productivity by altering their fish distribution and reproductive cycles. Higher concentrations of CO_2 in the atmosphere are affecting the sea's pH level, and higher ocean temperatures are causing fish to migrate to deeper zones. Sea level rise and intense storm activities are expected to impact the breeding grounds of some key species, such as prawns. Climate change can directly affect the livelihood of the fishermen's communities by melting glaciers, leading to sea level rise and an increase in tidal surges (Mitra, 2013).

Sea level rise (SLR), Sea surface temperature (SST), and alteration in salinity levels are closely associated with climate change (Jakimavičius et al., 2018; Datta et al., 2024). Rising temperatures due to global warming disrupt homeostasis in fish, causing metabolic imbalances that make them vulnerable to stress and disease. Temperature plays an important role in physiological phenomena such as feeding, respiration, osmoregulation, growth, and reproduction (Ninawe et al., 2018). A Sea Surface Temperature rise of 1.04 °C was reported across the tropical Indian Ocean from 1950 to 2015 due to the increase in the concentration of greenhouse gases (IPCC, 2019). A temperature increase of only a few degrees can increase the temperature of large water masses such as oceans, seas, and lagoons as well as create hydrological changes within the ecosystem, which alter the physical and chemical characteristics of the water bodies and affect the life cycle, physiology, and behaviour of the aquatic animals (Ninawe et al., 2018).

It is projected that sea surface temperature will increase by 4 °C towards the end of the twenty-first century in India. Makwana and Patnaik (2021) reported that a 1 °C rise in sea surface temperature may result in a nearly 9% decrease in total fish catch. Most fish species thrive within a specific temperature range that supports their fundamental metabolic processes and the availability of their food sources; a tiny change may alter their metabolism, distribution, and life processes (Cheung et al., 2016). Pauly and Cheung (2018) reported that an increase in Sea surface temperature results in a reduction of dissolved oxygen in the water, which compensates for the oxygen requirement by raising their metabolic rate through diverse activities, resulting in decreased body size of these fish.

Climate change affects the fish's feeding, migration, and breeding behaviour by changing their physical environments, which indirectly affects their growth, mortality, and reproduction (Brander, 2010; FAO, 2014). Perry et al. (2005) reported that fish migrate to deeper zones in the North Sea in response to the rising sea surface temperature (SST). An increase in SST is linked with an increase in cyclonic storms that inhibit the vertical mixing of Oceans, resulting in a lack of nutritional availability for aquatic creatures, which results in their migration to the colder and deeper regions (Makwana & Patnaik, 2021). Makwana and Patnaik (2021) reported that severe degrading conditions of marine phytoplankton due to rising SST and severe cyclonic storms are responsible for lower fish productivity and a reduction in fishing days. There will be a range of shifts in different fish species due to climate change in the future (Perry et al., 2005).

Apart from the direct effects, climate change indirectly affects the fishing resources by changing the water quality, shelter, and breeding grounds of fish, and disrupting the plankton (food for fish) community (Mitra, 2013). As a result of range shifts, fish species will face losses due to the arrival of some invasive species into their waters (Cheung et al., 2016). Iwasaki et al. (2009) reported that an increase in weed species hinders the movement of fishing boats, resulting in a reduced fish catch. These shifts can directly impact catches, negatively affecting small-scale fishermen who are often tied to a particular fishing territory or region (Barange et al., 2014). Amin et al. (2024) reported that due to high temperatures and irregular rainfall, drought is the most harmful natural disaster for aquaculture systems in Bangladesh. Barua et al. (2020) reported that precipitation, salinity, and turbidity significantly affect the fish productivity in the Bakkhali River in Bangladesh. Gills et al. (2024) reported a 51–69% reduction in fish availability due to climate change in Karnataka. This reduction in fish populations has some harmful economic and social implications for the fishing-dependent communities (Fig. 8.1).

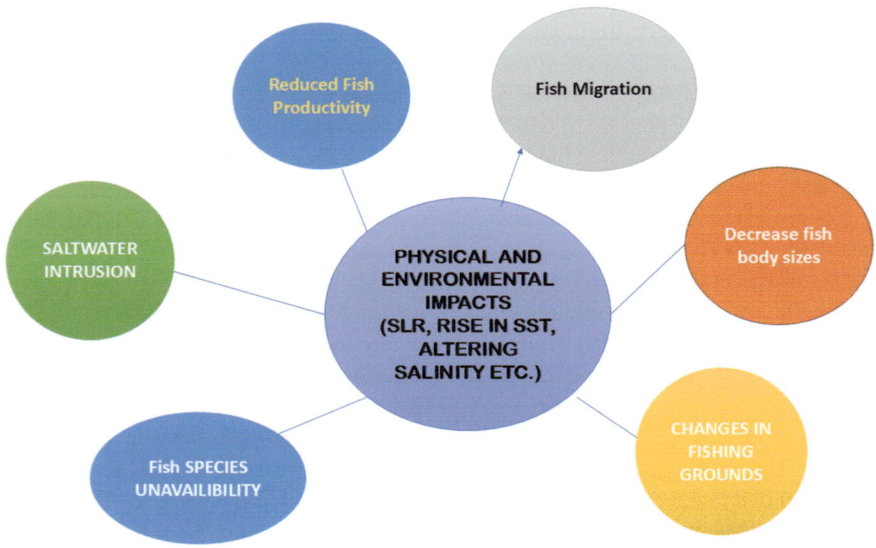

Fig. 8.1 Physical and environmental impacts of climate change on the fisheries

8.3.2 Economic Impacts

Fisheries support the Global economy with an estimated revenue of US$80–85 billion annually (Sumaila et al., 2011). Nearly 500 million people derive their income from fisheries worldwide, supporting the livelihood of 10–12% of the world's population (Barua et al., 2020). In South Asia alone, 5.5 million fishermen are employed in the fisheries sector (FAO, 2024). Most fishery-dependent people reside in developing countries where there is low income, food scarcity, and few livelihood opportunities, making them vulnerable to climate change (Barua et al., 2020; Islam et al., 2014). Small-scale fishermen are quite vulnerable to climate change and natural disasters because of their small boats and fragile equipment. Variability in the climatic conditions restricts small-scale fishermen fishing activities as it will expose their small vessels to strong winds and waves, which ultimately reduce the fishing days and, hence, the income (Colgan & Merill, 2008). Central Marine Fisheries Research Institute recorded a 36% reduction in fishing days in India due to extreme weather in 2019 (Kajal et al., 2020). Gills et al. (2024) reported an average loss of 31, 39, and 22 fishing days in the 03 marine districts of Karnataka due to Cyclone Tauktea in 2021. Climate change adversely impacts fish habitats, which forces fishermen to look for new fishing locations, resulting in longer fishing trips; they have to bear extra costs for fuel and ice as well (Salagrama, 2013).

Fish processing is also quite susceptible to climate change, as traditional fish processing depends upon the consistency of temperature and rainfall. Any variability in temperature and rainfall hampers the fish processing process, which leads to increased costs (Badjeck et al., 2010; Islam et al., 2014). Reduced catches and an increase in processing costs eventually influence the income, employment, and livelihood of the fishing-dependent communities (Badjeck et al., 2010). Islam et al. (2014) reported that sea level rise, temperature & rainfall variability, and land erosion hampered the fish drying process in Bangladesh. Sea level rise increases the coastal soil salinization and salinity level increase in groundwater aquifers, which ultimately leads to reduced availability of economically important fish from coastal ecosystems. Around 60 million people live near high-risk coastal flooding zones in South Asia alone. Extreme weather events often destroy the physical infrastructure (homes, schools, roads, etc.), which results in increased financial burden on the already economically backward fishing communities. Gills et al. (2024) highlighted that there is a significant loss of income, infrastructure, housing, and fishing equipment in Karnataka due to Cyclone Tuktae in 2021.

Franco et al. (2020) reported that global warming is adversely impacting the socioeconomic life of fishermen by changing seasons, reducing yield, and income. Disruptions in the biophysical characteristics of the aquatic environment and frequent natural disasters adversely affect fish ecosystems, which ultimately affect food security in various ways, such as extinction and migration of some fish species, resulting in less catch, which leads to loss of income and livelihood (Mahammad & Eraguchi, 2013). A sharp change in the environmental conditions can severely impact the income of the fishing-dependent communities. For instance, Iwasaki et al. (2009)

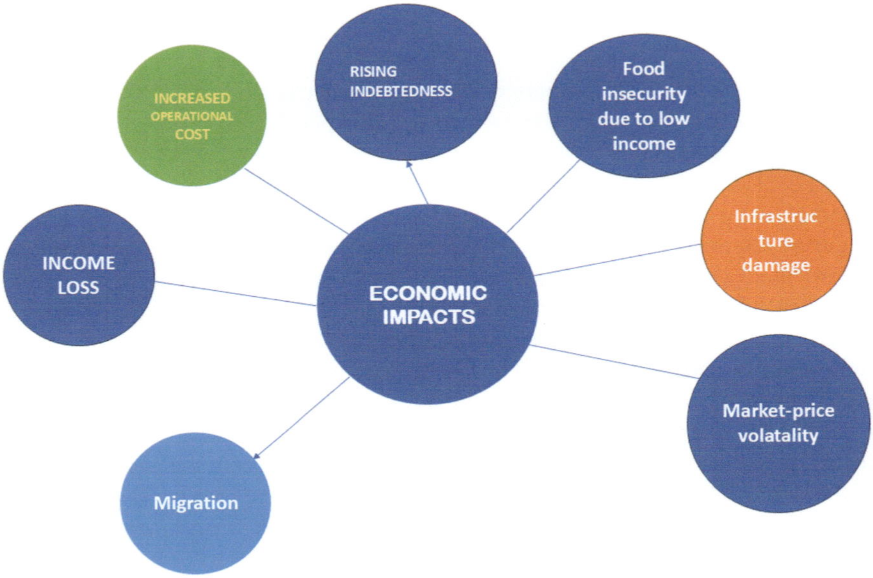

Fig. 8.2 Economic impacts of climate change on the small-scale fishermen of South Asia

reported that fishermen of Chilika lagoon face significant loss of income in the rainy season because they couldn't go fishing due to intense rainfall and winds. Coulthard (2008), reported that in the Pulicat lagoon of Tamil Nadu, the Dhoniveru fishermen community is now earning one-tenth of what they used to earn from fishing. Climate change has significantly impacted fish quantity and quality, further diminishing fishing activities (Rijnsdorp et al., 2009). The informal borrowing from the local merchants at high interest rates, to meet the expenses during the time of natural disasters, makes the fishermen fall into indebtedness, which exacerbates poverty and the debt trap (Islam et al., 2014). The loss of income, increased operational costs, lack of proper market access, and volatile fish selling prices are negatively affecting the economic well-being of the small-scale fishermen communities (Fig. 8.2).

8.3.3 Social Impacts

Globally, 1.3 billion people reside in tropical coastal regions, and the majority of them rely on fishing activities for their livelihoods and food security (Mozumder et al., 2023). The majority of small-scale fishermen are socioeconomically vulnerable due to low income, poverty, and limited occupational opportunities. Mozumder et al. (2023) reported that most of the fishing communities of southern Bangladesh live in houses made of thatch with their large families. Most of the fishermen don't have proper education, and they don't even own a boat. They live in precarious conditions without proper education and healthcare facilities. Barua et al. (2020) reported that fishing people are significantly affected by environmental change and climatic hazards like storm surge in the Bakkhali river area in Bangladesh due to their inability to cope with the changing environmental conditions. Senapati and Gupta (2017) reported that the Coastal fishing community's livelihoods are heavily impacted by the pollution of coastal waters due to the overuse of coastal resources, which leads to health issues, loss of wetlands, loss of mangroves, and corals.

Hossain et al. (2024) reported that due to saltwater intrusion, the coastal households are unable to get freshwater in nearby areas. Therefore, they have to go far to source freshwater, which disrupts their daily life and puts a financial burden on them. The reliance on tubewells for freshwater makes them susceptible to various kinds of water-borne diseases, and most of them suffer from hypertension in these areas. During our field visit to the Chilika lagoon, most of the fishermen reported having suffered from hypertension and high blood pressure due to reduced catches and excessive salt intake. Iwasaki et al. (2009) reported that due to extreme rainfall-induced water pollution, the fishermen of Chilika have suffered from various waterborne diseases such as diarrhoea, cholera, etc.

Climate change affects different aspects of the well-being of the fishing-dependent communities. High tidal waves due to climate change result in overflow of water into the low-elevation village areas, which ultimately results in displacement of affected households (Hossain et al., 2024). In the Gwadar region of Pakistan, most of the fishermen families have been displaced from their traditional places as a result of development projects such as sea port construction (Ullah et al., 2023). This climate induced displacement of fishermen communities has led to the loss of the traditional knowledge of fishing (Ullah et al., 2023). Due to their low income and low savings, small-scale fishermen have been borrowing from different informal resources to cope with the effects of natural disasters.

Due to environmental change and other Government regulations, the fishing communities of Sundarban have to borrow money from informal money lenders to continue their operations, which ultimately leads to a state of indebtedness (Hossain et al., 2024). Due to frequent natural disasters in the coastal areas, the fishermen are migrating to other places in search of employment opportunities, which critically hampers their social well-being (Islam et al., 2014). Reduction in fish availability due to climate change sometimes creates community conflicts. Iwasaki et al. (2009) reported that in the Chilika lagoon, fishermen communities tend to compete with each other for scarce resources. Coulthard (2008) reported that due to the reduction in fish productivity, community conflicts have risen in the Pulicat Lagoon of Tamil Nadu, India. The children of Chilika, due to their lack of money, are leaving formal schooling to start fishing at a very young age (Iwasaki et al., 2009). Notably, climate change has significantly affected the social life of fishermen through forced displacement from their roots, migration, escalating community conflicts, adverse health impacts, and loss of traditional knowledge (Fig. 8.3).

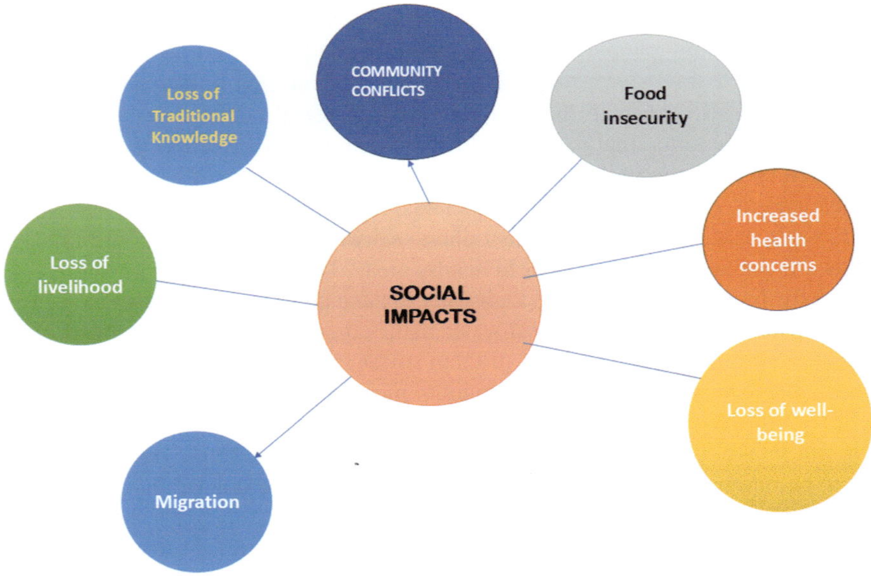

Fig. 8.3 Social Impacts of climate change on the small-scale fishermen of South Asia

8.3.4 Food Security

Fisheries are a cornerstone of global food security, supplying essential protein and nutrients to billions of people, particularly in coastal and low-income communities. Globally, there are approximately 2 billion people are deemed to be food insecure. Fishery plays a key role in assuring the food security of millions of people for their social, economic, and nutritional benefits. The role of fisheries is highly critical in ensuring food and nutritional security for people living near the coasts because these regions are particularly vulnerable to extreme weather events. Over 3.3 billion people depend on fish for at least 20% of their animal protein intake (FAO, 2018). It is higher for some countries and communities as it accounts for 50% of total protein intake in small island states (FAO, 2014). Fish is an important part of the diet of the people living in coastal areas of South Asian countries such as India, Sri Lanka, Bangladesh, and the Maldives. Fish is a rich source of animal protein, vitamins, and micronutrients, which are not available in such vast quantities and diversity in any other crops or animal products (Lauria et al., 2018). Recently, the world's per capita fish consumption has almost doubled from 9.9 kg in the 1960s to above 20 kg in 2016 (FAO, 2018). Allison et al. (2009) have reported that those fishermen who are dependent on marine resources for their dietary protein intake will be severely hit by the changing climate, especially in Asia and Africa.

Fisheries is an important contributor to food supply, food security, and livelihoods both at the local and global level (FAO, 2014). Marine fisheries also provide part-time and full-time jobs to approximately 260 million people worldwide, a large proportion of whom are engaged in small-scale fisheries (Teh & Sumaila, 2013). Food and Nutrition security is a serious threat globally, as 735 million people remain undernourished worldwide (FAO et al., 2024; Vijayan et al., 2025). Marine fisheries make a crucial contribution to food security both directly and indirectly. Directly as the source of protein, fatty acids, and micronutrients, and indirectly via generating employment and income for food purchases. Hossain et al. (2024) reported that the fishermen are now reducing their daily household food intake during the time of natural disasters. In this way, changing climatic conditions influence the food security of the fishery-based households in South Asia (Table 8.3).

8.3.5 Adaptive Responses

Climate change, coupled with anthropogenic activities, has a significant impact on the livelihoods of people, especially in developing countries (Iwasaki et al., 2009; Naveen et al., 2024). Occupational groups whose livelihoods are dependent on environmental conditions are most likely to be victims, including fishermen. Climate Change significantly affects the socio-economic life of the fishermen (Shaffril et al., 2017). Efforts should be made to strengthen the ability of the fishermen to cope with the inevitable natural changes. There is an urgent need for climate adaptation to protect nature and people. Adaptation is the fundamental response to climate change.

Climate Change adaptation refers to the actions that help to reduce the vulnerability to the current and future effects of climate change, like extreme weather events, natural disasters, sea level rise, biodiversity loss, and food and water insecurity. Adaptation can be anticipatory or reactive (Mitra, 2013). In the former case, the system adjusts before the final effect takes place, but in the latter case, the change is introduced to reduce the effect of the impacts. In most cases, local adaptation strategies are reactive and short-term (Islam et al., 2014). Barange et al. (2014) reported that the least developed countries are particularly vulnerable to climate change because of their high dependence on fish and fewer available resources to invest in climate change adaptation. In South Asia, almost all countries are lower-income countries with low adaptive capacity, which makes them quite vulnerable to the frequent natural disasters like floods, cyclones.

Coastal areas that are quite vulnerable to multiple disasters like cyclones, tsunamis, soil erosion, etc., preparedness and adaptation are extremely important (Mitra, 2013). Adaptation to climate change in fisheries has been gaining importance recently because of its impact on catch composition, catch potential, and fishery revenues. Fishermen are particularly vulnerable to the changing climate due to their low income, limited knowledge about the climatic conditions, and institutional and governance issues in the fisheries sector. The lack of occupational flexibility among

Table 8.3 Climate change impacts on the small-scale fishermen of South Asia

Impact type	Climate change impacts	Adaptative responses	Sources
Physical and environmental impacts	Sea level rise Coastal erosion Coastal floods High tidal waves Salt water intrusion Ocean Acidification Rise in Invasive Species	Mangrove restoration Community-based resource management	Mozumdar et al. (2023) Basnayake and De Silva (2025) FAO (2014) Islam et al. (2025)
	Decrease in fish body sizes Fish migration & changes in species composition Reduction in fish availability	Cage aquaculture Trap pond management system	
	Damage to the boat and fishing equipment	Borrowing	Gills et al. (2024) Roy and Sen (2024)
	More frequent cyclones and extreme weather events	Early warning systems Climate literacy	Basnayake and De Silva (2025)
Social impacts	Food insecurity	Small-scale homestead pens Engage in alternative income-generating activities	FAO (2014)
	Loss of livelihood Migration and displacement Increased School Dropouts Increased Health Risks Loss of Indigenous Knowledge	Livelihood diversification Cage aquaculture Promotion of local employment	Hossain et al. (2024) FAO (2014), Ullah et al. (2024)
	Community conflicts	Padu system Building strong social connections	Iwasaki (2014) Coulthard (2008) Mozumdar et al. (2023)
Economic impacts	Loss of income due to reduced catches Increased operational costs	Adoption of new fishing gears & equipment New generation electronic fishing devices	FAO (2014) Ullah et al. (2024)
	Rising indebtedness	Engage in alternative income-generating activities Family support Migration	Hossain et al. (2018)
	Market price volatility Migration due to loss of income	Livelihood Diversification	Hossain et al. (2024)
	Infrastructure damage	Mangrove restoration Building cyclone-resistant houses	Mozumdar et al. (2023)

Table 8.3 (continued)

Impact type	Climate change impacts	Adaptative responses	Sources
	Food insecurity due to loss of income	Reducing spending on food items Honey collection	Roy and Sen (2024) Hossain et al. (2018)

small-scale fishermen makes them vulnerable to climate change because they rely entirely on fishing activities for income (Brashares et al., 2004).

Mozumder et al. (2023) reported that fishermen are focusing on disaster risk reduction techniques as part of their social adaptation strategies, such as early warning systems, timely seasonal weather predictions, and catastrophe recovery programs to reduce climate vulnerabilities. These fishermen are also focusing on mangrove restoration because it helps in reducing the hazardous effects of the tropical cyclones and also helps in shrimp, crab, and mullet harvesting. To tackle high waves and extreme water conditions during floods, Bangladeshi fishermen are now using motorised boats instead of wooden boats. Mozumder et al. (2023) also reported that Bangladeshi fishermen emphasized the importance of building strong social connections to enhance resilience, overcome stress, speed up asset transfers, relocation of members, and rebuilding infrastructures that were destroyed due to cyclones.

Classically, adaptation can be classified into 'hard' (for instance, infrastructure and institutional changes), 'soft' (policy and governance changes), and 'ecosystem-based' (conservation and management) (Mozumder et al., 2023). Adaptation can take various forms and shapes according to the setting of the community or the region. For capture fisheries, adaptation involves keeping fishing activities to sustainable limits. Governments and authorities should impose catch limits based on changes in growth, survival, and reproduction (FAO, 2014). Several other adaptation strategies include: shift to culture-based fisheries, adjustments in insurance markets, and using improved vessels to reduce fuel cost (FAO, 2014). In Bangladesh, trap pond management is being practiced on seasonally flooded lands by building traps and modifying bamboo traps to reduce per-unit household effort (FAO, 2014). Mozumder et al. (2023) reported that Bangladeshi fishermen are focusing on strengthening conservation techniques, using indigenous knowledge, changing fishing technology, location, and target, and collaboration as part of their ecological adaptation strategies.

In Nepal, cage aquaculture is being practiced by the local dam-displaced communities. Hossain et al. (2018) reported that Bangladeshi fishermen are practicing family support and migration as key adaptation strategies during the events of natural disasters. Basnayake and De Silva (2025) reported that in Sri Lanka, artisanal fishermen are focusing on climate literacy, informal education programmes, and community-based resource management as the adaptation measures for the smooth production of lobsters. Iwasaki et al. (2014) reported that in the Negombo lagoon of Sri Lanka, the local fishermen are practicing the Padu system for the sustainable management of fishery resources. Coulthard (2008) reported that fishermen in the Pulicat lagoon of Tamil Nadu in India are using different kinds of nets to catch various fish species in

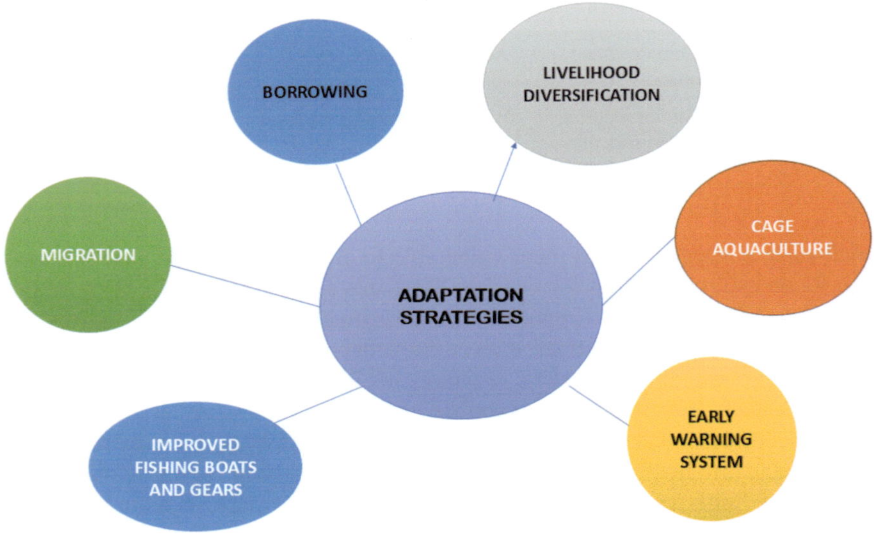

Fig. 8.4 Adaptation Strategies practiced by the small-scale fishermen of South Asia

diverse fishing seasons. Islam et al. (2025) reported that in the Haor region of northeastern Bangladesh, fishermen are using cage aquaculture as the primary adaptation strategy to fight climate change.

Hossain et al. (2024) reported that the fishermen households in Sundarban are focusing on livelihood diversification as one of the main adaptation decisions. They are engaged in shrimp frying, crab collection, and honey collection throughout the year to meet their expenses. In Karnataka, most of the fishing-dependent people are taking credit to meet the financial loss (Gills et al., 2024). Ullah et al. (2023) reported that Fishing communities of the Gwadar region of Pakistan are using new-generation electronic fishing devices even on small boats to increase the catch. Roy and Sen (2024) reported that the fishermen of West Bengal are practicing borrowing money from informal sources, migration, and working as daily labourers as some of their coping strategies (Fig. 8.4).

8.4 Discussion

The literature consistently highlights that climate change has had a profound impact on the socioeconomic conditions (income, livelihood, food security, health, displacement, and well-being) of the small-scale fishermen in South Asia, particularly through declining fish stocks, erratic weather patterns, and coastal erosion. Several studies from India and Bangladesh report increased frequency of cyclones and rising sea levels, which have directly affected the number of fishing days and damaged fishing infrastructure (Islam et al., 2014; Senapati & Gupta, 2017). One of the main findings across countries was the decline in fish stocks, hence income among small-scale fishermen in the coastal areas of India, Bangladesh, and Sri Lanka (Islam et al., 2014; Makwana & Patnaik, 2021). Ullah et al. (2024) reported the importance of traditional knowledge in enhancing the overall well-being of the small-scale fishermen communities in the Gwadar region of Pakistan, but due to climate change-induced displacements, the traditional knowledge is not being passed to future generations. Additionally, several studies reported that the small-scale fishermen are selling their catch through the middlemen as they don't have proper knowledge and access to the market and storage facilities. This results in a rise in exploitation by the middlemen; hence, the fishermen become poorer. South Asian fishermen employ a variety of adaptation strategies to cope with environmental and economic challenges. Some of these—such as mangrove restoration, cage aquaculture, the use of motorized boats, and early-warning systems—are broadly sustainable, whereas others—like taking high-interest loans, engaging in casual labour, migrating in search of work, and using gill nets—tend to undermine long-term resilience and are therefore unsustainable. Shortly, the fishery-based livelihoods are expected to be greatly impacted by climate change. In the absence of proper adaptation, the rise in frequency and intensity of natural disasters such as cyclones and floods may result in more fatalities in the coastal zones and the sea, greater loss of livelihood assets, destruction of fishing equipment, and loss of fishery-based income (Islam et al., 2014). As migration emerged to be one of the well-practiced adaptation strategies in the coastal regions of India and Bangladesh, it greatly affects the livelihood and cultural identity of the fishermen communities, as it leads to the loss of traditional knowledge and weakening of communal relationships. Displacement due to development projects in the Gwadar region led to the loss of traditional knowledge, as the knowledge couldn't be passed on to future generations (Ullah et al., 2024).

Climate change and other environmental disruptions make the affected communities' food insecure. Roy and Sen (2024) reported that the fishermen communities of West Bengal are reducing their household food consumption to save money. This results in malnutrition of the people affected. Most of the Bangladeshi fishermen are selling their fish catch because the price of the fish has gone up so high, it is better to sell than to consume. Hossain et al. (2024) reported that saltwater intrusion due to floods and high tidal waves adversely affects the health of the household. Climatic changes affect the overall well-being of the small-scale fishing communities by reducing catches, inducing displacement, creating several health problems, accelerating migration, and creating conflicts among the communities. Coulthard (2009) reported that community conflict is increasing in the Pulicat lagoon of Tamil Nadu due to a reduction in catches and competition among the fishermen communities for scarce resources. Most of the studies focused on the impact of climate change on the fishermen's households, but very few studies focused on the gender side of it. Females play an important role in the coping and adaptation process by engaging in daily labour, collecting honey and wax, etc. Furthermore, most of the studies are being done in India and Bangladesh, but there are very few studies in other South Asian countries like Sri Lanka, Pakistan, and the Maldives. Studies need to be done on these two areas to develop proper adaptation strategies. While most of the studies focused on the economic and social impacts of climate change, very few studies have discussed the mental health and psychological well-being of the small-scale fishermen communities due to less catch, conflicts, and food insecurity. This is a pivotal aspect of a community to develop; hence, more studies need to be done on this topic. Lastly, most of the studies ignored the pivotal role of indigenous knowledge for the growth of the fishery sector, so more literature needs to be undertaken on the role of indigenous knowledge to enhance the community's well-being and to achieve peace and harmony.

8.5 Conclusion and Policy Implications

The study has examined the multifaceted impacts of climate variability on the socioeconomic conditions of small-scale fishermen in South Asia. The results show that a rise in sea surface temperature, changes in salinity levels, ocean acidification, and climate-induced sea level rise have profoundly impacted the fish availability and migration patterns, which ultimately affect the socioeconomic life and well-being of fishermen through income disruptions, loss of livelihoods, rising indebtedness, and exacerbating food insecurity. The composite effects of reduction in fish catch, loss of fishing days, destruction of fishing gears and boats, and rising operational costs have forced the small-scale fishermen to adopt unsustainable, insecure sources of income. The paper highlighted the need to provide alternative income-generating opportunities, better market access to sell their catch, and the creation of a value chain to enhance their income. Most of the small-scale fishermen sell their catch through middlemen because of their lack of proper access to the market and storage facilities; as a result, they often get exploited. There is an urgent need for the Government to provide them with storage and market facilities to sell their catch.

Furthermore, the results show that while some of the small-scale fishermen communities have adopted sustainable adaptation strategies such as mangrove restoration, livelihood diversification, Internet-based communication systems, early warning systems, etc., others are trying to cope with disasters and change by adopting unsustainable measures like selling assets, migration, and informal borrowing. This discrepancy in adaptation measures highlights the urgent need for Governments to implement proper training and capacity-building programmes to enhance the capability of the small-scale fishermen communities to cope with fragile environmental conditions.

Notably, in South Asia, a disaster-prone region, the fishing equipment of small-scale fishermen often gets destroyed by frequent disasters like cyclones and floods. Governments should focus on implementing early warning systems, internet-based communication systems, providing jetties, and building proper cyclone rescue centres to minimize the damage. The Government also needs to provide proper life insurance for the small-scale fishermen as well as insurance for their fishing equipment (such as nets, boats, etc.), in the event of natural disasters. There is also a need to develop local and regional institutions to govern the fishing activities and provide solutions in case of conflicts and damage. As fish is a perishable product, special emphasis must be given to the post-harvest technology for fish processing. Overfishing is a significant problem in South Asia because of its high population density, so policies should be made to prohibit the destructive and illegal fishing techniques. It is crucial to arrange training programmes for alternative income-generating activities, provide proper sanitation facilities, and implement micro credit schemes to uplift the socioeconomic well-being of the fishermen in South Asia.

Declaration of Competing Interest The authors declare that they have no conflict of interest.

References

Ahmed, M., Saha, S. M., Hossain, M. E., Khan, M. A., & Prodhan, M. M. H. (2021). Assessment of livelihood and food poverty status of the floating fishermen in riverine system of Bangladesh. *Social Sciences & Humanities Open, 4*(1), Article 100219.

Allison, E. H., et al. (2009). Vulnerability of national economies to the impacts of climate change on fisheries. *Fish and Fisheries, 10*, 173–196.

Badjeck, M. C., Allison, E. H., Halls, A. S., & Dulvy, N. K. (2010). Impacts of climate variability and change fishery-based livelihoods. *Marine Policy, 34*(3), 375–383. https://doi.org/10.1016/j.marpol.2009.08.007

Barange, M., Fernandes, J. A., Kay, S., Hossain, M. A., Ahmed, M., & Lauria, V. (2018). Marine ecosystems and fisheries: trends and prospects. In *Ecosystem services for well-being in deltas: Integrated assessment for policy analysis* (pp. 469–488).

Barange, M., Merino, G., Blanchard, J. L., Scholtens, J., Harle, J., Allison, E. H.,... & Jennings, S. (2014). Impacts of climate change on marine ecosystem production in societies dependent on fisheries. *Nature climate change, 4*(3), 211–216.

Barua, P., Rahman, S. H., Barua, S., & Rahman, I. M. (2020). Climate change vulnerability and responses of fisherfolk communities in the South-Eastern coast of Bangladesh. *Water Conservation and Management, 4*(1), 20–31.

Basnayake, R., & De Silva, A. (2025.) Institutional innovations towards sustainable small-scale fisheries: an assessment of co-management approach in lobster and giant freshwater prawns Fisheries, Sri Lanka. *Research on World Agricultural Economy, 6*(1), 624–638. https://doi.org/10.36956/rwae.v6i1.1547

Bindoff, N. L., Cheung, W. W., Kairo, J. G., Arístegui, J., Guinder, V. A., Hallberg, R., et al. (2019). Changing ocean, marine ecosystems, and dependent communities. https://www.ipcc.ch/srocc/chapter/chapter-5/

Brander, K. (2010). Impacts of climate change on fisheries. *Journal of Marine Systems, 79*(3–4), 389–402.

Brashares, J. S., Arcese, P., Sam, M. K., Coppolillo, P. B., Sinclair, A. R., & Balmford, A. (2004). Bushmeat hunting, wildlife declines, and fish supply in West Africa. *Science, 306*(5699), 1180–1183.

Chefaoui, R. M., Duarte, C. M., & Serrão, E. A. (2018). Dramatic loss of seagrass habitat under projected climate change in the Mediterranean Sea. *Global change biology, 24*(10), 4919–4928.

Cheung, W. W., Reygondeau, G., & Frölicher, T. L. (2016). Large benefits to marine fisheries of meeting the 1.5 C global warming target. *Science, 354*(6319), 1591–1594.

Cinner, J. E., Adger, W. N., Allison, E. H., Barnes, M. L., Brown, K., Cohen, P. J., et al. (2018). Building adaptive capacity to climate change in tropical coastal communities. *Nature Climate Change, 8*(2), 117–123.

Colgan, C. S., & Merrill, S. B. (2008). The effects of climate change on economic activity in Maine: Coastal York County case study. *Maine Policy Review, 17*(2), 66–79 (2008). https://digitalcommons.library.umaine.edu/mpr/vol17/iss2/10

Coulthard, S. (2009). 16 adaptation and conflict within fisheries: insights for living with climate change Sarah Coulthard. In *Adapting to climate change: Thresholds, values, governance* (p. 255).

Coulthard, S. (2008). Adapting to environmental change in artisanal fisheries—Insights from a South Indian Lagoon. *Global Environmental Change, 18*(3), 479–489.

FAO. (2014). *The state of world fisheries and aquaculture 2014* (223 pp.).

FAO. (2020). The State of World Fisheries and Aquaculture 2020. Sustainability in action. Rome.

FAO. (2022). *The state of world fisheries and aquaculture 2022. Towards blue transformation.* FAO. https://doi.org/10.4060/cc0461en

FAO, Ifad, UNICEF, WFP, & WHO. (2024). *The state of food security and nutrition in the world 2024—Financing to end hunger, food insecurity and malnutrition in all its forms.* https://doi.org/10.4060/cd1254en

Food and Agriculture Organization of the United Nations. (2018). *The state of world fisheries and aquaculture 2018: Meeting the sustainable development goals.* FAO. https://doi.org/10.4060/i9540en

Franco, B. C., Defeo, O., Piola, A. R., Barreiro, M., Yang, H., Ortega, L.,... & Möller, O. O. (2020). Climate change impacts on the atmospheric circulation, ocean, and fisheries in the southwest South Atlantic Ocean: a review. *Climatic Change, 162*(4), 2359–2377.

Gills, R., Sulochanan, B., Padua, S., Nakhawa, A. D., Abdul Azeez, P., Edward, L., et al. (2024). Impact of cyclone Tauktae on fishermen's livelihood in Karnataka: A climate change perspective analysis. *Indian Journal of Fisheries, 71*(3), 132–139.

Giri, S. S. (2024). Sustainable aquaculture practices in South Asia: A comparative analysis of feed formulation and utilization. *Animal Frontiers, 14*(4), 6–16. https://doi.org/10.1093/af/vfae020

Hossain, M. A., Ahmed, M., Ojea, E., & Fernandes, J. A. (2018). Impacts and responses to environmental change in coastal livelihoods of south-west Bangladesh. *Science of the Total Environment, 637*, 954–970.

Hossain, M. A., & Shrestha, R. B. (2019). Fisheries in South Asia: Trends, challenges and policy implications. *Agricultural Policy and Program Framework: Priority Areas for Research & Development in South Asia*, 332.

Hossain, M. Z., Rahman, M. A. U., Rahaman, K. R., et al. (2024). Investigating critical relationships among vulnerability, livelihoods, and non-migration strategies at the fishing communities in the Sundarbans. *Environment, Development and Sustainability, 26*, 29129–29168. https://doi.org/10.1007/s10668-023-03857-y

IPCC, 2019: IPCC Special Report on the Ocean and Cryosphere in a Changing Climate [H.-O. Pörtner, D.C. Roberts, V. Masson-Delmotte, P. Zhai, M. Tignor, E. Poloczanska, K. Mintenbeck, A. Alegría, M. Nicolai, A. Okem, J. Petzold, B. Rama, N.M. Weyer (Eds.). Cambridge University Press, Cambridge, UK and New York, NY, USA, (pp. 755). https://doi.org/10.1017/9781009157964.

Islam, M. M., Sallu, S., Hubacek, K., & Paavola, J. (2014). Vulnerability of fishery-based livelihoods to the impacts of climate variability and change: Insights from coastal Bangladesh. *Regional Environmental Change, 14*, 281–294.

Islam, Sheikh & Sanjida, Laila & Sabit, Mohammad & Hossain, Bhuiya Md Tamim Al. (2025). Assessing Climate Vulnerability in the Haor Wetland Region of Bangladesh: Policy and Governance Implications for Livelihood Resilience and Ecosystem-Based Adaptation. *International Journal of Agricultural Economics.* https://doi.org/10.11648/j.ijae.20251006.11.

Iwasaki, S. (2014). Driving forces of the long-enduring institutional mechanism of Padu system in Negombo Lagoon, Sri Lanka. *Marine Policy, 50*, 190–196.

Iwasaki, S., Razafindrabe, B. H. N., & Shaw, R. (2009). Fishery livelihoods and adaptation to climate change: A case study of Chilika lagoon. *India. Mitig Adapt Strateg Glob Change, 14*, 339–355. https://doi.org/10.1007/s11027-009-9167-8

Jakimavičius, D., Kriaučiūnienė, J., & Šarauskienė, D. (2018). Impact of climate change on the Curonian Lagoon water balance components, salinity and water temperature in the 21st century. *Oceanologia, 60*(3), 378–389.

Kajal, K., 2020. Low Fish Catch Along India's Western Coast Hints at Impacts of Climate Change. Mongabay, Retrieved from: https://india.mongabay.com/2020/08/low-fish-catch-along-indias western-coasthints-at-impacts-of-climate-change/

Lam, V. W., Allison, E. H., Bell, J. D., Blythe, J., Cheung, W. W., Frölicher, T. L., et al. (2020). Climate change, tropical fisheries and prospects for sustainable development. *Nature Reviews Earth & Environment, 1*(9), 440–454.

Lauria, V., Das, I., Hazra, S., Cazcarro, I., Arto, I., Kay, S., et al. (2018). Importance of fisheries for food security across three climate change vulnerable deltas. *Science of the Total Environment, 640*, 1566–1577.

Makwana, M., & Patnaik, U. (2021). Vulnerability of marine fisheries to sea surface temperature and cyclonic events: Evidences across coastal India. *Regional Studies in Marine Science, 48*, Article 102002.

Mitra, A. (2013). *Sensitivity of mangrove ecosystem to changing climate* (Vol. 62, pp. 143–157). Springer.

Mohammed, E. Y., & Uraguchi, Z. B. (2013). *Impacts of climate change on fisheries: Implications for food security in Sub-Saharan Africa* (pp. 114–135). Nova Science Publishers Inc.

Mozumder, M. M. H., Schneider, P., Islam, M. M., Deb, D., Hasan, M., Monzer, M. A., & Nur, A.-A.U. (2023). Climate change adaptation strategies for small-scale Hilsa fishers in the coastal area of Bangladesh: Social, economic, and ecological perspectives. *Frontiers in Marine Science, 10*, 1151875. https://doi.org/10.3389/fmars.2023.1151875

MRAG. (2011). *Fisheries and livelihood*. Fisheries Management Science Programme (FMSP), Marine Resources Assessment Group (MRAG), and Department for International Development (DFID). www.mrag.co.uk/Documents/PolicyBrief4_Livelihoods.pdf. Accessed 25 October 2011.

Ninawe, A. S., Indulkar, S. T., & Amin, A. (2018). Impact of climate change on fisheries. In *Biotechnology for sustainable agriculture* (pp. 257–280). Woodhead Publishing.

Pauly, D., & Cheung, W. W. (2018). Sound physiological knowledge and principles in modeling shrinking of fishes under climate change. *Global change biology, 24*(1), e15–e26.

Perry, A. L., Low, P. J., Ellis, J. R., & Reynolds, J. D. (2005). Climate change and distribution shifts in marine fishes. *Science, 308*(5730), 1912–1915.

Rijnsdorp, A. D., Peck, M. A., Engelhard, G. H., Möllmann, C., & Pinnegar, J. K. (2009). Resolving the effect of climate change on fish populations. *ICES Journal of Marine Science, 66*, 1570–1583.

Rilov, G. (2016). Multi-species collapses at the warm edge of a warming sea. *Scientific reports, 6*(1), 36897.

Rother, E. T. (2007). Systematic literature review X narrative review. *Acta Paulista De Enfermagem, 20*, v–vi.

Roy, A., & Sen, A. (2024). Importance of indigenous small fish to boost livelihood and nutritional security: Evidences from Indian Sunderbans. In *Perspectives and applications of indigenous small fish in India: An introduction* (pp. 583–593). Springer Nature Singapore.

Salagrama, V. (2013). Climate change and livelihoods: perspectives from small-scale fishing communities in India. *Food Chain (2046–1887), 3.*

SAARC. (2008). Coastal and Marine Risk Mitigation Plan for South Asia - Proceedings of SAARC Regional Workshop. March 27–28, 2008, Goa, India.

Senapati, S., & Gupta, V. (2017). Socio-economic vulnerability due to climate change: Deriving indicators for fishing communities in Mumbai. *Marine Policy, 76*, 90–97.

Shaffril, H. A. M., Samah, A. A., & D'Silva, J. L. (2017). Adapting towards climate change impacts: Strategies for small-scale fishermen in Malaysia. *Marine Policy, 81*, 196–201.

Sivakumar, M. V., & Stefanski, R. (2010). Climate change in South Asia. *Climate change and food security in South Asia* (pp. 13–30). Springer, Netherlands.

Sumaila, U. R., Cheung, W. W., Lam, V. W., Pauly, D., & Herrick, S. (2011). Climate change impacts on the biophysics and economics of world fisheries. *Nature Climate Change, 1*(9), 449–456.

Teh, L. C., & Sumaila, U. R. (2013). Contribution of marine fisheries to worldwide employment. *Fish and Fisheries, 14*(1), 77–88.

Townhill, B. L., Radford, Z., Pecl, G., van Putten, I., Pinnegar, J. K., & Hyder, K. (2019). Marine recreational fishing and the implications of climate change. *Fish and Fisheries, 20*(5), 977–992.

Ullah, S., Khan, U., Begum, A., Han, H., & Mohamed, A. (2024). Indigenous knowledge, climate change and transformations of Gwadar fishing community. *International Journal of Climate Change Strategies and Management, 16*(3), 298–317. https://doi.org/10.1108/IJCCSM-06-2022-0069

Ullah, S., Khan, U., Begum, A., Han, H., & Mohamed, A. (2023). Indigenous knowledge, climate change and transformations of Gwadar fishing community. *International Journal of Climate Change Strategies and Management, 16*(3), 298–317.

Vijayan, V., Suresh, A., Sabu, S. S., & Padi, H. P. (2025). Fish for food and nutrition security in India: A comprehensive framework analysis. *Food Security*, 1–17.

Datta, P., Behera, B. and Rahut, D.B., 2022. Climate change and Indian agriculture: A systematic review of farmers' perception, adaptation, and transformation. Environmental Challenges, 8, p.100543.

Chettri, D., Datta, P. and Behera, B., 2024. Climate change and household food security in the Himalayas: A systematic review of the challenges and household adaptative measures. Environmental Development, 51, p.101019.

Datta, P., Behera, B. and Rahut, D.B., 2024. Climate change and water-related threats in the Indian Sundarbans: food security and management implications. International Journal of Water Resources Development, 40(3), pp.323-344.

Naveen, N., Datta, P., Behera, B. and Rahut, D.B., 2024. Climate-smart agriculture in South Asia: Exploring practices, determinants, and contribution to sustainable development goals. Mitigation and Adaptation Strategies for Global Change, 29(4), p.31.

Chettri, D., Datta, P., Rahut, D.B., Behera, B. and Sonobe, T., 2025. Factors Influencing the Household Food Security of Forest Villagers in Eastern Sub-Himalayan India. Food and Energy Security, 14(2), p.e70052.

Chapter 9
Services Sector and Its Interactions with Climate Change and Environmental Degradation: An Analysis in the Indian Context

Anurag Krishna Vippala

Abstract As a result of great progress in various developmental targets, developing nations are progressively investing more heavily into the rapid growth of their respective services sectors. However, there is very little academic research being conducted on the growing role of the services sector, and its sensitivity to climate change, even as economies become more reliant on its growth and well-being. Given the growing significance of the services sector, this paper aims to bridge this gap. This study focuses on the business process outsourcing services sector with an analysis of the global marketplace, and India's place in it. It is imperative to establish an understanding of the sensitivity of the industry, and thus the economy at large, to climate and macro-economic shocks. The paper finds that the health of the services sector in developing nations is closely linked to monetary and foreign trade policy. Climate change adds an additional stress to an already tenuous balance between developmental and trade policy. The nature of the services sector itself further aggravates the sensitivity of players to climatic and economic shocks, along with policy uncertainty. Thus, a comprehensive evaluation of the sector, India's place in the global services marketplace, and the potential aggravation and sensitivity of the sector in relation to climate change establishes that things cannot continue laissez-faire. There is an urgent need for policy to improve local demand for the expertise being developed by business process outsourcing. India must also embrace the opportunity to set the tone on AI legislation. Thus, a reduction on the reliance of the sector on foreign outsourcing, paired with strong legislation on climate change, can position India uniquely to tackle the challenges of development with more certainty.

Keywords Global services Industry · Global trade and policy · Industrial analysis · Macro-economic sustainability · Climate change · Climate resilience · Artificial intelligence · AI policy · Business process outsourcing · Monetary policy

A. K. Vippala (✉)
Livelihoods and Natural Resources Management Institute, Hyderabad, India
e-mail: anuragvippala@gmail.com

P. Datta et al. (eds.), *Climate Change and Human Well-being*, Advances in Global Change Research 83, https://doi.org/10.1007/978-3-032-25479-5_9

9.1 Introduction

In an increasingly inter-connected global economy, governments face increasing complexity when it comes to framing economic and foreign policy. Terms of trade and the sensitivity of economies to global forces form the basis of strong policy. Historically, trade has facilitated the transfer of resources in order to meet imbalances in natural resource re-distribution. Despite technological progress, and the development of a highly complex and robust trade ecosystem, trade continues to be focused on the same fundamental natural resources viz., raw materials, forest produce, etc. While this largely continues to be the case, complex governance systems and production needs have transformed this process. It is now composed of the populations of nations who produce these goods and services on a scale that allows for trade to happen freely, but remain unable to rely directly on the proceeds from the goods and services produced alone.[1] Maintaining a flourishing economy, is thus far more challenging, and policy makers in developing nations now find themselves at the helm of a more complicated mechanism, with more sensitive participants to protect, while maximising the return from production activities. This is especially so in the absence of smooth a economic transition from primary to tertiary sector.

India's economic development has missed on this smooth transition, as it jumped to tertiary sector without achieving maturity of manufacturing or industry sector as per Rostow's take-off model (Bade, 2003). This is mainly due to the nature of economic reforms India adopted to overcome economic crisis during early 1990s. Though it is a relatively new entrant to the economic development discourse, India has done well in terms of economic performance. India achieved high growth (9%) in service sector when compared to global growth (4%) between 1996 and 2022 (Pant et al., 2024). This also helped India to achieve high overall GDP growth rates. India's investment in education and development of local technology and infrastructure paid off, enabling the development of a robust services industry. Consequently it continues to be a hub for the technology services industry, continuing to leverage its human resource base to produce high value services for western corporations, while also collaborating with them to further develop local products and technologies.

This development came at considerable cost—to both the natural and human resources of these countries. Pollution levels climbed rapidly, with India now home to some of the most polluted cities in the world.[2] Aggressive agricultural intensification, combined with weak environmental and labour protection legislation boosted food security and economic productivity, but at the cost of exploitation of both natural resources (Reddy et al., 2025). It is easy to assume that the services sector would have a lower impact on the environment and climate change accelerators like greenhouse gases. However, studies have found that while the environmental impacts of the services sector are lower at the discretized level, given the scale of operations of the

[1] Trade balance of developing and developed countries continues to converge, UNCTAD statistics show | UN Trade and Development (UNCTAD) 2014.

[2] https://www.iqair.com/in-en/?srsltid=AfmBOoq0WRyJ6aMVqrzsAerJvF3JdTG3RmCN3Kg9lO_s1x3l1Xs6DDod.

services sector, the net environmental impact of the services sector becomes significant even when compared to the manufacturing sector (Rosenblum et al., 2000). This is mainly due to: (i) expansion of urbanisation and the associated land use change and ecosystem services, (ii) high energy requirements, (iii) expansion supply chains to service the service sector and its impact on energy and natural systems. In fact, the energy requirements are increasing multi-fold with the advent of Artificial Intelligence (AI). Thus, the services sector is not a silver bullet for sustainable/climate conscious economic growth—it still comes with important environmental costs and can exacerbate climate change on a scale comparable to the manufacturing sector. At the same time, climate change, and especially increased frequency of extreme events like floods could impact the continuity of services, which can threaten the viability of India as an outsourcing destination. This paper explores the climate risks that impact the service sector and vice versa.

9.2 Rationale of the Study and Literature Review

This paper dives into the India case-study covering the evaluation of service sector, its stability, and consequently its contribution to climate change and sensitivity to climate shocks. The focus of this study however, is the sub-sector with the greatest share in India, namely, business process outsourcing (BPO). BPO services focus on the provision of globalized services to multinational corporations incorporated outside of India. It includes the sub-sectors of information technologies, financial services and various business operations such as customer support and vendor management. The services sector, and BPO services in particular, are largely left to open market supply and demand forces. Thus, the dynamics at play within the sector are more aligned to modern business strategy, and see very little research and literature from academic research and publishing. Thus, there is very little academic literature on the services sector. Most research on the sector is typically a direct result of players within this space (such as consulting firms) producing reports on the development and state of the sector, and its potential growth prospects. Such research is typically focused on assessing business opportunity, and therefore falls short on analysis on long term sustainability and inter-linkages with allied sectors. There is no academic work conducting a critical analysis of the space as a whole, and this study's primary goal is to bridge this gap. A more comprehensive approach is required in order to understand the sustainability of the growth path the services sector is on globally, and within the Indian context as well. Policy and development of this sector is not traditionally driven by independent research, but rather through industry engagement and lobbying. This is problematic as policy tends to be shaped by an incomplete understanding and heavily influenced by vested interests. When this is further contextualized with the understanding that BPO services have far reaching impacts in terms of global trade and monetary policy as well, it is imperative that independent academic research form the basis of policy making. The current approach places a lower focus on sustainability and the protection of vulnerable sections who interact with the sector. Thus, more independent work needs to be done in this space, especially considering how reliant modern economies are on the services industry in particular—this is an area of great importance, flying under the radar of research on more financially vulnerable populations in developing nations.

9.2.1 The Services Sector and India: A Story of Growth

The Indian economy is largely driven by a robust services sector, accounting for nearly 50% in 2023 (Fig. 9.1). While the manufacturing sector fluctuates in its contribution to the GDP, it continues to be the second largest contributor to GDP output. Agriculture continues to employ the largest population within the country (Fig. 9.2). However, labour markets are shifting towards increased growth in the services sector, and given the significance to the economy at large, there is considerable value in evaluating the state of the services sector within India.

The workforce is gradually transitioning towards the manufacturing and services industries, assisted by improved access to both higher levels of education and technical training. Poor urbanization rates and a declining capacity of the agriculture sector to provide a sustainable and comfortable quality of life further incentivize the workforce to shift away from agrarian to other livelihoods. This is also visible in a shift in the geographic distribution of the population as well (Fig. 9.3). Urban

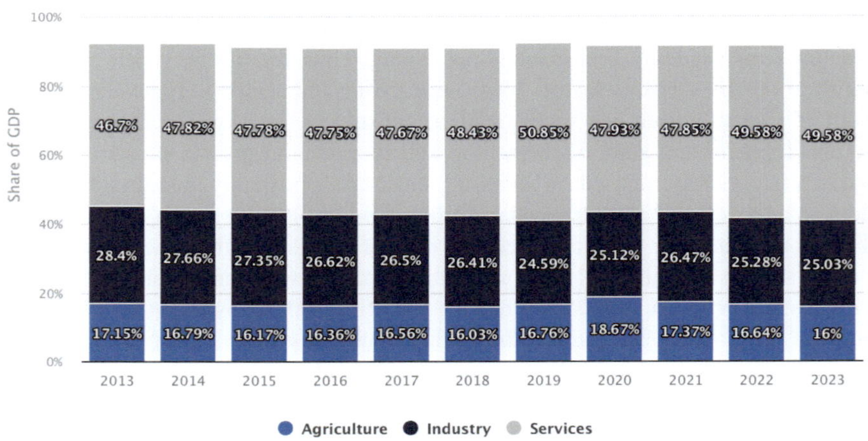

Fig. 9.1 India: Distribution of gross domestic product (GDP) across economic sectors from 2013 to 2023. *Source* https://www.statista.com/statistics/271329/distribution-of-gross-domestic-product-gdp-across-economic-sectors-in-india/

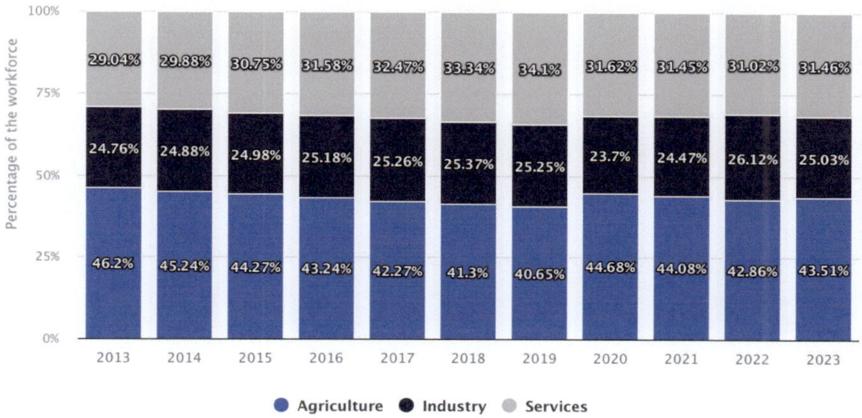

Fig. 9.2 India: Distribution of the workforce across economic sectors from 2013 to 2023. *Source* https://www.statista.com/statistics/271320/distribution-of-the-workforce-across-economic-sectors-in-india/

and peri-urban communities also provide even further improved access to modern infrastructure and education systems, with the promise of even further improved livelihoods and socio-economic outcomes for the children of migrating labour, who would benefit from more robust education and infrastructure (Ghose, 2021). As populations shift to urban centres, the agrarian sector's contribution to employment levels will decrease, and the manufacturing and services sectors will eventually become more significant in terms of population employed as well.

India in particular is also highly vulnerable to climate change. While research has been largely focused on the impact of climate change will have on food systems and the agriculture sector within India, these vulnerabilities cascade across sectors and have a major impact on the services and manufacturing sectors. The shift in the workforce from agrarian to the professional services sector means that the demographics making up the agrarian workforce shift towards even more vulnerable populations, as the activity will now be taken up by those without access to the high skill (and high cost) technical education required in cutting edge services jobs (Reddy et. al., 2025).

Thus, the stage is set for the significance of the service sector to India's economy, and the value of an independent, critical analysis of its sensitivity to climate change. Disruptions in access to critical systems due to challenges with climate or urban planning have already resulted in significant financial losses for various stakeholders in the services space.[3]

[3] Bengaluru sees massive traffic jam, thousands left stranded on flyover for around 3 h. 2024; Livemint (2023); Bengaluru Loses Nearly Rs 20,000 Crore Due To Traffic Mess: Study 2023.

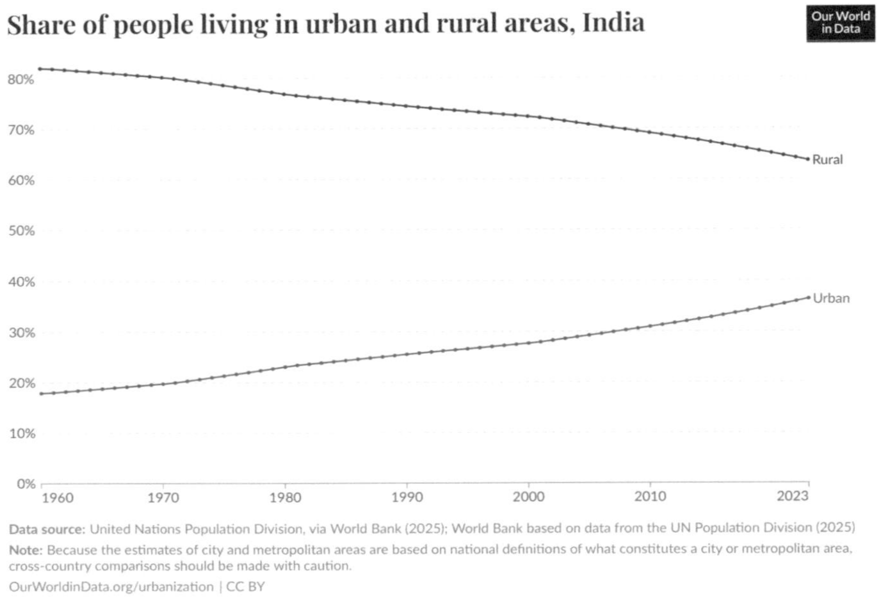

Fig. 9.3 Share of Rural and Urban Population in India, 1960–2023. *Source* https://ourworldindata.org/urbanization

9.2.2 Climate Change and the Service Sector

Fundamentally, the services sector is recognized to have substantially lower environmental impact than the manufacturing sector. Studies show that the services sector is correlated with lower levels of emissions through two major pathways. Firstly, liberalisation of the services sector facilitates increased trade of technology and knowledge enabling improvements in manufacturing processes. These improvements allow for lower emissions and more efficient manufacturing, leading to lower emissions as a result. Furthermore, growth in the services sector is also correlated to more stringent environmental norms. Thus, as the more polluting manufacturing sector continues to get more rigorously regulated, activity in the services sector as a result is further incentivized, moving labour and activity away from more environmentally degrading activity in the manufacturing sector (Cosimo Beverelli et al., 2025).

Compared to the manufacturing sector, the services sector undoubtedly produces less waste, and has lower emissions, paired with lower energy requirements as well, especially when contextualized in terms of emissions/cost per unit of value produced. However, these savings in emissions and energy consumption can quickly be erased when the scale of operation exceeds a critical mass (Roberts et al., 2021). When the services sector becomes a majority driver of GDP, it is found that the aggregated energy requirements and indirect impacts of the sectoral activity becomes significant once again. The services sector's major contributions to emissions emerges from logistics and transportation, relying on increased penetration of supply chains to support services activity (Rosenblum et al., 2000). “Tertiarization” was found to have further knock on effects in terms of shifts in consumption behaviour as a result of income and lifestyle shifts. Growth of the services sector can be associated with significant increases in emissions, land use, and water consumption (Greenford et al., 2020).

India’s focus on economic growth has meant that policy around the services sector is more focused on encouraging growth, rather than environmental sustainability. The rapid growth of the sector has been fuelled by environmental degradation. A similar study focused on Thailand (whose services sector grew under similar circumstances) found that when the benefits of the services sector are weighed against the combined cost of growth for the sector, lax policy that focuses on the development of the sector tend to offset the positive outcomes through associated degradation and resource stress (Sutthichaimethee & Sawangdee, 2016). A lot of the literature on the overall environmental impact of the services sectors considers all sub-sectors such as transportation, logistics, and so on. However, in the Indian context, the largest contributor to the GDP continues to be BPO services, which greatly differ from these more conventionally environmentally stressful services. However, there are higher indirect costs as a result of the un-regulated manner in which the sector has and continues to grow.

In summary, this study is an evaluation of the sustainability of the growth path India continues to invest so heavily into. It is an assessment of what the future of India’s economy stands to look like, and therefore it is important to evaluate and recognize the risks facing the services sector. This paper aims to conduct a preliminary analysis of the current status of Indian economy, with the services sector as a focal point in order to identify areas of vulnerability, and provide policy guidance in order to protect the economy from external shocks that threaten to slow down or even destabilize the economy. It further aims to explore the interactions of these forces with climate change, while assessing the potential impact of unchecked climate change on systems supporting the industry as well as the industry’s impact on climate change as well.

9.3 Methodology and Data

The study is split into 2 sections—an overall industrial analysis that establishes the nature of the globalized services sector as a whole, followed by an internal competitiveness analysis of India as a player in the globalized services sector. The study is primarily supported by a comprehensive review of existing literature, coupled with secondary data analysis in order to support the industrial analysis.

For the purposes of the analysis, the paper adopts the approach of treating India’s services sector as a business. However, rather than restricting the scope to a particular industry within the country, the idea is to contextualize the services sector of India relative to the global services sector, and analyse its capacity to continue to compete in the global services marketplace. Furthermore, the study also aims to evaluate the impact of climate change on the overall sustainability of an economy this heavily invested in the services sector, as well as the nature of services that are driving economic growth.

The study leverages various publicly available sources of data. Data from various data aggregators such as "our wold in data", "Statista" and "Investing.com" have been combined with an extensive review of published studies sourced from aggregators such as "sciencedirect.com", "researchgate.com" and "Elsevier". "Our world in data" is a notable data aggregator that compiles data from various reputable sources such as the World Bank, United Nations, UNESCO, and so on. Statista is also a trusted aggregator, and is one of the leading aggregators of statistics on the internet. "Investing.com" is a leading aggregator of financial data sourced from various market sources and validated by the financial industry.

9.4 Theoretical Frameworks

The corporate finance and investment banking industry has a robust toolkit in order to objectively assess and evaluate the strengths, weaknesses, opportunities and threats for any industry. Proposed by Michael Porter in 1980, the Competitive Forces Model (CFM) is used to objectively assess the salient forces impacting the performance of an industry.

9.4.1 Porter's Competitive Forces Model (CFM)

Michael Porter hypothesized that while businesses instinctively keep a close eye on their competitors in order to assess and evaluate business strategy, it is vastly more beneficial to evaluate the external forces shaping and influencing the business environment. Porter identified 5 forces that apply directive pressure on business strategy. While this model continues to be widely adopted, there are certain limitations to the version of the model proposed in the original paper—the model works best in a perfectly competitive environment, and is not as well-suited to markets where competitors aren't purely rivals. It also falls short of fully assessing the impacts of a dynamic and rapidly evolving market. The CFM also serves best as a starting point for a broad macro level analysis and needs to be supplemented with a more in-depth analysis.

Porter's CFM consists of 5 forces: 'Competitive Rivalry' focuses on examining the nature and intensity of competition in the marketplace (Fig. 9.5). 'Threat of New Entrant' evaluates the barrier to entry for new players, and the market's capacity to absorb new players. Bargaining 'Power of Suppliers' analyses the role of incumbent players in solidifying market dynamics. Bargaining 'Power of Buyers' does the same, but from a demand perspective. Finally, 'Threat of Substitution' dissects the nature of

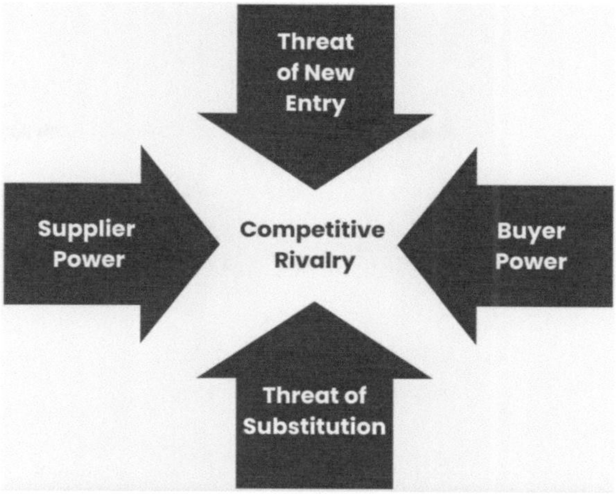

Fig. 9.5 Porter's five forces model. *Source* https://www.oxfordcollegeofprocurementandsupply.com/wp-content/uploads/2019/08/Porters-5-Forces.jpg

the product itself, and analyses the likelihood of the product itself being transformed meaningfully, and effectively being replaced by a better or alternative product.

SWOT Analysis is also an industry standard tool used to evaluate the internal Strengths, Weaknesses, Opportunities, and Threats (SWOT) of an organization. While Porter's model focuses on external forces influencing the internal strategies and decisions of an organization, through indirect market action, SWOT analysis is a tool used to analyse the internal forces directing strategies for navigating the competitive environment (Fig. 9.6).

9.4.2 SWOT Analysis

Climate change has an amplifying effect on the industry as a whole, and has the potential to introduce new challenges (threats) or exacerbate the weaknesses inherent to the India's services sector. Climate change threatens to escalate the costs of building resilient infrastructure, which forms a central pillar for a healthy services industry. Adverse climate events also threaten the overall willingness of businesses to move processes into countries like India, as the stresses placed on socio-economic infrastructure may result in wider scale instability from a policy standpoint. Finally, India is especially exposed to climate risk. A world Bank report has found that the major outsourcing hubs in India—Hyderabad, Kolkata, Mumbai, Delhi, Chennai, Pune and Bangalore—are particularly exposed to adverse climate risk.[4]

[4] https://climateknowledgeportal.worldbank.org/sites/default/files/country-profiles/15503-WB_India%20Country%20Profile-WEB.pdf.

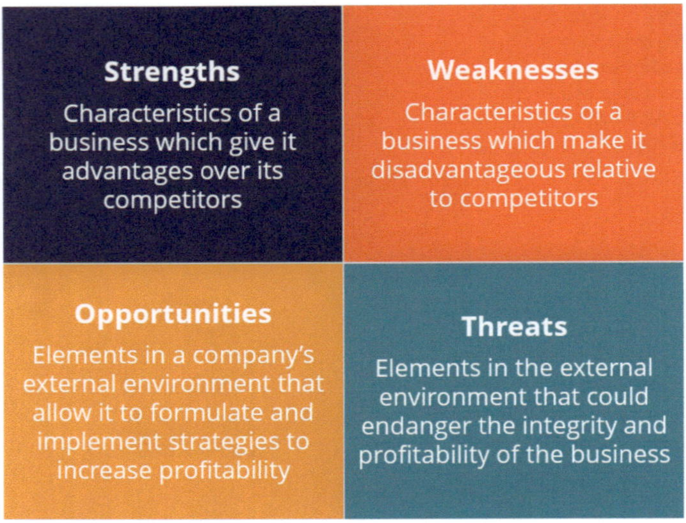

Fig. 9.6 4 dimensions of SWOT analysis. *Source* https://corporatefinanceinstitute.com/resources/management/industry-analysis-methods/

9.5 Analysis

9.5.1 Background

Globalisation and the prevalence of the internet in the early 2000s unlocked the potential for the world to be more inter-connected than ever before (Borcuch et al., 2012). Paired with liberalisation of markets in much of Asia, this meant that communications were simplified, and the west could now develop hitherto inaccessible markets and resources. It also meant that lower value activities could now be moved out of western economies into comparatively much cheaper eastern labour market.[5] Manufacturing was the first to go, as stronger labour and environmental policies in the west escalated the cost of manufacturing, and corporations were intent on keeping these costs as low as possible. This meant that intellectual property (design and innovation) and as a result, services, were now being developed within the west as a primary activity of the workforce, shifting away from manufacturing and hard labour (Fig. 9.7).

Eventually, the pursuit of perpetually increasing growth of balance sheets year on year meant that healthy top line growth was no longer sufficient to meet shareholder expectations. The gains in the bottom line would have to be bolstered through cost savings as well. This was the beginning of an increasingly aggressive approach to cost management—what started as a control on excessive spending and reduction of inefficiencies, quickly progressed into a search for finding the lowest cost way to

[5] Low-cost country sourcing (cheap labor) | EBSCO Research Starters 2025.

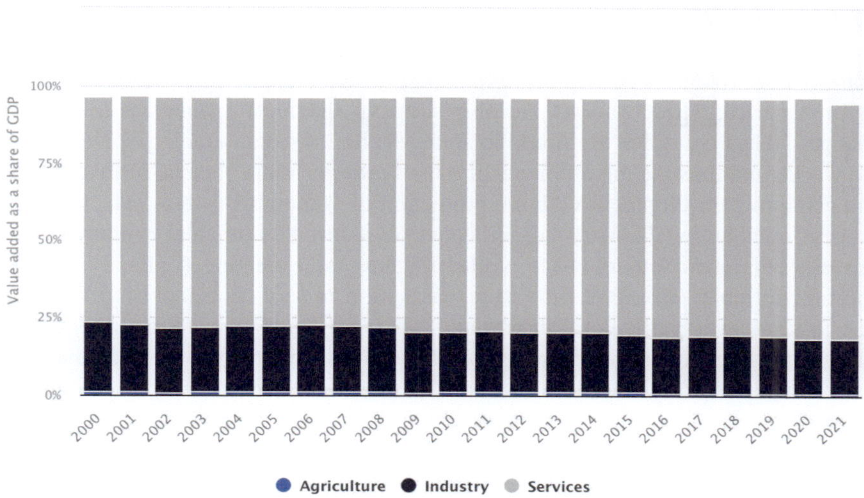

Fig. 9.7 Sectoral share of GDP in the USA, 2000–2021. *Source* https://www.statista.com/statistics/270001/distribution-of-gross-domestic-product-gdp-across-economic-sectors-in-the-us/

achieve similar levels of output, particularly in terms of support services (Martin, 2024). Customer support, low level software support, vendor management, human resource management—gradually, more departments were being moved offshore, in pursuit of cost savings. Nearly two decades later, this aggressive cost cutting approach has culminated in the increasing adoption of technologies that remove human intervention altogether. AI is now being marketed as the next big step, not just in improved quality or experiences, but cost reductions. However, most AI products, while being clearly inferior to direct human interaction, are still being adopted in order to cut costs further, despite a worse end user experience for all stakeholders.[6]

This is the context in which the services industry operates. It went from a value add for all stakeholders—an offloading of certain functions, allowing for a more focused approach to business management and development—to now a race to the bottom, an endeavour to maintain a bare-minimum level of service at the lowest cost possible. This is what made Asian economies competitive in the first place, and as can be seen, the market place is now facing a potential disruption which will pose new problems to economies highly reliant on this industry. Thus, we can preface the analysis with this baseline analysis of the "product" itself—it is not specialized, or proprietary, and can be easily replaced.

[6] https://www.theverge.com/news/657594/duolingo-ai-first-replace-contract-workers; https://www.polygon.com/ai-artificial-intelligence/603216/duolingo-ai-language-lessons.

9.5.1.1 The Climate Perspective

The services industry is rapidly expanding in developing nations in Asia and Latin America. These nations are also some of the most vulnerable to climate change—these countries are home to some of the most vulnerable populations in the world, and thus the governments of these countries must balance infrastructural and economic development with the protection of vulnerable groups. The business process outsourcing industry is predicated on a few fundamental principles—(i) Processes that are labour intensive, while being low value creating, (ii) Easily scalable and transferrable, (iii) Have little to no disruption in service availability (uptime). The third tenet is particularly important, and forms the basis of a comprehensive analysis for target countries. Nations are evaluated for geographic, political and now, even climatic stability, in order to ensure that there is no disruption or hinderance to the access of critical services. Climate change threatens the reliability with which these services The services industry and those employed within it rely on a collection of supportive services and infrastructure—the services industry has surprisingly high energy requirements, both directly and indirectly. The carbon footprint from the energy requirements alone is significant (Charlotte Fourcroy, Faiz Gallouj, and Fabrice Decellas 2012). Urbanisation is also a key requirement for a strong services industry, which in itself has high environmental cost and contributes significantly to climate change.[7]

Thus, how does a nation reliant on the services sector, and mainly the BPO industry for continued economic growth remain resilient to economic, and climatic shocks? This is the basis for assessing India's economic resilience to evolving market and climatic conditions. Significant changes in these two factors could jeopardise the path of progress India has been on, since the early 2000s.

9.6 CFM Analysis of the Global Services Sector

9.6.1 Competitive Rivalry

Competitors influence the strategies businesses employ to get ahead in the marketplace. Various factors influence the intensity of rivalry and competition in the marketplace. Market size, product differentiation, demand pressure, and number of competitors. The Asia Pacific (APAC) region is home to some of the most lucrative targets to outsource business processes and services[8] (Fig. 9.8).

Beyond Asia, MEA and LATAM are also potential outsourcing destinations for business process outsourcing (Pena, 2024) (Fig. 9.9).

---

[7] https://www.ox.ac.uk/news/2024-01-29-expert-comment-urbanisation-s-role-climate-crisis-being-overlooked.

[8] India and services outsourcing in Asia | IIM Bangalore (2008).

APAC Outsourcing Selection Criteria

Criterion	Criteria Breakdown	Importance
Language	• English Proficiency • Local Language Proficiency • Accent and Clarity • Communication Skills	High
Government Support	• Tax Incentives • Ease of Doing Business • Infrastructure Development • Data Protection Laws	High
Time Zone	• Proximity to key markets (UK, USA, Australia etc) • 24/7 Support Capability • Follow-the-Sun Model	High
Political & Economic Environment	• Regional Political Tensions • Business-Friendly Regulations • Trade & Investment Policies	Medium
Educational System	• STEM Education • Industry-specific Training • Presence of Globally Recognized Institutions	Low

Fig. 9.8 APAC outsourcing criteria. *Source* https://www.invensis.net/blog/top-countries-for-outsourcing

Globally, the services sector employs nearly half the global workforce (Fig. 9.10). And despite outsourcing matrices accounting for various aspects such as political stability, legislative risk, and natural disaster risk, it is only now that climate risk is starting to make its way into outsourcing risk assessments. What is most interesting, is that the countries being chosen for outsourcing, are particularly vulnerable to climate change, and extreme adverse climate events leave the populations and economies of these countries especially vulnerable, through disruptions in infrastructure, food systems and socio-economic stability.

This means that as countries continue to compete to capture the growing outsourcing market, they must also keep a close eye on climate targets in order to offset emerging climate risks, which would eventually give them an edge over competitors who fail to achieve climate resilience. Mounting global pressure to take stronger climate action means that nations that can prioritize sustainable development and access to clean energy set themselves up for success in the long run. Being able to meet the growing needs of the services sector in a sustainable manner will be an edge over competitors.

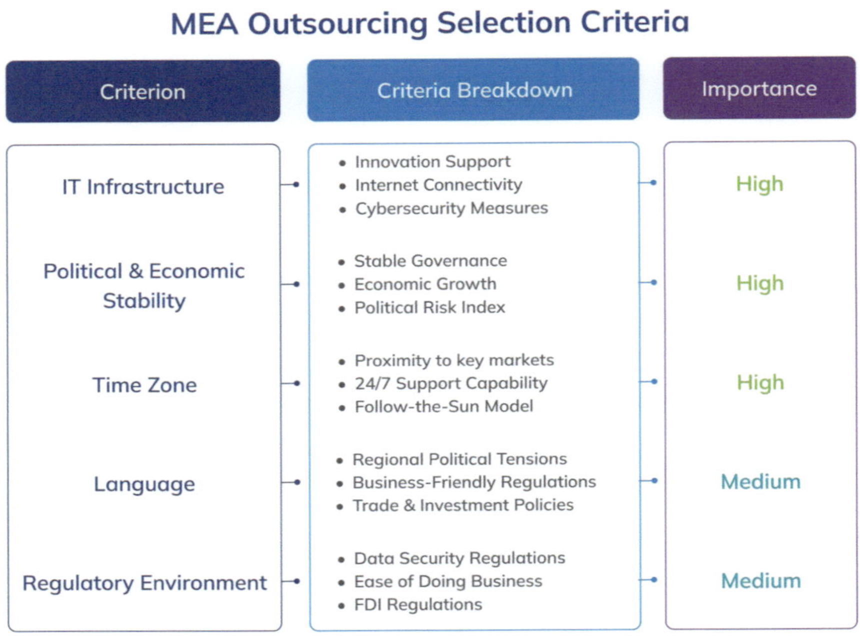

Fig. 9.9 MEA outsourcing criteria. *Source* https://www.invensis.net/blog/top-countries-for-outsourcing

Fig. 9.10 Global share of employment over time. *Source* https://ourworldindata.org/data-insights/the-service-sector-now-represents-about-half-of-employment-across-the-world

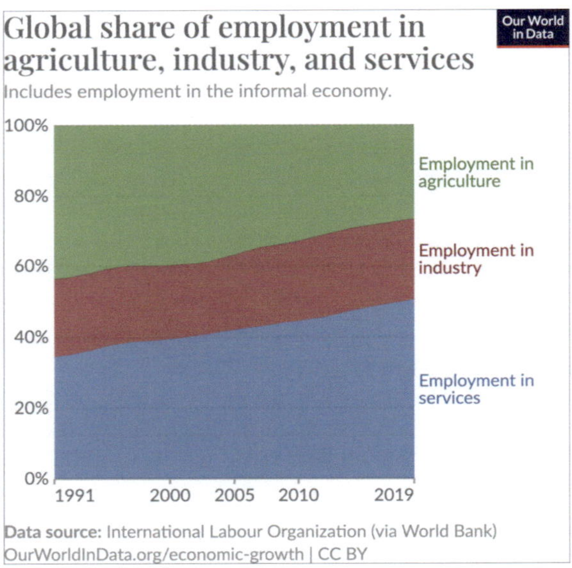

9.6.2 *Threat of New Entrants*

Corporations are value maximising, and are constantly weighing out cost–benefit metrics in order to determine which nations to expand their operations into (Jensen, 2000). Thus, the barriers to new entrants are largely a function of new entrant's willingness to participate in the global services market place, along with the endemic circumstances of the nation's demographics—penetration of language proficiency, skill development, population of young and motivated professionals, and finally trade barriers for goods and services.

Services tend to be homogenized as a product, with labour being highly mobile between organizations hosting similar services—across service types—Information Technology, Financial Services, or Operations. Initially, the barrier for new entrants were at the lowest, over time, as corporations continue to expand the portfolio of services being offshored, new barriers tend to be introduced, over time lowering the threat of new entrants in the traditional sense. The bigger threat however would be improvements to technology and processes that would make some of these services defunct, which will be touched upon in the threat of substitution section.

9.6.2.1 Bargaining Power of Suppliers and Bargaining Power of Buyers

The bargaining power of suppliers in the Industry is quite low—given the primary motivation for the industry is cost minimization, suppliers (of services) are incentivized to keep prices low, and the bargaining power is asymmetrically aligned in the favour of the corporations outsourcing the services to the target countries.[9]

The most that suppliers can do in order to court these businesses is to reduce costs further, or by expanding talent pools and making the case for corporations that the labour market is maturing, and developing further, enticing higher value processes to be offshored into these markets (Mahadevan et al., 2017). India for example has made great efforts to attract outside investment through the development of Special Economic Zones (SEZ), the most significant and ambitious being the setting up GIFT city, a special economic zone dedicated to the incentivization of the Financial Services Industry moving their operations to the state of Gujarat. In the Philippines, the Philippine Economic Zone Authority is tasked with managing SEZs aimed at providing tax incentives and streamlining of bureaucracy of setting up corporate entities within the Philippines.

[9] https://www.infosysbpm.com/blogs/customer-service/customer-satisfaction-through-outsou cring.html.

9.6.3 Threat of Substitution

Any industry structured around a highly specialized, unique and potentially proprietary product, will have high barriers to substitution placed around it. The development of proprietary products requires time, capital, and expertise. And any novel solution being developed thus must improve upon existing solutions, while offering up new advantages, whether significant or otherwise.

While services as a product per se cannot be substituted carte-blanche, it is more meaningful to evaluate the likelihood of substitution of the current "product" offering from incumbent players in the global market place in terms of "how" services are being rendered. And within this context, the threat of substitution poses the greatest threat to the industry as a whole. Technological progress threatens the competitive advantage being offered by the incumbents in the industry. Artificial Intelligence currently targets the lowest level of work, and already has many of the advantages that incumbents in the industry leverage—it can be done remotely, does not require highly specialized skills and has a very low floor for quality. AI as a solution offers significant advantages over the incumbents. It can be deployed at scale with moderate incremental cost, an extremely short implementation timeline (especially compared to the cost, effort and timelines for hiring personnel), can function essentially 24 × 7, and over time, promises to become better and more efficient, with costs eventually scaling further down.

9.6.4 The Climate Effect

With the BPO sector relying on low cost, highly skilled labour in order to be an attractive business strategy, the scale of operations can rapidly lead to aggregate costs and risks which are significant to business stability and reliability. Furthermore, the sector is highly centralized in urban areas, meaning significant supportive infrastructure—high speed internet, sufficiently public and private transport infrastructure, housing, healthcare, and education—must all scale as labour migration progressively concentrates in urban centres. The linkages between urban development and environmental degradation are well established. It is in fact recognized that much of the world's potential environmental challenges will stem from rampant and poorly planned/managed urbanization. The overall environmental impact of urbanization is well documented—ranging from shifting temperatures due to building materials and concentrated emissions. There is also significant impacts on land use patterns and thus the biomes around urban and peri-urban areas (Uttara et al., 2012). Urbanization is also said to impact humidity, winds, dispersion of air pollutant and storms (Qian et al., 2022). The UNEP also recognizes the impact urbanization has on climate change. The IPCC in 2022 found that urban areas are responsible for nearly 70% of global CO2 emissions (Lwasa et al., 2022). Despite strong research highlighting the clearly detrimental impact of urbanization on climate and the environment, the policy response to this information has been lax. There is research indicating a direct correlation between urbanization and various climate change indicators (Maheshwari et al., 2020). Currently, the BPO industry has achieved significant returns on investment by concentrating in urban centres. This has meant that these organizations have access to skilled urban workers, while also benefitting from the superior technological infrastructure made available in urban environments.

Policy however, has lagged behind in recognizing the risks of this concentration of capital, infrastructure, and workforce. Governments in developing nations are aware of the benefits of attracting foreign investment in the form of BPO development. The most common method of attracting organisations is through channels called "Special Economic Zones" or SEZs. These are strategic geographic zones in countries with special allowances on cost management and taxation policy, aimed at incentivizing global organizations to setup shop. While encouraging the growth of the services sector (and the economy), SEZs have an unintended consequence of concentration of employment opportunities. This leads to rapid and highly concentrated urban development in these areas, with population congestion issues leading to strained resources and infrastructure.

9.7 SWOT Analysis for India

India's performance in the services industry marketplace is a function of its intrinsic strengths, weaknesses, opportunities and threats. The SWOT analysis will provide a framework to evaluate these aspects specific to the India context.

9.7.1 Strengths

India has a wealth of strengths that has kept it as the preferred hub for outsourcing services since the 90s. India's core strength comes from its vast wealth of human resources. India has made massive strides in the overall literacy rate, starting at an estimated low point of 17.5% in 1950 and reaching 82% in 2023. During this same time frame, India's population has grown from 346.28 Million to 1.44 Billion (Fig. 9.11).

This has meant that India, across a period of 73 years, India has gone from a literate population of just 60.599 million in 1950, to 1.18 billion in 2023 (Fig. 9.12). While this is an encouraging metric as a standalone, it is much more important to understand what share of this population has at least a tertiary level education. Tertiary level education refers to a collegiate level education, meaning at least a bachelor's degree. This data is especially encouraging, despite not growing at quite the same rate as the primary education share. The share of population who have at least enrolled in a tertiary education was as low as 5% of the qualified population in 1971, which has reached 34.4% of the eligible population in 2024 (Fig. 9.13).

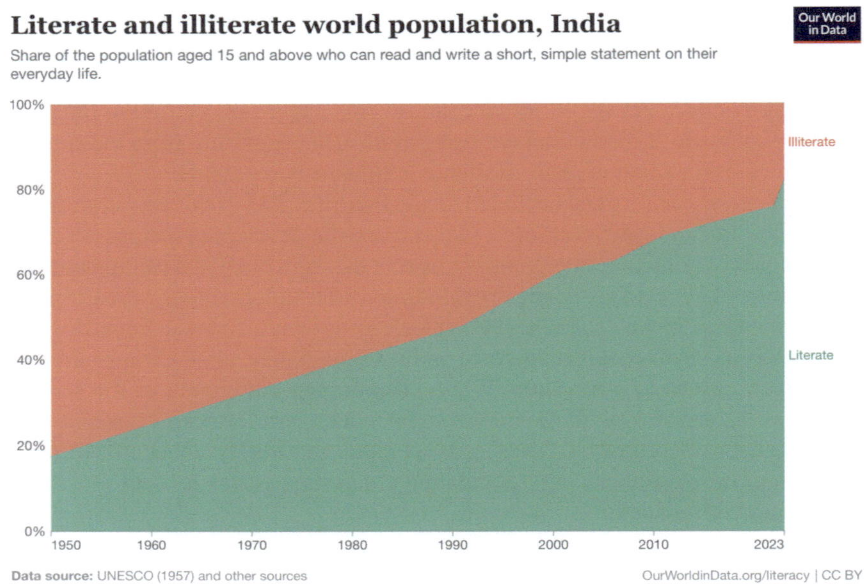

Fig. 9.11 Share of literate population for India—1950–2023. *Source* https://ourworldindata.org/grapher/literate-and-illiterate-world-population?time=1950..latest&country=~IND

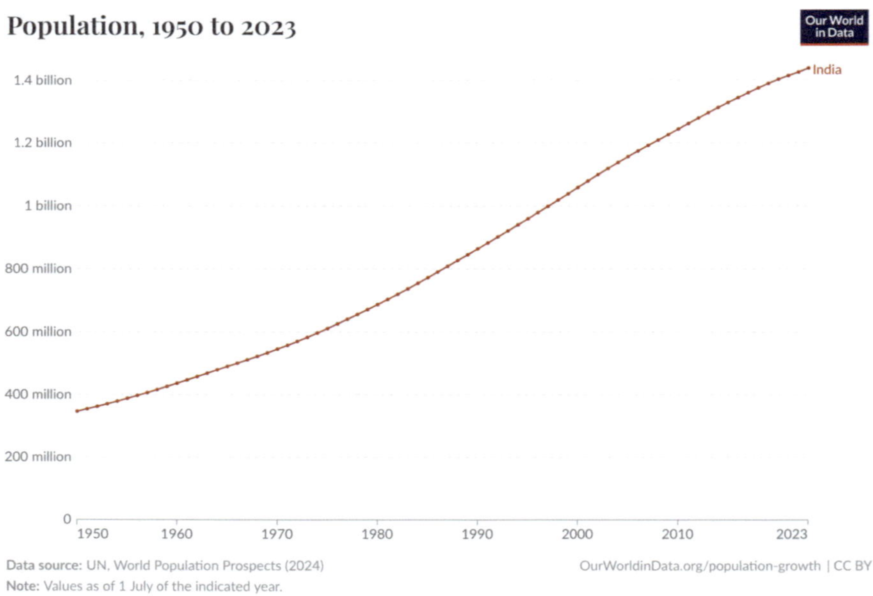

Fig. 9.12 Total population for India 1950–2023. *Source* https://ourworldindata.org/population-growth

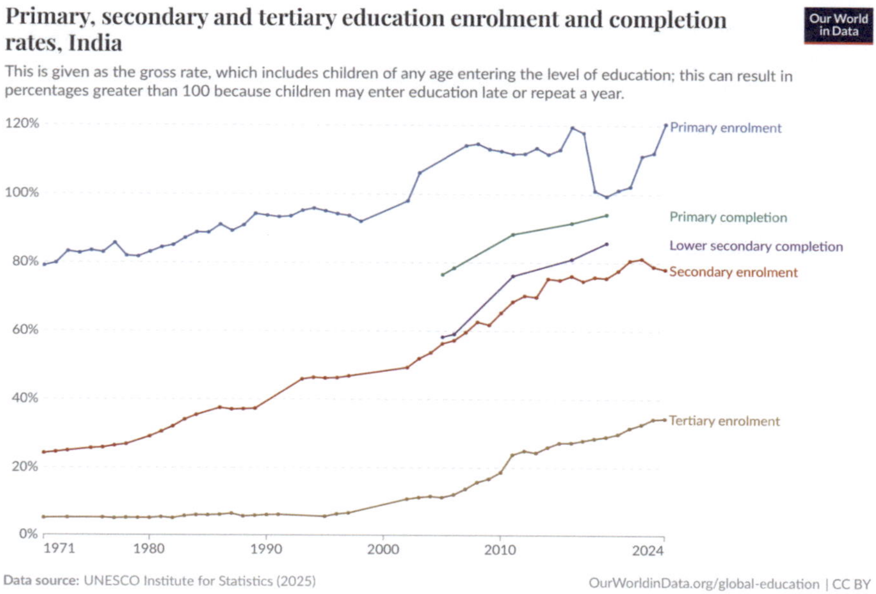

Fig. 9.13 Enrolment and completion rates for different levels of education for India. *Source* https://ourworldindata.org/global-education

Thus, a well-educated and qualified population captured the attention of a global marketplace looking for low-cost, highly qualified workers, bringing an influx of jobs in the services industry. This is further bolstered by the ongoing penetration of urban development, and the rapid expansion of urban infrastructure, although the case can be made that sectoral expansion is greatly outpacing the urban development necessary to support the industry.

India also has a significant policy strength—the government recognizes the importance of India's competitive advantage in the services marketplace, and drafts policy that keeps it competitive. In order to maintain cost leadership, the currency exchange rate with the dollar is carefully managed to keep cost of outsourcing low, despite global inflationary pressures. This has meant that over the years, the rupee has progressively gotten weaker against the dollar, while being balanced out by increased inflow of foreign currency compensation, leading to rising income levels (Fig. 9.14).

State governments across the country also continue to draft policy that incentivise businesses to move their processes to their states and cities. India also continues to have a robust education policy, with private education rapidly expanding access to highly specialized technical education, such as IT services, and software engineering. The expansion is in fact so rapid, that there is now a shortage of capable high quality faculty to meet the large demand for technical education.[10]

[10] https://www.indiatoday.in/education-today/news/story/iits-iims-nits-have-56-faculty-vacancies-says-parliamentary-report-2702244-2025-04-01.

Fig. 9.14 USD INR exchange rate 1975–2025. *Source* https://in.investing.com/currencies/usd-inr

9.7.2 Weaknesses

Most of India's service sector is reliant on the offshoring of business processes developed countries. There is very little internal demand for these services,[11] and the growth of the economy is also catalysed by the currency exchange rate gap. The monetary policy must be carefully managed in order to balance 3 forces—keep outsourcing to India lucrative from a cost standpoint, keep incomes from outsourced jobs at a healthy value (or in other terms, keep a stable value for the GDP contribution from services, instead of a gradual or continuous devaluation), and a firm grip on inflation in the country, as incomes climb thanks to the measures catalysing the first two forces.

An inherent weakness for the industry as a whole, is a lack of push from the supplier side. As such, nations providing the outsourced services keep vying for business from developed nations, and must continue to sweeten the deal in order to keep interest high. This means that Indian service providers need to make themselves more vulnerable to demand pressures, resulting in sub-optimal labour protections, especially in the case of contractual working hours, with over-time, and night shift allowances being weakly enforced in order to maintain the cost leadership advantage.

While the services industry has the greatest contribution to the GDP, the greatest share of the work force is still employed by the agriculture sector, which does not benefit directly from the growth of the services sector, relying on the trickle-down effect instead. However, in the case of the agriculture sector in particular, India continues to closely control food costs in order to protect vulnerable populations. What this means is that the inflationary effect that monetary policy protects the services sector which can cause considerable harm to the agriculture sector. That

[11] https://www.deccanherald.com/opinion/policy-pangs-india-s-services-sector-needs-internal-markets-3341854.

is weakening the real income of those operating in the agriculture sector. Such low food costs cannot be sustained, as it impacts the viability of the farm sector adversely. This limits India as an outsourcing destination, unless the more systemic problems of food security and the agriculture sector are solved. Climate risk further aggravates this risk by amplifying the vulnerability of the agriculture sector and the associated populations. Thus, policy must split its attention between protecting the vulnerable populations in the agriculture sector, and either pivoting development activity into more sustainable practices, or slowing down urban development.

Thus, the policies for nurturing the services sector may indirectly conflict with the policies and goals for continued sustainable development, poverty alleviation and improving food security, not to mention protecting the agriculture sector. The long term developmental goals being in conflict means that policy expansion, and the reliance on overseas businesses, and their continued well-being are significant weaknesses within India's services sector. India's ability to balance these challenging goals, while continuing to solve key poverty, food security, housing and sanitation challenges with its economic development through the services sector is a weakness given the size and scale of India's economy and population.

Finally, a major weakness India faces is in fact a consequence of the high concentration of the services industry in a handful of cities in the country. The services industry currently employs about 30% of the national workforce,[12] and this workforce is primarily concentrated in about 6 cities[13] (Agarwal & Behera, 2023). This puts an extreme strain on urban infrastructure, as systems run the risk of being overwhelmed by mass migration to urban centres of outsourcing. This stress on urban infrastructure and support systems become even more vulnerable during times of adverse climatic events.[14] Even without significant climate events, it is estimated that stressed public infrastructure is leading to large scale productivity losses in corporations concentrating their operations in highly resource stressed cities.

9.7.3 *Opportunities*

Despite the challenging weaknesses highlighted there exist significant opportunities for India's services sector. Despite such a strong presence from the tertiary sector, most of the revenue is generated through contracts with foreign corporations, and while there is a growing local demand for these services, it is greatly outpaced by external demand. Thus one major opportunity would be a direct result of developing increased local demand for these services, tapping into a relatively under-developed market. The challenge here, however, is that the cost advantage presented to foreign

[12] https://www.pib.gov.in/PressReleaseIframePage.aspx?PRID=2098048#:~:text=The%20service%20sector%20also%20provides,and%20post%2Dproduction%20value%20addition.

[13] https://timesofindia.indiatimes.com/india/6-indian-cities-among-top-outsourcing-hubs/articleshow/29226856.cms.

[14] https://thewire.in/urban/is-urban-infrastructure-resilient-to-the-growing-climate-change-risks.

corporations would not be as compelling a benefit for local corporations. This would mean that rather than banking purely on a cost leadership angle, the sector would greatly benefit from marketing itself as vendors of expertise. That is, leveraging the expertise gained from working on projects and solutions to foreign clients, and providing specialized services for local corporations as well. This outcome can be achieved by also nurturing India's growing manufacturing sector—existing policies are already incentivizing local production.

Policy can bridge some of the gaps between the manufacturing and services sectors by drafting policy that incentivize collaboration between the two sectors. While businesses are keen on keeping product design and development activity in-house in order to protect intellectual property (IP), stronger IP laws and guardrails against corporate espionage. This paired with tax benefits for firms that collaborate to drive businesses in both sectors would encourage the local demand for the services sector to grow.

9.7.4 Threats

The threats facing India's leadership in the global services marketplace stem from competitors who can exploit the inherent weaknesses of India's services industry. Particularly, any competitors who can afford to take a more aggressive cost approach, or have a greater base of highly trained, and technically skilled workers, who can present a more lucrative skill set for organisations looking to further outsource higher levels of work would be a strong threat to India's services sector. The Philippines and Mexico are rapidly emerging as significant competitors with the same strengths as India, and China continues to expand its strengths into the services sector at the same time.

However, the biggest threat to the Indian services sector, is one that is threatening services as an industry, globally—Artificial Intelligence (AI). It's already been discussed how the services industry has now culminated in a race to the bottom, and many firms continue to prioritize reduction in the cost that is associated with providing these services, both internally to the firm, and externally to clients. It is now quite evident that corporations are embracing AI as the next big technological leap that will help aggressively reduce costs (Yee, 2024). There is no real defence against the widespread adoption of AI solutions. Over time, it is estimated that complex AI systems can replace human intervention even in the most complex systems and processes. For instance, certain models have already shown great promise in helping identify critical illnesses such as cancer and Alzheimer's in patients, in some cases with a higher efficacy than even veteran doctors (Liao et al., 2023).

AI however, has significant climate impact. While AI positions itself to replace labour costs with a scalable solution, the energy costs for running the models is significant, and rapidly increasing. AI runs on data centres which require significant power and cooling infrastructure in order to remain functional. It is projected that energy use for data centres is expected to double by 2030, exceeding the current total use of energy in Japan today.[15] While nations are quick to develop AI strategies from the perspective of developing their own version of the technology, very little work is being done in global policy in order to assess the environmental impact of the technology, and heading off the climate risks it poses. Thus the advent AI is likely to exacerbate the climate impacts of the service sector.

9.8 Conclusions and Policy Recommendations

The BPO services industry has provided a pathway for economic development that is an attractive alternative to expansion of the manufacturing sector. It promises to be less capital intensive, with shorter gestation periods. It also presents itself as being more ecological through multiple pathways—the most significant being lower emissions from production activity, and the introduction of newer and improved technologies that aid in lowering emissions for other industries as well. However, there can also always be too much of a good thing. Currently, the scale at which the services industry operates, some of these benefits are now beginning to be outpaced by some of the challenges introduced by a significant services sector.

The viability of the sector as a whole is predicated on the ability of businesses to rapidly and reliably deploy certain activities remotely. There is a significant reliance on information technology infrastructure. The success of any outsourcing activity hinges on the target destination's ability to keep these services up and running consistently. The geographic separation of the consumers and producers of this services means that any disruptions would be challenging to navigate at the consumer stakeholder level, and would require remote intervention in order to address challenges. Corporations are left stranded in the case of any significant disruption in services due to circumstances outside of their control. While this risk is monitored and mitigated through industry standard practices such as business continuity plans and assessments, major geographic events always stand to overpower even the strongest planning. Climate change, and specifically adverse severe climate events, introduce significant risk to continuity planning and management. It is also important to note, that risk and cost escalation need not be a result of significant weather events alone. Infrastructure collapse brought about by abuse of resources and improper maintenance threatens the stability of services being rendered in over-burdened urban centres themselves. Mass health outbreaks, disruption in basic amenities, and challenges in access to workplaces or marketplaces due to mismanaged resources can lead to climbing cost of operations, as critical services stand to be delayed, or shutdown altogether. While the industry insists that the activities being offshored are non-essential and low value, progressively, more critical services are being outsourced as costs in home countries continue to rise. As the industry progresses in this direction,

[15] https://www.iea.org/news/ai-is-set-to-drive-surging-electricity-demand-from-data-centres-while-offering-the-potential-to-transform-how-the-energy-sector-works.

the acceptability of any disruption in services for any reason, especially climate, will rapidly decline.

Nations with significant climate and weather risk (typhoons, earthquakes, volcanos, floods and so on) will progressively find less favour with organisations looking to outsource their processes and services. Nations who neglect to keep their infrastructure updated and well maintained will also find less favour once failing infrastructure begins to add to the cost of low down time operation, offsetting some of the economic gains of outsourcing. Monetary policy and labour markets alone cannot keep a country competitive in the global outsourcing marketplace, without sufficient climate resilience and infrastructure development in place. With outsourcing activity (and policies that tend to attract such business) actively contributing to climate change, and in turn operational risk, the industry as a whole faces a catch 22—cutting too many costs can result in significant risk, which in turn increases other costs, such as regulatory or even market share, as customer experience gets impacted.

India now finds itself at the crossroads of sustainable development and growing its economy aggressively. India's base of well-educated, English speaking population positioned it advantageously when services started being outsourced. Indian infrastructure was also quick to react, recognizing the market's demand for technically trained professionals. The education sector and urban development exploded, rapidly expanding to meet the demand of a growing population of job seekers, who wished to work in the now rapidly growing services industry. While initially, the services sector was focused on providing bare-bones customer support services, such as call centres, this growth in highly qualified, technically trained workers meant that other higher level work was quick to follow.

India's key advantages are its robust workforce, and ability to establish itself as a cost leader, striking a delicate balance between maintaining a low cost of operation within India, while keeping incomes and the local economy healthy enough to benefit from the massive influx of foreign capital (albeit in the form of salaries, rather than direct investment for production or sales activity within the country). However, as a result, India's fate is inextricably linked with the prosperity of western corporations, and the overall health of the Western world in general—with outsourced services now forming nearly a third of India's GDP. Thus, it is especially important now to take a critical look at the continued expansion of the services sector, and form legislation that helps de-couple the success of this sector from the success of other businesses globally, in order to make it more resilient to global shocks.

India cannot rely on its resource base and cost leadership alone—other competitors are building similar competencies, and India's advantages need to extend beyond the historical relationship that corporations have with outsourcing in the country. Thus, there are two major areas of concern—an over-reliance on foreign corporations to continue to outsource work to India, and the emerging threat of AI. Both of these challenges can be addressed decisively with strong policy making. India's reliance on the international market to drive the demand for homegrown services means that a large portion of the talent pool finds itself creating value for external corporations, both directly and indirectly. India continues to be a hub for top tier services for the information technology industry, with the financial services industry following closely. However, it is unfortunate that the exchange rate deficit means that remuneration in multinational corporations handsomely outpaces the pay in indigenous corporations, meaning there is little motivation for the best in the country to be engaged directly in local corporations, adding significantly higher value through research and development of technology locally.

Indian legislators are empowered to significantly address this disparity. By incentivizing services and product firms within the country to collaborate more strongly, a dual purpose can be served. Helping reduce the reliance of the services industry on influx of external capital, and helping bolster local businesses to build better products and technologies that can compete and hopefully one day outpace the standards and technologies of corporations in Europe and the US.

As for tackling the AI threat, legislation is key. AI is now throwing regulators and legislators across the world off balance—by taking a strong stance with respect to AI, and defining clear regulatory boundaries for the technology, India can set the global tone and stance on managing this existential threat to employment across various industries. India also stands to be the trailblazer in terms of setting stronger terms of taxation and leveraging the economics of AI platforms towards ensuring increased job security for non-technical fields. As organizations start leveraging AI solutions to replace workers, strong guardrails need to be kept in place in order to ensure the labour market does not collapse. The growth of AI also comes with significant climate and intellectual costs. The computing power required to drive AI solutions is massive, and therefore requires significant energy and thermal management solutions in order to keep the platform functioning. If renewable, low impact solutions to the energy and thermal requirements of AI platforms are not built in at inception, then AI threatens to be the latest in a long list of climate change accelerators, while also harming the labour market, and eventually the economy.

India with its experience in managing the global services outsourcing market, paired with strong leadership in climate change, and developmental policy is uniquely positioned to make significant changes to strengthen itself against the risks plaguing the global services industry. It is crucial that policy and legislation shape the direction of market forces now, rather than the other way around, which is what has been happening so far, particularly in the case of the services sector within India. India has the distinct capacity of building a truly resilient services sector, by improving local demand for services, investing heavily in sustainable infrastructure, and aggressively encouraging de-centralization of urban development and job creation in the services sector. Strong policy and thoughtful leadership are critical to achieving this outcome. Further, there is very little work being done on the sustainability of India's economic development—more research on the space is crucial to framing well informed policy that can set a strong foundation for the future.

References

Agarwal, S., & Behera, S. R. (2023). Spatial concentration of Indian Service Industries in rural and urban areas: A micro-unit-level analysis. *Cogent Economics & Finance, 11*(2), 2270231. https://doi.org/10.1080/23322039.2023.2270231

Bade, K. J. (Ed.). (2003). Migration during the shift from agrarian to industrial societies. In *Science of the total environment. Migration in european history.* (1st ed., pp. 1–52). Wiley. https://doi.org/10.1002/9780470754658.ch1

Bengaluru sees massive traffic jam, thousands left stranded on flyover for around 3 hours. Watch viral video. (2024 October 24). *The Economic Times.*

Beverelli, C., Breitkopf, M., Fiorini, M., & Tresa, E. (2025). *Services trade and environmental sustainability*. https://cepr.org/voxeu/columns/services-trade-and-environmental-sustainability.

Biscontini, T. (2025). *Low-cost country sourcing (Cheap Labor).* https://www.ebsco.com/research-starters/social-sciences-and-humanities/low-cost-country-sourcing-cheap-labor.

Borcuch, A., Piłat-Borcuch, M., & Świerczyńska-Kaczor, U. (2012). *The influence of the internet on globalization process.*

Chanda, R. (2008). India and services outsourcing in Asia. *The Singapore Economic Review, 53*(03), 419–447.

Fourcroy, C., Gallouj, F., & Decellas, F. (2012). Energy consumption in service industries: Challenging the myth of non-materiality. *Ecological Economics, 81*, 155–164.

Ghose, A. (2021). Structural change and development in India. *Indian Journal of Human Development*. https://doi.org/10.1177/09737030211005496

Greenford, D., Horen, T., Crownshaw, C., Lesk, K. S., & Matthews, H. D. (2020). Shifting Economic activity to services has limited potential to reduce global environmental impacts due to the household consumption of labour. *Environmental Research Letters, 15*(6), 064019. https://doi.org/10.1088/1748-9326/ab7f63

Jensen, M. C. (2000). *Value maximization and the corporate objective function.*

Liao, J., Li, X., Gan, Y., Han, S., Rong, P., Wang, W., Li, W., & Zhou, L. (2023). Artificial intelligence assists precision medicine in cancer treatment. *Frontiers in Oncology, 12*, 998222. https://doi.org/10.3389/fonc.2022.998222

Livemint. (2023). Bengaluru: Massive Traffic Jam on Tech Corridor on September 27. *Today News.* https://www.livemint.com/news/india/bengaluru-massive-traffic-jam-on-tech-corridor-on-september-27-netizens-furious-2-km-in-1-hour-11695875022701.html.

Lwasa, S., Seto, K. C., Bai, X., Blanco, H., Gurney, K. R., Kılkış, Ş., Lucon, O., Murakami, J., Pan, J., Sharifi, A., & Yamagata, Y. (2022). *Urban systems and other settlements.* IPCC Sixth Assessment Report. https://www.ipcc.ch/report/ar6/wg3/chapter/chapter-8/

Mahadevan, B., Hazra, J., & Jain, T. (2017). Services outsourcing under asymmetric cost information. *European Journal of Operational Research, 257*(2), 456–467. https://doi.org/10.1016/j.ejor.2016.07.020

Maheshwari, B., Pinto, U., Akbar, S., & Fahey, P. (2020). Is urbanisation also the culprit of climate change?—Evidence from Australian cities. *Urban Climate, 31*, 100581. https://doi.org/10.1016/j.uclim.2020.100581

Martin, R. (2024). Racing to the bottom. https://rogermartin.medium.com/racing-to-the-bottom-ce2fc7bf2240.

Pant, S., Sharma, P., Gupta, A., & Agrawal, P. (2024). Identifying potential service sub-sectors: Insights from GVA, exports, and employment data. In *NITI Aayog working papers.*

Pena, D. (2024). *What US Tech Companies should consider when outsourcing to Latin America.* https://www.forbes.com/councils/forbesbusinesscouncil/2024/12/11/what-us-tech-companies-should-consider-when-outsourcing-to-latin-america/.

Peters, J. (2025). *Duolingo will replace contract workers with AI*. https://www.theverge.com/news/657594/duolingo-ai-first-replace-contract-workers

Qian, Y., Chakraborty, T. C., Li, J., Li, D., He, C., Sarangi, C., Chen, F., Yang, X., & Ruby Leung, L. (2022). Urbanization Impact on regional climate and extreme weather: Current understanding,

uncertainties, and future research directions. *Advances in Atmospheric Sciences, 39*(6), 819–860. https://doi.org/10.1007/s00376-021-1371-9
Reddy, V. R., Rahut, D., & Sonobe, T. (2025). *The future of smallholder farming in Asia.* Asian Development Bank Institute.
Rosenblum, J., Horvath, A., & Hendrickson, C. (2000). Environmental implications of service industries. *Environmental Science & Technology, 34*(22), 4669–4676. https://doi.org/10.1021/es9914083
Roberts, S. H., Foran, B. D., Axon, C. J., & Stamp, A. V. (2021). Is the service industry really low-carbon? Energy, jobs and realistic country GHG emissions reductions. *Applied Energy, 292.*
Sutthichaimethee, P., & Sawangdee, Y. (2016). Modeling environmental impact of service sectors to promote sustainable development of Thailand. *Ethics in Science and Environmental Politics, 16*(1), 11–17. https://doi.org/10.3354/esep00169
Thakur, A. (2023). *Bengaluru loses nearly Rs 20,000 crore due to traffic mess: Study.* https://www.ndtv.com/india-news/bengaluru-loses-nearly-rs-20-000-crore-due-to-traffic-related-hindrances-study-4275195
UN Trade and Development. (UNCTAD). (2014). Trade balance of developing and developed countries continues to converge. UNCTAD statistics show. https://unctad.org/news/trade-balance-developing-and-developed-countries-continues-converge-unctad-statistics-show
Uttara, S., Bhuvandas, N., & Aggarwal, V. (2012). Impacts of urbanisation on environment. *IJREAS, 2.*
Yee, L. (2024). *Commentary on AI.* http://ceros.mckinsey.com/commentary-ai-2024-2-lareina-yee

Suggested Reading

Bai, X., et al. (n.d.).*Chapter 8: Urban systems and other settlements.* Retrieved August 19, 2025, from https://www.ipcc.ch/report/ar6/wg3/chapter/chapter-8/
Das, S., Choudhury, M. R., Chatterjee, B., Das, P., Bagri, S., Paul, D., Bera, M., Dutta, S., Das, S., Choudhury, M. R., Chatterjee, B., Das, P., Bagri, S., Paul, D., Bera, M., & Dutta, S. (2024). Unraveling the urban climate crisis: Exploring the nexus of urbanization, climate change, and their impacts on the environment and human well-being—A global perspective. *AIMS Public Health, 11*(3), 963–1001. https://doi.org/10.3934/publichealth.2024050
Dwivedi, Y. K., Hughes, L., Kar, A. K., Baabdullah, A. M., Grover, P., Abbas, R., Andreini, D., Abumoghli, I., Barlette, Y., Bunker, D., Chandra Kruse, L., Constantiou, I., Davison, R. M., De', R., Dubey, R., Fenby-Taylor, H., Gupta, B., He, W., Kodama, M., … Wade, M. (2022). Climate change and COP26: Are digital technologies and information management part of the problem or the solution? An editorial reflection and call to action. *International Journal of Information Management, 63*, 102456. https://doi.org/10.1016/j.ijinfomgt.2021.102456
Gallé, J., Overbeck, D., Riedel, N., & Seidel, T. (2024). Place-based policies, structural change and female labor: Evidence from India's Special Economic Zones. *Journal of Public Economics, 240*, 105259. https://doi.org/10.1016/j.jpubeco.2024.105259
https://www.unep.org/explore-topics/resource-efficiency/what-we-do/cities-and-climate-change
https://www.ox.ac.uk/news/2024-01-29-expert-comment-urbanisation-s-role-climate-crisis-being-overlooked
Murshed, M., & Saadat, S. Y. (2018a). Effects of urbanization on climate change: Evidence from Bangladesh. In *SSRN Scholarly Paper No. 3247302.* Social Science Research Network. https://papers.ssrn.com/abstract=3247302
Murshed, M., & Saadat, S. Y. (2018b). Effects of urbanization on climate change: Evidence from Bangladesh. In *SSRN Scholarly Paper No. 3247302.* Social Science Research Network. https://papers.ssrn.com/abstract=3247302

Ortega, A. A., Acielo, J. M. A. E., & Hermida, M. C. H. (2015). Mega-regions in the Philippines: Accounting for special economic zones and global-local dynamics. *Cities, 48*, 130–139. https://doi.org/10.1016/j.cities.2015.07.002
Punyamurthy, C., & Bheenaveni, R. (2023). Urbanization in India: An overview of trends, causes, and challenges. *International Journal of Asian Economic Light*, 9–20. https://doi.org/10.36713/epra12473
Sethi, S. S., & Vinoj, V. (2024). Urbanization and regional climate change-linked warming of Indian cities. *Nature Cities, 1*(6), 402–405. https://doi.org/10.1038/s44284-024-00074-0
Sharma, V. (2025). Urbanization Induced climate change: Impacts and sustainable solutions. *Asian Research Journal of Arts & Social Sciences, 23*(5), 135–142. https://doi.org/10.9734/arjass/2025/v23i5688
Tafese, T., Lay, J., & Tran, V. (2025). From fields to factories: Special economic zones, foreign direct investment, and labour markets in Vietnam. *Journal of Development Economics, 174*, 103467. https://doi.org/10.1016/j.jdeveco.2025.103467
Wei, M. (n.d.). *Climate change and business continuity management: What should we be looking for?* Retrieved August 19, 2025, from https://blog.bcm-institute.org/blog/climate-change-bcm
Zarghamipour, M., & Malakooti, H. (2025). The projected effects of urbanization and climate change on urban Heat Island and thermal comfort over the Tehran metropolitan. *Science of the Total Environment, 992*, 179955. https://doi.org/10.1016/j.scitotenv.2025.179955

Part IV
Social and Cultural Dimensions

Chapter 10
Impact of Climate Change on the Traditional Practices of Changpa Tribal Community: A Case Study of Tsomoriri Lake, Leh (Ladakh)

Maansi Malik and S. C. Rai

Abstract Wetlands and their associated ecosystem services are severely threatened by the impact of climate change and global warming which can be detrimental for biodiversity and ethnic communities. The Tsomoriri Lake is a High-Altitude Wetland located at an altitude of more than 4500 m in the Changthang Plateau, the western extension of the Tibetan Plateau. It has been declared as a Wetland of International Importance and is protected under the Ramsar Convention due to it's unique and endemic faunal diversity. Changpas, the ethnic nomadic tribes have been inhabiting the Changthang Plateau since time unknown to human history. These communities practice pastoralism as their primary source of occupation and have been utilizing the lake pastures as fodder for their livestock. The risk of glacial melting and long-term fluctuations in temperature and rainfall is perceptible and experienced by these communities through changes in their migratory patterns, shortage of water, declining availability of fodder, herd-size composition etc. Recent developments such as expansion of tourism and accompanied infrastructural developments exaggerates these risks due to increased anthropogenic pressure. This paper is an attempt to explore this entwined complicated relationship and its long-term impacts on traditional practices which are a part of our intangible heritage.

Keywords Tsomoriri lake · Changpas · Climate change · Anthropogenic pressure

M. Malik (✉) · S. C. Rai
Department of Geography, Delhi School of Economics, University of Delhi, New Delhi, India
e-mail: malikmaansi@gmail.com

S. C. Rai
e-mail: raisc1958@rediffmail.com

P. Datta et al. (eds.), *Climate Change and Human Well-being*, Advances in Global Change Research 83, https://doi.org/10.1007/978-3-032-25479-5_10

10.1 Introduction

Climate change refers to a change in the state of climate that can be identified by changes in the mean values and/or variability of its properties and that persists for an extended period, typically decades or longer. Climate change may occur due to natural internal processes or external forcings such as modulations of the solar cycles, volcanic eruptions, and persistent anthropogenic changes in the compositions of the atmosphere or in land use (Allen et al., 2018). Moreover, the Framework Convention on Climate Change (UNFCCC) defines climate change with respect to direct and indirect human activity that alters the composition of global atmosphere in addition to the natural climate variability observed over comparable time periods. Climate change is anticipated to bring about irreversible losses to mountain regions and associated ecosystems, affecting livelihoods and cultural identity. Approximately 15% of the global population inhabits these regions with 90% of them concentrated in the developing countries (Kapruwan et al., 2025).

High-Altitude Wetlands (HAWs) are an important part of the Trans-Himalayas and some of the largest wetlands are located here. There is no credible definition for these High-Altitude Wetlands (HAWs) but a defining altitude limit has been applied to this category of wetlands. By definition, High-Altitude Wetlands as a generic term is used to describe areas of swamp, marsh, meadow, fen, peatland or water located at an altitude above 3000 m, whether natural or artificial, permanent or temporary (Panigrahy et al., 2012). The natural environment is harsh with extreme climate with precipitation and air temperature being integral abiotic factors in such ecosystems. They are effective carbon sinks and thus have other impressive ecological, cultural and spiritual, hydrological and provisioning functions such as flood regulation, harbouring valuable biodiversity, source of food and fodder, water etc. Ethnic communities inhabiting these areas depend on these systems for livelihood purposes. These wetlands act as headwaters for several major rivers and thus feed the essential water supply requirement of downstream regions (Chatterjee et al., 2010). These regions are also important areas of tourism due to their recreational and cultural value which provides an additional source of income for the local communities. Therefore "ecosystem services" provided by these wetlands add immense value and functionality. It is clearly evident that these ecosystems are extremely fragile and yet have several vital functions and their preservation and maintenance is integral to wetland conservation.

Climatic stress can pose a significant challenge to such high-altitude environments characterised by a delicate balance where a minor interference can have severe consequential implications. Anthropogenic interference, acts as the single most important challenge in these areas which infringes the ecological balance of these ecosystems. Expansion of tourism and related infrastructural developments compounds the severity of climatic stress and unfolds a complicated scenario requiring immediate attention. Numerous studies have highlighted the impacts of climatic stress in cryosphere environments of varied nature. However, several gaps continue to exist due to sparse network of stations providing climatic data which is inherent with variation across organisations. Gupta and Shukla (2016) conducted a land-use/cover change assessment of Tsomoriri Lake and analysed climatic parameters to corroborate the impact of climate change and anthropogenic activities in the region based on data derived from India Water Portal. They pointed a persistent warming trend in the lake region and a declining precipitation in Ladakh region. Chevuturi et al. (2018) utilized different climatic datasets and observed a similar trend with respect to temperature and fluctuating precipitation levels increasing somewhere and decreasing elsewhere. Chatterjee et al. (2010) highlighted adverse hydrological impacts on high-altitude lake ecosystem due to warming in the form of degraded ecosystem functioning and altered biochemistry. Barrett and Bosak (2018) integrated household survey with geo-spatial analysis to study climate change perception among the local population of Ladakh showing moderate to low awareness amongst the locals about the risks of climate change with strong evidences of erratic climate patterns, receding glaciers and altered seasonality and declining water availability in the region. Tashi and Sudan (2021) on the other hand raised a strong awareness among the local indigenous people about fluctuating temperature and rainfall which leads to land degradation, water crisis, low agricultural property, increased incidence of natural disasters, and decline in forest resources. This has impacted cropping patterns and techniques such as mixed farming, organic farming and more plantation of trees to adapt to the changing climate.

Mountain communities are therefore, particularly vulnerable to the impact of climate change and climate variability due to possession of limited resources to cope in extreme environments. Their livelihoods and everyday life are inextricably linked to the natural environment with nature playing a defining role in their occupational choices. The Changpa herders are one of the several communities inhabiting the hostile environment of the Changthang, the western extension of the Tibetan Plateau in India. This study is focused on the Changpas of Korzok village, located on the western bank of Tsomoriri Lake. These semi-pastoral communities depend on pastures surrounding the Tsomoriri Lake for grazing in the summer season and also use these for cultivation of potato, spinach, barley hay etc. which is essentially subsistent in nature. They do not use the lake water for consumption as the water is brackish is nature but instead depend on a perennial spring originated through melting of snow. Highland glaciers are therefore, intrinsic for meeting the water requirements of the local community. The Changpas practice several traditional adaptive strategies and mechanisms to efficiently utilize the scarce resource base in the region which has helped them sustain in the harsh and extreme environment of Changthang. However, several social, political, economic and ecological factors have altered the local ecological practices and ecology of the landscape from time to time. Several waves of transformation have altered the landscape and the Changpa community that has remained the historical custodians of the natural resources no longer enjoy the same position. The Indian Independence of 1947, annexation of Tibet by China in 1951, Sino-Indian conflict of 1962 and the more recent opening of Tsomoriri Lake to tourism in 1993–94 and an inroad to modern ideas and beliefs have completely modified the ecological-social-economic landscape of the region. The increased awareness about the threats of global warming and climate change to mountain communities has added to the vagaries of a system already experiencing a transition. This study is essentially focused on:

1. understanding the climatic stress in the region and it's impact on the local ecology with particular emphasis on hydrology and natural vegetation.
2. establish a relationship between changing socio-economic, political and ecological conditions in the region and it's impact on the traditional practices and customs of the Changpa tribal community.

There exists a significant gap here as limited studies on Tsomoriri Lake have focused on climate change, related stresses and related socio-ecological and economic implications for the ethnic communities. Lack of data, expensive travel, language barriers and seasonal accessibility are common factors that can inhibit research in such difficult terrains. This study is an attempt to highlight the exigencies of mountain environments and communities. These communities are under a constant state of change that is affecting their cultural heritage and local practices. This intangible heritage based on coexistence and harmony needs to be preserved rather than being abandoned under the influence of continuous change.

10.2 Study Area

10.2.1 Physical Characteristics

The Tsomoriri Lake is a High-Altitude Wetland located at an elevation of 4595 m above mean sea level and lies between 32°54′ N latitude and 78°18′ E longitude in Trans-Himalayas. Administratively it is located in the Rupshu Tehsil of Leh district in the Union Territory of Ladakh (Fig. 10.1). Geographically, the Tsomoriri Lake is located in the south eastern part of Ladakh which is also called the Changthang Plateau *(byang-thang)* or Northern Plains. It represents the south-western extension of Tibetan Plateau in India nestled in Eastern Part of Ladakh. Some of the most important High-Altitude Wetlands in the Himalayas—the Pangong Tso and Tso Kar Wetland complex are located here. The Tsomoriri Lake is depression bounded by rolling hills on all sides. It is bounded by the Korzok Range on its western side with Mentok Kangri I (6250 m) and Mentok Kangri II (6210 m) as the highest peaks. Studies point that the lake is tectonic in origin and is a remnant of the Ice Age formed due to melting of ice masses. It is a brackish oligotrophic lake spread over an area of 120 km^2 with a maximum depth of 40 m. It has a north–south length of 27 km and east–west extent of 5–7 km. The lake formerly had an outlet to the south but is now land-locked. It is fed by two major streams entering the lake—Gyoma from the north and the Phirse Phu in south-west (Fig. 10.2).

Average annual rainfall in the lake varies from 100 to 300 mm and temperature varies from 0 to 30 °C in the summers and − 40 to − 10 °C in the winters (Panigrahy, 2012) and is affected by both the Indian summer monsoons and winter westerlies.

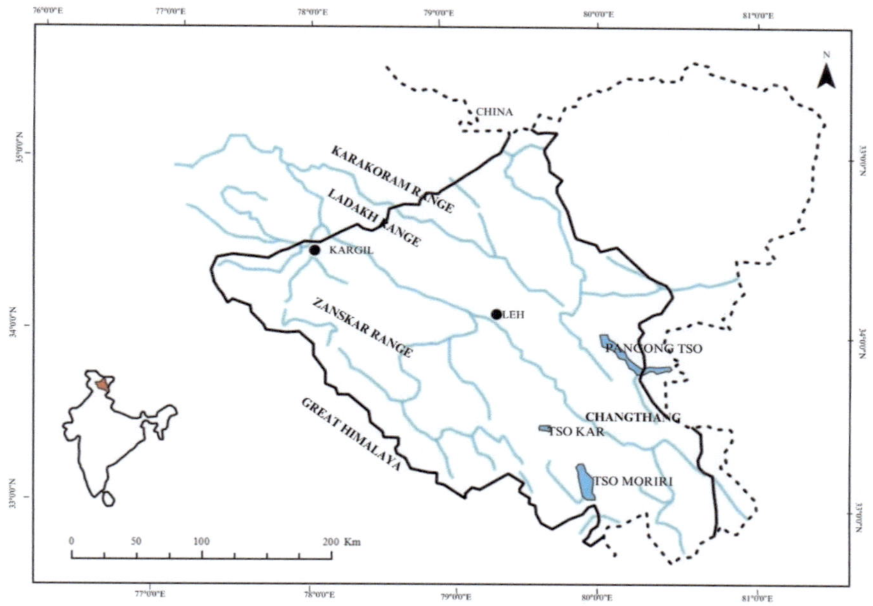

Fig. 10.1 Locational setting of Tsomoriri Lake in Leh

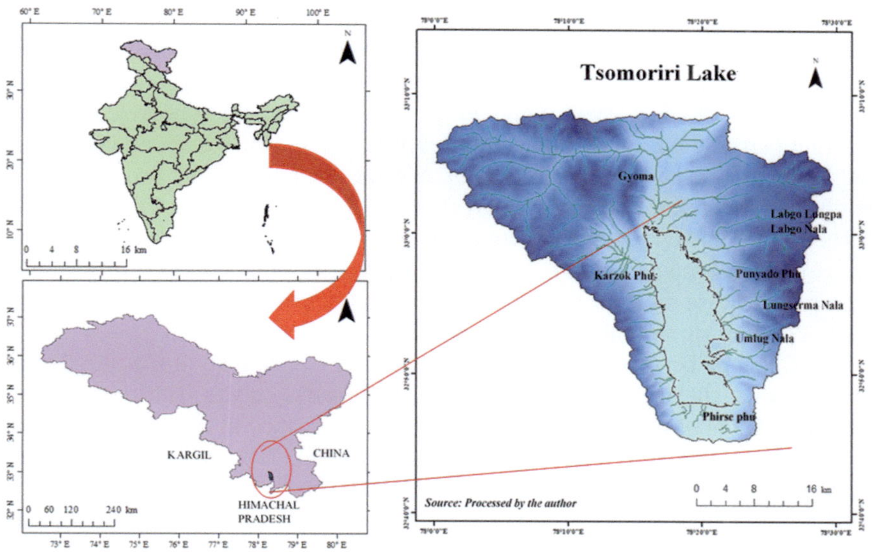

Fig. 10.2 Location and Administrative Division, Tsomoriri Lake Leh

The region is characterized by an arid steppe environment with low primary productivity and supports a scarce presence of grasslands. The Tsomoriri Lake has been declared as a Wetland of International Importance under the Ramsar Convention since 2002 due to its uniqueness being one of the highest in the world and it's unique faunal assemblages with high diversity, endemism and hosting a wide variety of flora and fauna including the Black-Necked Crane (*Grus nigricollis*), Bar-Headed Geese *(Anser indicus).* The Tsomoriri Lake is the only breeding ground for the Bahr-Headed Geese in India and the only breeding ground in the world for the Black-Necked Crane outside China. Other waterfowl species include Ruddy Shelduck *(Tadorna ferruginea)*, Brown Headed gull *(Larus brunnicephalus)*, Common Tern *(Sterno hirundo)*, Common Pochard *(Aythya fernia)*, Great Crested Grebe *(Podiceps cristatus)* Great crested Grebe *(Podisceps cristatus)* etc.Several species of mammals include the Blue Sheep or Bharal (*Pseduois nayur)*, Tibetan Argali or Nayan *(Ovis ammon hodgsoni*), Tibetan Wild Ass or Kiang *(Equus kiang kiang)*, Himalayan Marmot (*Marmota himalayana)*, Red Fox *(Vulpes Vulpes)*, and Tibetan Gazelle *(Procapra picticaudata)* (Anon, 2002).

10.2.2 Cultural Characteristics

The Changthang Plateau or the Northern Plains and the surrounding areas are inhabited by an indigenous tribal community called the Changpas or the Northerners also locally referred to as the *"drokpas"*. They are believed to be of Tibetan descent having migrated from the Zhang Zhung province of Tibet and have a nomadic lifestyle (Chaudhuri, 1988; Jina, 1999; Singh et al., 2013). In Ladakh they are confined to different part of the Rupshu valley with several of them being concentrated around the Tsomoriri-Tsokar Wetland Complex and the Pangong Lake, two most noteworthy High-Altitude Wetlands (HAWs) of the Trans-Himalayas. Their economy has been traditionally pastoral in nature and they rear a variety of livestock including sheep, yak, horses and pashmina or "*changra*" goats. Pashmina Goats (*Capra hircus*) are a prized possession for these herders as they yield what is also known as soft gold—"pashm" or pashmina fiber which is highly valued globally.

The Changpa herders brave extremely cold and arid climate to rear these goats which produce the finest quality of pashmina fiber native only to such extreme environments. Trade of pashmina has been a historical practice in the Changthang and it was initially traded to shawl makers from Kashmir via Ladakhi middlemen by Western Tibet. The pashmina derived from Ladakh was then considered inferior but the annexation of Tibet in 1951 completely sealed it's borders for any form of trade. Prior to this the Changpas were also dependent on the lakes in Tibet for salt. After the annexation, there was sudden surge in the demand for pashmina from South Eastern Ladakh which caused an exponential rise in the price of pashmina fibre.

The Changpa community practice transhumance and migrate to low-lying areas in the summer and move back to their villages in the highlands in winters. Herders around the Tsomoriri Lake are settled in the Korzok hamlet, which is one of the highest settlements in the world and is also known for it's 400-year old Korzok monastery. The other hamlets which constitute the Korzok village–Tegazong, Chummur, Angchuk and Sumdo continue to practice the migratory lifestyle. Herders from highland settlements Tegazong and Chummur all migrate to the regions in and around the Tsomoriri Lake and use the pastures around the Tsomoriri Lake for grazing purposes only to return back in winters. The indigenous Changpa tribe of the Korzok hamlet now lead a sedentary lifestyle and no longer migrate. They use pastures in and around the Tsomoriri for summer grazing and are dependent on stocked feed, government supplies or cultivated barley hay in the winters. Prior to the Sino-Indian Conflict in 1962 all Changpa herders of Korzok moved to the Skagjung pastures in winters located in Tibet which now lies in the Chinese territory. The war modified the migratory routes of the Changpas as they lost access these pastures which were lush green, and devoid of snow in winters as it was comparatively lower in altitude and was sufficient enough.

The Changpas have traditionally employed conscious mechanisms that have helped them cope the hardships of a pastoral life in an extreme environment This includes manipulation of resources through appropriate livestock management strategies via species diversification; raising a variety of livestock to ensure diversity availability of products for consumption and trade, herd disposal; slaughtering of animals that have outlived their productive potential. Species diversification ensures herd maximization i.e. realization full potential as they are not only a source of income but also provides several by-products. The traditional economic system of the herders was initially based on the principle of barter which was ecologically sustainable and ecologically tenable in the hostile Changthang environment. The Changpas shared stable trading relations with several partners in Tibet, Himachal Pradesh and the Zanskar Valley and exchanged products like butter, cheese, pashmina, wool, lambskin, hides, livestock, salt and small amounts of borax and sulphur for items that were not available in the region (Bijoor et al., 2024). Barter anthropologically has been a crucial adaptation strategy for mountain communities where one community is symbiotically linked to the other through goods available in their ecological zones. This helps prevent sabotage of resources and ensures resource maximization. The Changpas also possess and have a strong awareness about the rich biodiversity of the region. They possess a vast knowledge about the vegetation phenology of the area and determine the quality of the pasture based on it's composition. They also utilize these for their local health care practices. Gautam et al. (2007) reported the use of more than 300 varieties of medicinal plants in Ladakh for Tibetan medicine. According to the herders, the commonly found grasses found in Tsomoriri include *Bol, Shyol, Longma, Nyalo, Gyapshen, Burtze, Tolo, Trama* etc. with *Gyapshen* and *Trama* being the most abundant. *Nema* is the rarest variety of grass found in the region. Tolo is the most preferred variety and is also reportedly declining according to the Changpas. They attributed this qualitative decline to the fluctuating precipitation and vehicular movement on the pastures. Some particular species of grass found in the region have special characteristics according to the local tradition and consumed seasonally or completely avoided. The *chilse* is a typical variety found closer to the sweet water of the streams but it's consumption can be fatal for the livestock. *Lachu* is a variety grown in the month of April–May and can only be consumed after a flash of rainfall otherwise it can make the livestock blind. (Gautam et al., 2007). There are no scientific reasons to back these claims but are being practiced as a local tradition since ages.

The pastoral lifestyle of these communities is well suited to the unique ecology of the Tibetan Plateau and their age-old customs imitate the technological practices of the Western rangeland managers such as rest/rotation, deferred grazing, monitoring of plant and animal performance. The Changpa nomads with implicit knowledge of the local ecology and natural environment have survived harsh climate and have been managing their pastures with mobility governed by strict regulations enforced by the religious head or the *"goba"*. He is responsible for determining the rules of pasture usage and allocation to the Changpas for grazing, dates for movement and migration, fines, punishments etc. These rules are agreed upon by the Changpas and cannot be violated under any circumstances (Bijoor et al., 2024). Religion thus, had a strong influence on the customary pastoral practices of the Changpa nomads. The Changpas were followers of an animistic religion called *"Bon"* and later adopted Buddhism of the *Kargyud-pa* sect. These religions strongly promote the idea of harmony and co-existence with nature which is strongly reflected in cultural practices of the Changpas. These communities inhabit a harsh terrain with limited resources and have live in harmony with their natural environment without exerting any pressure. This strongly a reflection of resilience of Buddhist traditions which embraces sustainability amidst limited resources. Natural resources are thus considered sacred and revered as a matter of religious belief and custom. Slaughtering of animals which have outlived their productive value is however permitted as an adaptation mechanism to prevent wastage of resources.

Marriage and kinship in a Changpa household was traditionally polyandrous i.e. two brothers married the same women which led to large family sizes. This ensured consolidation of property and provided manpower essential to manage the herds. Herd population in a Changpa household have traditionally been large and required more people to take care of the herds. A polyandrous marriage prevented division of the herds and the household. A polyandrous marriage was thus, considered as a traditional method of population stabilisation in an area where resources were scant and had to be managed efficiently. Polyandry is no longer practiced by the younger generation and was also legally banned in 1941 by the Ladakh Buddhist Polyandrous Marriages Prohibition Act. More and more Changpa youth are adopting monogamy as they no longer associate marriage as a survival strategy but equates it with companionship (Bhasin, 2012). Correspondingly, the availability of new economic opportunities has created more disillusionment with polygamy amongst the youth (Namgail, 2015).

10.3 Database and Methodology

The study is based on an integrated statistical and geo-spatial analysis primarily focused on assessment of various climatic parameters and statistically validate the nature of climatic stress in the region. Satellite Data used to conduct the geo-spatial analysis mentioned in Table 10.1. The false colour composite images were created after appropriate stacking of select bands. This data obtained through different satellite sensors were then geo-rectified to a common projection and radiometrically corrected to enhance quality through Histogram Equalisation. Later in order to extract the watershed of the lake, to be used as the study area for analysis the Hydrology function of the Spatial Analysis Tool were used in Arc Map 10.1. SRTM-DEM data was used for this purpose.

This data was finally used to determine the extent of vegetation, snow cover and water body area in the region using spectral indices. A Spectral index is a multi-band method based on band-ratio approach also called the normalized index where usage of different band combinations highlights essential information about surface features as all Earth features respond differently to varied wavelengths of electromagnetic radiation. Band ratios such as Normalised Difference Vegetation Index (NDVI), Normalised Difference Water Index (NDWI) and Normalised Difference Snow Index (NDSI) have been used to map the vegetation, water body extent and snow cover in the region for three time periods 1990, 1998 and 2023. 1990 represents the year before Tsomoriri Lake was opened for tourism, 1998 represents the transitional year and 2023 is the recent year.

The NDVI is sensitive to sub-pixel vegetation and takes advantage of greater reflectance of vegetation in the infrared band (Townshend & Justice, 1986).

Table 10.1 Details of satellite data

S. No.	Satellite	Year	Sensor	Date of acquisition	Path/Row	Band Combinations
1	Landsat 4–5	1989	Thematic Mapper	02.08.1990	146/37	4,3,2
2	Landsat 4–5	1998	Thematic Mapper	24.08.1998	146/37	4,3,2
3	Landsat 4–5	1998	Thematic Mapper	31.08.1998	147/37	4,3,2
4	Landsat 8	2023	Operational Land Imager	29.08.2023	146/37	6,5,4,3
5	Landsat 8	2023	Operational Land Imager	04.08.2023	147/37	6,5,4,3
6	SRTM DEM	2014	Imaging Radar	23.09.2014	–	–

Source earthexplorer.usgs.gov

$$\text{NDVI} = (\text{NIR}-\text{Red})/(\text{NIR}+\text{Red})$$
$$\text{NDVI} = \text{B}(4)_{\text{TM}}-\text{B}(3)_{\text{TM}}/\text{B}(4)_{\text{TM}}+\text{B}(3)_{\text{TM}}$$
$$\text{NDVI} = \text{B}(5)_{\text{OLI}}-\text{B}(4)_{\text{OLI}}/\text{B}(5)_{\text{OLI}}+\text{B}(4)_{\text{OLI}}$$

The delineation of the lake area is based on the used on NDWI which aims to enhance the visibility of open-water areas in remotely sensed imageries taking advantage of high reflectance of water bodies in the green band (McFeeters, 1996).

$$\text{NDWI} = (\text{Green}-\text{NIR})/(\text{Green}+\text{NIR})$$
$$\text{NDVI} = \text{B}(2)_{\text{TM}}-\text{B}(4)_{\text{TM}}/\text{B}(2)_{\text{TM}}+\text{B}(4)_{\text{TM}}$$
$$\text{NDVI} = \text{B}(3)_{\text{OLI}}-\text{B}(5)_{\text{OLI}}/\text{B}(3)_{\text{OLI}}+\text{B}(5)_{\text{OLI}}$$

Snow cover estimation using Normalised Difference Snow Index (Riggs et al., 1994) is based on high reflectance of snow cover in the visible region and declining reflectance in the infrared region of the electromagnetic spectrum.

$$\text{NDSI} = (\text{Green}-\text{SWIR})/(\text{Green}+\text{SWIR})$$
$$\text{NDVI} = \text{B}(2)_{\text{TM}}-\text{B}(5)_{\text{TM}}/\text{B}(2)_{\text{TM}}+\text{B}(5)_{\text{TM}}$$
$$\text{NDVI} = \text{B}(3)_{\text{OLI}}-\text{B}(6)_{\text{OLI}}/\text{B}(3)_{\text{OLI}}+\text{B}(6)_{\text{OLI}}$$

NDVI Tool of Erdas Imagine Software (2014) with an inbuilt algorithm for used for the purpose of calculation of these spectral indices. Appropriate thresholding was done and extraction of vegetation (NDVI), waterbody (NDWI), and snow cover (NDSI) was performed using the Reclassify tool of ArcMap software (Version 10.1). Waterbody Area was calculated using the Raster Calculator.

This data was supplemented with gridded climate data obtained through NASA MERRA 2 for a period of 42 years from 1981 to 2022. MERRA 2 is a long-term record of global climatic analysis which began in 1980. Temperature data at 2 m (°C) and aggregated rainfall data (mm) was obtained for both annual and monthly cycle for statistical analysis. These hydro-meteorological variables were used to conduct a simple time series analysis using MS-Excel and establish a link between changing snow-cover dynamics, vegetation and lake area. Moving Average over a five-year period and a regression equation for moving average have been plotted to understand the temporal trend and R^2which represents the strength of the relationship between two variables has also been calculated which is also called the Co-efficient of Determination. Moving average has been calculated to smoothen the impact of short-term fluctuations in annual monthly average data. A regression equation is of the form

$$y = mx + c$$

where y and x, represent the dependent and independent variables

m = slope; determining the rate of change in a variable

c = intercept.

The statistical and geo-spatial analysis of climatic parameters was supplemented with a household survey conducted with 50 Changpa households in Korzok Village. Questions to the local community were mainly divided into four categories—Personal information, Occupational Structure and Traditional Practices and Climate Change Perception. The part of traditional practices and Climate change Perception has been analysed in this chapter of Traditional Ecological Knowledge. Informal open-ended interviews were also conducted with 20 rebo settlers habiting around the Tsomoriri Lake. Language was a persistent barrier while conducting the primary survey and assistance was sought from the village headman or "*Lumberdaar*" and a sub-Inspector from 37th Battalion, ITBP. Prior permission was sought from the village chief before accessing the households. In-depth interviews were conducted with the head-monk of the Korzok monastery and the village headman. The data obtained through the primary survey was supplemented through an extensive review of existing research on the traditional practices and adaptation strategies of the Changpas.

10.4 Results

10.4.1 Analysis of Climatic Parameters

The statistical analysis of temperature data over a 42-year period has been shown in Figs. 10.3 and 10.4. It represents average annual temperature data in three forms—raw plotting, moving average over a five-year period and a regression line has also been plotted with a calculated Coefficient of Determination (R^2).

10.4.1.1 Change in Temperature

The raw data indicates a marginal fluctuation in temperatures with a warming phase between 1998 and 2000 followed by a sharp decline in temperature after 2015. The Moving Average over a 5-year period smoothened the short = term fluctuations indication a more stabilized temperature till 2012 and a consistent downward shift after that. The regression equation as indicated in Fig. 10.3 is as follows:

$$y = -0.0284x - 6.3667$$

Here m = −0.0284 indicated a drop in the temperature at the rate of 0.028 °C/year which cumulates to a drop of1.16 °C (0.028 * 42) between 1981 and 2022. The Coefficient of Determination (R^2) value of 0.24 indicates a lack of consistency and a weak moderate trend in the variability of data.

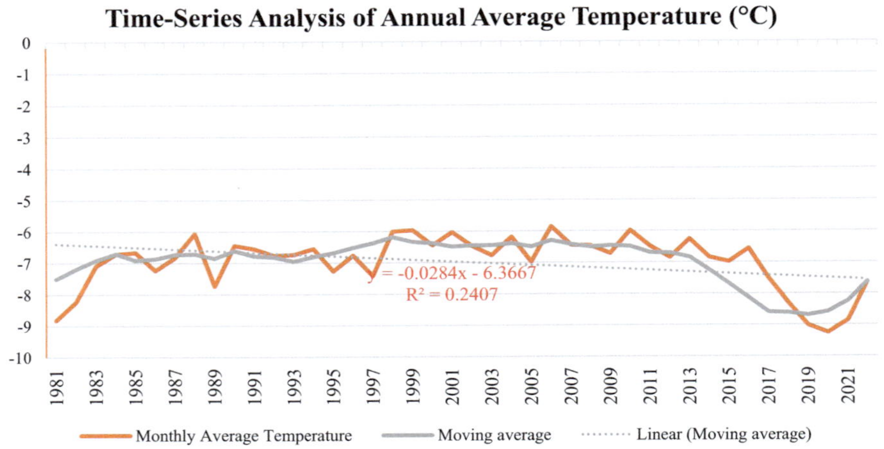

Fig. 10.3 Time-series analysis of temperature, Tsomoriri Lake (1981–2022). *Source* Processed by the author using MS-Excel

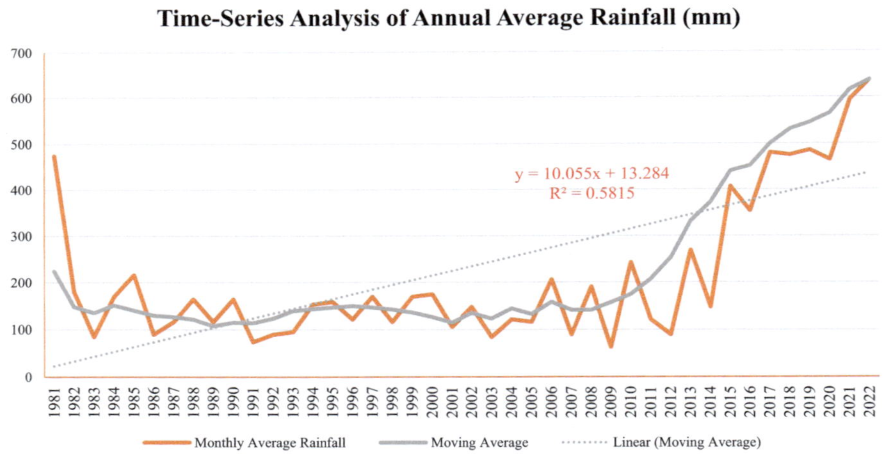

Fig. 10.4 Time-series analysis of rainfall, Tsomoriri Lake (1981–2022). *Source* Processed by the author using MS-Excel

The Tsomoriri Lake is a snow-fed basin located at an average elevation of more than 4000 m above mean sea level. The time series analysis of temperatures indicates a trend of atmospheric decoupling, meaning regional temperatures do not align with the global trends or continental trends of global warming and the declining temperatures of Tsomoriri Lake indicate a complex behavior of cryosphere environments. Ladakh as a region is thermally and hydrologically distinct from other parts of India. It lies in the rain shadow region of Himalayas which explains the extreme aridity and low humidity. Common reasons for this decoupling include a thinner troposphere at such altitudes with reduced surface-air interactions, a high albedo due to predominance of snow which reflects most incoming radiation and enhanced radiative cooling (at night) due to clear skies and low humidity reduces the net solar heating. The Tsomoriri Lake is also under the influence of katabatic winds reinforcing cold pulling in valleys and basins explaining the cooling trend. This leads to the creation of a unique micro-climate that is inconsistent with the global system. Comparing on a decadal scale, temperatures after 2019 indicate an upward trend which is almost a degree higher than the 1980s. We can therefore, infer that although a cumulative cooling trend exists average annual temperatures are comparatively higher which can have severe implications for a fragile ecosystem.

10.4.1.2 Change in Rainfall

Rainfall patterns show a comparatively more stable and upward trend especially after 2010. The regression equation as shown in Fig. 10.4 is as follows:

$$y = 10.55x + 113.284$$

Here m = 10.55 indicates an increase in the temperature at the rate of 10.5 mm/year which cumulates to an increase of 422 mm (10.55 * 42) between 1981 and 2022. The Coefficient of Determination (R^2) value of 0.585 indicates a strong positive trend with a consistent increment in rainfall over time. While the raw data shows more fluctuation, the moving average indicates a clear-cut upward trend especially after 2010 indicating a marked increase from 100 to 600 mm in 2022. Although a marked wetter phase in indicated in the region but a prior warming phase in 1998–2000 has more likely shifted the precipitation form to rain which may not be beneficial for cryospheric regimes. Increased rainfall impacts snow-covered areas in the form of reduced albedo, removal of latent heat from partially snow-covered or frozen areas. This reinforces the cooling trend as previously indicated. Indicatively, more rainfall in cryosphere environments is likely to drive a cooler temperature rather than contradicting each other.

10.4.1.3 Analysis of Spectral Properties

10.4.1.4 Snow-Cover Dynamics

The NDSI values shown in Table 10.2 reflect the changing snow-cover dynamics in the region. The maximum value for NDSI has declined from 0.78 in 1990 to 0.64 in 2023 suggesting a reduction in snow reflectance due to declined extent. The cryosphere is predominated by the presence of snow for a prolonged duration which remains frozen in the winter and determines the local ecology of the region. The visual analysis of land-use maps in Fig. 10.5 shows a notable decline in snow covered

Table 10.2 Spectral properties of land-use classes, Tsomoriri Lake (1990–2023)

Land-use class	Index		1990	1998	2023
Vegetation	NDVI	Min	– 0.39	– 0.45	– 0.28
		Max	0.36	0.35	0.46
Snow Cover	NDSI	Min	– 0.63	– 0.36	– 0.36
		Max	0.78	0.80	0.64
Waterbody	NDWI	Min	– 0.59	– 0.59	– 0.43
		Max	0.78	0.79	0.33
Waterbody Area (km^2)			137	137	132

Source Processed by the author

areas across the region from 1990 to 2023. Extensive snow-covered areas existed in the central and eastern portions in 1990 but have declined significantly in 1998 with only partial signs of visible recovery only in some high-altitude zones. Prior warming before 2010 induced melting of snow reducing accumulation but despite of a slight cooling after 2015 the snowpack recovery seems slow. This represents a climatic system in transition with altered snow-dynamics which can lead to long-term snow loss. The current phase of higher rainfall may not be favourable for snow regeneration with significant hydrological and ecological implication with long-term impacts on local livelihoods.

10.4.2 Changes in Vegetation

The maximum NDVI values show an increase from 0.35 in 1990 to 0.46 in 2023 indicating denser or comparatively productive vegetation patches in some area. This is clearly indicative of localized recovery of vegetation in some patches due to increased precipitation. Although changes in vegetation are less concerning the complex climatic dynamics in the region can impact the natural vegetation of the region which can threaten the local biodiversity and impact local livelihoods of the Changpa community who depend on the pastures for grazing their livestock. Eventually a situation of human-wildlife conflict due to shortage of pastures may befall.

The decline in snow cover can lower the retention of soil moisture and exacerbate soil freeze and thaw damaging the seeds and roots leading to pasture degradation. Rainfall in such areas is not as beneficial as it lacks the slow-release mechanism of melting snow. This aggravates soil erosion and diminished soil fertility as well as failure to replenish root-zone moisture. A cooling trend after 2010 indicates a shortening growing period for the vegetation which impacts accumulation of biomass, productivity and regeneration. It is easy to conclude that climate induced pasture

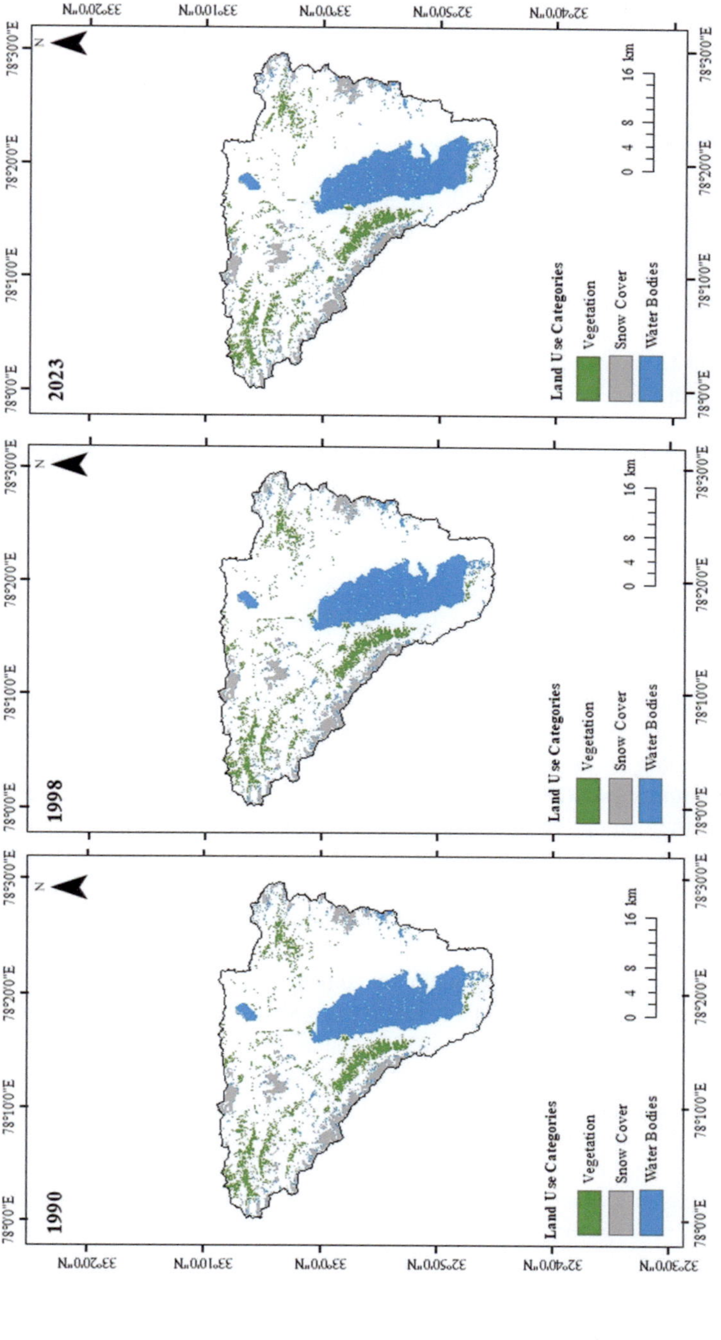

Fig. 10.5 Temporal Analysis of Vegetation, Snow-Cover and Waterbody

degradation in this region will not be stimulated due to water stress but more indicative of climatic mismatch—transition to rainfall from snow, early snow melt and lower temperature regimes that does not support full plant development.

10.4.3 Changes in Lake Area

The NDWI values indicated in Table 10.2 reflects a less pure water signature as maximum values decline from 0.78 to 0.33 indicating declining surface water quantity and quality. Lake area calculations show a decline of an area of 5 km^2 which can be attributed to several factors. High-altitude terrains are exposed to intense solar radiation with strong winds and low humidity which can accelerate the rate of evaporation from lake areas. The arid topography of the region leads to more runoff than infiltration of water. The lake is also heavily fed by snowmelt which is declining comparably due to prior warming phases and could have also shortened the snowmelt season creating a less stable lake inflow. The role of anthropogenic factors is currently less pervasive in this region but the expanding rate of tourism and related developments in the region presents an alarming situation.

10.4.4 Local Perception on Climate Change

The term "climate change" and scientific studies validating global warming and erratic climate patterns are unknown to the Changpas, yet there is a strong agreement about the changing weather conditions in Changthang which is making grazing a difficult community practice for livelihood. 70% of the respondents agreed about the change in the magnitude of snowfall and decreasing availability of water in the region (Fig. 10.6). There is therefore, a very strong positive perception about the variation in regional climate. The indigenous communities believe these harsher situations are also responsible for more and more people wanting to give up the age-old traditional occupation and lead a settled and easier life. The impact of climate change is also evident in the Rangeland management system of the Changpas. These communities recognize the limited availability of pasture and use the rangelands on a rotational basis with clear-cut seasonal demarcation. Mobility and an emphasis on co-operation through mutual support and reciprocity are strong coping strategies (Bijoor et al., 2024). Cumulatively, this has impacted the availability of pastures from time to time which has had a major bearing on the pastoral cycles. These cycles were comparatively longer and only 6 to 8 shifts were done yearly based on availability of water and pasture which has now increased to 8 to 15 times in a year (Bhasin, 2012, 2021). A more pertinent implication of climate change in the region has been decreased water availability as already reported through declining snow-cover in the region. The decline in snow-cover results in declining availability of snow-melt and glacier-melt in summers which feeds the local streams of the region which are

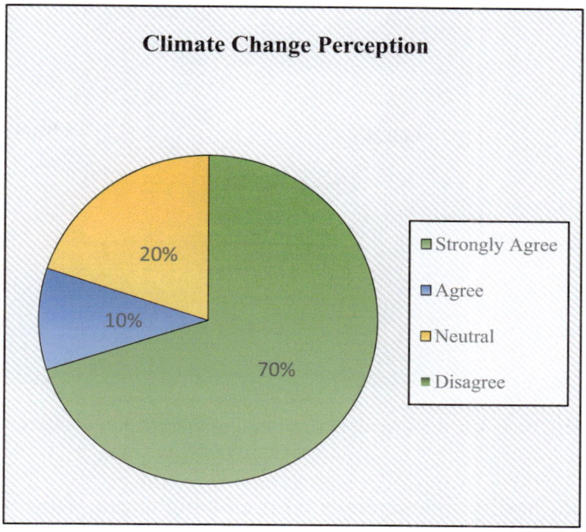

Fig. 10.6 Climate change perception

a primary source of water for the tribal community. Respondents also reported a diminished availability of water for consumption since they do not use the water of Tsomoriri Lake as holds sacred value for the local people and it has turned brackish with time. They did not reveal any specific reasons behind considering the lake sacred but associated it with the arrival of the Black-Necked Crane which is considered by them as the "*mani ratna*" according to the village chief.

10.4.5 Changes in Occupational Structure and Traditional Practices

The occupational structure of the inhabitants of Korzok was traditionally quite simplistic. Men practiced nomadic herding and women were entrusted with household responsibilities and were involved in handicraft work which they used and also sold for a living. The economy of the Changthang Plateau is traditionally based on livestock rearing and the trade of pashmina and wool produced by pashmina goats and sheep. 100% of the survey population practices livestock rearing as their primary occupation. Four amongst the surveyed households also practiced agriculture and owned lands near the village. Two amongst the surveyed households was operating a small restaurant and shop respectively in Korzok as they had a small herd-size which was not sufficient to meet the family requirements. The elder in the family had passed away and the younger son sold a significant proportion of the herd for monetary requirements. Though communities practice agriculture, it has largely remained subsistent in nature owing to the extreme terrain, low precipitation

Fig. 10.7 Changpa herders shearing pashmina fibre. *Source* Clicked by the author

and aridity and thereby a short growing season. However, these highlands seem suitable for grazing and the Changpas thereby depend on animal husbandry for basic necessities and for a living (Fig. 10.7).

An informal interaction with the village chief and the monastery monks provided several insightful details about the changing economic system of the Changpas which is no longer dependent exclusively on livestock rearing and is experiencing diversification. Introduction of tourism, expansion and proliferation of pashmina trade, education and introduction of military and sectoral jobs, transition to monogamy and role of climate change in impacting quality of pastures are commonly identified factors for abandoning local livelihoods. An important reason has been to supplement family income or shift to off-farm income. Korzok being a tourist spot being located on the banks of Tsomoriri is witnessing a comparatively rapid transformation. While families continue to pursue nomadic herding but several are supplementing their income through homestays, restaurants, shops, or working has porters and guides with the military and travel agencies respectively. The livestock composition with time has also become favourable to pashmina goats as they generate more income. Traditionally there was a large proportion of sheep in the herd because they were easier to tend in the hardy environment of Changthang. The village chief also revealed that most young children in Changpa families no longer wish to pursue nomadic herding and migrate to the cities for educational and employment purposes. This is however income driven as only wealthy Changpa herders can afford to send their children. Those who do not migrate continue to perform odd jobs in Korzok as mentioned above. Education and increased interaction with the modern ways of living is causing a sense of disillusionment amongst the youth with traditional mode of Changpa life. They no longer wish to pursue the hardships of nomadic life, disregard the traditional practices of the Changpas and want to lead a comfortable sedentary life with stable income. Discontinuation of polyandry is more an impact of changes in attitude and beliefs of the youth rather than the legal ban according to the village chief. The migration of the youth to the cities has completely altered the demographic composition of Korzok which is now mostly comprised of senile population.

A fundamental change that is occurring in Korzok according to the respondents is the diminishing authority of the "goba". The cultural tradition of Changpas consider the religious head or the "goba" as the supreme authority and the village chief although enjoys the position of the administrative head but is subservient to the "goba". However, with increasing administrative responsibilities, frequent travel the goba's involvement is declining towards pasture management which has also impacted his social status and violation of pastoral rules is common.

Accumulation of wealth has been an important economic consequence of augmented pashmina trade. This has enabled them to send their children for educational purposes to cities like Leh, Delhi, Chandigarh and Jammu. There is no statistical data available to support these claims but personal interaction and existing research supplements this information. With increased knowledge and skill and increased affinity with the modern way of life which is settled, comparatively luxurious, stable and has less hardships, the younger generation no longer wishes to pursue their traditional nomadic mode of living and have migrated to the cities with no intentions of returning. They consider the age-old traditions and customs of the ethnic communities as regressive and outdated. For instance, they prefer monogamous relationship based on companionship rather than pursuing polygamy, which according to them is outdated but has been an important adaptation mechanism to ensure population stabilization, efficient resource utilization and ensure availability of people to maintain herds and prevent herd disintegration. This has resulted in a demographic restructuring of Changpa villages which are now mostly inhabited by the elderly. With exacerbated hardships due to climatic hostility and geo-political changes the elderly population are more than willing to let their children adopt a modern lifestyle and give up nomadic herding but concerns about vanishing traditions and customs is equally worrisome. Many among those who do not possess the wealth to relocate take up odd jobs as labor with BRO, porters with the military and several unskilled jobs during the peak tourist season. Families who have completely given up livestock rearing now survive on meagre income from running shops and restaurants in Korzok which are only operational in the season. Limited skills and lack of education serve as a major impediment in such situations.

10.5 Discussion

The Tsomoriri Lake is a Wetland of International Importance with ecological, hydrological, spiritual and cultural values. It is of extreme importance to the local communities of the Korzok village, located on the western bank of the lake. It is also a site of cultural importance for the Buddhists as the 400-year old Korzok monastery is located here. The Changpas who have inhabited the region for several years, display a complex interface with nature that resonates in their customs, land-use, local knowledge and spiritual beliefs, and are currently threatened by changing socio-economic and anthropogenic climate threats (Kothari, 2019). Trade of pashmina has been the backbone of economic development in various parts of the Changthang Plains. It has contributed towards social upliftment of several large Changpa herders who have been able to accumulate massive wealth due to rising prices and demand of pashmina fiber in the global market and this region continues to the retain the monopoly of supplying the best quality fiber. The cold climate and extreme aridity where the pashmina or *"changra"* goats are reared is the primary reason for maintaining the finesse of the fiber. This well explains the importance of retaining the "natural environment" of the region in order to conserve local livelihoods and traditional practices. The traditional pastoral economy of Korzok is experiencing a monumental socio-economic transformation since the 1990s after the Tsomoriri Lake was opened for external access through tourism in 1993–94 (Ramsar Information Sheet, 2002). This not only provided an additional source of income for the local people through home-stays, small restaurants, shops etc. but also brought them in contact with the modern way of living. Anthropogenic influence, geo-political changes, increased wealth and risks of climate change today are driving more and more people out of livestock rearing particularly within the younger generation. A complex web of mutually reinforcing factors are leading to a loss of a traditional occupation being practiced since long as shown in Fig. 10.8.

Climate change and accompanied fluctuations of temperature and precipitation has had a multiplier effect on the rangelands of the Changthang. Unprecedented extreme events in the Changthang such the snow-storm in 2013 is resulting in a massive loss of local biodiversity, submergence and shrinkage of shallow lake islands are perceptive evidences of habitat destruction due to climate change (Kothari, 2019). The quality of grasslands has become extremely deteriorated and unable to meet the requirements of augmented livestock population. Pastures that were previously lush green are now degrading due to continuous usage and overcrowding after the influx of Tibetan refugees and loss of Skagjung winter pastures after the annexation of Tibet in 1951 by China and Sino-Indian War of 1962. Development of roads, infrastructure, military encampments have also encroached on the pasture lands. Expansion of tourism has exacerbated pasture degradation through unregulated development of camping sites, waste disposal and infrastructural development.

Climatic stresses have comparatively made livestock rearing difficult with challenges of pasture degradation, over-grazing in vulnerable areas, alteration of migratory routes, human-wildlife conflict which is consequent of declining pasture quality and quantity. Fluctuation in temperature and precipitation also affect the daily lives of the local people as advancing snowmelt period reduces early access to pastures. Changing temperature and precipitation regimes also impacts the herds leading to herd mortality with the younger population being more vulnerable. Harsher climate makes nomadic herding a difficult community practice. Cyclic rotational of pastures has become longer with more shifts in a year. As local people are dependent on glacial melt for water, degradation of snow cover poses a big risk to water security. Pasture degradation is also contributed by the change in species composition and growth of unpalatable plants which is detrimental for the pastoral communities as well as the overall health of the wetland ecosystem. Locust attacks on pastures are now a

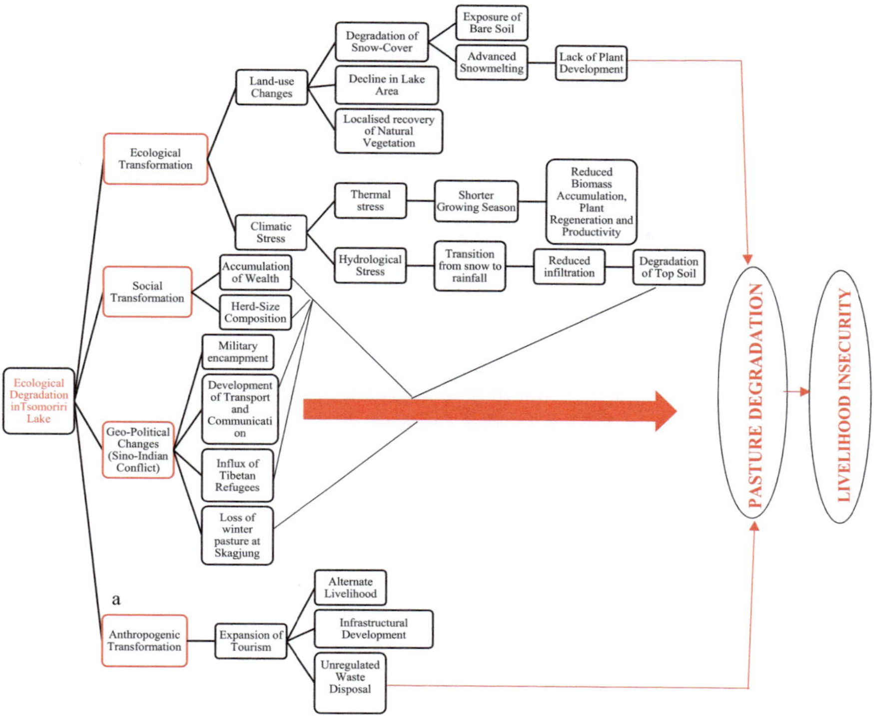

Fig. 10.8 Factors affecting ecological degradation in Tsomoriri Lake (self-generated)

common occurrence in the summer season leading to a massive shortage. Changpa herders also cultivate barley hay during the short growing season but yields are low and with lack of appropriate fodder management due to which disease outbreak, storage issues emerge causing a shortage particularly when the temperature drops aggravating livestock mortality especially neonates.

Competitive interaction between the endemic fauna and livestock is consequent of declining pastures and increased livestock population. This can pose a severe threat for the existence of endangered species like Tibetan Gazelle which are on the verge of extinction. As of today, less then hundred individuals of Tibetan Gazelle remain and require immense protection and conservation (Namgail et al., 2007). As a result, the Changpa community is becoming increasingly dependent on the supplies by the government which is also creating lethargy amongst the herders. Change in herd composition due to expansion in number of pashmina goats has also contributed towards the deterioration of rangelands as goats possess hooved feet which degrade the top soil. They dig deeper into the grasses and extract them from the root making regeneration difficult (Bijoor et al., 2024; Gaur et al., 2022; Namgail et al., 2010).

The impact of climate change has been exacerbated in the form of anthropogenic interferences through the expansion of tourism and military infrastructure due to augmented tension with China. Tourism in remote locations has been a key factor of change spurring infrastructure development from roads, housing, administrative and defense posts resulting in a complete transformation of the landscape. Construction of rest houses, restaurants and hotels due to expansion of tourism and military infrastructure has occupied the northern meadows of the Korzok village (Humbert-droz, 2023). Car Rallies deteriorate the top soil and deplete the quality of grasslands. Waste generation and solid waste disposal are common environmental challenges these pristine environments face. On the flip side a budding tourism economy today offers a great alternative economic opportunity. As of 2015, there were five seasonal camps, four campsites, four hotels and resorts, more than 20 home-stay guest houses and two café restaurants (Humbert-droz, 2023). This has been able to supplement the income of the Changpa communities effectively. This has also become a lucrative opportunity for the Changpa tribal to give away their traditional nomadic profits.

The marketisation of Pashmina and involvement of the government has transformed the traditional systems and have integrated the ethnic tribes into a market economy which has both positive and negative implications. The preference of pashmina goat for herding has increased overwhelmingly as "pashmina" fetches a much higher price than sheep wool. Sheep wool is used for either personal use or traded to neighboring communities in Spiti. An important outcome of the growing pashmina trade and increased income for the Changpa community has been a better lifestyle for the younger generation who can now access better food, education and health facilities in urban cities but it also means a loss of traditional practices and occupational system which were ecologically sustainable and culturally rich. Absence of the younger generation here also means a lack of assistance to manage the livestock which also inhibits the transfer of traditional knowledge and experience on grazing practices, local vegetation and knowledge of bio-physical conditions (Ladon et al., 2023). The inter-generational loss of traditional ecological knowledge and institutional mechanism is exacerbated by the introduction of centralized education systems since, children now spend more time in school than with the elders and parents to access their skill, knowledge and experiences required to practice pastoralism in Changthang. As more and more people from the younger generation migrate to Leh and other parts, the higher is the possibility that this traditional practice would become a lost tradition one day. Rate of such forms of emigration from Korzok are comparatively less in villages as compared to Samad and Kharnak where 90% of the population has migrated to a developing congregation in Leh called "Kharnakling" named after the village Kharnak whose inhabitants constitutes the most proportion of it's population (Ahmed, 2004). There are instances of families wanting to return back to their villages and continue the traditional occupation but their acceptance back in the community varies from village to village and also depends on the degree of association the migrated families maintained with their community post-migration in the form payment of taxes, fulfilling the rituals etc. Families who severed ties with their villages completely were not allowed to return back and increase the competition for pastures. This new group from Changthang settled in Leh have also transformed the pashmina trade by entering the trade market as new traders. They act as middle men for their family members or other Changpas and bring their produce to Leh to sell it to private traders from Leh or Kashmiris (Ahmed, 2004).

The proliferating demand of pashmina with halted trade with Tibet after its annexation promulgated several transformations in the herding practices of the Changthang which can be detrimental to the local ecology. The herd-size composition has changed with time suiting to changing conditions. Herd-sizes were traditionally large with each herder possessing 500–1000 livestock as pastures were abundant (Ahmed, 2004). Influx of refugees, loss of pastures on the Chinese side, declining pasture health with continued grazing increased the competition which drastically reduced the herd-sizes. As of today, only a few farmers possess more than 100 goats and most herders have an average herd-size of 50 goats (Livestock Census 2020–21, Sheep Husbandry Office, Korzok). Herd-size composition traditionally had more sheep as they were easier to rear and there was sufficient demand for wool but has now become more favourable to goats to cater to the demands of pashmina. While sheep to goat ratio was previously 3:1 it has now transformed to 1:3 with a larger number of goats in the pack (Bhasin, 2012, 2021; Singh et al., 2013). Cumulatively, where on one hand pashmina trade has proved profitable for the Changpas through accumulation of wealth, climatic hostility and hardships of nomadic life has discouraged the younger generation to continue with this traditional practice of nomadic herding. Possession of wealth has enabled children to gain education and get exposed to an urbanized way of life. Students who leave Korzok to gain education rarely come back to pursue the nomadic profession. Many among the older generation are more than willing to let their children pursue the age-old traditions but concerns about vanishing customary practices are equally bothering. The population composition of Korzok has undergone a shift and today only the elders of the family continue to remain back while the younger children have moved to the city life for better employment, health and education facilities. As more and more people withdraw from Changpa households, accompanies the loss of traditional livelihood, local knowledge and customary heritage of Changpa households. There is an emergent to need conserve and protect this intangible heritage and indispensable knowledge that has helped communities sustain in this harsh environment since ages to preserve the eco-cultural landscape of the Changthang.

10.6 Conclusion and Policy Recommendations

The current economic system in Changpa communities is experiencing a transition towards diversification of livelihoods which has impacted traditional livelihoods. The impacts of globalization led economic expansion has also been impacting the mountain ecosystems which is also precipitating shifts in pastoral strategies and practices interfering with the delicate balance of ecosystems such as High-Altitude Wetlands (HAWs). Nomadic herding as a practice has evolved over a long course of time with an acquired knowledge of the environment and strong social organization among the Changpa herders. The herders brave the extreme climate and undulating terrains of the

Changthang, and have well adapted to the environmental instability through a variety of socio-ecological mechanisms and herd-management strategies. The challenge of global warming interferes with the harmonious relationship between the Changpas and their natural environment presenting new challenges. The process of this change has been gradual but the last few decades have been quite transformative due to anthropogenic interference with significant impacts on the Himalayan ecosystem in terms of health, food and fiber, degradation of the land, livelihood and biodiversity. The analysis reveals a stark atmospheric decoupling in the thermally and hydrologically distinct region which is inconsistent with the global trends of continental warming. Yet, this is also indicative of a climatic stress in a cyrospheric environment with complex ecological regimes with long-term impacts on local hydrology, ecosystem functioning and water and food security. Erratic climate pattern is directly linked with the availability of pasture and water supply in the region which can have long term implications in the form of pasture degradation and water shortage. Pressures of herd overstocking, changing herd-size composition, expanding commercial development to suits the needs of military and tourism is detrimental for the ecology of the Changthang. Climate changes and its associated impacts on the grasslands, biodiversity exacerbates the risks of a threatened traditional livelihood. Sustainable land-use planning which integrates the concerns of local pastoral economy, biodiversity conservation and the rich cultural heritage of the region, integration of traditional adaptation strategies in new age policy making, extended government support and assistance to the herder community for increasing returns on trade to prevent mass exodus from Changpa villages and protection of snow-fed catchments are essential for enhancing the resilience of the ecosystem. Long-term monitoring of hydrological and climatic parameters, stringent monitoring of tourism related expansion, checking encroachment of pastures and degradation of the top soil and local awareness about the changing thermal dynamics is the need of the hour.

Declaration The authors confirm that the content is not AI generated.

Conflict of Interest The authors have no conflict of interest.

References

Ahmed, M. (2002). Living fabric. Weaving among the nomads of Ladakh Himalaya. Bangkok: Orchid Press. 192 pp. Hb.: £35.00. ISBN: 974-8304-87-6. *Social Anthropology*, *13*(1), 91–92. https://doi.org/10.1017/s0964028204211089

Ahmed, M. (2004). The politics of Pashmina: The Changpas of Eastern Ladakh. *Nomadic Peoples, 8*(2), 89–106. https://doi.org/10.3167/082279404780446041

Allen, M., Dube, O. P., Solecki, W., Aragón-Durand, F., Cramer, W., Humphreys, S., & Kainuma, M. (2018). Special report: Global warming of 1.5 C. *Intergovernmental Panel on Climate Change (IPCC), 677*(393), 49–92.

Anand, A., Chandan, P., & Singh, R. B. (2012). Homestays at Korzok: Supplementing rural livelihoods and supporting green tourism in the Indian Himalayas. *Mountain Research and Development, 32*(2), 126–136. https://doi.org/10.1659/MRD-JOURNAL-D-11-00109.1

Anon. (2002). Information Sheet on Wetlands. *Ramsar Secretariat. Ramsar Convention.*

Bahuguna, A., & Ramaswamy, G. P. (2019). Highland to high fashion: The development of a women cooperative in Ladakh. *Indian Journal of Public Administration, 65*(1), 152–170. https://doi.org/10.1177/0019556118814653

Bahuguna, A., & Ramaswamy, G. P. (2022). The Changpa and Ladakh Pashmina: Craft Revivalism, skill development and ecology in Ladakh, India. *Nomadic Peoples, 26*(2), 167–189. https://doi.org/10.3197/np.2022.260202

Barrett, K., & Bosak, K. (2018). The role of place in adapting to climate change: A case study from Ladakh, Western Himalayas. *Sustainability (Switzerland), 10*(4), 1–22. https://doi.org/10.3390/su10040898

Bhasin, V. (2012). Life on an edge among the Changpas of Changthang, Ladakh. *Journal of Biodiversity, 3*(2), 85–129. https://doi.org/10.1080/09766901.2012.11884738

Bhasin, V. (2013). Pastoral women of Himalayas. *Journal of Biodiversity, 4*(1), 1–23. https://doi.org/10.1080/09766901.2013.11884740

Bhasin, V. (2021). Livelihood, sustainability and change among Changpas of Changthang: Ladakh India. *Journal of Human Ecology, 63*(1–3), 34–73. https://doi.org/10.31901/24566608.2018/63.1-3.3118

Bhatnagar, Y. V., Wangchuk, R., Prins, H. H. T., Van Wieren, S. E., & Mishra, C. (2006). Perceived conflicts between pastoralism and conservation of the kiang Equus kiang in the Ladakh Trans-Himalaya, India. *Environmental Management, 38*(6), 934–941. https://doi.org/10.1007/s00267-005-0356-2

Bijoor, A., Khanyari, M., & Suryawanshi, K. (2024). Sustainable development in the high mountains—The case of Ladakh Pashmina. *Economic and Political Weekly, 59*(19), 59–64.

Chandan, P., Chatterjee, A., & Gautam, P. (2008). Management planning of Himalayan high-altitude Wetlands. A case study of Tsomoriri and Tsokar Wetlands in Ladakh, India. In *Proceedings of Taal 2007: The 12th World Lake Conference: 1446–1452, Management* (pp. 1446–1452).

Chandola, S. (2012). *An assessment of human-wildlife interaction in the Indus valley, Ladakh, Tran-Himalaya* (pp. 1–238).

Chatterjee, A., Blom, E., Gujja, B., Jacimovic, R., Beevers, L., O'Keeffe, J., Beland, M., & Biggs, T. (2010). WWF initiatives to study the impact of climate change on Himalayan high-altitude wetlands (HAWS). *Mountain Research and Development, 30*(1), 42–52. https://doi.org/10.1659/MRD-JOURNAL-D-09-00091.1

Chaudhuri, A. (1988). Change in changthang: To stay or to leave? *Economic and Political Weekly, 23*(44), 2244. http://about.jstor.org/terms

Chevuturi, A., Dimri, A. P., & Thayyen, R. J. (2018). Climate change over Leh (Ladakh), India. *Theoretical and Applied Climatology, 131*(1–2), 531–545. https://doi.org/10.1007/s00704-016-1989-1

Gaur, M. K., Goyal, R. K., Saha, D., Singh, N., Shekhar, S., Ajai, & Chauhan, J. S. (2022). The estimation of snow cover distribution using satellite data in the cold arid LEH region of Indian Himalaya. *Polish Journal of Environmental Studies, 31*(1), 63–73. https://doi.org/10.15244/pjoes/135606

Gautam, P., Chandan, P., Chatterjee, A., Vidya, S., & Areendran, G. (2007). *Management planning for Tsomoriri – Tsokar—A framework*

Geneletti, D., & Dawa, D. (2009). Environmental impact assessment of mountain tourism in developing regions: A study in Ladakh, Indian Himalaya. *Environmental Impact Assessment Review, 29*(4), 229–242. https://doi.org/10.1016/j.eiar.2009.01.003

Goodall, S. (2007). *From plateau pastures to urban fringe: Sedentarisation of nomadic pastoralists in Ladakh, North-West India.*

Gupta, S. K., & Shukla, D. P. (2016). Assessment of land use/land cover dynamics of Tso Moriri Lake, a Ramsar site in India. *Environmental Monitoring and Assessment, 188*(12). https://doi.org/10.1007/s10661-016-5707-3

Humbert-droz, B. (2023). *Environmental change and development in Ladakh, Indian Trans-Himalaya.* https://link.springer.com/https://doi.org/10.1007/978-3-031-42494-6

Jina, P. S. (1999). The chang Pa of the Ladakh Himalayas: Some observations on their economy and environment. *Nomadic Peoples, 3*(1), 94–105. https://doi.org/10.3167/082279499782409460

Kapruwan, R., Saksham, A. K., Ford, J. D., Petzold, J., McDowell, G., Rufat, S., Schneiderbauer, S., Tiwari, S., & Pandey, R. (2025). Climate change adaptation in global mountain regions requires a multi-sectoral approach. *Communications Earth and Environment, 6*(1). https://doi.org/10.1038/s43247-025-02444-5

Kothari, R. (2019). Nature-culture mapping in the trans-Himalayas. *Journal of World Heritage Studies*, 99–105. https://doi.org/10.15068/00157693

Ladon, P., Nüsser, M., & Garkoti, S. C. (2023). Mountain agropastoralism: Traditional practices, institutions and pressures in the Indian Trans-Himalaya of Ladakh. *Pastoralism, 13*(1), 1–14. https://doi.org/10.1186/s13570-023-00289-1

McFeeters, S. K. (1996). The use of the Normalized Difference Water Index (NDWI) in the delineation of open water features. *International Journal of Remote Sensing, 17*(7), 1425–1432. https://doi.org/10.1080/01431169608948714

Mishra, C., Bagchi, S., Namgail, T., & Bhatnagar, Y. V. (2009). Multiple use of trans-Himalayan Rangelands: Reconciling human livelihoods with wildlife conservation. In *Wild Rangelands: Conserving wildlife while maintaining livestock in semi-arid ecosystems, Mishra 2001* (pp. 291–311). https://doi.org/10.1002/9781444317091.ch11

Mishra, P. K., Prasad, S., Jehangir, A., Anoop, A., Yousuf, A. R., & Gaye, B. (2018). Investigating the role of meltwater versus precipitation seasonality in abrupt lake-level rise in the high-altitude Tso Moriri Lake (India). *Palaeogeography, Palaeoclimatology, Palaeoecology, 493*, 20–29. https://doi.org/10.1016/j.palaeo.2017.12.026

Namgail, S. (2015). Changes in the institution of family amongst Changpas. *Ladakh Studies, January 2015*, 4–17.

Namgail, T., Bhatnagar, Y. V., & Fox, J. L. (2011). Pastoral production and wildlife conservation in a trans-Himalayan wildlife reserve in Ladakh. *Sustainable Pastoralism in the Himalayas, January 2006*, 165–175.

Namgail, T., Bhatnagar, Y. V., Mishra, C., & Bagchi, S. (2007). Pastoral nomads of the Indian Changthang: Production system, landuse and socioeconomic changes. *Human Ecology, 35*(4), 497–504. https://doi.org/10.1007/s10745-006-9107-0

Namgail, T., van Wieren, S. E., & Prins, H. H. T. (2010). Pashmina production and socio-economic changes in the Indian Changthang: Implications for natural resource management. *Natural Resources Forum, 34*(3), 222–230. https://doi.org/10.1111/j.1477-8947.2010.01303.x

Nautiyal, S., Rao, K. S., Kächele, H., Raju, K. V., & Schaldach, R. (2013). *Knowledge systems of societies for adaptation and mitigation of impacts of climate change/Gesellschaftliche Wissenssysteme zur Anpassung und Abmilderung der Auswirkungen des Klimawandels.*

Pandit, A., Mir, M., Mir, M., Wani, Y., Bisati, I., Nisa, S. U., Khan, H. M., & Shah, R. A. (2024). Pastoralism in Changthang, Ladakh: Adaptations, challenges, and pathways for sustainability. *Mountain Research and Development, 44*(1), A1–A7. https://doi.org/10.1659/mrd.2023.00028

Panigrahy, J. G. P. P. (2012). National Wetland Atlas. High Altitude Lakes on India. In *Advances in remote sensing* (Vol. 18, Issue October).

Prins, H. H. T., & Namgail, T. (2017). Bird migration across the Himalayas: Wetland functioning amidst mountains and glaciers. In *Bird migration across the Himalayas: Wetland functioning amidst mountains and glaciers.* https://doi.org/10.1017/9781316335420

Riggs, G. A., Hall, D. K., & Salomonson, V. V. (1994). Snow index for the Landsat thematic mapper and moderate resolution imaging spectroradiometer. *International Geoscience and Remote Sensing Symposium (IGARSS), 4*, 1942–1944. https://doi.org/10.1109/igarss.1994.399618

Sabharwal, A. (2016). Contested affluence: Cultural politics of Pashmina wealth and wildlife conservation in Ladakh. *Research in Economic Anthropology, 36*, 77–113. https://doi.org/10.1108/S0190-128120160000036004

Singh, N. J., Bhatnagar, Y. V., Lecomte, N., Fox, J. L., & Yoccoz, N. G. (2013). No longer tracking greenery in high altitudes: Pastoral practices of Rupshu nomads and their implications for biodiversity conservation. *Pastoralism, 3*(1), 1–15. https://doi.org/10.1186/2041-7136-3-16

Singh, R., Sharma, R. K., & Babu, S. (2015). Pastoralism in transition: Livestock abundance and herd composition in Spiti, Trans-Himalaya. *Human Ecology, 43*(6), 799–810. https://doi.org/10.1007/s10745-015-9789-2

Tashi Dolkar, F. D. S., et al. (2017). To study the migration route of Changpa Nomads of Ladakh. *IMPACT: International Journal of Research in Humanities, Arts and Literature (IMPACT: IJRHAL), 5*(11), 79–84.

Townshend, J. R. G., & Justice, C. O. (1986). Analysis of the dynamics of African vegetation using the normalized difference vegetation index. *International Journal of Remote Sensing, 7*(11), 1435–1445. https://doi.org/10.1080/01431168608948946

Tashi, T., & Sudan, F. K. (2021). Perceived impacts of climate change on economic activities and adaptation strategies: A case study from cold desert of Ladakh, India. *Regional Economic Development Research*, 2(2), 193–210. https://doi.org/10.37256/redr.2220211068

Chapter 11
Threads of Resilience: Climate Change, Culture, and Community Wellbeing

Deepen Chettri, Bhagirath Behera, and Pritha Datta

Abstract Climate change is also recognized as a serious social and cultural shift that has been treated not only as an ecological disaster but as a profound crisis too. The review synthesizes the results of 51 research studies published since 2010 and has investigated a two-way relationship: how climate change compromises the social and cultural fabric of society with diminished livelihoods, displacement, and diminished traditions and, in its turn, how that same social relations fabric, cultural practices, and traditions are a critical storage of resilience to rise to these challenges. Such losses cannot be seen, and they lack an economic value. Nevertheless, the concept of community agency as emphasized in this review can also be of significance and in this case, the practices that are embedded in culture, the networks of trust, and traditional knowledge are the foundation of collective adaptation. The possibilities of dealing with climate-related challenges among communities are extremely diverse and include ecological stewardship practices and early warning systems, as well as constitutive knowledge frameworks. In line with this, climate adaptation policies have to be based on cultural realities and should be socially inclusive. It is, therefore, important to appreciate the intangible heritage; its traditions, knowledge and social relationships as basic infrastructure in sustaining long-term wellbeing and resilience in a community.

Keywords Climate change · Social cohesion · Cultural continuity · Traditional knowledge systems · Community resilience · Intangible cultural heritage

D. Chettri (✉) · B. Behera
Department of Humanities and Social Sciences, Indian Institute of Technology Kharagpur, Kharagpur, India
e-mail: deepenchettri8@gmail.com

P. Datta
Rekhi Centre of Excellence for the Science of Happiness, Indian Institute of Technology Kharagpur, Kharagpur, India

P. Datta et al. (eds.), *Climate Change and Human Well-being*, Advances in Global Change Research 83, https://doi.org/10.1007/978-3-032-25479-5_11

11.1 Introduction

Climate change is not simply an environmental problem but a rather a daunting social problem that can radically transform the organization structure, relationship among people, and culture of different societies (Adger et al., 2013; Datta et al., 2022). The social fabric is getting increasingly disrupted due to the growing frequency and scale of the repercussions of climate change, leading to discontinuous networks of trust, displaced populations, weakened cultural identities, and unequal resilience capacities (Feola, 2015; Scheffran, 2020; Vinyeta et al., 2016). These failures are not only dangerous to the stability of local systems, but they also tend to undermine the collective resilience that is necessary in order to approach an uncertain future.

The world is facing the ripple effects of climate change events such as floods, droughts, wildfires, and sea levels (IPCC, 2014; Datta and Bose, 2020). Although the physical destruction caused by these occurrences is well documented, current literature often overlooks their contribution towards weakening the intangible, but invaluable aspects that bind communities together: communication of shared culture, relationship based on trust and survival tactics. Social cohesion, which can be described as the degree of relationships, values and efficacy (Markus & Kirpitchenko, 2007), can be viewed as both a source of resilience and a victim of environmental stressors (Aall & Crocker, 2019; Townshend et al., 2015). Similarly, the danger of erosion of cultural practices, the foundation of which is based on traditional knowledge, rituals, land-based identities, and intergenerational relationships, is quite high in the context of ecological instability (Nnandi, 2015; Redvers et al., 2023).

Meanwhile, communities cannot be considered passive participants of climate change. To increase resilience, local stakeholders in many situations have mobilized a multiplicity of adaptive responses utilizing cultural knowledge, historical memory, and solidarity to mitigate the impact (Adger et al., 2013; Nyong et al., 2007). Based on the local circumstances, these strategies demonstrate the significance of local considerations, personal behavior, and cultural tradition in the development of the plans aimed at climate change adaptation.

With the increased awareness of the socio-cultural dimensions of climate change, however, new systematic measurements indicate that still there is a gap. Dembedza et al. (2022) state that climate change-related disasters damage food-related intangible cultural heritage (ICH) like rituals, culinary traditions, and traditional food systems. But there are not sufficient methods of quantifying these losses. At the same time, Higgins (2022) criticizes the international legal frameworks (e.g., UNESCO conventions) because they provide little protection of Indigenous ICH against being degraded by climate changes, and because they have little substantive inclusion of the Indigenous perspectives in policymaking.

Moreover, geographical as well as thematic diversities further weaken the evidence base. According to Orr et al. (2021), climate-heritage scholarship is Western-centric (seven out of ten entries represent Euro-North American), and focuses on tangible heritage, which does not give due focus to intangible practices. Pearson et al. (2021) support this claim, indicating that South American, African, and

Asian contexts were underrepresented as well as the empirical analysis of the existing (as opposed to projected) losses was limited. It is important to note that, although cultural resilience is an eminent part of the Pacific Island studies, McNamara et al. (2021) note that psychological ramifications (e.g., solastalgia) and intergenerational transmission of knowledge have not been sufficiently studied.

All in all, these reviews show three key gaps, namely, the imbalance in geography; the absence of focus on intangible heritage, particularly, the absence of study on how climate change influences social practices, oral traditions, and intergenerational knowledge transfer; and the resilience-action gap, which refers to the absence of information on how communities adapt using cultural resources.

This research question can be discussed as filling the knowledge gap by addressing the following questions: (1) to conduct a systemic review of the impact of climate change on social cohesion, cultural practices, and traditional knowledge under various contexts, and (2) to record the global example of how the strengthening or mobilization of these factors leads to the development of adaptive capacity and community resilience. Therefore, answering these aims, this review aims to offer a synthesized insight into the fact that climate change breaches social cohesion, cultural practices, and traditional knowledge in all parts of the world and to record and examine agentic community responses uniquely whereby, the same aspects are utilized as resilience building blocks.

11.2 Materials and Methods

This paper takes a Systematic Literature Review (SLR) methodology to analyze the effects of climate change on social cohesion, cultural continuity, and indigenous knowledge, and how they lead to community resilience. The Preferred Reporting Items of Systematic Review and Meta-Analysis (PRISMA) guidelines were used in the review to make sure that the review process is transparent, replicable and comprehensive (Chettri et al., 2024).

11.2.1 Search Strategy and Databases

Relevant literature was identified by use of two notable academic databases namely Scopus and Web of Science. The following Boolean string was used in the search:

((climat chang) OR (climat variability) OR (environmental chang)) AND ((social cohesion) OR (social capital) OR (community ties) OR (social networks) OR (social trust) OR (social resilien)) AND ((cultural practices) OR (cultural heritage) OR (traditional knowledge) OR (Indigenous knowledge) OR (rituals) OR (cultural identit*).

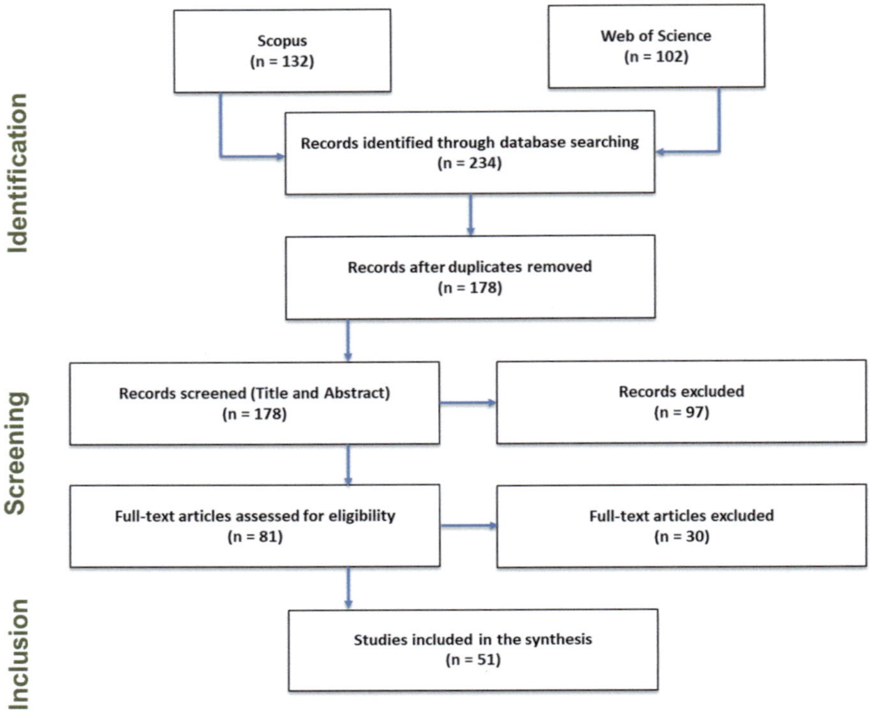

Fig. 11.1 Systematic process followed for the review

The search restricted to peer-reviewed journal articles in the period between January 2010 and May 2025. All the results were filtered by relevance and duplicates were filtered off (Fig. 11.1).

11.2.2 Screening and Eligibility Criteria

From a total of 234 records retrieved (Scopus = 132; Web of Science = 102), 56 duplicate records were identified and removed. The remaining 178 articles were screened for relevance based on the inclusion–exclusion criteria presented in Table 11.1.

After full-text screening, 51 articles were selected for inclusion in the final review.

Table 11.1 Inclusion–exclusion criteria for screening

	Inclusion	Exclusion
Type of study	• Only empirical evidence-based studies and book chapters presenting original fieldwork and case studies were included. These were selected to capture grounded and real-world insights into social and cultural responses to climate change	• Review articles, editorials, commentaries, or conceptual/theoretical papers without empirical data were excluded • Technical climate modelling studies or those focused solely on biophysical impacts (e.g., glacier melt, temperature projections, biodiversity loss) without linking to cultural or social dimensions were omitted
Topical focus	• Studies must explicitly address climate change or climate variability in relation to at least one of the following: – Social cohesion (e.g., communal solidarity, mutual aid, social networks) – Cultural practices and continuity (e.g., rituals, oral histories, traditional festivals) – Intangible cultural heritage (ICH) (e.g., local knowledge systems, traditional ecological knowledge) – The role of these elements in building resilience to climate-related stressors and shocks	• Studies focusing primarily on economic loss, infrastructure, or agricultural productivity without integrating aspects of social cohesion, Indigenous cultural heritage, or indigenous responses were excluded • Papers that mentioned cultural practices only peripherally or as background, without analyzing their role in climate adaptation or resilience building, were excluded

11.2.3 Data Extraction and Categorization

A thematic analysis (Braun & Clarke, 2006; Datta et al., 2024) was used to determine how climate change influences social cohesion and continuity of culture, and traditional knowledge systems (TKS), and subsequently, the degree to which these factors affect community resiliency. After conducting a full review of the 51 studies that matched the selection criteria for the full review, relevant data were coded by using a structured coding schema. The coded data were analyzed thematically according to the two objectives of the research. Each publication was analyzed in detail, to extract information under themes that included geography and socio-ecological context, climate change stressors, impacts to sociocultural domains, adaptive practices, and engagement with policy frameworks.

The analysis of 18 studies that addressed Objective 1 identified three major themes: (i) the Fracturing of Social Cohesion (i.e., weakening of kinship bonds and dislocation due to displacement, conflict over diminishing resources); (ii) Loss and Transformation of Cultural Practices (i.e., disruption of rituals, loss of festivals, abandonment of sacred places, and diminishing place-based identity); (iii) Erosion of Traditional Knowledge Systems (i.e., unpredictability of ecological systems and development of policies by higher authorities undermine the relevance and continuity of Aboriginal and local knowledge).

Similarly, the 39 studies used to address Objective 2 were analyzed for the purpose of identifying how social and cultural assets are utilized to enhance resilience within communities. Four interconnected themes were produced from the current study: social cohesion as a basis for collective adaptation; cultural practices as pathways to psychological and ecological durability; traditional knowledge systems as reference-specific adaptation tools. The first theme demonstrates how trust networks, mutual support, and inter-household cooperation create the foundation for collaborative action. The second theme highlights seven cultural practices (seasonal celebrations, land-based ceremonies, and collective memory systems) that reinforce psychological and ecological durability. The third exploded the use of indigenous knowledge systems; traditional weather forecasting; and sustainable land-use practices as examples of reference-specific adaptation tools.

This thematic analysis not only synthesized the diverse qualitative findings across global case studies but also provided insight into how intangible cultural dimensions, often overlooked in policy, play critical roles in shaping vulnerability and adaptation (Adger et al., 2013; Braun & Clarke, 2006; Redvers et al., 2023).

11.3 Results

11.3.1 General Description of the Reviewed Papers

This systematic review aggregated 51 empirical studies published between 2010 and 2025 focusing on the socio-cultural aspects of climate change, focusing on three areas: social cohesion, cultural continuity and traditional knowledge systems. The studies are geographically diverse and encompass areas vulnerable to climate-related disruptions (Fig. 11.2), including: Small island states such as Fiji, Micronesia, Vanuatu; coastal and delta surrounding Bangladesh and the Philippines; arid and semi-arid regions in Sub-Saharan Africa; highland areas in Nepal and the Indian Himalayas; and Indigenous lands in North and South America.

The temporal distribution of this literature (Fig. 11.3) reflects increased academic interest over the last ten years and a marked increase in the number of publications since 2015, likely coinciding with the Paris Agreement and the rise in global awareness of non-economic losses and damages. The studies examined used many different qualitative methodologies including but not limited to; ethnographic study, case studies, interview and participatory research which provided an understanding of Climate Disruption from the perspectives of the lived experiences of climate disruption and of adaptations by local communities to climate change.

Notably, the dataset reflects both geographic and thematic imbalances in the current literature. A significant proportion of studies focus on Asia and the Pacific, while fewer originate from Latin America and the Middle East. Furthermore, while

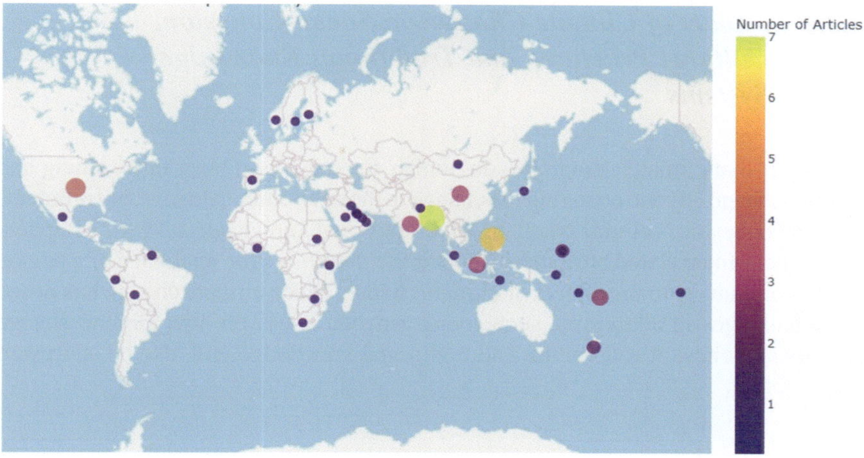

Fig. 11.2 Spatial distribution of reviewed papers (Map data © OpenStreetMap contributors. Visualization by Plotly)

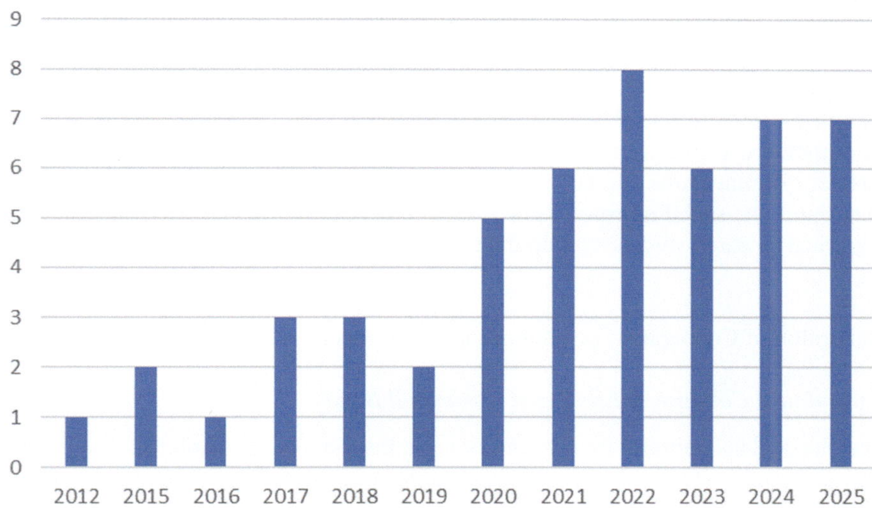

Fig. 11.3 Temporal distribution of reviewed studies

intangible cultural heritage and social cohesion are central themes, only a subset of studies explicitly frame their findings within the context of non-economic loss and damage or legal adaptation frameworks.

In general, the literature reviewed has strong evidence to inform the comprehension of how climate change is impacting not only the physical environment but also the intangible social and cultural fabrics that underpin community resilience.

11.3.2 Impact of Climate Change on Social Cohesion, Cultural Practices, and Traditional Knowledge Systems

The effects that climate change has on human societies have been far-reaching and profound that go beyond changing the change in the environment to severely influence the social cohesion, cultural practices, and traditional knowledge systems (Stephen, 2018). These interrelated attributes of social life determine the resilience of a person and the community to the effects of climate change. Since climate change has posed some challenges in addressing such social aspects, issues on how climate change alters social cohesion and culture and traditional knowledge will also be addressed in this section.

11.3.2.1 Impact on Social Cohesion: How Climate Change Fractures Communities

Social cohesion is explained as the bonds, trust, and common norms of conduct that cement the individuals in the community; social cohesion is a significant factor of resilience in human communities. One of the major disruptive forces is climate change, which causes the transformation of social relationships between individuals in the manifestation of using social networks, trust-building events, and the determination of the norms to follow in cooperative and reciprocal actions. This chapter examines how climate change causes direct and indirect consequences that undermine the social narrowness of a community across various ways in which 18 empirical research studies explain the issue.

Disruption of Cooperative Livelihoods and Reciprocity

Agricultural Collapse & Erosion of Communal Bonds

The physical and cultural foundations of agriculture are being displaced by environmental changes stemming from climate change. In coastal Bangladesh, rice fields are permanently flooded due to salinization (Assaduzzaman et al., 2023; Rakib et al., 2019) and erratic rainfall disrupts planting seasons in Ghana (Boafo et al., 2024) and Bolivia (McDowell & Hess, 2012). In addition, flood and riverbank erosion are destroying farmland in the char lands of Bangladesh (Al Mamun et al., 2023) and the Indian Himalayas (Joshi & Rawat, 2021). As farms are destroyed, people can no longer collaborate on farm-based efforts which is vital for social cohesion. For instance, *Nnoboa* (collective planting) is currently not feasible in Ghana due to erratic rainfall (Boafo et al., 2024) while the *Ashar Pandra* rice-transplanting ceremony is no longer celebrated in India since farmers are transitioning from paddy (rice) to commercial crops (Datta & Behera, 2024). In Bolivia, reciprocal labor exchange (*ayni, quri*) is declining as farmers prioritize their own cash needs over the tradition of community reciprocity (McDowell & Hess, 2012). Due to such changes, there is a general movement towards more individualistic livelihoods, such as commercial crop production (e.g. shrimp, French beans) or non-agricultural work, which results in the loss of valuable social bonds, and mutually dependent communal relationships which traditionally arose through the mutual cooperation of agricultural activities.

Resource Scarcity Fuels Conflict and Distrust

Crops destruction and climatic excesses enhance the pressure on scarce basic resources, which fracturing communities directly. In Bangladesh, the saline intrusion gives rise to intense competition over the limited fresh water, and this results in conflicts, refusal of access to ponds by neighbours and even secret night bathing (Budrudzaman et al., 2025). In Bolivia, land is limited, and scarce resources are at fault, as upstream residents disenfranchise downstream populations in the struggle to access the river (McDowell & Hess, 2012). On the same note, the system of land leasing in intensified agriculture in India leads to payment disagreement between tenants and landowners (Datta & Behera, 2024). Although fisheries in Oceania became weaker, which intensified competition, disruption of traditional sharing networks, which are needed in times of shortages (Tverskoi et al., 2024). Such widespread competition brings in distrust, ego conflicts, social distrust, intensifies tensions within the villages and disrupts social harmony, and instead of unity, it creates hostility and suspicion (Budrudzaman et al., 2025; Datta & Behera, 2024; Qaisrani et al., 2023). In such a way, the results demonstrate that resource scarcity is one of the leading stressors that transform the neighbours into competitors and disintegrate social networks.

Livelihood Shifts Exacerbate Inequality and Resentment

Livelihood transitions caused by climate change tend to create or exacerbate social fissures by disproportionately distributing the benefits and burdens of these transitions. In Bangladesh, the transition to shrimp farming (*gher*) is dominated by non-local investors which leads to huge power imbalances and economic inequalities that lead to crime and social instability (Rakib et al., 2019). Equally, inequality in Langtang Valley, Nepal also arises because tourism related households are more likely than the traditional agro-pastoral households to get humanitarian aid, which means the two groups will resent each other (Tuladhar et al., 2021). The ethnic minority groups in Bangladesh have been systemically disadvantaged in new opportunities (land rights, fair wages, and decisions), thus becoming trapped in high-risk areas, which exacerbates the old inequalities (Assaduzzaman et al., 2023). Some of the effects anticipated with the changing species distributions in the Pacific Islands include the fact that, these changes can become an obstacle to inter-tribal reciprocity and prestige networks, which are key stabilizers of social structures (Bond et al., 2019). Such unequal transitions disruption disturbs the more homogenous communities that have occurred and forms new divisions between the members of the community according to their access to the new economy or traditional peripheral positions.

Migration and Displacement

Forced Mobility Shattering Community Foundations

Study under review show that climate change directly contribute to displacement through processes such as riverbank erosion, where 25% of households in char regions have migrated more than three times over a ten-year period (Al Mamun et al., 2023). Cyclone-induced displacement has also led to family disintegration, as observed following Cyclone Aila (Malak & Lina, 2024). Similarly, cyclones and rising sea levels have contributed to displacement along coastal areas (Rakib et al., 2019). Livelihood collapse further contribute displacement, as reported by the agro-pastoral decline in Langtang, which forces residents to leave their home communities (Tuladhar et al., 2021). Displacement removes the physical and social anchors of community life, fracturing kinship networks as family members become dispersed (Al Mamun et al., 2023; Malak & Lina, 2024; Rakib et al., 2019). It also causes the loss or abandonment of shared community spaces where people can stay connected, and older adults are hit the hardest. They are often left behind by family members who move away, feel alone, or have trouble getting help (Malak & Lina, 2024). Collectively, these processes dissolve the geographic and relational fabric that holds communities together.

Trapped Populations and Social Isolation

Marginalized people have difficulty migrating due to numerous insurmountable and increasing barriers which ultimately increase the risk taken by the aforementioned social group(s) through repeated occurrences of isolating social experiences and circumstances that compound social inequality among the affected population; An example highlighted in the study of Munda, Bangladesh (Assaduzzaman et al., 2023) illustrates that marginalized populations are discriminated against through a lack of access to land rights, fair wage opportunities, and local governance, preventing successful migration due to the language barrier and discrimination issues. Consequently, this population is physically trapped (geographically) in high-risk salinized areas, hindering their ability to access a network of local community support and resource systems. At the same time, their exclusion during climate stress extenuates pre-existing inequalities and erodes any collective sense of community trust or resilience.

Youth Outmigration Draining Vitality and Continuity

Migration of young people seeking education or work elsewhere results in in the loss of vital human resources needed for community development. The Indian Himalayas saw a population decline of 34.7%—leaving these communities without the capability of working together as a group (Joshi & Rawat, 2021). There are severe labor shortages that are crippling communities' efforts to maintain critical shared infrastructure such as irrigation systems (Boafo et al., 2024; Joshi & Rawat, 2021; McDowell & Hess, 2012). Cultural festivals and communal ceremonies, such as the *Langshisha* festival and *Dhupa Chhiju* ceremony, no longer occur due to the lack of participants or leadership, as in the case of the Langtang area in Nepal (Tuladhar et al., 2021). This means that there will be significant impact upon intergenerational transfer of knowledge since elders now do not have someone to transfer their traditional agricultural, animal husbandry, environmental, and spiritual knowledge to (Joshi & Rawat, 2021; McDowell & Hess, 2012; Tuladhar et al., 2021). The loss of the younger generation also leads to cultural destruction and severely hinder the ability of the community to adapt in the future.

Erosion of Shared Spaces and Institutions

Destruction of Physical and Cultural Anchors

Climate-related disasters disrupts the physical connections that form the basis of community. In Bangladesh, cyclone destroy various market places and medical clinics that people depend on as places where they can conduct their daily lives and as sources of mutual help or cooperative action (Malak & Lina, 2024). In addition to these communities losing their physical infrastructure, climate change has also caused the communities to be cut off from the natural landscapes that are essential to their cultural identity and spiritual practices. For example, the cryosphere degradation in Nepal (Langtang Valley) cause disruptions to the connections between sacred mountains (Tuladhar et al., 2021) and the continuation of sea level rise and coastal erosion places culturally significant seascapes under threat in places like Oceania (*Hawai'i's Papahānaumokuākea*) and African American Churches located in vulnerable coastal wetlands (Friedlander, 2018; Hesed & Paolisso, 2015). This loss removes the physical venues for social bonding and the landscapes that embody shared heritage and meaning.

Collapse of Trust in Collective Governance

The issue of climate impacts unveils the flaws in the approach to responding to and fairly distributing resources through institutions and the loss of the social contract necessary to participate in collective action. Post disaster assistance may leave behind a perception of being left behind or inequitable distribution of aid (as in Crisfield, Maryland, in the aftermath of Hurricane Sandy) that may generate mistrust in both external agencies and the government agencies (Hesed & Paolisso, 2015). The systemic exclusion of African Americans from the decision-making process, whether through the inability to participate in discussions related to floodgate maintenance, disengaged residents of Char Land from interacting with local government, where 75–81% of residents did not seek assistance, or members of Munda communities from participating in local governance, demonstrates that these communities were left behind by institutions (Al Mamun et al., 2023; Assaduzzaman et al., 2023; Hesed & Paolisso, 2015). This type of "social-political isolation" generates feelings of resentment, discourages faith in collective responses to chronic issues, and significantly hinders local-level adaptation activities (Hesed & Paolisso, 2015).

Fracturing of Household and Social Fabric

Pressure from climate stress causes strain inside households, and disrupts the traditional social roles of men and women, especially in regards to gender. As a result of male outmigration (as much as 90% in some areas of Bangladesh) (Khalil et al., 2020), women are being forced into dual caretaking and income-earning positions that were previously occupied by men, are often experiencing social stigma for entering traditional "male" spheres such as fishing and construction (Assaduzzaman et al., 2023; Khalil et al., 2020), which is resulting in an alteration of the power dynamics within the household. Children (especially girls) will drop out of school or take on labor-intensive tasks such as collecting water to help their families, which is limiting their future opportunities (Khalil et al., 2020; Malak & Lina, 2024). In some cases, women are being burdened by climate-induced resource shortages (i.e., walking 3–6 h every day to collect saline water) (Assaduzzaman et al., 2023; Khalil et al., 2020; Nyahunda & Tirivangasi, 2022), which causes exhaustion and chronic health problems (i.e., urinary tract infections) (Budrudzaman et al., 2025), which in turn, is leading to increased domestic tension and reduced capacity for women to engage in community activities. Additionally, patriarchal norms (e.g., restrictive purdah) act to limit the mobility and collective action of women, which leads to a fragmentation of women's potential support networks (Assaduzzaman et al., 2023).

11.3.2.2 The Cultural and Knowledge Nexus: Underpinnings of Cohesion Under Threat

The erosion of social cohesion is intrinsically linked to the parallel loss of cultural practices and traditional knowledge systems:

Loss of Cultural Practices and Identity

Ritual Disruption & Identity Loss

When rituals that used to be done together no longer take place, they are actually removing part of the community's identity and uniting force. For instance, the salt levels in the water in Bangladesh is not only ruining the drinking water but also ruining the way *Wudu*—a daily activity that brings people together in their faith–is performed (Budrudzaman et al., 2025). In India, when farmers shift from planting rice to planting pineapples, they not only remove their ability to grow a critical crop but they are also removing the ability for people to participate in the *Ashar Pandra* and gather together at this time to share food, sing and work together (Datta & Behera, 2024). In addition, climate change causes certain species such as *néré* in West Africa, which is an important ingredient for making soumbala, to decline drastically; this will have a profound effect on the meaning, purpose and relationship that people have when performing their birth, marriage and death rites (Bond et al., 2019; Djoudi et al., 2022; Friedlander, 2018).

Livelihood and Dietary Shifts Undermining Self-sufficiency and Belonging

Culturally-based economic activities and food supply systems rely on community interdependence and self-sufficiency (Chettri et al., 2025). In the case of salinity in Bangladesh, saline intrusion is not only a problem for growing rice; it has also destroyed the ability of many rural people to maintain productive home gardens, which provides food for families, allows women to maintain their culturally-based roles and creates an important part of community identity. Furthermore, salinity has destroyed the freshwater fish farming system, eliminating both fish species that are an important part of people's diets and thousands of years of knowledge on aquaculture (Assaduzzaman et al., 2023; Budrudzaman et al., 2025; Rakib et al., 2019). The adoption of monoculture production has severed the linkages rural communities have to their land, traditional processing methods, and their right to breed their own food crops, and has caused to be replaced with market dependency on imports (Datta & Behera, 2024; Joshi & Rawat, 2021; McDowell & Hess, 2012; Rakib et al., 2019; Tuladhar et al., 2021). The change in Bolivia of producing *kaya* and *chuno* (processed freeze-dried potatoes), which used to provide for their families, to producing cash crops demonstrates the destruction of food-based cultural identity and buffers of risk (Rakib et al., 2019).

Severed Place-Based Connections

Communities are physically and spiritually disconnected from the landscape that serves as the basis for their identity and knowledge systems from the displacement and environmental degradation. Coastal erosion and sea level rise in Bangladesh and Oceania not only physically destroy homes, but they also destructively sever ties between people and their ancestral lands, their "cultural seascapes," and their sacred sites; these physical removals dissolve the geographic foundation of the individuals' sense of belonging and narrative (Friedlander, 2018; Qaisrani et al., 2023; Tverskoi et al., 2024). Additionally, youth becoming disengaged from the cryosphere (the mountains that are viewed as deities), as their primary source of income shifts from agriculture to tourism, exacerbates the spiritual and cultural disconnection of Nepal's Langtang community and the subsequent loss of place-based or indigenous knowledge (Tuladhar et al., 2021).

Traditional Knowledge System (TKS) Erosion: Breaking the Chain of Resilience

The disruption of TKS transmission and applicability directly weakens the adaptive capacity underpinning cohesion.

Broken Intergenerational Transmission

From the reviewed study, It has been identified that Climate driven migration among youth—for example, in the Himalayas, Bolivia, and Bangladesh—lack of interest in, and the death of elders due to climate change (i.e., Nepal earthquake) is contributing to the breaking down of the cultural transmission chain (Joshi & Rawat, 2021; McDowell & Hess, 2012; Rakib et al., 2019). Due to this breakdown, community members are no longer being taught how to select indigenous seeds, use environmental signals to predict changing weather patterns, traditional methods for controlling pests (e.g., *Vitex negundo* in India), harvest medicinal plants, understand natural disaster types (e.g., Avalanches in Nepal), and manage resources in specific locations (e.g., *Kuta* collection by *Maori*). The result is there is a decrease in potential skill sets within communities - "deskilling." (Al Mamun et al., 2023; Boafo et al., 2024; Bond et al., 2019; Budrudzaman et al., 2025; Datta & Behera, 2024; Joshi & Rawat, 2021; Malak & Lina, 2024; McDowell & Hess, 2012).

Outpaced by Unprecedented Change

Many TKS represent a diminished basis of reliability due to the rapid pace and unpredictability of the impacts of climate change. The unpredictability of rainfall patterns and changing temperatures in Ghana and Bolivia render traditional planting calendars unreliable, as do changing frost prediction methods (Boafo et al., 2024; McDowell & Hess, 2012). In Bangladesh, elders can no longer predict the occurrence/impact of cyclones using their traditional methods because the development of storms has become so rapid and intense, as was the case with Cyclone Aila (Rakib et al., 2019). The rates at which ocean acidification and coral bleaching are occurring in Oceania may exceed the limits to adaptive capacities within the lunar-calendar based fishing knowledge system (Friedlander, 2018).

Policy Marginalization Undermining Agency

The systematic exclusion of TKS and indigenous institutions from formal adaptation strategies actively weakens community agency and cohesion. Top-down strategies overlook useful activities such as Locally Managed Marine Areas in Oceania; customary water management in Porunto, Bolivia; mosaic agroforestry in West Africa; and subtle local flood pathways in African American settlements (Djoudi et al., 2022; Friedlander, 2018; Hesed & Paolisso, 2015; McDowell & Hess, 2012). Such ignorance of alternative ways of knowing does not only dismiss possible solutions but also undermines the knowledge of communities, which leads to the development of mistrust and dependence (Djoudi et al., 2022).

11.3.3 Resilience and Adaptation

Although the erosion story may have it, several studies also show the possibility of social cohesion, cultural practices, or traditional knowledge system being the source of resilience:

11.3.3.1 Role of Social Cohesion

The synthesis that follows, based on the literature, proves how social bonds make collective response to climate change possible.

Mutual Support and Resource Sharing

Food and Livelihood Security

Communities utilize social cohesion to create networks of familial and communal cooperation to support one another in times of crisis; for example, communities in Southern Africa use their support networks to borrow staple food (e.g. maize) when there is a drought or to collectively relocate livestock to communal grazing areas (called emlageni) when there is a shortage of pasture (Matandirotya et al., 2024). In Fiji, where the practice of solesolevaki, which is a component of traditional iTaukei culture, allows for the mobilization of labour to rebuild homes and restore farms in the aftermath of a cyclone ((Nakamura & Kanemasu, 2020). In the same way, women collectively prepare and distribute food and aid to those in need, rather than being given based on social status. In Kenya, women's self-help groups (e.g. Amani) are pooling their resources to enable members to borrow funds and diversify their livelihoods in times of food insecurity (Liru & Heinecken, 2021). Furthermore, the church provides a source of funding for members of different social classes through cross-class cooperatives.

Risk Pooling

The decentralization of social networks is an effective way to equalize resources amongst members of the community and provide additional protection against problems caused by large scale structural factors. In the Indigenous communities of Alaska, the use of reciprocity networks and so-called "super-households" keeps food production flowing from the community to the elders and/or vulnerable populations and ensures that food is plentiful throughout the entire community, even in times of shortages or disruptions (Scaggs et al., 2021). In Vanuatu, *rispek* (respect based relationships) help pool resources following destructive cyclones and kinship networks link people in urban areas with family members in the rural areas of the country to send remittances and provide temporary housing assistance (Granderson, 2017). In Micronesian cultures the concept of pooling risk through family obligations, especially through matrilineal kinship systems, provides for the basic needs (food/shelter) of displaced kin (Perkins & Krause, 2018). Additionally, the adoption of children is used strategically to redistribute labor and resources across islands. Finally, the *sawei* system helps formalize reciprocal forms of support across island groups, where individuals from the Main Islands (the more populated Islands) provide a parent like role to individuals from the Outer Islands (less populated Islands).

Collective Action for Disaster Response

The bayanihan spirit in the Philippines (Mabborang et al., 2022) is one such example of rapid mobilization through mutual aid, wherein communities come together to move light vernacular homes to higher ground during floods using a sound signal called the ammariaw to assemble workers and delegate tasks organically. Likewise, Micronesian cultures (Perkins & Krause, 2018) utilize matrilineal clan responsibilities, wherein scattered exogamous clans foster large kinship networks, to require shelter, food, and relocation assistance for affected members after a typhoon, in addition to cultural protocols such as the *sawei* system of reciprocal exchange.

On the other hand, Japanese fishermen in Soma (Mabon et al., 2020) showed self-organization in post-Fukushima recovery by building upon pre-existing local trust networks with fisheries extension officers, which enabled free communication, daily updates on radiation risk at fish markets, and joint trial fisheries for "safe" fishing, in direct contrast to the community's lack of trust in national institutions (e.g., TEPCO/government).

Trust Networks and Information Flow

Trust networks and information Flow appear as important resilience strategies, functioning through both decentralized governance and crisis communication. Decentralized Governance is based on localized, kinship, or customary systems to establish trust and facilitate adaptive decision-making. For example, Sámi reindeer herders (Horstkotte et al., 2022) use silda units to negotiate pasture entry during cold winters, while the Comcaac of Mexico (Martínez-Tagüeña & Renteria-Valencia, 2020) use fluid band organizations (*ihiizitim*) and sharing rituals (*kimosin*) for justifiable resource sharing. In a similar fashion, Micronesian societies (Perkins & Krause, 2018) rely on matrilineal clan commitments and adoption strategies to share labor and resources across islands during typhoons.

On the other hand, successful Crisis Communication requires pre-existing, trusted intermediaries to enable information flow during environmental disturbances. In

Tacana societies, Bolivia (Bauer et al., 2022), community leaders strategically coordinated external assistance from NGOs and the government during floods by connecting local and external networks. In a similar fashion, African American churches in Crisfield, USA (Hesed & Paolisso, 2015), functioned as trusted nodes to coordinate flood relief efforts and share information about available aid through social networks. Tibetan pastoralists (Hopping et al., 2016) also function in this manner, using village leaders to mediate knowledge, resolve grazing disputes, and contest state policies that are maladaptive.

Institutional and Governance Resilience

Forest-dwelling communities in Suriname (Smith et al., 2023) use their high level of social cohesion and shared values to facilitate balanced community decision-making for climate change adaptation, while Peruvian ecotourism groups (Gabriel-Campos et al., 2021) use democratic elections of community leaders to establish the trust and legitimacy required to implement collective actions such as lockdowns during the pandemic.

At the same time, the Comcaac of Mexico (Martínez-Tagüeña & Renteria-Valencia, 2020) developed hybrid systems that combine their original reciprocity networks (*kimosin*) with newly elected councils and elders' institutions to balance the pressures of the outside world and resist the incursion of extractive industries, showing adaptive resilience in their institutions that are rooted in, yet transforming, traditional social relationships.

Psychological and Emotional Buffering

In Fijian communities (Nakamura & Kanemasu, 2020), social cohesion facilitates community-based trauma processing strategies such as storytelling (tadanoa) and humor in response to climatic disasters, in addition to group prayers that consolidate community asylum and psychological resilience. In a similar manner, African American churches (Hesed & Paolisso, 2015) use their long-standing social relationships to offer spiritual support and hope in times of crisis such as floods, turning their shared religious beliefs into emotional buffers.

Examples of Hawaiian systems (McMillen et al., 2017) illustrate identity-based resilience, in which a place-based cultural identity, fostered by ahupuaʻa stewardship groups and values such as *aloha kekahi i kekahi* (love one another), bonds communities against climate change, embedding a sense of belonging that underpins collective resilience. These bonding social networks, therefore, act as key buffers, translating a shared cultural identity and emotional support into adaptive psychological resources during climatic disturbances.

But social cohesion has a radically limited role in resilience building in the face of social fragmentation and lack of policy congruence; examples include that rapid urbanization of Bangladesh (Assaduzzaman et al., 2023) leaves ethnic minorities out of central decision-making networks, disaggregating social networks, and that top-down development strategies tend to disrupt the adaptive social networks, such as the state boundaries dividing Sami pastoralist groups in Scandinavia (Horstkotte et al., 2022), or the traditional norms of reciprocity among Mongolian Pastoralist, that eventually reduce collective capacity (Chao et., 2023).

Social cohesiveness plays an important role in climate adaptation, which enables communities to self-organize, co-risk, and mobilize fast. Nevertheless, it can play the role of resilience building, but this depends on inclusive governance and protection against external disruptions.

11.3.3.2 Role of Cultural Practices

Cultural practices act as key processes towards the development of adaptive capacity by engaging in ritualized environmental stewardship, passing of intergenerational knowledge as well as place-based innovations. Key findings include:

Rituals and Ceremonies as Risk Management Tools

The rituals and ceremonies form a significant element of risk management tools that bring proactive adaptation and psychological resilience to the lives of communities. As a case in point, in Vanuatu (Bryant-Tokalau, 2018), collective reaction to imminent risk can be facilitated by disaster preparedness rituals like pre-storm roof-tying, or in Bangladesh (Datta & Kairy, 2024), by pre-cyclone dry food storage. Post-disaster rites like the *Monu ain ba lulik* of Timor-Leste restore the social bonds, as well as encouraging environmental stewardship through the symbolic reconciliation with nature (Hiwasaki et al., 2014). The agricultural rituals, like the *kahilamun* field-clearing rituals by the Ifugao, support ecological reciprocity (e.g., offering to ensure productive harvests) and also place the community in the context of cultural identity, thereby boosting psychological resiliency to climate change (Delina et al., 2025). These ceremonies translate abstract climatic dangers into events that can be culturally handled, combining preparedness with spiritual and social empowerment.

Identity-Preserving Practices

The practices that are identity sustaining play an important role in instilling resilience through connection with their heritage in times of crises. The revival of the *Soma Nomaoi* samurai festival in Fukushima, Japan, provides an example because it is necessary to restore the sense of community purpose and overcoming trauma as it constructs social cohesion and continuity in the disaster aftermath (Mabon et al., 2020). In addition, the restoration of *Jincang* embroidery in Quanzhou, China, is important in ensuring that intangible cultural heritage is adapted into climate-resilient livelihoods (such as tourism and design sectors), which are able to adapt economically during a transition to a knowledge economy while maintaining intergenerational skills and place-based identity (Shen et al., 2022). At the same time, the matrilineal *tabinaw* land tenure protocols of Yap's matrilineal society maintain resource access rights and social responsibilities in the face of urbanization pressures, which ensures fair distribution during climate-related disasters through culturally embedded governance and kinship structures (Perkins & Krause, 2018).

Stewardship-Embedded Livelihoods

Tabu systems (e.g., temporal/spatial fishing bans in Vanuatu and the Philippines) are community-based ecological insurance that limits the use of resources during critical periods to ensure recovery and availability, directly conserving biodiversity and protecting against climate-driven scarcity (Granderson, 2017; Maliao et al., 2025).

Moreover, dynamic architectural cultures, such as the Philippine *bayanihan* (mutual aid) tradition, facilitate rapid and inexpensive avoidance of disasters; the community-wide relocation of light vernacular dwellings using communal labor reduces flood and landslide damage while fostering community cohesion during disasters (Mabborang et al., 2022).

Moreover, the gendered knowledge systems like women networks of seed conservation and exchange in Kenya actively conserve the essential crop diversity in the event of droughts; the social system ensures food sovereignty and adaptive agriculture in situations where foreign inputs fail (Liru & Heinecken, 2021).

Value Transmission Mechanisms

Values that are instilled in oral cultures, including Lun Bawang folk tales (*beras dan harta*), include the readiness to face the unknown, the need to preserve the transmission of responsive attitudes across the generations (Hosen et al., 2020). Place-based ecological knowledge ceased by economic shocks is fashioned through ritual apprenticeships in Hawaiian fishing or planting practices which bridges the gap between ancient and contemporary challenges (McMillen et al., 2017). Adaptive behaviors are normalized through the conceptualization of resilience as a key element of cultural identity, which is proverbial wisdom, as in Tagalog metaphors (e.g., bamboo as the flexibility) (Shah et al., 2017). Storytelling, apprenticeship and symbolic language, respectively, are part of these mechanisms, and, when combined, enable the communities to remain strong against disturbances in the environment.

Cultural practices are thus not fossilized things, but systems. Their surpassing abilities to resist decline disconnected to land, generational perpetuation, and shared government. Resilience then has to be culturally grounded living infrastructure (Delina et al., 2025; Turner et al., 2025).

11.3.3.3 Role of Traditional Knowledge System (TKS)

By facilitating local ecological knowledge of climate threats, TKS can become the main pillars of adaptive capacity and resilience as they allow communities to identify, interpret, and manage climate threats independently. Key mechanisms include:

Early Warning Systems and Preparedness

Droughts, floods, storms, and seasonal changes are predicted contextually specific observations of both biotic (bird migration, frog callings, nests moved by hornets, and unusual fruiting) and abiotic (moon halos, wind direction, mist patterns, and tides) indicators, which communities use to make predictions (Hosen et al., 2020; Matandirotya et al., 2024; Nakamura & Kanemasu, 2020; Perkins & Krause, 2018). This enables proactive measures to be implemented, including food storage, livestock moving, changed planting dates, evacuation, and allocating resources (Bauer et al., 2022; Maliao et al., 2025; Nyahunda & Tirivangasi, 2022).

Historical memory of TKS, which was preserved through oral history, rituals, and landmarks (tsunami shrines in Japan and narratives of resilience in the Philippines, which is a yearly cycle tied to the tsunami threat), is an important component of disaster resilience planning in communities since it is necessary to know the past effects of disasters, recovery strategies, and patterns of risks over time to make informed decisions (Mabon et al., 2020; McMillen et al., 2017; Shah et al., 2017).

Ecosystem-Based Adaptation & Sustainable Resource Management

TKS supports the ecological wisdom of sustainable resource management practice including sustainable harvesting practice, including controlled fishing practice based on season indicational practices (Maliao et al., 2025; Wongbusarakum et al., 2021) and innovation in water resource management practice, including excavation of deeper wells during dry conditions (Matandirotya et al., 2024) or conservation of ancient gravity-fed irrigation systems (*falaj/aflaj*) in arid networks (AlShalan and Al-Saud, 2025). Besides, TKS accelerates ecological recovery and conservation, such as the planting of degraded habitat with native, climate-tolerant species (Gabriel-Campos et al., 2021) and the imposition of the conservation of important habitats like mangroves and forests with the help of traditional regulations and taboos (Hiwasaki et al., 2014; Rizki & Asteria, 2023).

Above all, TKS guides complex spatial resource management activities such as agroforestry practices that maximize the use of watersheds by using zone-based agroforestry (Perkins & Krause, 2018), pasture rotation (*otor*) practices that maintain the health of a grassland (Chao et al., 2023), and maintaining a diverse plot with different elevations (Gabriel-Campos et al., 2021; McMillen et al., 2017).

Livelihood Diversification and Food Security

TKS supports diversification approaches that are essential for building resilience, for example, through the adoption of drought-, flood-, and salt-tolerant crops (Khalil et al., 2020; Liru & Heinecken, 2021), intercropping, seasonal foraging of wild foods (Matandirotya et al., 2024; Scaggs et al., 2021), and the incorporation of new sources of income (e.g., handicrafts, ecotourism) based on local resources and knowledge (Gabriel-Campos et al., 2021; Khalil et al., 2020).

In addition, traditional food storage and preservation methods (e.g., elevated rice stores *silap padi*, dried breadfruit) ensure food availability during times of scarcity (Granderson, 2017; Hosen et al., 2020; Scaggs et al., 2021).

Autonomous Adaptation and Self-reliance

By drawing on local resources and local expertise, TKS supports low-cost and resource-saving practices, including the shift of planting seasons based on ecological indicators (e.g., bird life or moon phase), the development of disaster-resistant homes based on bamboo or rattan, or the development of natural pesticides based on local plants (Khalil et al., 2020; Mabborang et al., 2022; Nyahunda & Tirivangasi, 2022; Smith et al., 2023).

Such strategies not only reduce the levels of reliance of communities on external support, but also make sure that adaptation is ecologically and culturally situated and that self-determination is upheld. As an example, Pacific countries are encouraging nature-based solutions like multipurpose planting of pandanus before imported infrastructures, and Inner Mongolian herders are using nature-based methods of managing pastures through rituals-based taboos as opposed to the top-down controls, like grazing prohibitions, which ignore local conditions (Chao et al., 2023; Datta & Kairy, 2024; Turner et al., 2025).

Dynamic Knowledge Evolution and Hybrid Systems

TKS is evolving; it is changing according to new challenges. As an example, the Comcaac youth in Mexico combine scholarly education in biology and law with the ancestral species monitoring with the development of new TKS applications (Martínez-Tagüeña & Rentería-Valencia, 2020). Likewise, to react to current ecological stressors, Mongolian herders create new environmental regulations in *Ovoo* rituals, including the prohibition of hunting marmots or using cars (Chao et al., 2023). In the Philippines, fishers have preserved the main principles and introduced new materials to the traps such as PET bottles as weaving knowledge reduces (Maliao et al., 2025).

More importantly, integrating TKS and science can be most effective to increase adaptive capacity, as observed in Micronesian societies that require integrated information to manage fisheries (Wongbusarakum et al., 2021), *Maori* communities integrating *matauranga* and geospatial technology in the *Ki Uta Ki Tai* framework (Reid et al., 2024), and Tacana communities in Bolivia who need to have the local observations validated by meteorological information (Bauer et al., 2022). Nevertheless, such integrations can remain biased toward TKS, such as Sami reindeer herders, in which subtle snow classifications and sustainability measures have been disregarded by the state-imposed carcass-weight measures in Norway (Horstkotte et al., 2022).

Foundation for Culturally-Grounded Solutions

TKS lays the necessary basis of the culturally-grounded climate adaptation solutions. These systems provide the ecological knowledge local to the place - such as the knowledge of pasture rotation by Inner Mongolian herders using the term of *baigal/baidal* (ecological balance) (Chao et al., 2023) or the Maori *matauranga* used to understand holistic management of the so-called *ki uta ki tai* (mountains-to-sea) watershed (Reid et al., 2024), which is rooted in the local worldviews and values like reciprocity with nature such as, Mongolian *ayni* The resulting social acceptability and long-term sustainability are ensured by this cultural grounding, such as in the practices such as the climate-active vernacular architecture made of local bamboo in the Philippines (Mabborang et al., 2022) or the so-called *tabu* systems that preserve marine resources in Vanuatu (Granderson, 2017).

Moreover, TKS tend to produce more cost-effective and ecologically competent solutions, including locally available materials to disastrous housing (Mabborang et al., 2022) or biotic indicators of an upcoming disaster (e.g., the behavior of birds, the season of plants) (Hosen et al., 2020; Nyahunda & Tirivangasi, 2022), which makes TKS inherently more feasible than technologically imposed solutions.

Despite its high effectiveness, TKS is threatened by land dispossession, state policies in support of scientific methods, such as the reindeer carcass measurements in Norway superseding Sami knowledge (Horstkotte et al., 2022), and migration by young people. Effective resilience is achieved by legalizing the rights of Indigenous (e.g., ICCAs) and co-management systems that integrate TKS with scientific data (Datta & Kairy, 2024; Smith et al., 2023).

11.4 Discussion

Our review demonstrates that climate change is not merely a problem of the environment or economy; it is a highly social one that can tear society right away. This review discovers that there is an interwoven web of stories at work here: the story of discontinuity and destruction, and the story of stability and adaptation which is founded on cultural continuity, social integration, and cultural knowledge. These, in their turn, challenge the mainstream techno-centric climate adaptation discourses.

11.4.1 Climate Change as a Driver of Social Disintegration

The findings of this review indicate that climate change is the factor that raises the social vulnerabilities and that it forms new forms of disintegration. Disintegration is caused by the breakdown of co-operative systems of livelihood, displacement, conflict of livelihood resources and inequalities in livelihood. All these are not chances, but rather systemic in nature, indicating a trend under which climate change stressors unveil and amplify the cracks in the systems and structures of governance in communities. These results are consistent with previous research (e.g., Adger et al., 2013; Feola, 2015) that has linked climate change stress to social vulnerability, but this review goes further to describe in detail how this erosion takes place in the breakdown of systems of reciprocity, ritual participation, and place-based identities.

It is also important to note that this review points out that climate change migration, whether it is forced, voluntary, or aspirational, is a factor that changes the demographic structures of society, increases the rate of intergenerational disconnection, and results in the loss of critical cultural knowledge. The loss of youth from mountain and coastal areas, for example, is a factor that not only deprives society of labor but also deprives society of the bearers of cultural knowledge and future leadership. This aspect of climate change migration has not been adequately explored in the climate change policy literature, which tends to view migration as a strategy of adaptation rather than a factor of socio-cultural disintegration.

11.4.2 Cultural and Knowledge Erosion: Invisible Losses with Lasting Impacts

More than physical displacement, loss of intangible cultural heritage, rituals, language, cuisine and spiritual attachment to land has been shown as one of the key themes. They are usually unseen losses which are not measured or even included in climate adaptation planning, yet their impact is far reaching. The discontinuity between ancestral lands and the loss of ceremonies and the traditional knowledge systems destroys the sense of identity and emotional strength that is the core of long-term adaptive potential.

This is in line with the accumulating literature that questions mainstream adaptation strategies as failing to take cultural factors into consideration (Redvers et al., 2023; Tschakert et al., 2017). Unless cultural knowledge is actively included in the process of adaptation, it may be disempowering, alienating, or even counterproductive, in particular, to Indigenous and marginalized communities.

11.4.3 Culture, Cohesion, and Knowledge as Foundations of Resilience

Nevertheless, the strong counter-narrative also emerges in this review, which states that culture and social cohesion, despite the disruptions, are not only endangered by climate change, but it is also seen as a power source. Communities with social cohesion mobilize faster, risk-share more, and their group morale does not decline during trauma. Likewise, the environmental stewardship and preparedness are institutionalized in rituals and place-based knowledge systems. These findings are in line with the speculations of other researchers like Aldrich and Meyer (2015) and Adger et al. (2013) who perceive social capital and cultural identity as a key resource in adaptive governance.

Interestingly, the review cites the fact that TKS does not only offer historical knowledge, but it also offers dynamic and context-specific responses. TKS enhances self-sufficiency and low-cost adaptation, such as biotic warning systems through early warning indicators and taboo-based conservation systems and climate-resistant vernacular architecture. Nevertheless, one of the most significant adaptation barriers is its lack of concern by the state and scientific systems.

11.4.4 Hybrid Models and the Challenge of Integration

Despite the potential of hybrid knowledge systems, the integration of traditional and scientific knowledge systems, the review also cautions about the perils of mere integration. The power balances are often not eliminated and TKS is relegated into second instead of first position. The cases of such situations are Sami reindeer herders and Mongolian pastoralists who have been subject to state-imposed models that ignore their epistemologies (Chao et al., 2023; Horstkotte et al, 2022). The fundamental changes needed include legal recognition of Indigenous rights, governance, and epistemic justice in climate science, to affect the effective integration and implementation.

11.4.5 Implications for Policy, Practice and Future Research

The implications of the results are that the climate policy and programming should change its orientation towards culturally informed and socially inclusive strategies. These are the recommendations:

To begin with, the adaptation planning practice should focus on cultural loss as a critical non-economic form of loss and destruction. Secondly, enhancing local institutions and the customary governance systems as spaces of participatory adaptation. Thirdly, the investment in intergenerational knowledge transfer in the form of education, mentorship and revitalization of rituals. Fourthly, safeguarding of shared land ownership and access to sacred places in order to maintain cultural continuity and place-based resilience. And, fifth, the value co-creation of TKS and scientific methodologies that are equitable and also manage power imbalances through the development of hybrid adaptation models. The limitation of this review is that it uses peer-reviewed English-language literature, which could not represent some of the geographies or oral knowledge systems. Also, although it is a synthesis of empirical case-studies, further comparative and longitudinal studies are required to determine how cultural and social resilience changes with time and compounding stressors. It is also worthwhile to examine in future how digital tools, youth engagement, and urban migration is transforming the transmission and transformation of traditional knowledge and cultural identity, in terms of climate change.

11.5 Conclusion

This review has noted that climate change is indeed an ecological crisis but a profound social and cultural crisis, and that this is tearing apart the social glue and is weakening the cultural continuity and excluding the traditional knowledge systems that have long held people together. The overall result is the destruction of the pillars that constitute collective resilience and whole being during times of adversity.

Nevertheless, even in a state of fragmentation, strong currents of resilience are also discerned in the literature review. Communities are no longer passive receivers but they are actively engaging their social networks, cultural practices and traditional knowledge to adapt and innovate to face the climate change. These culturally specific resilience practices provide both scaleable and context-focused climate adaptation opportunities, including support networks, ecological stewardship practices and knowledge systems.

Nevertheless, to be fair and effective, adaptation must transcend infrastructure and technocratic ways of engagement extending into the lived experiences, identities, and agency of communities. The adaptation policies should see culture as an alive system of strength and not a fixed part of communities that can be pushed aside. Rather they should foster inclusive governance that recognizes the various ways of how communities understand and react to climate change. Ultimately, a resilience strategy that looks forward must make these aspects the core of such resilience, so that adaptation is not just sustainable but well-founded, rooted, and meaningful. The future of human-kind and earth is in danger.

References

Aall, P., & Crocker, C. A. (2019). Building resilience and social cohesion in conflict. *Global Policy, 10*(S2), 68–75. https://doi.org/10.1111/1758-5899.12681

Adger, W. N., Barnett, J., Brown, K., Marshall, N., & O'Brien, K. (2013). Cultural dimensions of climate change impacts and adaptation. *Nature Climate Change, 3*(2), 112–117.
Al Mamun, A., Islam, A. R. M. T., Alam, G. M. M., Sarker, M. N. I., Erdiaw-Kwasie, M. O., Bhandari, H., & Mallick, J. (2023). Livelihood vulnerability of char land communities to climate change and natural hazards in Bangladesh: An application of livelihood vulnerability index. *Natural Hazards, 115*(2), 1411–1437. https://doi.org/10.1007/s11069-022-05599-y
Aldrich, D. P., & Meyer, M. A. (2015). Social capital and community resilience. *American Behavioral Scientist, 59*(2), 254–269. https://doi.org/10.1177/0002764214550299
AlShalan, M. S., & Al-Saud, N. T. (2025). Weaving resilience: The enduring threads of GCC communities. In A. Arora, F. Belaïd, & S. Lechtenberg-Kasten (Eds.), *Climate-resilient cities. Contributions to economics*. Springer. https://doi.org/10.1007/978-3-031-73090-0_9
Assaduzzaman, M., Filatova, T., Lovett, J. C., & Coenen, F. H. J. M. (2023). Gender-ethnicity intersectionality in climate change adaptation in the coastal areas of Bangladesh. *Sustainability (Switzerland), 15*(4). https://doi.org/10.3390/su15043744
Bauer, T. N., de Jong, W., & Ingram, V. (2022). Perception matters: An Indigenous perspective on climate change and its effects on forest-based livelihoods in the Amazon. *Ecology and Society, 27*(1). https://doi.org/10.5751/ES-12837-270117
Boafo, J., Yeboah, T., Guodaar, L., Stephanie, Y., & Nyantakyi-Frimpong, H. (2024). Understanding non-economic loss and damage due to climate change in Ghana. *Climate and Development, 16*(2), 109–118. https://doi.org/10.1080/17565529.2023.2183074
Bond, M. O., Anderson, B. J., Henare, T. H. A., & Wehi, P. M. (2019). Effects of climatically shifting species distributions on biocultural relationships. *People and Nature, 1*(1), 87–102. https://doi.org/10.1002/pan3.15
Braun, V., & Clarke, V. (2006). Using thematic analysis in psychology. *Qualitative Research in Psychology, 3*(2), 77–101. ISSN 1478-0887
Bryant-Tokalau, J. (2018). Indigenous knowledge systems and urbanization: Relocation, planning and modern disasters. In *Indigenous Pacific approaches to climate change. Palgrave studies in disaster anthropology*. Palgrave Pivot. https://doi.org/10.1007/978-3-319-78399-4_5
Budrudzaman, M., Bärnighausen, K., Selim, S., & Irfanullah, H. M. (2025). Water, water, everywhere, nor any drop to drink: Qualitatively unraveling the nuances of non-economic loss and damage in Bangladesh. *Local Environment*, 1–20. https://doi.org/10.1080/13549839.2025.2496158
Chao, O., Li, X., & Reyes-García, V. (2023). Faith, reciprocity, and balance: Inner Mongolian Ovoo offering ritual and its contribution to climate change adaptation. In *Routledge handbook of climate change impacts on indigenous peoples and local communities* (pp. 348–357). https://doi.org/10.4324/9781003356837-26
Datta, R., & Kairy, B. (2024). Decolonizing climate change adaptations from indigenous perspectives: Learning reflections from Munda indigenous communities, coastal areas in Bangladesh. *Sustainability (Switzerland), 16*(2). https://doi.org/10.3390/su16020769
Datta, P., & Behera, B. (2024). Assessing farmers' maladaptation to climate change in a sub-Himalayan region of India. *Environment, Development and Sustainability, 26*(12), 30621–30638. https://doi.org/10.1007/S10668-023-03925-3/TABLES/1
Delina, L. L., Fuerzas, I., Dulay, M. J., Ludovice, N. P., Tam, K. P., & Salamanca, A. (2025). Affective relationships as non-material capital assets in building resilience within the rice-terraced cultural landscapes of the Philippine Cordillera. *Applied Geography*, 179. https://doi.org/10.1016/j.apgeog.2025.103631
Dembedza, V. P., Chopera, P., Mapara, J., & Macheka, L. (2022). Impact of climate change-induced natural disasters on intangible cultural heritage related to food: A review. *Journal of Ethnic Foods, 9*(1), 1–11. https://doi.org/10.1186/S42779-022-00147-2/FIGURES/1
Djoudi, H., Locatelli, B., Pehou, C., Colloff, M. J., Elias, M., Gautier, D., Gorddard, R., Vinceti, B., & Zida, M. (2022). Trees as brokers in social networks: Cascades of rights and benefits from a Cultural Keystone Species. *Ambio, 51*(10), 2137–2154. https://doi.org/10.1007/S13280-022-01733-Z/FIGURES/7

Feola, G. (2015). Societal transformation in response to global environmental change: A review of emerging concepts. *Ambio, 44*(5), 376–390. https://doi.org/10.1007/S13280-014-0582-Z/FIGURES/2

Friedlander, A. M. (2018). Marine conservation in Oceania: Past, present, and future. *Marine Pollution Bulletin, 135*, 139–149. https://doi.org/10.1016/j.marpolbul.2018.05.064

Gabriel-Campos, E., Werner-Masters, K., Cordova-Buiza, F., & Paucar-Caceres, A. (2021). Community eco-tourism in rural Peru: Resilience and adaptive capacities to the Covid-19 pandemic and climate change. *Journal of Hospitality and Tourism Management, 48*, 416–427. https://doi.org/10.1016/j.jhtm.2021.07.016

Granderson, A. A. (2017). The role of traditional knowledge in building adaptive capacity for climate change: Perspectives from Vanuatu. *Weather, Climate, and Society, 9*(3), 545–561. https://doi.org/10.1175/WCAS-D-16-0094.1

Hesed, C. D. M., & Paolisso, M. (2015). Cultural knowledge and local vulnerability in African American communities. *Nature Climate Change, 5*(7), 683–687. https://doi.org/10.1038/nclimate2668

Higgins, N. (2022). Changing climate; Changing life—Climate change and indigenous intangible cultural heritage. *Laws, 11*(3). https://doi.org/10.3390/laws11030047

Hiwasaki, L., Luna, E., Syamsidik, & Shaw, R. (2014). Process for integrating local and indigenous knowledge with science for hydro-meteorological disaster risk reduction and climate change adaptation in coastal and small island communities. *International Journal of Disaster Risk Reduction, 10*, 15–27. https://doi.org/10.1016/j.ijdrr.2014.07.007

Hopping, K. A., Yangzong, C., & Klein, J. A. (2016). Local knowledge production, transmission, and the importance of village leaders in a network of Tibetan pastoralists coping with environmental change. *Ecology and Society, 21*(1). https://doi.org/10.5751/ES-08009-210125

Horstkotte, T., Heikkinen, H. I., Næss, M. W., Landauer, M., Forbes, B. C., Risvoll, C., & Sarkki, S. (2022). Implications of norms and knowledge in customary reindeer herding units for resource governance. In *Reindeer husbandry and global environmental change: Pastoralism in Fennoscandia* (pp. 133–149). https://doi.org/10.4324/9781003118565-11/IMPLICATIONS-NORMS-KNOWLEDGE-CUSTOMARY-REINDEER-HERDING-UNITS-RESOURCE-GOVERNANCE-TIM-HORSTKOTTE-HANNU-HEIKKINEN-MARIUS-WARG-N

Hosen, N., Nakamura, H., & Hamzah, A. (2020). Adaptation to climate change: Does traditional ecological knowledge hold the key? *Sustainability (Switzerland), 12*(2). https://doi.org/10.3390/su12020676

IPCC (2014). Climate Change 2014: Synthesis report. In R. K. Pachauri, L.A. Meyer, & Core Writing Team (Eds.), *Contribution of working groups I, II, and III to the fifth assessment report of the Intergovernmental Panel on Climate Change* (151 p.). IPCC.

Joshi, N. C., & Rawat, G. S. (2021). An integrated approach for the identification and prioritization of areas based on their livelihood vulnerability index: A case study of agro-pastoral community from Western Indian Himalaya. *Mitigation and Adaptation Strategies for Global Change, 26*(6). https://doi.org/10.1007/s11027-021-09962-5

Khalil, M. B., Jacobs, B. C., McKenna, K., & Kuruppu, N. (2020). Female contribution to grassroots innovation for climate change adaptation in Bangladesh. *Climate and Development, 12*(7), 664–676. https://doi.org/10.1080/17565529.2019.1676188

Liru, P., & Heinecken, L. (2021). Building resilience: The gendered effect of climate change on food security and sovereignty in kakamega-kenya. *Sustainability (Switzerland), 13*(7). https://doi.org/10.3390/su13073751

Mabborang, Ma. H., Nozaleda, B. M., Maguddayao, R. N., Udaundo, L., Laggui, N., Martin, Engr. B., & Sibal, C. (2022). Vernacular house architecture and climate change adaptation: Lessons from the indigenous peoples of Cagayan, Philippines. *Journal of Climate Change, 8*(4), 25–33. https://doi.org/10.3233/jcc220027

Mabon, L., Kawabe, M., Huang, Y. C., Moller, L., Gu, J., Wakamori, D., Narita, K., Ito, T., Matsumoto, A., Niizeki, K., Suzuki, S., & Watanabe, M. (2020). Inherent resilience, major marine environmental change and revitalisation of coastal communities in Soma, Fukushima

Prefecture, Japan. *International Journal of Disaster Risk Reduction, 51*, Article 101852. https://doi.org/10.1016/J.IJDRR.2020.101852

Malak, M. A., & Lina, N. K. (2024). Tropical cyclones and the mobility of older persons: Insights from coastal Bangladesh. Sustainable Development Goals Series, Part F3368 (pp. 17–33). https://doi.org/10.1007/978-981-97-3234-0_2/TABLES/1

Maliao, R. J., Cahilig, R. C., Cahilig, R. R., & Jaspe, B. T. (2025). Climate change, traditional ecological knowledge, and riverine biodiversity conservation: A case in Aklan, Central Philippines. *Environment, Development and Sustainability, 27*(2), 4745–4767. https://doi.org/10.1007/S10668-023-04096-X/TABLES/2

Markus, A., & Kirpitchenko, L. (2007). Conceptualising social cohesion. *Social Cohesion in Australia*, 21–32. https://doi.org/10.1017/CBO9780511481574.004

Martínez-Tagüeña, N., & Rentería-Valencia, R. F. (2020). Social cohesion and environmental governance among the Comcaac of Northern Mexico. *Springer Climate* (pp. 321–335). https://doi.org/10.1007/978-3-030-22464-6_18/FIGURES/2

Matandirotya, N. R., Filho, W. L., Mahed, G., Dinis, M. A. P., & Mathe, P. (2024). Local knowledge of climate change adaptation strategies from the vhaVenda and baTonga communities living in the Limpopo and Zambezi River Basins, Southern Africa. *Inland Waters, 14*(3), 268–280. https://doi.org/10.1080/20442041.2024.2437932

McDowell, J. Z., & Hess, J. J. (2012). Accessing adaptation: Multiple stressors on livelihoods in the Bolivian highlands under a changing climate. *Global Environmental Change, 22*(2), 342–352. https://doi.org/10.1016/j.gloenvcha.2011.11.002

McMillen, H., Ticktin, T., & Springer, H. K. (2017). The future is behind us: Traditional ecological knowledge and resilience over time on Hawaii Island. *Regional Environmental Change, 17*(2), 579–592. https://doi.org/10.1007/s10113-016-1032-1

McNamara, K. E., Westoby, R., & Chandra, A. (2021). Exploring climate-driven non-economic loss and damage in the Pacific Islands. *Current Opinion in Environmental Sustainability, 50*, 1–11. https://doi.org/10.1016/J.COSUST.2020.07.004

Nakamura, N., & Kanemasu, Y. (2020). Traditional knowledge, social capital, and community response to a disaster: resilience of remote communities in Fiji after a severe climatic event. *Regional Environmental Change, 20*(1). https://doi.org/10.1007/s10113-020-01613-w

Nnandi, F. (2015). The impact of climate change on African traditional religious practices. *Journal of Earth Science & Climatic Change, 5*(15), 2225–0484.

Nyahunda, L., & Tirivangasi, H. M. (2022). Adaptation strategies employed by rural women in the face of climate change impacts in Vhembe district, Limpopo province, South Africa. *Management of Environmental Quality: An International Journal, 33*(4), 1061–1075. https://doi.org/10.1108/MEQ-09-2021-0207

Nyong, A., Adesina, F., & Osman Elasha, B. (2007). The value of indigenous knowledge in climate change mitigation and adaptation strategies in the African Sahel. *Mitigation and Adaptation Strategies for Global Change, 12*(5), 787–797. https://doi.org/10.1007/S11027-007-9099-0/METRICS

Orr, S. A., Richards, J., & Fatorić, S. (2021). Climate change and cultural heritage: A systematic literature review (2016–2020). *Historic Environment: Policy and Practice, 12*(3–4), 434–477. https://doi.org/10.1080/17567505.2021.1957264

Pearson, J., Jackson, G., & McNamara, K. E. (2021). Climate-driven losses to knowledge systems and cultural heritage: A literature review exploring the impacts on Indigenous and local cultures. *The Anthropocene Review, 10*(2), 343–366. https://doi.org/10.1177/20530196211005482. (Original work published 2023).

Perkins, R. M., & Krause, S. M. (2018). Adapting to climate change impacts in yap state, federated states of Micronesia: The importance of environmental conditions and intangible cultural heritage. *Island Studies Journal, 13*(1), 65–78. https://doi.org/10.24043/isj.51

Qaisrani, S. A., et al. (2023). Climate change and social concerns. In W. N. Jatoi, M. Mubeen, M. Z. Hashmi, S. Ali, S. Fahad, & K. Mahmood (Eds.), *Climate change impacts on agriculture*. Springer. https://doi.org/10.1007/978-3-031-26692-8_3

Rakib, M. A., Sasaki, J., Pal, S., Newaz, M. A., Bodrud-Doza, M., & Bhuiyan, M. A. H. (2019). An investigation of coastal vulnerability and internal consistency of local perceptions under climate change risk in the southwest part of Bangladesh. *Journal of Environmental Management, 231*, 419–428. https://doi.org/10.1016/j.jenvman.2018.10.054

Redvers, N., Aubrey, P., Celidwen, Y., & Hill, K. (2023). Indigenous peoples: Traditional knowledges, climate change, and health. *PLOS Glob Public Health, 3*(10), Article e0002474. https://doi.org/10.1371/journal.pgph.0002474

Reid, J., Challies, E., Tau, T. M., & Awatere, S. (2024). Adapting to climate change through nature-based solutions and indigenous knowledge: the case for landscape-scale ecosystem regeneration in the Rokohouia Delta. *Kōtuitui: New Zealand Journal of Social Sciences Online, 20*(2), 249–267. https://doi.org/10.1080/1177083X.2023.2299364

Rizki, M. N., & Asteria1, D. (2023). Local ecological knowledge and it's benefit to conservation programs in Indonesia. *International Journal of Conservation Science, 14*(4), 1527–1548.

Scaggs, S. A., Gerkey, D., & McLaughlin, K. R. (2021). Linking subsistence harvest diversity and productivity to adaptive capacity in an Alaskan food sharing network. *American Journal of Human Biology, 33*(4). https://doi.org/10.1002/ajhb.23573

Scheffran, J. (2020). Climate extremes and conflict dynamics. In *Climate extremes and their implications for impact and risk assessment* (pp. 293–315). https://doi.org/10.1016/B978-0-12-814895-2.00016-1

Shah, S. H., Angeles, L. C., & Harris, L. M. (2017). Worlding the intangibility of resilience: The case of rice farmers and water-related risk in the Philippines. *World Development, 98*, 400–412. https://doi.org/10.1016/j.worlddev.2017.05.004

Shen, J., Chou, R. J., Zhu, R., & Chen, S. H. (2022). Experience of community resilience in rural areas around heritage sites in Quanzhou under transition to a knowledge economy. *Land, 11*(12). https://doi.org/10.3390/land11122155

Smith, G., Chowenga, M., & Karsters, J. (2023). Local knowledge matters: Understanding the decision-making processes of communities under climate change in Suriname. *Frontiers in Climate, 5*, 1294271. https://doi.org/10.3389/FCLIM.2023.1294271/BIBTEX

Stephen, K. (2018). Societal Impacts of a Rapidly Changing Arctic. *Current Climate Change Reports, 4*(3), 223–237. https://doi.org/10.1007/S40641-018-0106-1/METRICS

Townshend, I., Awosoga, O., Kulig, J., & Fan, H. Y. (2015). Social cohesion and resilience across communities that have experienced a disaster. *Natural Hazards, 76*(2), 913–938. https://doi.org/10.1007/S11069-014-1526-4/FIGURES/5

Tschakert, P., Barnett, J., Ellis, N., Lawrence, C., Tuana, N., New, M., Elrick-Barr, C., Pandit, R., & Pannell, D. (2017). Climate change and loss, as if people mattered: Values, places, and experiences. *Wiley Interdisciplinary Reviews: Climate Change, 8*(5), Article e476. https://doi.org/10.1002/WCC.476;JOURNAL:JOURNAL:17577799;PAGE:STRING:ARTICLE/CHAPTER

Tuladhar, S., Pasakhala, B., Maharjan, A., & Mishra, A. (2021). Unravelling the linkages of cryosphere and mountain livelihood systems: A case study of Langtang, Nepal. *Advances in Climate Change Research, 12*(1), 119–131. https://doi.org/10.1016/j.accre.2020.12.004

Turner, H., Rogers, B., Kneebone, S., Ramirez, D., French, M., Sawailau, M. J., Volavola, F., Baran, S., Matavesi, K., Newton, O., Luveniyali, M. B., Tela, A., & Vakarewa, I. (2025). Nothing about us without us: Harnessing local voices in shaping community-based adaptation in the Pacific. *Sustainability Science, 20*, 877–902. https://doi.org/10.1007/s11625-025-01638-2

Tverskoi, D., Ember, C. R., Gelfand, M. J., Jones, E. C., Skoggard, I., Toutée, L., Gavrilets, S., Waring, T. M., & Muthukrishna, M. (2024). Cultural tightness and resilience against environmental shocks in nonindustrial societies. *Proceedings of the National Academy of Sciences of the United States of America, 121*(49), Article e2403386121. https://doi.org/10.1073/pnas.2403386121

Vinyeta, K., Whyte, K., & Lynn, K. (2016). Climate change through an intersectional lens: Gendered vulnerability and resilience in indigenous communities in the United States. https://papers.ssrn.com/abstract=2770089

Wongbusarakum, S., Gorstein, M., Pomeroy, R., Anderson, C. L., & Mawyer, A. (2021). Mobilizing for change: Assessing social adaptive capacity in Micronesian fishing communities. *Marine Policy, 129*, Article 104508. https://doi.org/10.1016/J.MARPOL.2021.104508

Datta, P., Behera, B. and Rahut, D.B., 2022. Climate change and Indian agriculture: A systematic review of farmers' perception, adaptation, and transformation. Environmental Challenges, 8, p.100543.

Chettri, D., Datta, P. and Behera, B., 2024. Climate change and household food security in the Himalayas: A systematic review of the challenges and household adaptative measures. Environmental Development, 51, p.101019.

Datta, P., Behera, B. and Rahut, D.B., 2024. India's approach to agroforestry as an effective strategy in the context of climate change: An evaluation of 28 state climate change action plans. Agricultural Systems, 214, p.103840.

Chettri, D., Datta, P., Rahut, D.B., Behera, B. and Sonobe, T., 2025. Factors Influencing the Household Food Security of Forest Villagers in Eastern Sub-Himalayan India. Food and Energy Security, 14(2), p.e70052.

Datta, P. and Bose, S., 2020. Assessing the changes in climate extremes over Karbi Anglong district of Assam, North-East India. Spatial Information Research, 28(5), pp.547-558.

Chapter 12
Climate Change and Social Inequality: A Systematic Review of Differential Vulnerabilities and Adaptation Challenges

Neelam Singh, Dibakar Sahoo, and Prasanta Moharaj

Abstract Climate change exacerbates existing social and economic inequalities, disproportionately impacting vulnerable populations based on gender, age, ethnicity, and socio-economic status. This study systematically examines how climate-related vulnerabilities shape differential access to adaptation and mitigation strategies, thereby influencing human well-being. Following the PRISMA framework, this review synthesizes literature from Scopus and Web of Science, drawing from empirical research across diverse global contexts. Findings reveal that low-income and marginalized groups face disproportionate climate risks while having limited access to the resources and systems necessary for adaptation. These populations are often situated in environmentally vulnerable areas, such as floodplains or drought-prone zones, and endure compounded insecurities related to food, water, and health. Socio-economic constraints, including income poverty, poor housing, inadequate infrastructure, and limited access to healthcare, reduce their capacity to anticipate, absorb, and recover from climate-induced shocks. Furthermore, social barriers such as lack of education, language marginalization, and restricted mobility undermine their ability to engage with adaptation planning and early warning systems. The disparity is also reflected in global emissions patterns: while affluent groups are responsible for a disproportionate share of carbon emissions, vulnerable populations bear the brunt of environmental degradation. These patterns underscore that vulnerability is not merely a matter of physical exposure but is shaped by intersecting inequalities and structural exclusions that determine who adapts, how, and with what support. This review highlights the urgent need for intersectional climate policies to address systemic

N. Singh · D. Sahoo (✉)
Department of Economics, Faculty of Social Sciences, Banaras Hindu University, Varanasi, Uttar Pradesh, India
e-mail: dibakar@bhu.ac.in

N. Singh
e-mail: neelam.singh@bhu.ac.in

P. Moharaj
School of Law, Dayananda Sagar University, Bangalore, India
e-mail: moharaj.prasant@uohyd.ac.in

P. Datta et al. (eds.), *Climate Change and Human Well-being*, Advances in Global Change Research 83, https://doi.org/10.1007/978-3-032-25479-5_12

inequalities in vulnerability and adaptation. Aligning with SDG 3 (Good Health and Well-being), SDG 5 (Gender Equality), SDG 10 (Reduced Inequalities), SDG 11 (Sustainable Cities and Communities), and SDG 13 (Climate Action), targeted interventions must ensure equitable climate resilience and inclusive governance for marginalized populations.

Keywords Climate change · Inequality · Differential vulnerability · Systematic review · Adaptation

12.1 Introduction

Climate change is a defining challenge of the twenty-first century, manifesting as a global environmental crisis with profound and unequal social consequences. While the physical effects of global warming, such as rising temperatures, sea-level rise, extreme weather events, and altered precipitation patterns, are increasingly well-documented, the social dimensions of these changes remain inadequately understood, particularly regarding how existing inequalities shape differential vulnerabilities and adaptation capacities (IPCC, 2022; Olsson et al., 2014). Across the globe, marginalized groups such as women, children, the elderly, ethnic minorities, and low-income households are disproportionately exposed to climate risks while simultaneously having the least access to resources, information, and institutional support systems that enable adaptation (Islam & Winkel, 2017; Tschakert et al., 2013). This layered inequality exacerbates socio-economic precarity and entrenches existing injustices, posing serious challenges to inclusive and sustainable development.

The uneven impacts of climate change are not simply a matter of geography or hazard exposure but are embedded in broader socio-political and economic structures. Vulnerable populations are often concentrated in climate-sensitive sectors (e.g., agriculture, fisheries), occupy precarious or informal housing in environmentally hazardous areas, and lack the means to relocate or invest in adaptation measures (Rufat et al., 2015; Schlosberg et al., 2017). For instance, in South Asia and Sub-Saharan Africa, women and landless farmers face reduced crop yields due to climate-induced droughts yet lack the formal land rights or access to irrigation technologies necessary to adapt (Djoudi et al., 2016; Sugden et al., 2014). Similarly, urban poor populations in Latin America and Southeast Asia live in informal settlements vulnerable to flooding and heatwaves but are excluded from urban planning and social protection systems (Hardoy & Pandiella, 2012; Revi et al., 2014).

While global frameworks like the Paris Agreement and the UN Sustainable Development Goals (SDGs) advocate for inclusive and equitable climate action, implementation remains uneven and often fails to integrate social justice concerns into adaptation and mitigation planning (Roberts & Pelling, 2021). Adaptation policies, in particular, tend to adopt a technocratic lens, emphasizing infrastructural or ecosystem-based interventions while neglecting the social, political, and cultural barriers that prevent marginalized communities from benefiting from these measures

(Erickson, 2022; Nightingale, 2017). This disconnect between climate policy and ground realities reinforces a cycle of vulnerability, where the most affected are also the least empowered to influence adaptation strategies.

Scholarly research over the past decade has responded to this gap by highlighting the concept of *differential vulnerability*, which recognizes that the ability to anticipate, cope with, and recover from climate impacts is unevenly distributed across and within societies (Tschakert et al., 2013; Wisner, 2010). This concept builds on longstanding insights from feminist, critical development, and environmental justice literatures, which emphasize that social identity markers such as gender, caste, ethnicity, class, and age interact to produce complex layers of disadvantage. For example, studies in Bangladesh reveal that female-headed households are less likely to receive early warning information during floods and have fewer relocation options due to gendered mobility restrictions and economic dependence (Akter et al., 2017; Thompson-Hall et al., 2016). In the United States, racial minorities are more likely to live near hazardous industrial zones or in flood-prone areas, yet are less likely to receive equitable disaster relief or insurance coverage (Mohai et al., 2009).

Despite the growing interest in the intersection of climate change and social inequality, the literature remains fragmented across disciplines, sectors, and regions. Much of the existing research is either focused on sector-specific risks (e.g., agriculture, health) or on singular dimensions of inequality (e.g., gender or income), thereby overlooking the intersectional nature of vulnerability. Moreover, previous reviews have largely concentrated on high-income countries or emphasized biophysical risks, paying insufficient attention to how structural factors such as governance failures, historical marginalization, and colonial legacies shape climate outcomes in the Global South (Nagoda, 2015; Paavola, 2008). There is thus a critical need to systematically examine how empirical research has conceptualized and documented social inequality in the context of climate change, especially in low- and middle-income countries where both climate risks and social disparities are most pronounced. Within the broader framework of climate justice, the principle of *Common but Differentiated Responsibilities* (CBDR) offers a crucial lens to understand why vulnerability to climate change is unequally distributed. Although climate risks are global, the responsibilities for causing and addressing them are historically uneven. CBDR acknowledges these asymmetries by highlighting that those least responsible for emissions, particularly marginalized groups in the Global South, are often the most affected and least equipped to adapt (Okereke & Coventry, 2016; UNFCCC, 1992).

This paper seeks to fill this gap by conducting a systematic review of 108 peer-reviewed empirical studies, selected through a rigorous screening process from Scopus and Web of Science. The review focuses on how social inequalities, particularly those based on gender, socio-economic status, ethnicity, age, and geography, influence exposure to climate risks and shape differential adaptation responses. It explores the following core questions: Who is most vulnerable to climate change and why? How do structural inequalities restrict access to adaptation resources? What kinds of policy interventions have been proposed or implemented to reduce these disparities, and with what outcomes? Drawing on insights from development economics, feminist theory, political ecology, and climate justice literature, the paper aims to identify key themes, gaps, and future directions for research and policy.

We argue that understanding and addressing climate-related inequalities requires moving beyond vulnerability as an outcome to vulnerability as a *process,* one that is historically rooted, politically mediated, and socially reproduced (Eakin & Luers, 2006). This processual view compels researchers and policymakers to consider not only the distribution of climate impacts but also the power dynamics and institutional arrangements that govern adaptation planning and resource allocation. Our analysis highlights how adaptive capacities are constrained by factors such as unequal land tenure, lack of formal education, limited political representation, and exclusion from climate information systems. These constraints are often magnified by global economic structures and neoliberal policy prescriptions that promote privatized, market-based adaptation measures, which may be inaccessible to those most in need (Erickson, 2022).

Furthermore, we adopt an intersectional approach to vulnerability, emphasizing that climate risks do not affect individuals in isolated categories but through interlocking systems of oppression. For instance, in India's Eastern Gangetic Plains, low-caste women farmers face simultaneous disadvantages due to gender, caste, and landlessness, making them among the least resilient to droughts despite living in one of the world's most flood- and drought-prone regions (Bhusal et al., 2021; Sugden et al., 2014). Similarly, indigenous communities in the Amazon and the Arctic not only face ecological degradation but also cultural displacement and loss of identity as their traditional lands become uninhabitable (Ford et al., 2020). By highlighting these multi-dimensional forms of inequality, this review aims to provide a more nuanced and grounded understanding of climate resilience.

The remainder of the paper is structured as follows. Section 12.2 outlines the methodology of the systematic review, including search strategy, inclusion criteria, and coding procedures. Section 12.3 presents an overview of the studies incorporated. Section 12.4 deals with the results and discussion. Section 12.5 presents the research gaps, and finally, Sect. 12.6 presents the conclusion and policy implications.

12.2 Methodology

Systematic reviews are becoming increasingly valued in the social sciences because they offer a clear, structured, and transparent way of bringing together existing research (Petticrew & Roberts, 2006; Siddaway et al., 2019). They are particularly useful for identifying gaps in the literature and helping to shape future research directions (Müller-Bloch & Kranz, 2015). In conducting our review, we followed the four key principles proposed by Victor (2008): ensuring the review is comprehensive, maintaining high standards of quality, being transparent in how data is synthesised, and applying methodological rigor throughout.

To gather relevant studies, we used both the Scopus and Web of Science databases and included the journal articles with open access and limited to the English language only. The subject area was limited to social sciences, Gender studies, Climate Change studies, Economics, Econometrics and Finance. The process we followed for searching and screening the literature is shown in the PRISMA Flow Diagram (Fig. 12.1), and the full search strategy is detailed in Table 12.1. In the Web of Science, we used the TS (Topic) field tag along with Boolean operators (AND, OR, NOT) to build our search queries. The **TS** tag retrieves records where terms appear in the title, abstract, author keywords, or Keywords Plus. In Scopus, the equivalent search field, TITLE-ABS-KEY, searches the title, abstract, and author keywords. In both databases, keywords were organised into four thematic categories to ensure a comprehensive and reproducible search strategy: (i) terms relating to climate change as the primary environmental driver; (ii) terms reflecting social inequality and disparity; (iii) terms addressing adaptation and resilience challenges; and (iv) terms identifying vulnerable social groups. Table 12.1 lists the search strings we used, while Table 12.2 outlines the criteria we applied to decide which articles to include or exclude from our review. In the initial screening for the studies, we included the papers published after the year 2000, but subject to our filtration criteria of subject area, document type, and content on full text reading, there were no articles included till 2013, so our final filtered articles are from the year 2013–2025. Table 12.3 presents the main information of the studies included in the review.

12.3 Overview of the Studies

This section presents an overview of the studies incorporated in this review. Table 12.3 presents the bibliometric profile of the 108 studies included in this review, covering the period from 2013 to 2025. The dataset spans 69 sources, showing a rapid annual growth rate of nearly 30%, with an average of 20.97 citations per document. Authorship patterns indicate strong collaborative engagement, averaging 3.68 co-authors per paper, although international co-authorship remains relatively limited at just over 8%.

This map in Fig. 12.2, generated in R, visualizes the geographic distribution of studies across countries. Darker red shades indicate higher values, with South Asia (22) and several African Nations (21) showing the greatest counts, while lighter shades represent lower values. Countries in white have no literature included in the review.

The reviewed literature comprises 108 studies published between 2013 and 2025, reflecting an increasing research interest in gendered dimensions of climate change, vulnerability, and adaptation. In line with the growing recognition of climate change as both an environmental and social justice challenge, the volume of scholarly work in this field has expanded significantly over the past decade. Figure 12.3 on annual scientific production shows that from 2013 to 2025, research on climate change and social inequality has grown steadily, culminating in a sharp surge in 2022, followed by a brief slowdown in 2023 and renewed growth thereafter. This trend reflects how the topic has gained momentum over time, with post-COVID years marking

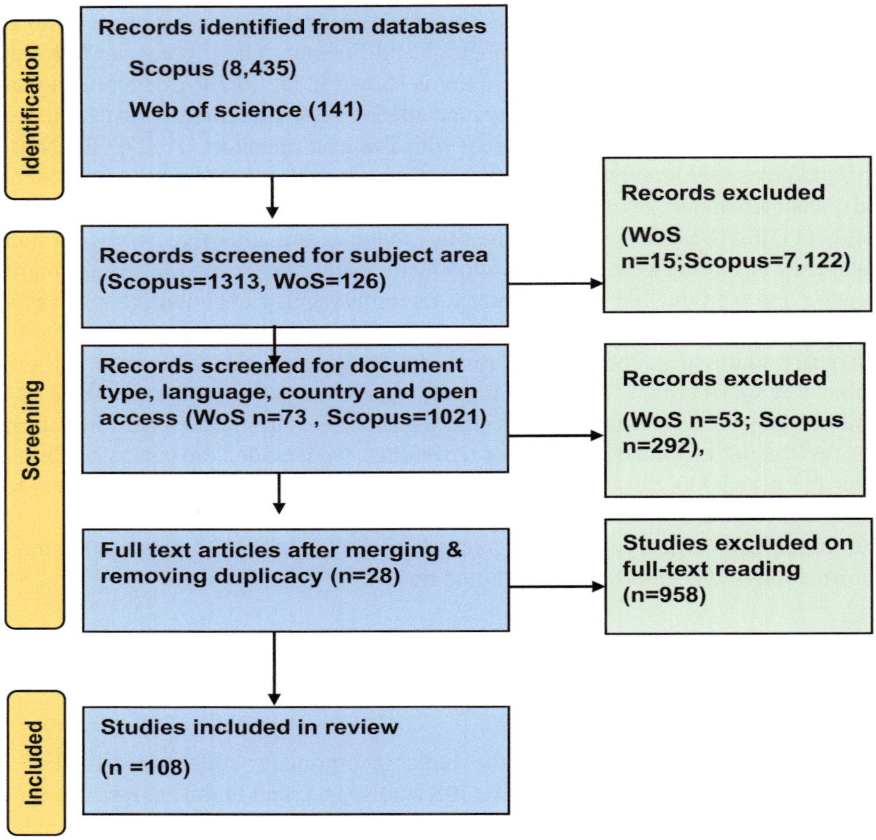

Fig. 12.1 The PRISMA flow

Table 12.1 Search strings used

Database	Search string
Scopus	TITLE-ABS-KEY((('Climate change' OR 'global warming') AND ('social inequality' OR 'social disparities' OR 'income inequality' OR 'differential vulnerability' OR 'vulnerable groups' or 'marginalized communities')) AND (('adaptation' OR 'resilience' OR 'climate mitigation') AND ('gender' OR 'women' OR 'female' OR 'children' OR 'elderly' OR 'ethnic minorities' OR 'socioeconomic status')))
Web of Science	Ts = ((('climate change' OR 'global warming') AND ('social inequality' OR 'social disparities' OR 'income inequality' OR 'differential vulnerability' OR 'vulnerable groups' OR 'marginalized communities')) AND (('adaptation' OR 'resilience' OR 'climate mitigation') AND ('gender' OR 'women' OR 'female' OR 'children' OR 'elderly' OR 'ethnic minorities' OR 'socioeconomic status')))

Table 12.2 The inclusion and exclusion criteria

Parameters	Inclusion	Exclusion
Language	Articles published in English	Articles published in languages other than English
Duplication	Unique studies are considered	Imitated studies or studies presenting replicated or redundant information are excluded
Content	Studies that explicitly examine differential vulnerability, social inequality, or equitable access to adaptation and resilience (gender, age, caste/ethnicity, socioeconomic status, marginality) in the context of climate change	Studies focused solely on technical/biophysical hazards without social or inequality analysis
Subject area	Articles categorized under social sciences, gender studies, climate change studies, economics, econometrics, and finance domains	Articles primarily from unrelated domains such as environmental sciences, engineering, or molecular biology
Year	Studies published from 2000 onwards	Studies published before 2000

Table 12.3 Main information of the studies incorporated in the study

Description	Results
Main information about data	
Timespan	2013:2025
Sources (journals, books, etc.)	69
Documents	108
Annual growth rate %	29.86
Document average age	2.74
Average citations per doc	20.97
Document contents	
Keywords plus (ID)	476
Author's keywords (DE)	421
Authors	
Authors	383
Authors of single-authored docs	13
Authors collaboration	
Single-authored docs	13
Co-authors per doc	3.68
International co-authorships %	8.333
Document types	
Article	108

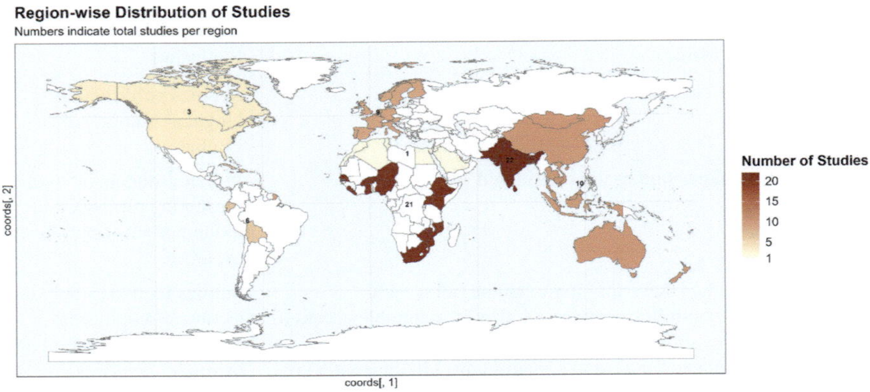

Fig. 12.2 Distribution of the studies region-wise (Author's compilation using R software)

a heightened awareness of the ways climate vulnerability and inequality intersect, prompting greater scholarly attention.

Geographically, the studies span diverse contexts, with the highest representation from South Africa (6), Bangladesh (6), India (5), Nigeria (4), Ethiopia (4), Kenya (4), Ghana (4), and other countries such as China (3), Pakistan (3), Nepal (3), USA (3), Vietnam (2), Indonesia (2), Austria (2), Ecuador (2), Bolivia (2), Malaysia (2), and a set of single-country entries (e.g., Uganda, Burkina Faso, Italy, Sri Lanka, Cambodia, São Tomé and Príncipe). Several studies adopt a multi-country focus, particularly in regions such as the Hindu Kush Himalaya or the West African Monetary Zone.

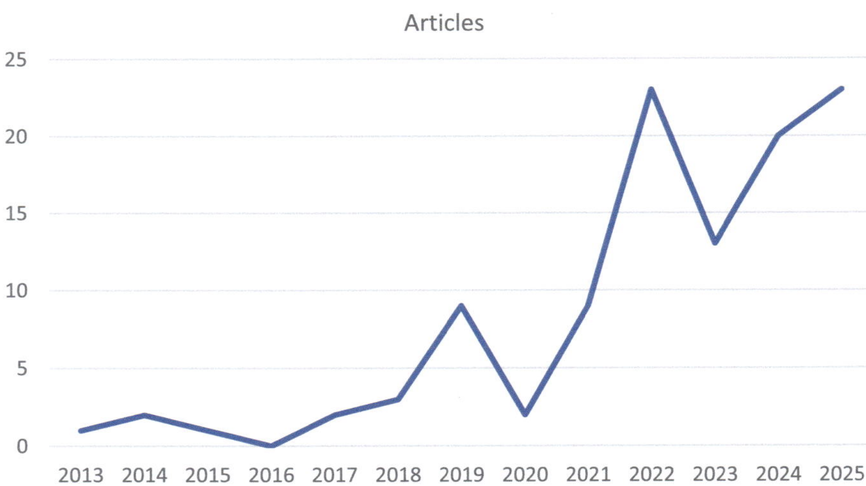

Fig. 12.3 Annual scientific production

Regionally, Sub-Saharan Africa accounts for the largest share (31 studies), followed by South Asia (19), Asia–Pacific (9), Latin America & Caribbean (4), Europe (4), East Asia & China (3), and North America (2). In addition, there are multi-country/regional (3) and global (5) studies, alongside entries with general or unspecified African or MENA regional scope.

Methodologically, the sample is diverse. Quantitative methods dominate (31 studies), encompassing regression models, econometric analyses, panel data approaches, and vulnerability indices such as LVI and LVI-IPCC. Mixed methods approaches are used in 19 studies, integrating surveys with interviews, FGDs, or spatial analysis to bridge statistical measurement with socio-political interpretation. Qualitative designs appear in 18 studies, particularly for exploring gendered vulnerabilities, local perceptions, and indigenous knowledge systems. Theoretical and conceptual analyses account for 7 studies, while modelling and simulation techniques (10 studies) apply tools such as CGE models, agent-based modelling, and flood risk simulations to explore systemic climate risks. Spatially explicit approaches, including GIS and remote sensing, are embedded in 8 studies, while case studies and literature reviews together contribute to in-depth contextual and policy-focused insights.

Sample size reporting varies, with clear quantitative detail in survey-based studies but many qualitative or theoretical works omitting numeric reporting. Where stated, sample sizes range from < 100 respondents in focused qualitative surveys to over 1000 participants in large-scale household surveys.

Thematically, these studies converge around gendered and intersectional inequalities as key determinants of vulnerability and adaptive capacity. Vulnerability is not merely shaped by climatic exposure, but is embedded in structural social, economic, political, and institutional inequalities that dictate who adapts, how, and with what resources. Women particularly in rural, indigenous, and low-income contexts are frequently identified as disproportionately exposed to climate hazards, while facing barriers to adaptation resources and decision-making platforms (Basiru et al., 2022; Dickin et al., 2021; Dilshad et al., 2019). In contexts such as Nigeria, Ethiopia, and Bangladesh, women's roles in water procurement, food provision, and caregiving increase their exposure to drought, flooding, and resource scarcity (Devonald et al., 2024; Tadesse et al., 2025), while restrictive social norms and limited mobility further constrain their participation in adaptation planning (Ferdous & Mallick, 2019; Hlahla et al., 2022).

Region-specific patterns emerge: in Sub-Saharan Africa, vulnerability often intersects with persistent gendered inequities and weak institutional support; in South Asia, caste, landlessness, and ethnicity compound the vulnerability of female-headed households; in Southeast Asia and the Pacific, urban informality and infrastructure deficits exacerbate flood risk; in Latin America & the Caribbean, adaptation measures are undermined by income inequality and insecure tenure.

Methodologically, the studies illustrate a balance between the precision of quantitative assessments useful for identifying 'hotspots' of vulnerability and the depth of qualitative and participatory approaches, which reveal lived realities and institutional blind spots. Mixed-method studies provide integrative perspectives, though some risk privileging numerical outputs over social narratives. A smaller but influential subset adopts theoretical and critical lenses, challenging dominant 'resilience' framings and emphasizing the political nature of adaptation (Eriksen et al., 2019; Mikulewicz, 2019). Remote sensing, spatial econometrics, and modelling approaches are also present, offering macro-scale insights but requiring triangulation with place-based perspectives for contextual validity.

Overall, this diverse body of literature captures the multiplicity of contexts, methodological approaches, and thematic priorities in the study of gender, vulnerability, and climate change, while also revealing gaps in sample size transparency, geographic coverage in under-represented regions, and the integration of locally defined resilience measures into policy frameworks.

12.4 Results and Discussion

Drawing from a rich body of 108 empirical studies, this systematic review reveals an increasingly urgent and multi-scalar pattern of climate vulnerability shaped by intersecting dimensions of social inequality. These findings, derived from diverse geographies, methodologies, and populations, demonstrate that climate impacts are not simply biophysical phenomena but are deeply embedded in structural, historical, and political contexts. Three major themes emerge: (1) the differentiated exposure and sensitivity to climate risks across gender, age, caste, class, and ethnicity; (2) the constraints and opportunities surrounding adaptive capacities; and (3) the institutional and policy-level dynamics that reinforce or challenge vulnerability. The identified themes have been summarised in the Table 12.4.

12.4.1 The Thematic Map

The thematic map (Fig. 12.4) reveals a field structured around four distinct clusters of research focus. In the motor themes quadrant, inequality, poverty, and development appear as both conceptually mature and highly interconnected. These themes consistently emerge in the literature as core determinants of climate vulnerability, framing adaptation within broader socio-economic and structural inequalities (Basiru et al., 2022; Devonald et al., 2024; Reimann et al., 2024). For example, Basiru et al. (2022) show that Nigerian women are significantly more vulnerable across six of seven LVI components due to poverty, gendered labour roles, and limited access to resources, while Reimann et al. (2024) demonstrate that education and gender income gaps are global predictors of flood vulnerability. Devonald et al. (2024) add that adolescent girls in Ethiopia face climate-related water and food insecurity compounded by social norms, showing how development deficits and inequality intersect to deepen vulnerability.

Table 12.4 Key thematic areas

Theme (n, %)	Keynote	Key studies
Differential exposure & sensitivity (42; 38.9%)	Socially marginalised groups by gender, age, income, or ethnicity face higher exposure and impacts	Basiru et al. (2022), Ferdous and Mallick (2019), Nunes (2018), Hagedoorn et al. (2021), Reimann et al. (2024)
Constraints on adaptive capacity (33; 30.6%)	Barriers include limited finance, weak social protection, poor access to technical information, high energy costs, and gender-blind policies	Hlahla et al. (2022), Devonald et al. (2024), Mohammed and Favretto (2025), Dilshad et al. (2019)
Institutional & policy dimensions (20; 18.5%)	Governance and institutional arrangements influence adaptation outcomes; coordinated policies and targeted interventions improve equity	Reimann et al. (2024), Mohammed and Favretto (2025), Eriksen et al. (2019), Akter et al. (2017), Omotayo and Omotoso (2025)
Intersectionality & compounded vulnerability (9; 8.3%)	Vulnerability arises from overlapping disadvantages such as gender, ethnicity, age, and class	Dickin et al. (2021), Omotayo and Omotoso (2025), Hagedoorn et al. (2021), Ferdous and Mallick (2019), Basiru et al. (2022)
Livelihood diversification & local adaptation strategies (cross-cutting)	Women adopt diverse livelihood strategies; resilience depends on access to infrastructure, insurance, and markets	Mohammed and Favretto (2025), Tadesse et al. (2025), Zaman et al. (2025), Asokan et al. (2025), Omotayo and Omotoso (2025)
Transformational/ cross-cutting approaches (4; 3.7%)	Adaptation framed as part of systemic transformation; multi-scalar governance needed to avoid maladaptation	Stringer et al. (2020), Asokan et al. (2025), Thunshirn et al. (2025), Riepl et al. (2025)

The niche themes quadrant contains well-developed but less widely integrated topics such as gini coefficient, income inequality, natural disasters, and context-specific clusters like Ethiopia, COVID-19, and financial inclusion. Notably, climate finance appears here as a critical enabler of adaptation. Hagedoorn et al. (2021) highlight that lower-income households in Vietnam express strong preferences for ecosystem-based adaptation benefits but have limited influence over project design due to resource constraints. Tadesse et al. (2025) show that participation in village savings and loan groups in Mozambique improves household assets and reduces hunger scores, directly linking access to finance with enhanced adaptive capacity. Similarly, Mohammed and Favretto (2025) document how women farmers in Nigeria adopt diverse livelihood strategies but face persistent financial and institutional barriers. Together, these studies suggest that while climate finance is conceptually robust, it remains under-integrated into community-led adaptation frameworks.

The basic themes quadrant includes climate change, vulnerability, climate change adaptation, perception, and Nepal concepts that are central to the literature yet less

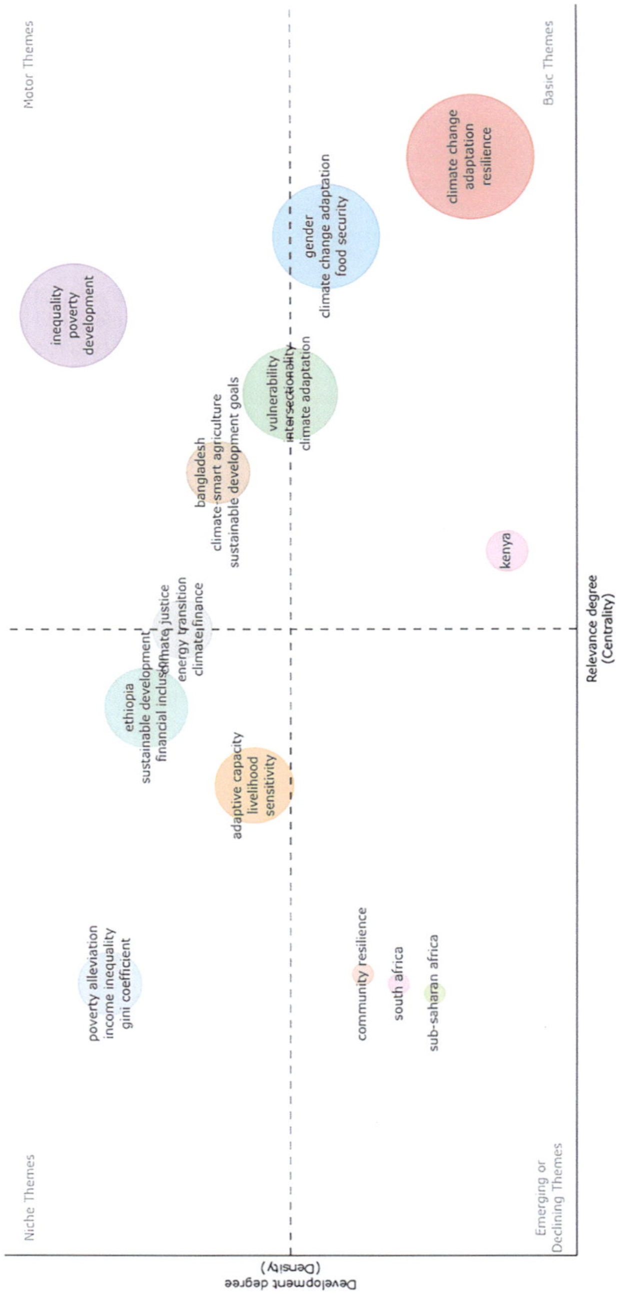

Fig. 12.4 The thematic map (Author's compilation using Bibliometrix R)

internally developed. Dilshad et al. (2019) in Pakistan's Teesta Basin show how gender, caste, and landlessness interact to shape vulnerability, while Ferdous and Mallick (2019) in Bangladesh emphasise the role of early marriage and illiteracy in limiting adaptation options. Dickin et al. (2021) in Burkina Faso reveal that water insecurity is compounded by gendered and ethnic norms, underlining that vulnerability remains a foundational but complex theme. These works demonstrate that while the concepts are widely used, deeper integration of intersectional analysis into vulnerability and adaptation frameworks is still needed.

The emerging or declining themes quadrant features sub-Saharan Africa, South Africa, climate variability, livelihood, adaptive capacity, and importantly, community resilience. Evidence from multiple contexts underscores its importance despite its peripheral map position. In Uganda, Eriksen et al. (2019) show how climate-smart agriculture can challenge existing authority relations when local actors are empowered. Escalante and Maisonnave (2023) in Dominica find that integrating community knowledge into urban planning improves both legitimacy and responsiveness. Hlahla et al. (2022) in South Africa demonstrate that urban poor women face barriers to resilience due to gender-blind policies, while Omotayo and Omotoso (2025) find that access to climate-smart technologies improves food security for female-headed smallholder households, provided they have institutional support. Collectively, these studies show that community resilience is a promising but fragmented theme, with its potential constrained by limited policy integration, lack of financing, and absence of scalable models.

Overall, the thematic mapping confirms the centrality of socio-economic inequality in shaping climate vulnerability, while identifying climate finance and community resilience as strategically important but underemphasised areas. The literature suggests that scaling community resilience requires targeted financial mechanisms (Tadesse et al., 2025), participatory governance (Escalante & Maisonnave, 2023), and gender-sensitive frameworks (Hlahla et al., 2022), alongside broader structural reforms to tackle inequality (Basiru et al., 2022; Reimann et al., 2024). Without deliberate efforts to integrate these into mainstream adaptation policy, both themes risk remaining peripheral despite their demonstrated potential to transform local and systemic resilience outcomes.

12.4.2 Differential Exposure and Sensitivity

Across the dataset, a recurring theme is that climate change disproportionately affects those already marginalized. Gender is the most prominent axis of analysis, featuring in over 60% of the studies. Women especially those from low-income, rural, or indigenous backgrounds emerge as the most vulnerable demographic, facing higher exposure and sensitivity to climate-induced stressors such as floods, droughts, and food insecurity (Basiru et al., 2022; Dickin et al., 2021). In Nigeria, for example, women were significantly more vulnerable across six of the seven components assessed using the Livelihood Vulnerability Index (LVI), including food security, health, and water access (Basiru et al., 2022).

Age and life stage also structure vulnerability. Devonald et al. (2024) reveal how Ethiopia's adolescent girls face compounded vulnerabilities due to climate-induced water scarcity and food insecurity, often exacerbated by early marriage and reduced mobility. Similarly, Nunes (2018) in Portugal finds that older adults, especially those living independently, experience limited adaptation options during temperature extremes due to declining health, inadequate housing, and social isolation. Both ends of the life spectrum youth and elderly are found to suffer unique vulnerabilities rarely addressed in mainstream adaptation policies.

Class and caste-based exclusion is also pervasive. Studies from South Asia, such as those by Dilshad et al. (2019) and Ferdous and Mallick (2019), underscore that lower-caste women in flood- and drought-prone areas like the Teesta Basin are constrained not only by climatic events but also by patriarchal norms, illiteracy, and lack of land tenure, which limit their access to adaptation resources. This aligns with Reimann et al. (2024), who find that education is the strongest determinant of vulnerability in their global model of social vulnerability to floods, while gender income gaps and elderly dependency ratios further compound risk.

Urban–rural divides further complicate the geography of exposure. Hlahla et al. (2022) demonstrate how urban poor women in KwaZulu-Natal, South Africa, are doubly burdened by poor infrastructure and a lack of gender-responsive policies, leading to insufficient coping mechanisms. In contrast, rural communities, while closer to nature-based livelihoods, face limited institutional support and declining ecological buffers (Dilshad et al., 2019; Tadesse et al., 2025). In informal settlements across Yogyakarta, Indonesia, gender and age further shaped evacuation patterns, where physical and cognitive capacities influenced the choice and feasibility of safe routes (Irsyad & Hitoshi, 2022).

12.4.3 Constraints on Adaptive Capacity

The review finds widespread evidence that adaptive capacity though theoretically expandable is in practice severely constrained for marginalized groups. Multiple studies reveal that women, youth, the elderly, and informal workers are often excluded from decision-making platforms, extension services, credit institutions, and formal adaptation programs. In Central Vietnam, Hagedoorn et al. (2021) find that women and lower-income groups show strong preferences for ecosystem-based adaptation (EBA) benefits like flood protection and food security, yet have little influence over their design or implementation. The gap between preference and access highlights systemic exclusion.

Financial inclusion and asset ownership emerge as pivotal for strengthening adaptation. Tadesse et al. (2025) find that participation in village savings and loan groups (VSLGs) significantly reduced food insecurity among low-income women in Mozambique, primarily by increasing household assets and reducing hunger scores.

Similarly, Nunes (2018) shows that among elderly populations, tangible and intangible assets (e.g., income, housing, social networks) are directly linked to their capacity to adapt to extreme temperatures.

Livelihood diversification is another frequently cited strategy, particularly among women. Mohammed and Favretto (2025) find that women farmers in flood-prone Jigawa State, Nigeria, engage in crop rotation, livestock rearing, and off-farm income generation as resilience strategies. However, barriers such as limited credit access, institutional inertia, and intra-household inequalities persist. Without supportive enabling environments, these strategies risk remaining coping mechanisms rather than transformative solutions.

In South Africa, Omotayo and Omotoso (2025) show that female-headed smallholder farmers benefit significantly from adopting climate-smart agricultural technologies (CSAT), with measurable improvements in food security, water access, and dietary diversity. Nevertheless, access to extension services, climate information, and market infrastructure remains uneven, especially for women and the landless, reflecting wider structural imbalances (Eriksen et al., 2019).

12.4.4 Institutional and Policy Dimensions of Vulnerability

One of the most critical insights from the review is that climate adaptation is often impeded not by lack of knowledge or will, but by institutional inertia, policy misalignment, and a failure to address systemic inequality. The technocratic bias in many climate interventions privileges infrastructural and economic efficiency over equity and participation. Mikulewicz (2019) critiques donor-driven adaptation in São Tomé and Príncipe for ignoring local needs, advocating for a justice-based approach. Similarly, Natarajan et al. (2019) in Cambodia show how migrant brickworkers displaced by both climatic and economic forces face precarity due to their invisibility in policy discourses.

The gender-blind nature of many climate policies is repeatedly emphasized. Hlahla et al. (2022) argue that the lack of gender-sensitive frameworks in South African urban policy has sidelined women from adaptation planning. Devonald et al. (2024) stress that adolescent girls are rarely recognized as distinct stakeholders in Ethiopia's climate strategy, despite facing specific and acute vulnerabilities. This absence of youth engagement reflects a broader institutional gap in recognizing intersectional identities within climate governance.

Some studies, however, do provide glimpses of more inclusive frameworks. In Dominica, Escalante and Maisonnave (2023) document efforts to integrate community knowledge into urban planning, suggesting that participatory governance can enhance legitimacy and responsiveness. Similarly, in Uganda, Eriksen et al. (2019) highlight how certain climate-smart agriculture initiatives can contest existing authority relations when local actors are empowered to challenge top-down definitions of 'progressive' development.

There is also an urgent need to rethink adaptation metrics and tools. Many of the reviewed studies employ the LVI and IPCC-LVI frameworks (e.g., Basiru et al., 2022; Jatav, 2024), which offer structured insights into vulnerability but may obscure local, subjective understandings of risk. While useful for comparison, such indices must be supplemented by qualitative, participatory, and intersectional analyses to ensure that policy interventions are grounded in lived realities.

12.4.5 Intersectionality and Compounded Vulnerability

Perhaps the most compelling insight from the review is the power of intersectional analysis to illuminate climate vulnerability as a socially constructed process. Vulnerability is not reducible to singular categories like gender or poverty; rather, it emerges from the co-constitution of multiple disadvantages. For instance, in the Hindu Kush Himalaya region, Dilshad et al. (2019) show how poor, ethnic minority women living in geographically remote areas face the compounded effects of climate hazards, infrastructural neglect, and social exclusion. Similarly, Ferdous and Mallick (2019) document how early marriage, illiteracy, and patriarchal norms amplify the vulnerability of women in rural Bangladesh.

In South Asia, caste and gender overlap to produce especially acute vulnerabilities. Jatav (2024) demonstrates that low-caste, landless, illiterate women in India's Eastern Gangetic Plains have the highest vulnerability scores, a result of their minimal asset base and social exclusion. Meanwhile, Dickin et al. (2021) show that water insecurity in Burkina Faso disproportionately affects women not only due to gendered roles but also because of ethnic norms (e.g., Mossi vs. Peul practices), illustrating that ethnicity, culture, and gender interact in complex ways. Livelihood diversification appeared both as a stand-alone theme and embedded in other categories. Women engaged in activities such as petty trade, livestock rearing, and non-timber forest product collection (Zaman et al., 2025), but often faced market access and capital barriers (Mohammed & Favretto, 2025). In Mozambique, Tadesse et al. (2025) found that Village Savings and Loan Groups significantly reduced household hunger through asset building, suggesting that financial inclusion mechanisms can enhance diversification's resilience benefits. However, without complementary infrastructure, insurance, and policy support, such diversification may remain a coping strategy rather than a pathway to transformation.

Finally, a small group of studies (3.7%, $n = 4$) approached adaptation from a transformational perspective, warning against maladaptation risks from technology adoption and advocating for multi-level governance. Stringer et al. (2020) emphasised that agricultural adaptation must be embedded in sustainability transitions to avoid reinforcing inequities. Asokan et al. (2025) showed how prolonged drought in Kenya's Turkana region forced shifts away from pastoralism, with water access as the critical determinant of livelihood choice.

These findings validate the need for intersectional adaptation planning that considers the cumulative and mutually reinforcing effects of different axes of disadvantage. Strategies that address only one dimension such as economic status are unlikely to reduce vulnerability unless they are embedded in broader efforts to dismantle systemic inequality and empower marginalized voices.

Figure 12.5 presents the conceptual framework flowchart, derived directly from the key insights of this systematic review, synthesizing the interplay of climate stressors (e.g., extreme heat, floods, droughts) with vulnerability factors such as age, gender, poverty, social norms, location, and governance. This leads to differential vulnerability manifested through variations in exposure, sensitivity, and adaptive capacity, modulated by assets and capacities including tangible (financial, physical) and intangible (human, social capital) elements. Barriers like resource gaps, high costs, discriminatory norms, and poor infrastructure hinder progress, while enablers policy support, education, extension services, social protection, and financial inclusion facilitate adaptive responses such as livelihood diversification, technology adoption, ecosystem-based adaptation, and community networks. Ultimately, the model culminates in positive outcomes, including enhanced resilience, reduced vulnerability, improved health and well-being, and secured food and water resources, underscoring the need for intersectional interventions to transform structural inequalities into equitable resilience pathways.

12.4.6 Bridging the Gap: Climate Finance, Community Resilience, and Technology for Equitable Adaptation

The synthesis of the reviewed literature reveals that the convergence of climate finance, community resilience, and technological innovation holds transformative potential for addressing the deep-rooted social inequalities that shape climate vulnerability. Climate finance, when designed with accessibility, affordability, and equity at its core, can move adaptation beyond short-term coping strategies toward long-term resilience. Evidence from community-based financial initiatives, such as village savings and loan groups, demonstrates that targeted resource flows can enhance household assets, improve food security, and expand livelihood options, particularly for women and low-income households. However, finance that is decoupled from social contexts and governance mechanisms often fails to reach the most marginalised, reinforcing existing disparities rather than mitigating them.

Community resilience, similarly, emerges not merely as a product of local coping capacities but as an outcome of sustained institutional support, participatory governance, and the recognition of local knowledge as a legitimate basis for decision-making. Initiatives that embed resilience-building within community structures such as participatory urban planning, locally driven ecosystem restoration, or gender-responsive adaptation programming illustrate that adaptive capacity is strongest when solutions are co-created with affected populations. Yet, such initiatives remain

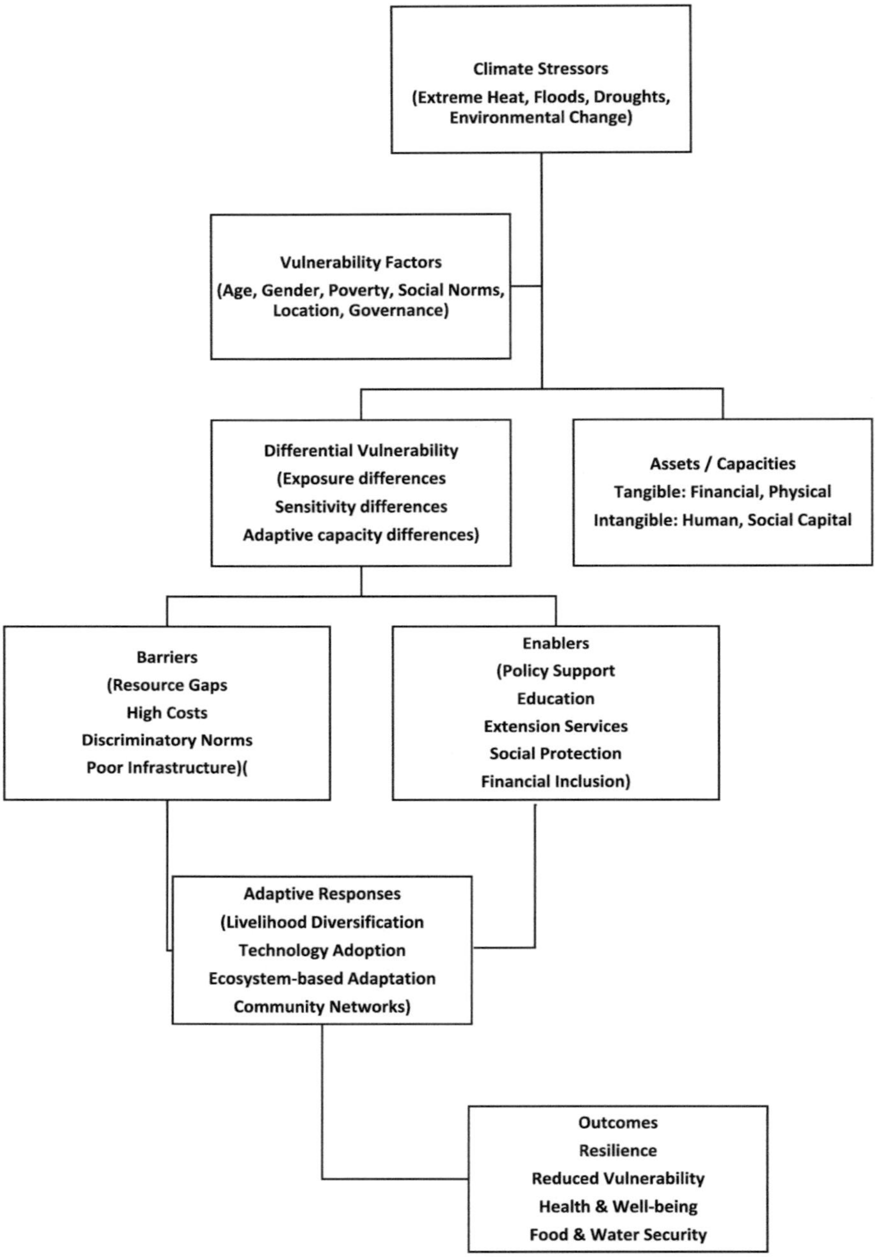

Fig. 12.5 Conceptual framework of the studies included in the review

unevenly implemented and are frequently limited in scale due to insufficient policy alignment and funding.

Technological innovation, including climate-smart agricultural practices, early warning systems, and decentralized renewable energy, can be powerful enablers of equitable adaptation. When deployed alongside inclusive knowledge systems that integrate scientific research with traditional and experiential knowledge, technology can support both livelihood security and environmental sustainability. However, unequal access to extension services, markets, and digital infrastructure means that benefits often accrue disproportionately to already advantaged groups, leaving vulnerable populations behind.

Taken together, the evidence suggests that adaptation strategies are most effective when climate finance supports community-identified priorities, when community resilience initiatives are institutionally and financially sustained, and when technological solutions are co-developed with and for those most at risk. This triad finance, resilience, and technology requires deliberate integration into adaptation governance at multiple scales. Without such integration, adaptation risks becoming a fragmented set of interventions that address symptoms rather than the structural drivers of vulnerability. Conversely, when these elements are strategically combined within equity-oriented policy frameworks, they can help shift adaptation from reactive measures toward transformative pathways that reduce inequality and foster long-term climate justice.

12.5 Research Gaps and Future Research Prospects

Despite the growing body of scholarship on climate vulnerability and adaptation, significant conceptual and methodological gaps limit its ability to inform inclusive climate policy. Gender remains a central focus, yet is often treated in overly generalised terms, obscuring how it intersects with caste, ethnicity, age, marital status, and other social markers to produce varied and compounded risks. Vulnerability is frequently conceptualised as static, with cross-sectional studies overlooking the dynamic processes through which repeated climate shocks erode coping capacity or reshape livelihoods and social networks over time. There is also a disproportionate emphasis on adaptation, with limited examination of how mitigation measures such as afforestation, carbon markets, or renewable energy transitions may exacerbate inequality when implemented through top-down, technocratic approaches that neglect local rights and knowledge. While community resilience is increasingly recognised as central to equitable adaptation, most studies focus on isolated local successes without exploring how these can be scaled or embedded within supportive governance and financing systems. Research also tends to prioritise household- and community-level analysis, underplaying how national policies, global donor priorities, and transnational climate finance mechanisms determine access to resources and shape adaptation pathways. Power dynamics within adaptation programming remain insufficiently scrutinised, with little interrogation of who defines risks, sets priorities, or controls resources, and how these processes may reinforce exclusion. Furthermore, prevailing monitoring frameworks often measure outputs like yield gains or reduced hazard exposure, without assessing whether interventions shift structural inequalities, enhance agency, or achieve socially transformative outcomes. Addressing these gaps requires a shift toward intersectional, process-oriented, and multi-scalar analyses that explicitly link local adaptation and community resilience initiatives to broader structural reforms, equitable financing mechanisms, and participatory governance, ensuring that resilience is both inclusive and sustainable over time.

12.6 Conclusion

This review of 108 empirical studies shows clearly that climate change does not affect everyone equally. Its impacts fall hardest on those who are already marginalised women in rural and low-income settings, elderly people, indigenous communities, low-caste groups, and others whose livelihoods depend on fragile ecosystems. These groups often live in areas most exposed to floods, droughts, and other hazards, yet have the least access to land, finance, education, and decision-making spaces. Their ability to adapt is not just limited by resources, but by the deeper structures of inequality governance systems that overlook them, public services that do not reach them, and social norms that restrict their choices. This review also highlights that climate finance and community resilience hold great promise, but they remain underutilised unless paired with inclusive governance and fair access to resources.

At the same time, there are important gaps in the current body of research. Gender is often treated as a single category, without recognising how caste, ethnicity, marital status, or age can combine to create very different experiences of vulnerability. Most studies provide only a snapshot in time, missing the way risks and capacities shift with repeated climate shocks. Research on how mitigation measures such as carbon offset projects or renewable energy transitions affect social equity is scarce. There is also limited attention to how national and global governance processes shape who gets support and which strategies are promoted. Few studies ask hard questions about who sets adaptation priorities, whose knowledge is valued, and whether interventions change the underlying power dynamics that drive inequality.

The evidence points to a simple truth: adaptation will only succeed if it is rooted in justice, inclusion, and accountability. That means designing policies that are informed by the lived experiences of those most affected, guaranteeing equal access to adaptation finance, protecting rights to land and resources, and strengthening essential public services like health care, education, and agricultural extension. Building resilience is not just about withstanding the next flood or drought it is about creating fairer systems that give all people the means and the voice to shape their futures. Aligning these efforts with the Sustainable Development Goals especially SDG 5 (Gender Equality), SDG 10 (Reduced Inequalities), SDG 11 (Sustainable Cities and Communities), and SDG 13 (Climate Action) offers a pathway to resilience that is not only climate-ready but socially just. In doing so, climate action can become a force for transforming the inequalities that make people vulnerable in the first place.

References

Akter, S., Krupnik, T. J., & Khanam, F. (2017). Climate change skepticism and index versus standard crop insurance demand in coastal Bangladesh. *Regional Environmental Change, 17*, 2455–2466. https://doi.org/10.1007/s10113-017-1174-9

Asokan, S. M., Kweyu, R. M., Kalibbala, M. M., & Obando, J. A. (2025). Prolonged drought and governance challenges in Turkana County, Kenya-Access to water and livelihood changes. *Environmental Development, 55*, Article 101193.

Basiru, A. O., Oladoye, A. O., Adekoya, O. O., Akomolede, L. A., Oeba, V. O., Awodutire, O. O., et al. (2022a). Livelihood vulnerability index: Gender dimension to climate change and variability in REDD+ piloted sites, Cross River state, Nigeria. *Land, 11*(8), 1240. https://doi.org/10.3390/land11081240

Bhusal, P., Kimengsi, J. N., & Awasthi, K. R. (2021). What drives environmental (non-) migration around the Himalayan Region? Evidence from rural Nepal. *World Development Perspectives, 23*, Article 100350. https://doi.org/10.1016/j.wdp.2021.100350

Devonald, M., Jones, N., Iyasu Gebru, A., & Yadete, W. (2024). Rethinking climate change through a gender and adolescent lens in Ethiopia. *Climate and Development, 16*(3), 176–186. https://doi.org/10.1080/17565529.2022.2032568

Dickin, S., Segnestam, L., & Sou Dakouré, M. (2021). Women's vulnerability to climate-related risks to household water security in Centre-East, Burkina Faso. *Climate and Development, 13*(5), 443–453. https://doi.org/10.1080/17565529.2020.1790335

Dilshad, T., Mallick, D., Udas, P. B., Goodrich, C. G., Prakash, A., Gorti, G., et al. (2019). Growing social vulnerability in the river basins: Evidence from the Hindu Kush Himalaya (HKH) Region. *Environmental Development, 31*, 19–33. https://doi.org/10.1016/j.envdev.2018.12.004

Djoudi, H., Locatelli, B., Vaast, C., Asher, K., Brockhaus, M., & Basnett Sijapati, B. (2016). Beyond dichotomies: Gender and intersecting inequalities in climate change studies. *Ambio, 45*, 248–262.https://doi.org/10.1007/s13280-016-0825-2

Eakin, H., & Luers, A. L. (2006). Assessing the vulnerability of social-environmental systems. *Annual Review of Environment and Resources, 31*(1), 365–394. https://doi.org/10.1146/annurev.energy.30.050504.144352

Erickson, K. A. (2022). Filling the protection gaps for climate change and disaster-induced migrants. *Human Rights Brief, 25*(2), 13.

Eriksen, S. H., Katherine, L., Ingvild, V., & Thornton, P. (2019). Can climate interventions open up space for transformation? Examining the case of climate-smart agriculture (CSA) in Uganda. *Frontiers in Sustainable Food Systems, 3*. https://doi.org/10.3389/fsufs.2019.00111

Escalante, L. E., & Maisonnave, H. (2023). Assessing the impacts of climate change on women's poverty: A Bolivian case study. *Journal of International Development, 35*(5), 884–896. https://doi.org/10.1002/jid.3711

Ferdous, J., & Mallick, D. (2019). Norms, practices, and gendered vulnerabilities in the lower Teesta basin, Bangladesh. *Environmental Development, 31*, 88–96. https://doi.org/10.1016/j.envdev.2018.10.003

Ford, J. D., King, N., Galappaththi, E. K., Pearce, T., McDowell, G., & Harper, S. L. (2020a). The resilience of indigenous peoples to environmental change. *One Earth, 2*(6), 532–543. https://doi.org/10.1016/j.oneear.2020.05.014

Hagedoorn, L. C., Bubeck, P., Hudson, P., Brander, L. M., Pham, M., & Lasage, R. (2021). Preferences of vulnerable social groups for ecosystem-based adaptation to flood risk in Central

Vietnam. *World Development, 148*, Article 105650. https://doi.org/10.1016/j.worlddev.2021.105650

Hardoy, J., & Pandiella, G. (2012). Urban poverty and vulnerability to climate change in Latin America. In *Adapting cities to climate change* (pp. 225–250). Routledge. https://doi.org/10.1177/0956247809103019

Hlahla, S., Simatele, M. D., Hill, T., & Mabhaudhi, T. (2022). Climate–urban nexus: A study of vulnerable women in urban areas of KwaZulu-Natal Province, South Africa. *Weather, Climate, and Society, 14*(3), 933–948. https://doi.org/10.1175/wcas-d-20-0180.1

IPCC. (2022). *Climate change 2022: Impacts, adaptation, and vulnerability*. Contribution of Working Group II to the Sixth Assessment Report of the Intergovernmental Panel on Climate Change. Cambridge University Press. https://doi.org/10.1017/9781009325844

Irsyad, H. A., & Hitoshi, N. (2022). Flood disaster evacuation route choice in Indonesian urban riverbank kampong: Exploring the role of individual characteristics, path risk elements, and path network configuration. *International Journal of Disaster Risk Reduction, 81*, Article 103275. https://doi.org/10.1016/j.ijdrr.2022.103275

Islam, N., & Winkel, J. (2017). *Climate change and social inequality*. https://doi.org/10.18356/2c62335d-en

Jatav, S. S. (2024). Farmers' perception of climate change and livelihood vulnerability: A comparative study of Bundelkhand and Central regions of Uttar Pradesh, India. *Discover Sustainability, 5*(1), 11. https://doi.org/10.1007/s43621-024-00193-7

Mikulewicz, M. (2019). Thwarting adaptation's potential? A critique of resilience and climate-resilient development. *Geoforum, 104*, 267–282. https://doi.org/10.1016/j.geoforum.2019.05.010

Mohai, P., Pellow, D., & Roberts, J. T. (2009). Environmental justice. *Annual Review of Environment and Resources, 34*(1), 405–430. https://doi.org/10.1146/annurev-environ-082508-094348

Mohammed, U. K., & Favretto, N. (2025). Livelihood diversification among women farmers in Nigeria's flood-resilient drylands. *Climate Risk Management, 48*, Article 100708. https://doi.org/10.1016/j.crm.2025.100708

Müller-Bloch, C., & Kranz, J. (2015). A framework for rigorous qualitative research synthesis: Lessons from systematic literature reviews. In *European Conference on Information Systems (ECIS) Proceedings* (pp. 1–15). https://aisel.aisnet.org/ecis2015_cr/49

Nagoda, S. (2015). New discourses but same old development approaches? Climate change adaptation policies, chronic food insecurity and development interventions in Northwestern Nepal. *Global Environmental Change, 35*, 570–579.

Natarajan, N., Brickell, K., & Parsons, L. (2019). Climate change adaptation and precarity across the rural–urban divide in Cambodia: Towards a 'climate precarity' approach. *Environment and Planning e: Nature and Space, 2*(4), 899–921. https://doi.org/10.1177/2514848619858155

Nightingale, A. J. (2017). Power and politics in climate change adaptation efforts: Struggles over authority and recognition in the context of political instability. *Geoforum, 84*, 11–20. https://doi.org/10.1016/j.geoforum.2017.05.011

Nunes, A. R. (2018). The contribution of assets to adaptation to extreme temperatures among older adults. *PLoS ONE, 13*(11), Article e0208121. https://doi.org/10.1371/journal.pone.0208121

Okereke, C., & Coventry, P. (2016). Climate justice and the international regime: Before, during, and after Paris. *Wiley Interdisciplinary Reviews: Climate Change, 7*(6), 834–851. https://doi.org/10.1002/wcc.419

Olsson, P., Galaz, V., & Boonstra, W. J. (2014). Sustainability transformations: A resilience perspective. *Ecology and Society, 19*(4). https://doi.org/10.5751/es-06799-190401

Omotayo, A. O., & Omotoso, A. B. (2025). Climate-smart agricultural technology and gender-differentiated food, and water security: Evidence from smallholder sunflower (*Helianthus annuus* L.) farmers. *Agricultural Water Management, 308*, 109276. https://doi.org/10.1016/j.agwat.2024.109276

Paavola, J. (2008). Livelihoods, vulnerability and adaptation to climate change in Morogoro, Tanzania. *Environmental Science & Policy, 11*(7), 642–654. https://doi.org/10.1016/j.envsci.2008.06.002

Petticrew, M., & Roberts, H. (2006). *Systematic reviews in the social sciences: A practical guide.* Blackwell Publishing. https://doi.org/10.1002/9780470754887

Reimann, L., Koks, E., de Moel, H., Ton, M. J., & Aerts, J. C. (2024). An empirical social vulnerability map for flood risk assessment at global scale ('GlobE-SoVI'). *Earth's Future, 12*(3), e2023EF003895. https://doi.org/10.1029/2023ef003895

Revi, A., Satterthwaite, D., Aragón-Durand, F., Corfee-Morlot, J., Kiunsi, R. B., Pelling, M., et al. (2014). Towards transformative adaptation in cities: the IPCC's Fifth Assessment. *Environment and Urbanization, 26*(1), 11–28.https://doi.org/10.1177/0956247814523539

Riepl, T., Schaffartzik, A., Grabow, S., & Banabak, S. (2025). Living well with the foundational economy: Assessing the spatial accessibility of foundational infrastructures in Vienna and the relationship to socio-economic status. *Ecological Economics, 232*, 108558. https://doi.org/10.1016/j.ecolecon.2025.108558

Roberts, E., & Pelling, M. (2021). Loss and damage: An opportunity for transformation? In *The third pillar of international climate change policy* (pp. 105–118). Routledge. https://doi.org/10.4324/9781003132271-10

Rufat, S., Tate, E., Burton, C. G., & Maroof, A. S. (2015). Social vulnerability to floods: Review of case studies and implications for measurement. *International Journal of Disaster Risk Reduction, 14*, 470–486.https://doi.org/10.1016/j.ijdrr.2015.09.013

Schlosberg, D., Collins, L. B., & Niemeyer, S. (2017). Adaptation policy and community discourse: Risk, vulnerability, and just transformation. *Environmental Politics, 26*(3), 413–437.https://doi.org/10.1080/09644016.2017.1287628

Siddaway, A. P., Wood, A. M., & Hedges, L. V. (2019). How to do a systematic review: A best practice guide for conducting and reporting narrative reviews, meta-analyses, and meta-syntheses. *Annual Review of Psychology, 70*, 747–770. https://doi.org/10.1146/annurev-psych-010418-102803

Stringer, L. C., Fraser, E. D., Harris, D., Lyon, C., Pereira, L., Ward, C. F., & Simelton, E. (2020). Adaptation and development pathways for different types of farmers. *Environmental Science & Policy, 104*, 174–189.

Sugden, F., Maskey, N., Clement, F., Ramesh, V., Philip, A., & Rai, A. (2014). Agrarian stress and climate change in the Eastern Gangetic Plains: Gendered vulnerability in a stratified social formation. *Global Environmental Change, 29*, 258–269. https://doi.org/10.1016/j.gloenvcha.2014.10.008

Tadesse, A., Li, K., Helton, J., Huang, J., & Ansong, D. (2025). The links between community-based financial inclusion and household food availability: Evidence from Mozambique. *Foods, 14*(2), 212. https://doi.org/10.3390/foods14020212

Thompson-Hall, M., Carr, E. R., & Pascual, U. (2016). Enhancing and expanding intersectional research for climate change adaptation in agrarian settings. *Ambio, 45*, 373–382. https://doi.org/10.1007/s13280-016-0827-0

Thunshirn, P., Ettwein, F., & Höferl, K.-M. (2025). Assessing the digital divide in the energy transition: Surveying the social factors influencing home energy management systems in Austria. *Energy Research & Social Science, 120*, 103941. https://doi.org/10.1016/j.erss.2025.103941

Tschakert, P., Van Oort, B., St. Clair, A. L., & LaMadrid, A. (2013). Inequality and transformation analyses: A complementary lens for addressing vulnerability to climate change. *Climate and Development, 5*(4), 340–350. https://doi.org/10.1080/17565529.2013.828583

UNFCCC. (1992). *United Nations framework convention on climate change.* United Nations. https://unfccc.int/resource/docs/convkp/conveng.pdf

Victor, L. (2008). Systematic reviewing in the social sciences: Principles and practices. *Social Science Update, 63*, 1–4.

Wisner, B. (2010). Climate change and cultural diversity. *International Social Science Journal, 61*(199), 131–140. https://doi.org/10.1111/j.1468-2451.2010.01752.x

Zaman, M., Jabeen, A., Waheed, M., Haq, S. M., Hashem, A., Almutairi, K. F., et al. (2025). Gendered ethnobotanical practices and their influence on livelihoods: Non-Timber Forest Product collection around Ayubia National Park. *Trees, Forests and People, 19*, Article 100752.

Part V
Adaptation and Resilience

Chapter 13
Assessing the Role of Home Gardens in Carbon Sequestration and Food Security in Assam, India

Anwesha Sarma and Pritha Datta

Abstract The present research is aimed at studying the carbon sequestration potential and socio-economic value of home gardens in Kamrup Metropolitan and Kamrup Rural Districts of Assam, India. As Assam is one of the most climate sensitive areas, the study seeks to determine the role home gardens can play in terms of carbon storage, food security as well as providing a knowledge base of sustainable agriculture. The study opted for a cross-sectional mixed-method approach, which combines a quantitative (biomass assessments) and qualitative approach (in-depth interviews). The results indicate that home gardens especially in Tatimora have vast carbon stocks that can be as high as 133.30 tC/ha. The amount of carbon in the Khetri and Aam Goan Bonda stocks is lower, still, home gardens are worth noting, they supply fresh produce in the homes on a regular basis. The study has emphasized the importance of home gardens in boosting food security and economic stability because people in the study mentioned that they found less need to buy food from the market and earn more income through the food they produced in the garden. Such results highlight the importance of advancing home gardens as a tool of climate adaptation, sustainable development, and resilience at the community level, and promoting the relevant policies that can facilitate the better management and integration of home gardens into the overall environmental initiatives.

Keywords Carbon sequestration · Home gardens · Food security · Assam · Sustainable agriculture · Socio-economic benefits

---

A. Sarma (✉)
Department of Policy and Management Studies, TERI School of Advanced Studies, New Delhi, India
e-mail: anweshasarma17@gmail.com

P. Datta
Rekhi Centre of Excellence for the Science of Happiness, Indian Institute of Technology Kharagpur, Kharagpur, India

P. Datta et al. (eds.), *Climate Change and Human Well-being*, Advances in Global Change Research 83, https://doi.org/10.1007/978-3-032-25479-5_13

13.1 Introduction

Climate change is becoming one of the most pressing issues in the world, and its effects are experienced in the ecosystems, economies, and human societies (Calvin et al., 2023; Bhatti, 2024; Datta et al., 2022; Chettri et al., 2024). We are witnessing a gradual increase in world temperature that, according to the Sixth Assessment Report of IPCC (AR6) (2023), has been causing the world to be warmed considerably by the difference of 1.07 °C between the pre-industrial period (1850–1900) and the more recent period of 2010–2019 (Calvin et al., 2023). Human activities, especially the combustion of fossil fuels, deforestation, and other industrial activities, have been found to be the major contributors to this unprecedented warming (Bhatti, 2024). Such activities have increased the natural greenhouse effect, causing more frequent and more severe climate-related phenomena (heatwaves, droughts, floods, and rising sea levels) (IPCC, 2021).

With the effects of climate change becoming more pronounced than ever, there has never been an urgent need to seek out an effective means of mitigating the effects of the issue (Datta & Behera, 2023). Carbon sequestration is one of such strategies that have attracted a lot of attention and this is a process of capturing carbon dioxide (CO_2) in the atmosphere and storing it in a number of natural and artificial reservoirs (Laikre et al., 2010). Carbon sequestration is important in the reduction of climate change through minimizing the quantity of CO_2 which is a significant greenhouse gas in the atmosphere. It may be achieved in the form of a natural process or by a biological method, which involves photosynthesis in plants and carbon sequestration in soils, forests, and oceans, or by an artificial method, which is carbon capture and storage (CCS) technologies (Sharma et al., 2014). Ecosystems such as forests, wetlands, and agro forest systems are considered to be important carbon sinks among the natural methods (Nzabarinda et al., 2024; Choudhary et al., 2024; Nair, 2011; Jose, 2009).

In this context, home-gardens have become a significant yet frequently ignored aspect of carbon sequestration, especially in rural and semi-urban areas (Korpelainen, 2023; Santos et al., 2022). Home-gardens are small-scale, controlled pieces of land, usually close to households, where various plant species are growing, such as vegetables, fruits, medicinal plants, and trees (Ferdous et al., 2016). Alongside the capacity to capture carbon, these gardens are appreciated due to their numerous socio-economic and cultural advantages (Ferdous et al., 2016; Santos et al., 2022). Home-gardens are not a new concept in the Indian rural environment, especially in places such as Assam, where farmers have been using home-gardens to ensure an uninterrupted flow of fresh food and medicinal products, as well as to supplement the family income (Das & Das, 2015). They play a crucial role especially in food and nutritional security, and provide a formidable and varied source of food amid climatic, economic uncertainties.

The contribution of home-gardens towards carbon sequestration is complex. To begin with, the variety of plant species in these gardens serve as carbon sinks as they absorb the CO_2 during their photosynthetic activity and store this in the biomass; both above and below ground (Kumar, 2006). The trees and perennial plants also play a major role in long term storage of carbon in the woody biomass and root systems (Jose, 2009; Nair, 2011). Moreover, the additional carbon content of soils by the decomposition of plant wastes improves the total carbon sequestration power of those systems (Ghale et al., 2022). This renders home-gardens an important asset in the effort on a global scale to reduce climate change particularly in areas that might not be able to undertake large scale reforestation or afforestation.

Home gardens are also essential in improving food and nutritional security besides their environmental advantages, which is relevant in the context of adaptation (Santos et al., 2022; Datta et al., 2024a; Datta et al., 2024b). The home gardens offer direct access to households to a range of fresh, nutritious foods, especially in areas where the markets may be inaccessible or food insecurity is a burning issue (Galhena et al., 2013). Home gardens make the households less susceptible to climate-related crop failures and economic shocks by diversifying the nature of crops that they produce (Galhena et al., 2013). In addition, they generate income to the household by selling excess produce hence sustaining livelihoods and promoting economic stability (Issahaku et al., 2023).

The state of Assam is one of the most climate-sensitive area in India with significant socio-economic challenges and widespread shifting cultivation activities (Hazarika et al., 2024; Government of Assam, 2013; Datta & Bose, 2020). These issues underscore the need to have sustainable solutions that can enhance protection of the environment and food security. Home gardens can be an excellent option in the management of climate change and adaptation in Assam where agriculture is deeply embedded in the cultural and economic life of the society. These gardens are an example of sustainable use of the land, a combination of the wisdom of the past, and the insights of the ecological theory, and an inspiration of community-based climate action. Policymakers and development practitioners can leverage their potential by promoting the use of home gardens and the improvement of these gardens as a strategy of climate mitigation (as well as socio-economic development). However, there is a substantial research gap on the potential of home gardens to sequester carbon, so their role in climate resilience is poorly studied.

In this context, the present study aims to assess the (i) role of home gardens in carbon sequestration and (ii) their impact on food and nutritional security in Assam, India. Through this study, we seek to deepen the understanding of how these small-scale agricultural systems can be harnessed to address some of the most pressing challenges of our time, offering insights that can inform policy and practice at both local and global levels.

13.2 Study Area

The study was carried out in two Assam districts in the villages of Khetri, Aamgaon and Tatimora in Kamrup Metropolitan district and Khatapara, Nangapam and Panbari in Kamrup Rural district. Kamrup Rural falls between 25.46–26.49 N Latitude and 90.48–91.50 E Longitude and Kamrup Metropolitan is located between 25 43–26 51

N Latitude and 90 36–92 12 E Longitude (Government of Assam, 2022). Such areas are typified by sub-tropical climate, semi dry summers, cold winters and humidity (average of 76%). The annual rainfall in these regions fall between 1500 and 2600 mm (Assam State Portal, 2022). The overall effect of this particular climatic profile on vegetation growth and biomass accumulation is significant, and it affects the carbon sequestration capacity of the area. High humidity coupled with large amount of rainfall provides good environment to the thick vegetation and strong plant growth, which is of paramount importance when it comes to increasing the carbon storage capacity of the home-gardens of the region. The interaction of these climatic conditions and carbon sequestration needs to be understood in order to maximize the contribution of home-gardens to mitigation of climate change in Assam.

13.2.1 Socio-Economic Condition of the Study Area

The socio-economic background of Kamrup Metropolitan and Kamrup Rural are different. The 2011 Census has shown that the population of Kamrup Metropolitan is 1.25 million with high literacy rate of 88.66, and is rapidly urbanizing with improvement in infrastructure and tertiary sector based on education, services, and trade. Kamrup Rural District on the other hand contains 1.51 million population and the literacy level is lower at 72.81. It is a majorly agricultural economy with the paddy, vegetables, and fruits as the major crops. The majority of households depend on smallholder agriculture, domestic gardening and other related practices like livestock and poultry. Scheduled Tribe (ST) and Scheduled Caste (SC) populations are also quite high in the district, most of whom live traditional lifestyles and rely on forest and land resources (Census of India, 2011).

13.3 Data and Methodology

The research design used in the study was cross-sectional, where the data were collected at a given time to determine the potential of the carbon sequestration capacity and socio-economic advantages of randomly sampled 30 home-gardens (15 in each district) in Kamrup Metropolitan and Kamrup Rural Districts of Assam, India. The biomass evaluation and the socio-economic information were obtained through primary data collection by interviewing the selected respondents through semi-structured schedules. This gave an insight into the home-garden management, the type of crops being grown, the income, food security, and environmental awareness. To estimate the potential of carbon sequestration, the analysis of data was carried out based on standardized guidelines and formulas in secondary sources, which are the IPCC guidelines, Training Manual on Capacity Building of Assam Forest Department on Carbon Stock Assessment of Forests (2019), FSI volume equations, and a resource manual by ICFRE (2020). The calculations were done using MS Excel and the mapping using Google earth.

13.3.1 Biomass Assessment

13.3.1.1 Sample Plot Layout

During the study, sample plots were laid out in three out of four selected home-garden sites (Fig. 13.1), following purposive sampling based on accessibility, diversity of species, and willingness of participants. The measurements from these plots were used to calculate carbon stock using standard IPCC and FSI methodologies. Ropes and measuring tapes were used for accurate layout demarcation, and data were recorded in field sheets. The sample plot layout is illustrated in Fig. 13.2.

For field data collection, one square plot sample of size 0.1 ha (31.62 m × 31.62 m) is established and then sub-plots of sizes 1 m × 1 m, 3 m × 3 m and 5 m × 5 m as

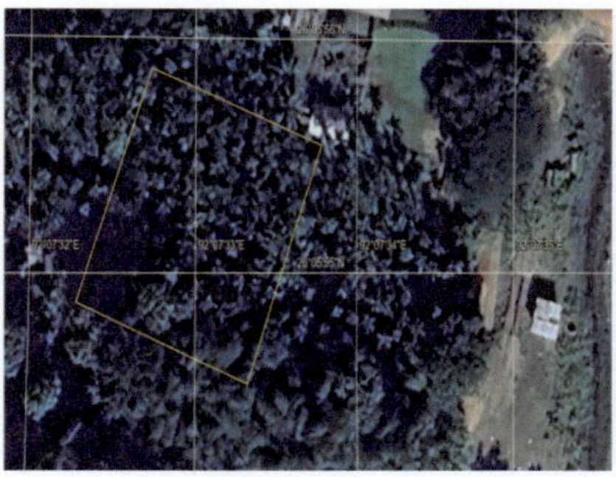

Plot 1: Khetri, (Source: Google Earth, 2024)

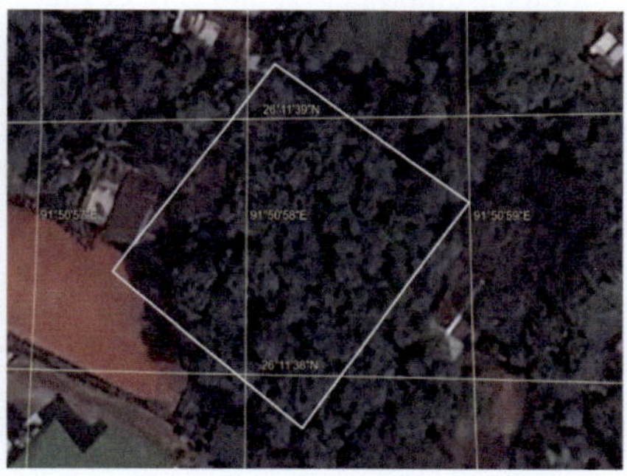

Plot 2: Aamgaon, (Source: Google Earth, 2024)

Plot 3: Tatimora (Source: Google Earth, 2024)

Fig. 13.1 Sample home-garden sites selected for this study

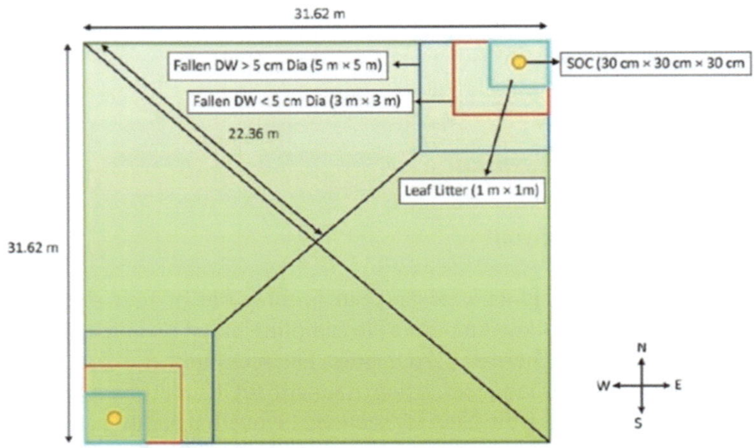

Fig. 13.2 Sample plot layout (*Source* GIZ & TERI)

quadrats are laid at NE and SW corners of the main plot (Fig. 13.1) to estimate the Above Ground Biomass (AGB), Below Ground Biomass (BGB), deadwood biomass, litter biomass and Soil Organic Content (SOC). They are essential for carbon stock measurements. The latitude and longitude are also measured at all four corners as well as at the centre of the plot (ICFRE, 2020). For laying the plot ropes of the aforementioned sizes are used.

13.3.1.2 Above-Ground Biomass (AGB)

Firstly, we measure the AGB. According to IPCC, AGB is "all living biomass above the soil, including stem, stump, branches, bark, seeds, and foliage". To calculate it we begin by measuring Circumference at Breast Height (CBH) of all the trees available inside the plot (usually at 1.37 m height) using a measuring tape in order to derive the Diameter at Breast Height (DBH). The formula to find DBH is:

$$\text{DBH} = \text{CBH}/\Pi \tag{13.1}$$

where, DBH = Diameter at Breast Height, CBH = Circumference at Breast Height $\Pi = 3.14$ (value of pi)

Height of the tree is measured using an altimeter. Using the altimeter, we find the angle of elevation (θ) to the top of the tree from our eye level.

The following formula is used then to derive the height:

$$\text{Tree Height(h)} = (\text{B} \times \tan\theta) + \text{h1} \tag{13.2}$$

where, B = horizontal distance from the tree to the observer (base of the triangle) θ = angle of elevation to the top of the tree measured by the altimeter h1 = height of the observer's eyes above the ground.

The height of the observer should be added from the calculated height to get the actual height of the tree.

Volume of the trees are calculated using volume equations or allometric equations provided by Forest Survey of India's report FSI volume equations for India, Nepal, and Bhutan (1996) for respective tree. FSI has provided a total of 753 volume equations for 198 species spread across India. Many states and UTs including Assam were covered through their study. For rest of the species in Assam a standard volume equation is followed which is $V = 0.11079 - 1.81103D + 11.4132D2 + 0.38528D3$, where D is the diameter. These standard equations differ according to regions. To find the volume of the trees we need DBH (Eq. 13.1) and height (Eq. 13.2), hence these values are calculated.

Finally, after computing the above factors, we find the AGB, formula for which is as follows:

$$
\begin{aligned}
&\text{Plant Weight(W)} = \text{Plant Volume (V)} \times \text{Wood Density(WD)} \\
&\text{Above Ground Biomass(AGB)} = \text{Plant Weight(W)} \\
&\qquad \times \text{Biomass Expansion Factor (BEF)} \\
&\text{Above Ground Biomass(AGB)} = \text{Volume} \times \text{Wood Density(WD)} \\
&\qquad \times \text{Biomass Expansion Factor(BEF)}
\end{aligned}
\tag{13.3}
$$

The wood density is different for each tree which has been provided in Training Manual for Capacity Building of the Assam Forest Department on Carbon Stock Assessment of Forests (2019). The Biomass Expansion Factor (BEF) values vary as per the density of the forest. The BEF values given in the training manual are:

- Open Forest: 1.14
- Moderately Dense Forest: 2.5
- Dense Forest: 3.4.

To measure the carbon stock density in a sampling plot, the weights (plant volume × wood density) of all the plants in the plot are added and the total weight is estimated in metric ton. The total weight is divided by the area of the plot, which is 0.1 ha, to find the biomass stock density in ton/ha. It is then multiplied with the value of BEF to obtain the AGB. Since the area is in a tropical and subtropical region, the AGB is converted into carbon stock density by multiplying it by the IPCC (2006) default carbon fraction of 0.47 and obtain the value in tC/ha.

13.3.1.3 Below Ground Biomass (BGB)

Every living biomass made up of roots are included in BGB. But less than 2 mm in diameter roots are typically disregarded because they frequently blend in with soil organic matter or litter. The Intergovernmental Panel on Climate Change (IPCC) recommends estimating BGB as 27% of AGB as it is a simplified approach. The formula to calculate BGB is as follows:

$$\text{BGB} = \text{AGB} \times 0.27(\text{IPCC Default Value}) \tag{13.4}$$

where, BGB = Below Ground Biomass, AGB = Above Ground Biomass

13.3.1.4 Dead Wood Biomass

Everything woody and non-living that isn't in the litter. They could be on the ground, or they could be standing. Dead roots and stumps with a diameter of at least 10 cm are also considered dead wood. Dead wood biomass is those whose diameter are above 5 cm and is collected from the 5 m × 5 m sub-plots. To estimate the biomass of deadwood, we start by obtaining the dry weight from laboratory measurements. The dry biomass weight is calculated separately for each plot which is kept in an oven for 24–48 h to achieve consistency and then converted to a per hectare basis. The biomass per unit area can be calculated using the following formula:

$$\text{DW} = \text{W}_{\text{fresh}}/\text{A} \times \text{W}_{\text{dry}}/\text{W}_{\text{fresh}} \times 1/100 \tag{13.5}$$

where, DW = Dead wood biomass, W_{fresh} = weight of fresh field sample, A = Size of the area W_{dry} = Weight of oven dry sample

13.3.1.5 Leaf Litter Biomass

All dead and present non-living biomass above the organic soil in different states of decomposition with a diameter of less than 5 cm (for FSI) is leaf litter biomass (FSI, n.d). Leaf Litter biomass is also collected from the sub-plots of size 1 m × 1 m. The fresh samples are weighed on-site, and a mixed sub-sample is placed in a marked bag. This sub-sample is then taken to the lab, where it is oven-dried until it reaches a constant weight to measure the water content. The formula to measure leaf litter is:

$$\text{L} = \text{W}_{\text{fresh}}/\text{A} \times \text{W}_{\text{dry}}/\text{W}_{\text{fresh}} \times 1/100 \tag{13.6}$$

where, L = biomass of litter, W_{fresh} = weight of fresh field sample, A = Size of the area, W_{dry} = Weight of oven dry sample

13.3.1.6 Soil Organic Content (SOC)

SOC incorporates uniformly applied organic carbon applied to a depth of 30 cm (for FSI) in both mineral and organic soils (FSI, n.d.). Soil samples are collected at 0-

30 cm depth. Samples of exactly 200gm are taken and transferred to pre-weighed sampling bags. Subsequently, samples are transported to the laboratory and oven dried (70° C) until constant weight to determine water content. Formula to estimate SOC is:

$$\text{SOC} = \text{rb} \times \text{d} \times \%\text{C} \tag{13.7}$$

SOC = soil organic carbon stock per unit area (t/ha), rb = soil bulk density (g/cm3) – Default value is 1.2, d = total depth at which sample is taken (cm), %C = carbon concentration.

13.3.1.7 Quantification of the Carbon Sequestration Potential

The total carbon stock for each sample plot was calculated by aggregating the carbon values derived from all the individual carbon pools assessed in the previous sections. These include Above Ground Biomass (AGB), Below Ground Biomass (BGB), Dead Wood Biomass, Leaf Litter Biomass, and Soil Organic Carbon (SOC). The sum of these components provides the total carbon stock per hectare for each site, as per the following formula:

$$\begin{aligned}\text{Total Carbon Stock} = {} & \text{AGB} + \text{BGB} + \text{Dead Wood Biomass} \\ & + \text{Leaf Litter Biomass} + \text{SOC}\end{aligned} \tag{13.8}$$

The above values in the graph have been derived by using data collected from the respective study areas in formulae given by the Forest Survey of India.

13.3.2 Household Socio-Economic Data Collection

For the socio-economic data collection, we conducted household surveys (n = 30) and interviews with the heads of households. If the head was unavailable, the next senior member was approached for the interview. Informed consent was obtained, and all respondents were briefed on the study's objectives. Data was collected from households about their basic information (such as the gender and age of the household head, religion, and education level), as well as details regarding the use and management of home gardens, the types of crops cultivated, income from sales, food security status, and their awareness of the environmental benefits associated with home gardens.

13.4 Results and Discussion

13.4.1 General Characteristics of the Households

Table 13.1 presents the socio-economic and demographic characteristics of the households surveyed for the study on *"Assessing the Role of Home Gardens in Carbon Sequestration and Food Security in Assam, India."* Most of the households were male-headed (93.33%), while only 6.67% were headed by widowed female members. A substantial proportion of the households were identified as Hindu (86.67%), followed by Muslim households (13.33%), with no representation from Christian families in the sample.

In terms of social grouping, Scheduled Tribe (ST) communities constituted the largest segment of the sample (36.67%), followed by Scheduled Castes (30.00%) and Other Backward Classes (20.00%). General category households accounted for a smaller share (13.33%). The high representation of ST and SC households indicates

Table 13.1 Household characteristics of the respondents. (*Source* Household Survey)

Particulars	Category	Count	Percentage (%)
Gender of the household head	Male	28	93.33
	Female	2	6.67
Religion	Hindu	26	86.67
	Muslim	4	13.33
	Christian	0	0.00
Community	Scheduled tribe	11	36.67
	Scheduled caste	9	30.00
	Other backward class	6	20.00
	General	4	13.33
Marital status of the household head	Married	27	90.00
	Widowed	2	6.67
	Unmarried	1	3.33
House construction type	Cemented	14	46.67
	Non-cemented	16	53.33
Particulars	**Mean**	**Maximum**	**Minimum**
Age of the respondent (years)	45	70	29
Education of the household head (years)	3.5	10	0
Household size	5.2	8	3
Landholding size (ha)	0.72	1.4	0.2
Farming experience (years)	28.1	55	12
Years of maintaining home garden	17.4	30	5

the predominance of socially and economically marginalized communities within the study area.

Regarding the marital status of household heads, 90% were married, while 6.67% were widowed and 3.33% were unmarried. Housing conditions varied, with 53.33% of households residing in non-cemented (kutcha) houses, and the remaining 46.67% in cemented (pucca) structures, indicating a moderate level of housing development in the area.

The average age of household heads was 45 years, with the oldest being 70 and the youngest 29 years. Educational attainment was low across the sample, with an average of 3.5 years of formal schooling, ranging from no education to a maximum of 10 years. Most households had a family size of around five members, with household sizes ranging from 3 to 8 individuals. The average landholding size was 0.72 ha, reflecting a predominance of smallholder farming systems.

The respondents reported an average of 28.1 years of farming experience, suggesting well-established engagement in agriculture. Additionally, households had been maintaining home gardens for an average of 17.4 years, with some continuing the practice for as long as 30 years. This indicates the enduring relevance of home gardens in supporting household-level food security and contributing to local ecological sustainability.

13.4.2 Overview of the Sample Home Gardens

The dominant tree species in most of the surveyed households is the betel nut (*Areca catechu*), typically cultivated alongside betel leaf (*Piper betle*), both of which are culturally significant in Assamese tradition and serve as major cash crops. Home-gardens also feature a variety of other tree species, including coconut (*Cocos nucifera*), lemon (*Citrus limon*), jackfruit (*Artocarpus heterophyllus*), banana (*Musa* spp.), wood apple (*Limonia acidissima*), lychee (*Litchi chinensis*), and mango (*Mangifera indica*). In smaller quantities, trees and plants such as drumstick (*Moringa oleifera*), tamarind (*Tamarindus indica*), and turmeric (*Curcuma longa*) are also grown. In addition, seasonal vegetables and edible wild herbs, commonly referred to as *xak*—such as *Alternanthera sessilis*, *Centella asiatica*, and *Marsilea minuta*—are integral to these gardens, contributing to household nutrition and dietary diversity.

The sizes of the home-gardens recorded in the study varied substantially, ranging from 1.5 to 180 kotha, with the majority of gardens being less than one hectare in area. This variation in garden size directly influences both the diversity and density of vegetation, and consequently, the potential for carbon sequestration and household-level food production.

When discussing irrigation methods, all respondents reported relying exclusively on rainwater for sustaining their home-gardens. Assam's high annual rainfall, particularly in the valley regions, provides adequate natural moisture, reducing dependence on artificial irrigation systems. While most households do not use any structured irrigation infrastructure, manual watering is occasionally practiced for seasonal vegetables and seedlings during dry spells or early stages of plant growth. Mature woody perennials, such as betel nut (*Areca catechu*), jackfruit (*Artocarpus heterophyllus*), and mango (*Mangifera indica*), require minimal intervention due to their deep root systems and the high-water table in the region. This reliance on rainfed and groundwater-supported systems aligns with traditional low-input farming practices and indicates a climate-resilient dimension of home-garden agriculture.

In terms of organic waste management, villagers demonstrate a diverse range of practices for handling leaf litter and plant debris. Some households actively clear and burn the litter, primarily to keep the garden tidy or to reduce pest habitats. However, others adopt more ecologically beneficial practices by leaving the organic matter on the soil surface, allowing it to naturally decompose and contribute to soil nutrient cycling. This not only improves soil organic matter content but also enhances moisture retention and carbon sequestration. A few respondents also reported burying decomposable waste in shallow pits within the garden, mimicking pit composting methods. While these practices reflect a mix of traditional knowledge and practical convenience, the absence of standardized or community-level composting systems limits large-scale optimization of biomass recycling. Additionally, due to the informal and unrecorded nature of these methods, it was not feasible to quantify the proportion of households practicing each approach. Nevertheless, the varied responses underline the potential for awareness-based interventions that can promote more sustainable and carbon-positive waste management strategies in home-gardens.

13.4.3 Quantification of Carbon Sequestration Potential of Home Gardens

The results of the carbon sequestration potential assessment reveal significant variations among the home gardens studied. Tatimora, Chandrapur, demonstrated the highest carbon stock at 133.30 tc/ha, indicating that well-managed home gardens can play a substantial role in carbon sequestration (Fig. 13.3). In contrast, Khetri and Aam Goan Bonda recorded lower carbon stocks of 77.90 and 71.51 tc/ha, respectively (Fig. 13.3). Although these values are not as high as those found in forested areas of Assam, they underscore the potential of home gardens to contribute to carbon sequestration, albeit on a smaller scale.

Several factors help explain this variation. Home-gardens with higher species richness and denser planting, such as those in Tatimora, tend to accumulate more biomass and soil organic carbon (SOC). This observation is consistent with Kumar (2011), who found that smaller and intensively managed homegardens in Kerala

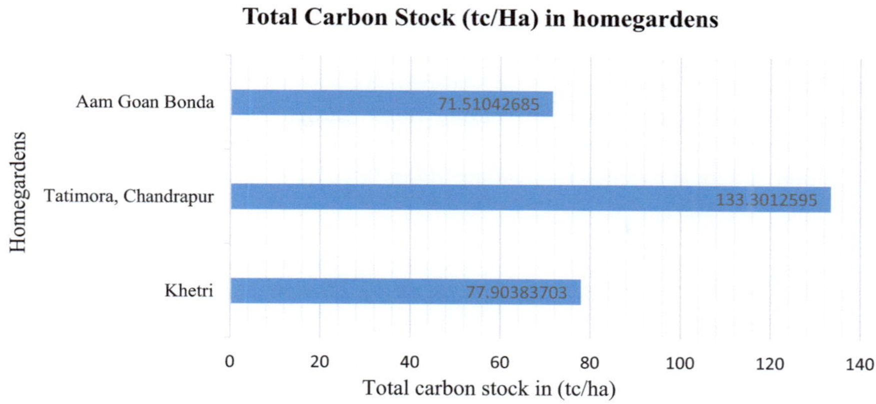

Fig. 13.3 Total carbon stock (tC/Ha) in home-gardens (*Source* Household Survey)

stored more carbon per unit area than larger holdings. Similarly, Nair (2011) emphasized that site characteristics, management intensity, and floristic diversity critically shape carbon sequestration potentials (CSPs). In contrast, less intensively managed sites like Aamgaon showed lower carbon stocks, reflecting the heterogeneity of homegarden systems.

Our findings align with broader tropical estimates. Reported CSPs for Javanese and Sumatran home-gardens range between 55.8 and 162.7 Mg C/ha, which is comparable to our results for Tatimora but higher than values from Khetri and Aamgaon (Kumar, 2006). While below forest levels (192–276 Mg C/ha in primary forests), home-garden carbon stocks are often comparable to those of secondary forests or mature agroforests (Kumar, 2006, 2011). This supports the notion that home-gardens function as human-managed analogues of forest systems, maintained in an early successional state but with considerable carbon storage capacity.

An important dimension is carbon permanence. Respondents reported that biomass is rarely fully removed from home-gardens, with tree residues and litter typically recycled back into the soil. This aligns with Kumar (2006, 2011), who stressed that such practices enhance SOC and ensure long-term stability of carbon storage. Indeed, many respondents linked soil fertility and stability directly to their garden vegetation, noting reduced erosion and improved moisture retention—observations supported by literature on home-gardens as stable, resilient carbon sinks (Nair, 2011).

At the community level, awareness of carbon sequestration was high; nearly all respondents associated their gardens with ecological benefits, including improved local climate regulation. They also observed negative impacts of surrounding deforestation, citing changes in rainfall, higher temperatures, and shorter winters. This local knowledge reinforces the ecological significance of home-gardens as buffers against broader environmental degradation.

Overall, these results confirm that while home-gardens cannot match forest ecosystems in biomass density, they represent a cost-effective, decentralized strategy for carbon sequestration with strong co-benefits for soil health and community resilience. Strengthening and expanding home-garden systems could therefore complement forest conservation efforts and contribute meaningfully to both climate mitigation and local livelihoods (Kumar, 2006, 2011; Nair, 2011).

13.4.4 Impact of the Home Gardens of Household Food Security

Most of the households interviewed have home-gardens, which they mostly use to meet their household food consumption (Fig. 13.4). About 63% of households use the production for self- consumption while 26% sell it in the market. Only 11% of them use it for both the purposes. This pattern highlights the dual role of home-gardens in ensuring subsistence food supply and generating supplementary income. Respondents consistently emphasized that home-gardens reduce dependence on markets, particularly during price hikes and seasonal scarcities. Similar observations have been reported in other developing-country contexts, where home-gardens serve as a cost-effective buffer against shocks in food systems (Galhena et al., 2013; Issahaku et al., 2023).

A key contribution of home-gardens is improved food availability and dietary diversity. The cultivation of vegetables, fruits, and herbs provides households with inexpensive and reliable sources of nutrition. This supports earlier findings that home-gardens supplement staple-based diets with essential vitamins and minerals, thereby reducing hidden hunger and micronutrient deficiencies (Ferdous et al., 2016; Galhena et al., 2013). In our study, respondents also reported that culturally important indigenous crops such as Dillenia indica and traditional leafy greens are often more

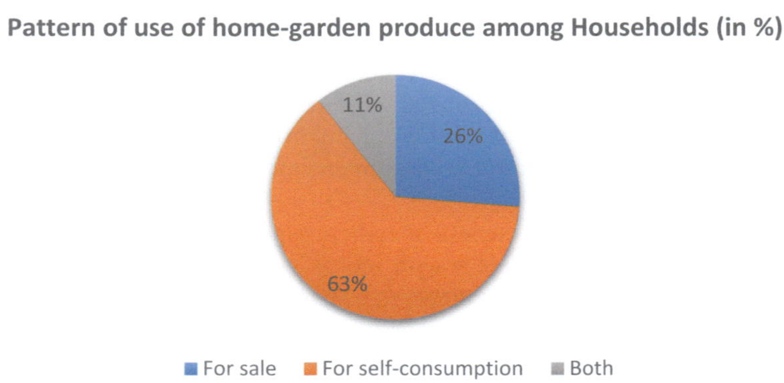

Fig. 13.4 Pattern of use of the home-garden produce among the households (*Source* Household Survey)

accessible through home-gardens than in local markets, reinforcing both dietary and cultural benefits.

The economic role of homegardens, while secondary to subsistence, remains significant. Respondents reported annual earnings of Rs. 7000–10,000 from surplus sales, which were used for education, healthcare, and household expenses. These findings are consistent with studies from South Asia and Sub-Saharan Africa that highlight home-gardens as a low-cost, low-input livelihood option that contributes to household income and reduces food expenditure shares (Ferdous et al., 2016; Issahaku et al., 2023).

Home-gardens also offer social and environmental co-benefits. Women, who are the primary managers of these gardens, gain decision-making autonomy and recognition in the household economy, echoing findings from Bangladesh and Rwanda (Ferdous et al., 2016; Issahaku et al., 2023). Environmentally, respondents noted that the diversity of plants in their gardens helps conserve soil fertility and reduces erosion, consistent with broader evidence of homegardens as repositories of agrobiodiversity and traditional knowledge (Galhena et al., 2013).

Overall, the results affirm that home-gardens contribute simultaneously to food availability, dietary diversity, income generation, and ecological stability. Their role as a multi-functional livelihood strategy is especially valuable in the context of increasing climate and market uncertainties. Strengthening support for home-garden practices, through access to quality planting material and extension services, could therefore enhance food security and resilience in the study area, in line with global findings (Galhena et al., 2013; Issahaku et al., 2023).

13.4.5 Seasonal Availability of Home-Garden Crops and Their Nutritional Benefits

Figure 13.5 illustrates the seasonal availability of various fruits, vegetables, and herbs commonly cultivated in home gardens across Assam. The chart uses three colour codes to denote availability status: light brown indicates general availability, dark brown represents peak harvest periods, and blank cells denote months when the crop is unavailable.

Perennial crops such as betel nut (*Areca catechu*) and betel leaf (*Piper betle*) show availability throughout the year, reflecting their continuous harvest nature. Crops like coconut (*Cocos nucifera*) and lemon (*Citrus limon*) peak in the pre-monsoon and early monsoon periods (May–July), while jackfruit (*Artocarpus heterophyllus*) and mango (*Mangifera indica*) are primarily available from April to July.

Seasonal vegetables display more defined patterns. Drumsticks (*Moringa oleifera*) and litchi (*Litchi chinensis*) show peak availability during April–June, while cauliflower, cabbage, and carrot, typical winter crops, are most abundant from November to February. Leafy vegetables like coriander and lady's finger are more frequently available during summer and early winter months.

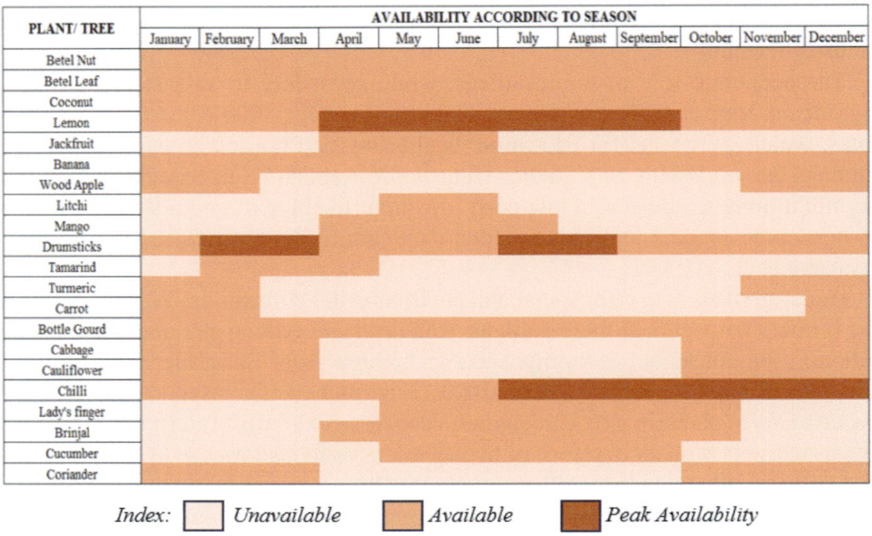

Fig. 13.5 Seasonal availability of home-garden crops (*Source* Household Survey)

This availability calendar reflects the diverse cropping cycles within home-gardens, contributing significantly to year-round food security. It also highlights the ecological resilience of home-garden systems, which integrate both perennial and seasonal crops to ensure a steady supply of produce.

Table 13.2 outlines the nutritional value of various crops commonly grown in home-gardens. These gardens provide households with a steady supply of vegetables, fruits, and herbs rich in essential nutrients. For instance, drumsticks and leafy greens (xak) are good sources of iron and calcium, while fruits like banana, jackfruit, and mango offer vitamins and fibre. Lemon ensures vitamin C intake and boosts immunity.

With diverse crops available across seasons, households enjoy year-round access to fresh, nutritious food, reducing reliance on market purchases. This is particularly important during price hikes, natural disasters, or supply chain disruptions, when home-grown food serves as a critical buffer. Moreover, several plants such as turmeric, tulsi, and wood apple have medicinal properties and are traditionally used to treat minor ailments, reducing the need for external healthcare.

When respondents were asked about the contribution of their home-gardens to overall food security, 100% reported a positive impact. Home-gardens serve as a vital source of fresh, seasonal produce, including fruits, vegetables, and herbs, which contribute significantly to daily household nutrition. This partial self-sufficiency not only enhances food availability but also reduces household dependence on market purchases, especially during times of price fluctuations or limited access to markets due to natural disasters or transport disruptions. The resulting cost savings provide economic relief and improve overall household stability. Similar findings from other

Table 13.2 Home-garden crops and their nutritional benefits. (*Source* Longvah et al., 2017)

Plants/crops	Nutritional value
Betel Nut	Alkaloids
Betel Leaf	Vitamin C, Calcium
Coconut	Manganese, Copper, Selenium
Lemon	Vitamin C
Jackfruit	Vitamin C, Potassium, Magnesium
Banana	Potassium, Vitamin C, Vitamin B6
Wood Apple	Vitamin C, Calcium, Phosphorus
Litchi	Vitamin C, Potassium
Mango	Vitamin C, Vitamin A
Drumsticks	Vitamin C, Vitamin A, Calcium, Potassium
Tamarind	Magnesium, Potassium, Iron
Turmeric	Iron, Manganese
Carrot	Vitamin A, Vitamin C, Potassium
Bottle Gourd	Vitamin C, Calcium, Potassium
Cabbage	Vitamin C, Vitamin K, Fiber
Cauliflower	Vitamin C, Vitamin K, Fiber
Chilli	Vitamin C, Vitamin A
Lady's Finger	Vitamin C, Vitamin A
Brinjal	Vitamin C, Vitamin K
Cucumber	Vitamin C, Potassium, Magnesium
Coriander	Vitamin C, Vitamin A, Iron

regions highlight the role of home-gardens as repositories of indigenous species and traditional knowledge (Kumar, 2011; Galhena et al., 2013; Datta et al., 2024c).

Beyond economic benefits, home-gardens also offer access to culturally appropriate foods that are either expensive or unavailable in local markets. Respondents mentioned growing traditional items like elephant apple (*Dillenia indica L*), manimuni (*Centella asiatica*), and other wild edible herbs or ferns (locally known as *xaak*) which are integral to Assamese cuisine and valued for their nutritional and medicinal properties. These leafy greens and indigenous crops are often foraged or home-grown and hold cultural significance, especially during festivals, fasting days, or traditional meals.

In addition to consumption, many households also sell surplus produce, earning approximately Rs. 7000 to Rs. 10,000 annually. Commonly sold items include betel nut (*Areca catechu*), betel leaf (*Piper betle*), banana (*Musa* spp.), lemon (*Citrus limon*), papaya (Carica papaya), and seasonal leafy vegetables (*xaak*) such as manimuni (*Centella asiatica*), dhekia xaak (*Diplazium esculentum*), lai xaak (*Brassica juncea*), etc. These items are either sold directly in nearby weekly markets (haats) or to local traders and middlemen who visit villages to purchase produce in small quantities. They spend that income to meet their household needs such as fooding, clothing, education for their children and other basic necessities. Their income from home- gardens is less as their production and marketing efforts are not very professional. While these earnings are modest, they supplement household stability and reduce reliance on market purchases, confirming the economic co-benefits of home-gardens (Ferdous et al., 2016; Issahaku et al., 2023).

By combining perennials with annuals, and integrating traditional herbs with fruit and vegetable crops, home-gardens in Assam mirror broader global evidence of their ecological resilience and socio-economic importance (Galhena et al., 2013; Kumar, 2011; Ferdous et al., 2016; Issahaku et al., 2023).

13.5 Conclusion and Policy Implications

The findings of this study underscore the significant potential of home gardens in contributing to carbon sequestration and enhancing food security within the communities of Kamrup Metropolitan and Kamrup Rural Districts of Assam. The highest carbon stock observed in Tatimora highlights the effectiveness of well-managed home gardens in sequestering carbon, while even the lower carbon stocks in other locations demonstrate the valuable role these gardens play in mitigating greenhouse gas emissions on a smaller scale. Furthermore, the diverse range of crops cultivated in these gardens not only provides essential nutritional benefits but also reduces households' dependence on external markets, offering a buffer against economic fluctuations and environmental shocks.

In light of these insights, several policy implications emerge. First, policymakers should recognize and promote home gardens as a viable strategy for climate change mitigation and food security enhancement. Supporting local agricultural practices through targeted programs can help foster sustainable land-use practices and enhance the resilience of communities. This includes providing training and resources for effective garden management, as well as facilitating access to markets for surplus produce.

Additionally, raising awareness about the environmental benefits of home gardens, including their role in carbon sequestration and biodiversity preservation, can encourage more households to adopt such practices. Integrating home gardening into broader agricultural and environmental policies will be crucial in promoting sustainable livelihoods and fostering community resilience in the face of climate change. Finally, further research should be encouraged to quantify the potential of home gardens in diverse socio-economic contexts, helping to build a more comprehensive understanding of their role in both environmental sustainability and community well-being.

References

Allen, M. (2021, August 26). The 6 environmental and health benefits of growing your own food. *The Garden Continuum.* https://www.thegardencontinuum.com/blog/the-6-environmental-and-health-benefits-of-growing-your-own-food

Bhatti, U. A., Bhatti, M. A., Tang, H., Syam, Awwad, E. M., Sharaf, M., & Ghadi, Y. Y. (2024). Global production patterns: Understanding the relationship between greenhouse gas emissions, agriculture greening and climate variability. *Environmental Research, 245*, 118049. https://doi.org/10.1016/j.envres.2023.118049

Calvin, K., Dasgupta, D., Krinner, G., Mukherji, A., Thorne, P. W., Trisos, C., Romero, J., Aldunce, P., Barrett, K., Blanco, G., Cheung, W. W., Connors, S., Denton, F., Diongue-Niang, A., Dodman, D., Garschagen, M., Geden, O., Hayward, B., Jones, C., & Ha, M. (2023). *IPCC, 2023: Climate Change 2023: Synthesis Report. Contribution of Working Groups I, II and III to the Sixth Assessment Report of the Intergovernmental Panel on Climate Change [Core Writing Team, H. Lee and J. Romero (eds.)].* IPCC. https://doi.org/10.59327/ipcc/ar6-9789291691647

CEEW. (n.d.). *Agroforestry.* https://www.ceew.in/agroforestry

Census of India. (2011). Office of the Registrar General & Census Commissioner, India

Chettri, D., Datta, P., & Behera, B. (2024). Climate change and household food security in the Himalayas: A systematic review of the challenges and household adaptative measures. *Environmental Development, 51*, 101019.

Choudhary, B., Dhar, V., & Pawase, A. S. (2024). Blue carbon and the role of mangroves in carbon sequestration: Its mechanisms, estimation, human impacts and conservation strategies for economic incentives. *Journal of Sea Research, 199*, 102504. https://doi.org/10.1016/j.seares.2024.102504

Das, T., & Das, A. K. (2015). The nature and function of traditional home gardens in Assam, Northeast India: A review. https://www.researchgate.net/profile/Shri-Tripathi/publication/341481463_Biodiversity_in_Tropics_1/links/5ec3b31c92851c11a87454bd/Biodiversity-in-Tropics-1.pdf#page=375

Datta, P., & Bose, S. (2020). Assessing the changes in climate extremes over Karbi Anglong district of Assam, North-East India. *Spatial Information Research, 28*(5), 547–558.

Datta, P., Behera, B., & Rahut, D.B. (2022). Climate change and Indian agriculture: A systematic review of farmers' perception, adaptation, and transformation. *Environmental Challenges, 8*, p.100543

Datta, P., & Behera, B. (2023). Climate change adaptation through agroforestry: Empirical evidence from Indian Eastern Himalayan Foothills. In *Strategizing agricultural management for climate change mitigation and adaptation* (pp. 167–181). Springer.

Datta, P., Behera, B., Rahut, D. B. (2024a). India's approach to agroforestry as an effective strategy in the context of climate change: An evaluation of 28 state climate change action plans. *Agricultural Systems, 214*, 103840.

Datta, P., Behera, B., Rahut, D. B. (2024b). Assessing the role of agriculture-forestry-livestock nexus in improving farmers' food security in South Asia: A systematic literature review. *Agricultural Systems, 213*, 103807.

Datta, P., Behera, B., Rahut, D. B., & Timsina, J. (2024c). Achieving sustainable development through agriculture-forestry-livestock nexus in Bangladesh: Synergies and trade-offs. *Agricultural Systems, 215*, 103854.

Deutsche Gesellschaft für Internationale Zusammenarbeit (GIZ) GmbH & The Energy and Resources Institute. *Training manual for capacity building of Assam forest department on carbon stock assessment of forests.* GIZ. https://forest.assam.gov.in/sites/default/files/swf_utility_folder/departments/pccf_lipl_in_oid_4/latest/giz_assam_training_emanual.pdf

Ferdous, Z., Datta, A., Anal, A. K., Anwar, M., & Khan, A. M. R. (2016). Development of home garden model for year round production and consumption for improving resource-poor household food security in Bangladesh. *NJAS - Wageningen Journal of Life Sciences, 78*(1), 103–110. https://doi.org/10.1016/j.njas.2016.05.006

Galhena, D. H., Freed, R., & Maredia, K. M. (2013). Home gardens: A promising approach to enhance household food security and wellbeing. *Agric & Food Secuity, 2*, 8. https://doi.org/10.1186/2048-7010-2-8

Ghale, B., Mitra, E., Sodhi, H. S., Verma, A. K., & Kumar, S. (2022). Carbon sequestration potential of agroforestry systems and its potential in climate change mitigation. *Water, Air and Soil Pollution/Water, Air & Soil Pollution, 233*(7). https://doi.org/10.1007/s11270-022-05689-4

Government of Assam. (2013). *Assam State Action Plan on Climate Change (2015–2020)*. https://moef.gov.in/uploads/2017/08/ASSAM-SAPCC.pdf

Hazarika, A., Nath, A. J., Reang, D., Pandey, R., Sileshi, G. W., & Das, A. K. (2024). Climate change vulnerability and adaptation among farmers practicing shifting agriculture in the Indian Himalayas. *Environmental and Sustainability Indicators, 23*, 100430. https://doi.org/10.1016/j.indic.2024.100430

ICFRE. (2020). *Resource manual: Measurement of forest carbon stocks for capacity building of state forest departments*. Indian Council of Forestry Research and Education, Dehradun (INDIA). https://moef.gov.in/uploads/2019/06/05-Forest-Carbon-Stocks-Measurement-RESOURCE-MANUAL-in-English.pdf

IPCC. (2021). *Climate change widespread, rapid, and intensifying*. IPCC

Intergovernmental Panel on Climate Change. (2006). *IPCC guidelines for national greenhouse gas inventories*. IPCC National Greenhouse Gas Inventories Programme. https://www.ipcc-nggip.iges.or.jp/public/2006gl/index.html

Issahaku, G., Kornher, L., Islam, A. H. M. S., & Abdul-Rahaman, A. (2023). Heterogeneous impacts of home-gardening on household food and nutrition security in Rwanda. *Food Security, 15*(3), 731–750. https://doi.org/10.1007/s12571-023-01344-w

Jose, S. (2009). Agroforestry for ecosystem services and environmental benefits: An overview. *Agroforestry Systems, 76*(1), 1–10. https://doi.org/10.1007/s10457-009-9229-7

Korpelainen, H. (2023). The role of home gardens in promoting biodiversity and food security. *Plants, 12*(13), 2473. https://doi.org/10.3390/plants12132473

Kumar, B.M. (2006). Carbon sequestration potential of tropical homegardens. In B. M. Kumar & P. K. R. Nair (Eds.), *Tropical homegardens. Advances in agroforestry* (Vol. 3). Springer. https://doi.org/10.1007/978-1-4020-4948-4_11

Kumar, B. M. (2011). Species richness and aboveground carbon stocks in the homegardens of central Kerala, India. *Agriculture Ecosystems & Environment, 140*(3–4), 430–440. https://doi.org/10.1016/j.agee.2011.01.006

Kumar, A., & Jasheena, J. C. (2016). Kudumbashree: Promoting the self-help group model of empowerment through women entrepreneurship in Kerala—A study. *Prabandhan Indian Journal of Management, 9*(6), 20. https://doi.org/10.17010/pijom/2016/v9i6/94959

Laikre, L., Schwartz, M. K., Waples, R. S., & Ryman, N. (2010). Compromising genetic diversity in the wild: Unmonitored large-scale release of plants and animals. *Trends in Ecology & Evolution, 25*(9), 520–529. https://doi.org/10.1016/j.tree.2010.06.013

Lal, R. (2020). Home gardening and urban agriculture for advancing food and nutritional security in response to the COVID-19 pandemic. *Food Security, 12*(4), 871–876. https://doi.org/10.1007/s12571-020-01058-3

Longvah, T., Ananthan, R., Bhaskarachary, K., & Venkaiah, K. (2017). *Indian food composition tables 2017*. National Institute of Nutrition, Indian Council of Medical Research. https://www.nin.res.in/ebooks/IFCT2017.pdf

Nair, P. K. R. (2011). Carbon sequestration studies in agroforestry systems: A reality-check. *Agroforestry Systems, 86*(2), 243–253. https://doi.org/10.1007/s10457-011-9434-z

NITI Aayog. (2019). *Transforming nutrition in India: Poshan Abhiyaan a progress report*. https://www.niti.gov.in/sites/default/files/2020-02/POSHAN_Abhiyaan_first_progress_report_6_Feb_2019.pdf

Nzabarinda, V., Bao, A., Tie, L., Uwamahoro, S., Kayiranga, A., Ochege, F. U., Muhirwa, F., & Bao, J. (2024). Expanding forest carbon sinks to mitigate climate change in Africa. *Renewable and Sustainable Energy Reviews, 207*, 114849. https://doi.org/10.1016/j.rser.2024.114849

Santos, M., Moreira, H., Cabral, J. A., Gabriel, R., Teixeira, A., Bastos, R., & Aires, A. (2022). Contribution of home gardens to sustainable development: Perspectives from a supported opinion essay. *International Journal of Environmental Research and Public Health, 19*(20), 13715. https://doi.org/10.3390/ijerph192013715

Sharma, P., Dhanwantri, K., & Mehta, S. D. K. M. (2014). Bamboo as a building material. *International Journal of Civil Engineering Research, 5*(3), 249–254.

Vuong, Q.-H., Sari, N. P. W. P., La, V.-P., & Minh-Hoang, N. (2025). *Exploring the health benefits of home gardens: Biological, psychological, and therapeutic perspectives*. PhilArchive. https://philarchive.org/archive/VUOETH

Chapter 14
Navigating Water Scarcity: Salinization, Well-Being, and Climate Resilience in Coastal Communities of Bangladesh

Md Lokman Hossain, Khusnur Jahan Shapna, Ashik Rubaiyat, Md. Humayain Kabir, Rakhi Baidya, Jianfeng Li, and Derrick Y. F. Lai

Abstract Salinization in coastal regions is a global challenge intensified by rising sea levels. Severe salinization in the coastal regions of Bangladesh forces communities to rely on stored rainwater, highlighting the need to assess its quality and storage challenges. This research adopted a mixed-method approach, incorporating water quality tests, household surveys, focus group discussions (FGDs), and in-depth

M. L. Hossain (✉) · K. J. Shapna · J. Li · D. Y. F. Lai (✉)
Department of Geography and Resource Management, The Chinese University of Hong Kong, Shatin, New Territories, Hong Kong, China
e-mail: lokmanbbd@gmail.com

D. Y. F. Lai
e-mail: dyflai@cuhk.edu.hk

K. J. Shapna
e-mail: k.jahan@link.cuhk.edu.hk

J. Li
e-mail: jianfengli@cuhk.edu.hk

M. L. Hossain
Department of Environment Protection, German University Bangladesh, Gazipur, Bangladesh

K. J. Shapna · J. Li
Centre for Environmental Policy and Resource Management, The Chinese University of Hong Kong, Shatin, New Territories, Hong Kong, China

A. Rubaiyat
Centre for International Studies, Ohio University, Athens, OH, USA
e-mail: ar585024@ohio.edu

Md. H. Kabir
Wegener Center for Climate and Global Change, University of Graz, Graz, Austria

Department for Knowledge and Communication Management, Danube University Krems, Krems an der Donau, Austria

Md. H. Kabir
e-mail: mh.kabir@uni-graz.at

P. Datta et al. (eds.), *Climate Change and Human Well-being*, Advances in Global Change Research 83, https://doi.org/10.1007/978-3-032-25479-5_14

interviews (IDIs) in the salinisation-affected region of Bangladesh. Twenty-three water samples from three types of water storage facilities were tested, 42 households were interviewed, and 4 FGDs and 4 IDIs were conducted to investigate rainwater quality, households' challenges and adaptive practices, and stakeholders' perceptions of drinking water management. The levels of lead (0.03–0.33 mgL^{-1}), faecal coliform (12–71 N/100 ml), and total coliform (81–444 N/100 ml) in all sources exceeded the permissible limits of national and international standards. Household surveys exhibited diverse water storage methods, with 33% of households relying on commercial water tanks and 26% using a combination of earthen pots and plastic drums. However, 69% of households reported consuming untreated water, leading to widespread health issues, including diarrhea (79%), vomiting (67%), and skin irritation (57%). FGDs identified water quality and contamination as core challenges, with proposed solutions including rainwater harvesting, desalination facilities, and pond sand filters. The analysis of top keywords obtained from IDIs, such as '*Water Salinity*' and '*Rainwater Harvesting*', underscores the urgency of addressing water scarcity and contamination. These findings provide valuable insights into adaptive water management strategies, health risks, and sustainable solutions that can be applied to salinization-affected regions worldwide.

Keywords Coastal Bangladesh · Climate change adaptation · Drinking water salinity · Rainwater harvest · Salinization-affected region · Water quality

14.1 Introduction

The scarcity of freshwater has become a serious global issue due to population growth, increased agricultural and industrial activities, and the unprecedented impact of global climate change (Raimondi et al., 2023). The Intergovernmental Panel on Climate Change (IPCC) Sixth Assessment Report (AR6) underscores that climate change is affecting both water availability and water quality, particularly freshwater for drinking purposes (Caretta et al., 2022). Globally, approximately four billion people experience severe water scarcity for at least one month per year due to climatic and non-climatic factors (Mekonnen & Hoekstra, 2016). The occurrence of droughts has increased by 29% since 2000, and more than three-quarters of the global population could suffer from water stress by 2050. Shifting precipitation patterns, rising temperatures, and frequent extreme weather events (e.g., droughts and periods of heavy rainfall) deteriorate surface water quality, increase dissolved organic carbon levels, and reduce the efficiency of traditional water treatment methods (e.g., coagulation and filtration) (Ma et al., 2022). These climatic perturbations have been reported to intensify the risks of water availability worldwide. For example, Leveque

R. Baidya
Dustha Shasthya Kendra (DSK), Dacope, Khulna, Bangladesh
e-mail: rakhibaidya87@gmail.com

et al. (2021) found that extreme weather events intensify risks to water quality and quantity in the northern regions of Quebec, Canada. Moreover, Boholm and Prutzer (2017) identified systematic risks to drinking water provisioning are strongly linked to climate change, such as microbial contamination from increased temperatures, and sea level rise-induced saltwater intrusion. Rising sea levels and intense storm surges contribute to saltwater intrusion into freshwater systems, especially in low-lying coastal and deltaic regions, compromising drinking water sources (Hock et al., 2019; Meredith et al., 2022). These impacts are particularly acute in South and Southeast Asia (Barbour et al., 2022; Mamun et al., 2024), where salinity and flooding heighten health risks (Rakib et al., 2019) and limit access to drinking water (Ward et al., 2013).

Saltwater intrusion into surface and subsurface water bodies has become a severe threat to safe drinking water in many coastal and low-lying countries (IPCC, 2023), including India and Bangladesh (Hoque et al., 2016). Over 25 million people in the Bengal, Mekong, and Red River deltas are exposed to saline drinking water due to storm surges, sea-level rise, and land subsidence (Hoque et al., 2016). These trends are especially worrying for Bangladesh (Rahman et al., 2014), where over 25% (42 million) of the total population live in delta areas and shortages of both surface and groundwater are worsening (Hossain & Li, 2024), especially in the southwest of the country (Abdullah et al., 2024; Afsari et al., 2022). Because of the shortage of freshwater flow from upstream and the intrusion of saline water into the land, nearly 37% of coastal agricultural land in the country has been affected by moderate to severe salinity (Chen & Mueller, 2018; Islam et al., 2010).

The southwestern coastal region of Bangladesh, the most climate-vulnerable regions, is increasingly affected by salinity intrusion in surface and groundwater sources due to sea-level rise, tidal flooding, excessive groundwater extraction, and reduced upstream freshwater flow (Barbour et al., 2022; Hossain & Hossain, 2012; Saha et al., 2024). Nearly 53% of the coastal area is already affected by salinity, with projections indicating further expansion under changing climate (CEGIS, 2006; SRDI, 2019). The issue is compounded by anthropogenic factors such as shrimp farming, reduced freshwater flows, and poorly planned infrastructure (Hossain & Li, 2024). A growing body of evidence has reported elevated salinity (electrical conductivity) levels in pond water (13.19–13.99 dSm^{-1}), canal water (65–78 dSm^{-1}), and ground water (0.98–14.16 dSm^{-1}) in the southwestern coastal region of the country (Hossain & Li, 2024; Rakib et al., 2020; Shaibur et al., 2021; Shapna et al., 2024; Tsai et al., 2024), indicating severe challenges for drinking and irrigation use. This has led to increased reliance on untreated or saline water, contributing to health problems such as hypertension, gastrointestinal illnesses, pre-eclampsia, and dehydration—particularly affecting women, children, and the elderly (Abdullah et al., 2024; Afsari et al., 2022; Asma & Kotani, 2021; Saha et al., 2024; Shapna et al., 2024). Recent evidence (Frisbie et al., 2024) demonstrates that climate change-induced sea level rise is likely to intensify arsenic contamination in the drinking well water of the country.

Given the heightened salt concentration in surface and ground water (Hossain et al., 2012; Rakib et al., 2020; Shaibur et al., 2021), people in climate-vulnerable communities in salinization-affected regions of the country are forced to choose alternative drinking water sources like rain-fed ponds, rainwater harvesting, and locally treated water (e.g., pond sand filters [PSF], and reverse osmosis systems) (Rakib et al., 2020; Tsai et al., 2024). However, many of these alternatives are either prone to salinity intrusion (e.g., salt deposition in ponds) or require high energy as well as operational and maintenance costs (e.g., reverse osmosis systems) (Rakib et al., 2020). To address these challenges, rainwater harvesting systems—an ancient practice—is considered a climate adaptation strategy worldwide, including in Germany, India, and Australia-, particularly in rural and developing regions (Raimondi et al., 2023; Ruiz Martínez & Cornejo Tueros, 2022). In Bangladesh, where the average annual rainfall exceeds 2000 mm (Shahid, 2010), the monsoon season (June–September) makes it easy to collect rainwater. As the quality of conventional drinking water decreases, rainwater harvesting is now often seen as an effective way to solve the water crisis in the southwestern regions of the country (Abdullah et al., 2024). Rainwater harvesting systems help households and communities collect, store, and use rainwater from their roofs or other types of surfaces for everyday tasks like household chores and gardening. These systems reduce reliance on contaminated groundwater and allow families to spend less time walking or traveling long distances to fetch water (Rodrigues de Sá Silva et al., 2022). To address water scarcity, rainwater harvesting systems have been adopted in Bangladesh by the Department of Public Health Engineering (DPHE) and non-government organizations (NGOs) (Saha et al., 2024). Most household systems in salinization-affected communities in the country feature rainwater harvesting facilities (e.g., plastic drums, earthen pots, and commercial water tanks) (Hossain & Li, 2024; Rakib et al., 2020).

Concerns about whether collected rainwater is safe or not to drink are still exist (Mostafa et al., 2020; Mukarram et al., 2023). Rainwater can become contaminated when it comes into contact with dirty roofs, leaves, bird droppings, or storage tanks that are not properly maintained (Raimondi et al., 2023; Saha et al., 2024). Inadequate treatment and maintenance can lead to the growth of unwanted bacteria, especially during warmer months when water is stored in tanks for longer periods (Ghosh & Ahmed, 2022). Field assessments in a sub-district (Dacope) in the southwestern region of the country show that many people use their rainwater harvesting systems during the monsoon and dry seasons. However, about 24% of the storage systems were found to be malfunctioning because of a lack of maintenance skills within the community (Ghosh & Ahmed, 2022). Ghosh and Ahmed (2022) also reported that only 3.3% of surveyed households were aware of any treatment methods, and less than 10% had knowledge of simple water quality tests.

Previous research has primarily focused on household-level harvesting systems, particularly tank sizing, water quality, and reliability, and multi-criteria decision-support tools (Kabir et al., 2016; Saha et al., 2024). Karim et al. (2015) analyzed the adequacy and reliability of household rainwater harvesting systems and found that under average climatic conditions, a maximum reliability of 70–85% is achievable with tanks up to 3000 L. Recently, Saha et al. (2024) developed a comprehensive, data-driven decision support system to evaluate the performance and suitability of community-managed rainwater harvesting systems in salinization-affected regions of the country, which reported that while 40% of systems performed well and 24% showed very high suitability, 36% were moderately suitable due to challenges in maintenance and low community engagement. In another study in the extremely salinization-affected sub-district (Shyamnagar) of the country, Afsari et al. (2022) evaluated the effectiveness of rainwater harvesting systems and concluded that rainwater harvesting at both household and community levels is a viable and widely accepted solution for drinking and cooking needs, especially during the monsoon season. However, multiple barriers, including high initial costs, lack of public awareness, limited training, weak institutional support, poor system design due to inadequate hydrological data, lack of coordination among government institutions, and insufficient subsidies, have been highlighted in previous studies (Afsari et al., 2022; Hossain & Li, 2024). Proper storage systems are crucial to maintaining better water quality after harvesting. Significant differences exist in how well tanks made from plastic, reinforced cement concrete (RCC), or earthy materials preserve water quality over the years (Abdullah et al., 2024). Several systems are prone to contamination as they lack features such as tightly fitting lids, first-flush waste diversion mechanisms, or water overflow controls (Saha et al., 2024). Poor maintenance, such as failing to clean roof catchments or allowing water to stagnate, can compromise water quality and reduce the long-term popularity of the systems (Rodrigues de Sá Silva et al., 2022).

Despite numerous qualitative and quantitative studies on rainwater quality assessment, community challenges, and adaptive mechanisms in drinking water management in the salinization-affected region of the country (Afsari et al., 2022; Hossain & Li, 2024; Saha et al., 2024; Shapna et al., 2024), a holistic understanding of drinking water management in the affected communities is still lacking. Specifically, it is crucial to examine (i) which water storage facilities (e.g., private, shared, and RCC tanks) provide better water quality, (ii) the challenges households face and the adaptive mechanisms they adopt to address water scarcity, and (iii) the perceptions of stakeholders and communities regarding drinking water management. In this study, by combining water quality assessment, household surveys, focus group discussions (FGDs) and in-depth interviews (IDIs) conducted in a severely salinization-affected coastal region of Bangladesh, we aim to provide valuable empirical evidence of drinking water quality, household challenges in meeting their water demands and stakeholders' perceptions of drinking water management. Such insights are essential for ensuring sustainable water management practices for the affected communities.

The novelty of our study lies in its holistic approach to drinking water management in a severely salinization-affected region, integrating water quality assessment of various water storage facilities, household surveys, FGDs, and IDIs. Unlike previous studies that focus on isolated aspects (e.g., water quality assessment or household adaptive mechanisms), our research provides a comprehensive understanding of (i) the comparative water quality of different storage facilities (private, shared, and RCC tanks), (ii) the challenges households face in addressing water scarcity and their adaptive strategies, and (iii) the perceptions of both stakeholders and communities regarding sustainable water management. By integrating these dimensions, our study offers valuable empirical evidence that bridges knowledge gaps and provides actionable insights for improving water management and climate change adaptation practices in coastal regions of Bangladesh and similar regions worldwide.

14.2 Data and Methodology

14.2.1 Study Area

The study was carried out in a salinization-affected region of Bangladesh (Fig. 14.1). Located near the Sundarbans mangrove forest, the study area, Kamarkhola Union, in the Dacope subdistrict of Khulna, Bangladesh, is susceptible to salinity intrusion (Hossain & Li, 2024). Salinity influences both soil and water in the area, especially in the dry season (November–May), when surface water bodies, including canals and ponds, become too saline for use (Chen & Mueller, 2018). For example, in our study region, elevated salinity levels have been observed in pond water (13.19–13.99 dSm^{-1}), and canal water (65–78 dSm^{-1}) (Hossain & Li, 2024). The drinking water sources in this union have been significantly impacted, with groundwater extracted from tubewells becoming increasingly saline (Tsai et al., 2024). The electrical conductivity (EC) of groundwater samples in the coastal region has been recorded between 980 and 14,160 $\mu S\ cm^{-1}$ (Rakib et al., 2020), where the maximum permissible limit is 300 $\mu S\ cm^{-1}$ (WHO, 2011). This growing salinity in water supplies has raised serious health concerns for the local communities (Hossain & Li, 2024). Thus, residents in this region must mostly rely on rainwater and unsafe surface water bodies (Shapna et al., 2024).

Kamarkhola Union spans 29.20 km^2 and is home to a population of 15,407 (Bangladesh National Portal, 2025). It is bordered by Dhaki and Vodra to the north, Suterkhali Union to the south, Vodra to the east, and Dhaki to the west. More than one-third of the land area is uncultivated because of high salinity (Hossain et al., 2025). Low-income farmers and fishermen make up the bulk of the population in this union. The literacy rate of this union is 47%, and the primary livelihoods include farming (60%), fishing (5%), and wood collection (2%) (Bangladesh National Portal, 2025). The region is traversed by two rivers and 26 canals. The area has 184 freshwater ponds and 43 PSFs, though only 16 of these PSFs are currently functional. To support disaster preparedness, the union is equipped with six cyclone shelters (Bangladesh National Portal, 2025). The study region is an ecologically important area, as it is located beside the mangrove ecosystem of the Sundarbans. This region plays a crucial role in agriculture and fisheries, supporting both local livelihoods and biodiversity. Its unique ecological significance stems from its rich natural resources and its vital role in maintaining environmental balance in the coastal zone.

This region experiences an annual precipitation of 2000–2300 mm (BMD, 2025; Shahid, 2010), with average temperatures ranging from a mean minimum of 20.8 °C to a mean maximum of 34.3 °C (BMD, 2025). Based on its climatic patterns,

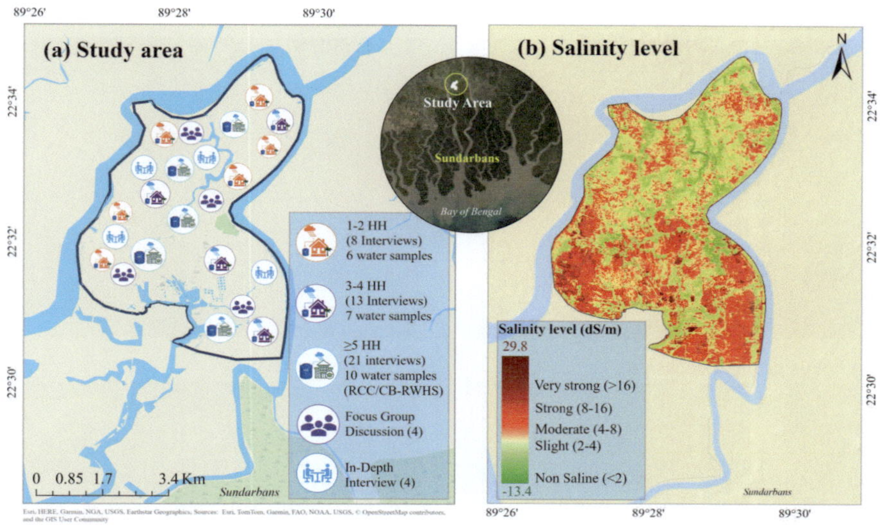

Fig. 14.1 Study area map showing the location of the household surveys, focus group discussions, in-depth interviews, and water sample collection points (**a**). A salinity gradient map of the study area is shown on the right side of the figure (**b**)

the area is divided into four distinct seasons: winter (December–February), pre-monsoon (March–May), monsoon (June–September), and post-monsoon (October–November) (Shahid, 2010). Apart from regular climate hazards (e.g., cyclones, storm surges, and tidal flooding), catastrophic events like Cyclone Sidr in 2007 and Aila in 2009 have worsened saltwater intrusion, destroyed infrastructure, and disturbed livelihoods (Chen & Mueller, 2018). Rising sea levels aggravating salinity over time provide difficulties for the community in maintaining agriculture and fisheries, as well as surface and subsurface water (Barbour et al., 2022).

14.2.2 Data Sources

The study used a mixed-methods approach, including household surveys, FGDs, IDIs, water quality tests, and climatic information. Household surveys gathered quantitative data, while FGDs and IDIs produced qualitative insights. These data were collected from December 2023 to January 2024.

Household survey: Using a semi-structured questionnaire, 42 households were surveyed (Fig. 14.2a and b). Participants were chosen by purposive sampling to ensure representation from various communities and socioeconomic levels. To reflect several points of view on water access and livelihood difficulties, the selected household respondents (27 female and 15 male) comprised farmers, fishermen, wood

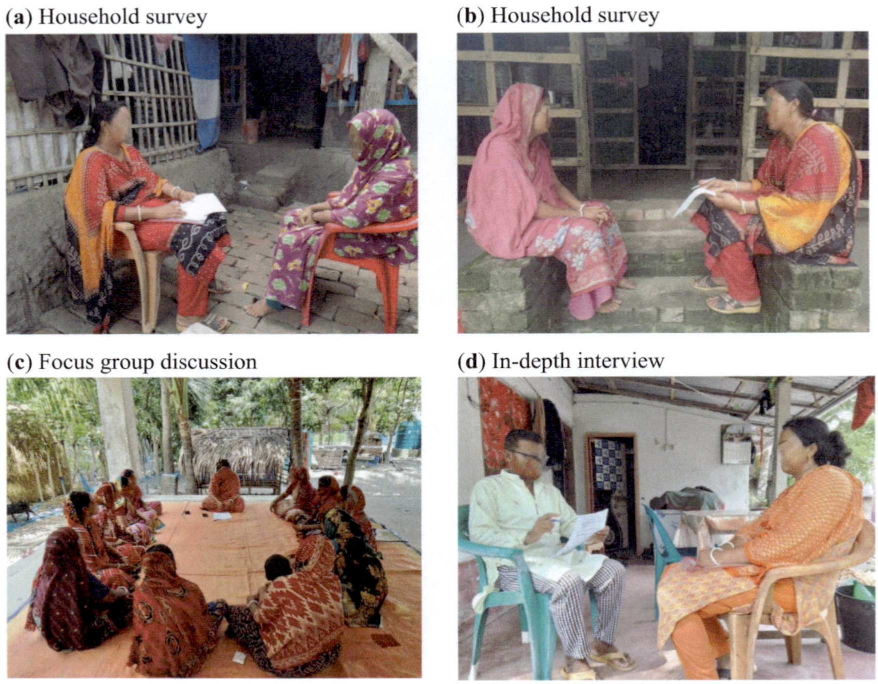

Fig. 14.2 Data collection from the field under Dacope sub-district: household surveys (**a** and **b**), focus group discussion (**c**), and in-depth interview (**d**) using a semi-structured questionnaire. © The authors

collectors, day laborers, businessmen, teachers, and development workers. Key questions, including water sources and salinity levels, rainwater collection methods, water storage, health effects connected to water quality, and household adaptation strategies, were covered in the questionnaire.

Focus group discussion and in-depth interview: We conducted 4 FGDs, comprising distinct groups of farmers, fishers, health workers, NGO workers, day laborers, students, teachers, religious leaders, and local leaders (Nyumba et al., 2018). With 8–11 participants in each group, each FGD investigated community views of shifting salinity patterns, availability of freshwater bodies, water management difficulties, rainwater harvesting and its associated challenges (e.g., storage capacity, treatment facility), adaptation and mitigation strategies for dealing with water salinity (Fig. 14.2c). To obtain institutional and expert viewpoints on the issues (Fig. 14.2d), four IDIs were also conducted with key informants (schoolteacher, healthcare provider, NGO representative, and Union Parishad Chairman) (Akhter, 2022). The research team was trained before collecting data to ensure consistent application of the approach. Five households outside the study area were used for pilot testing survey questionnaires, which were then polished in response. Throughout the process, ethical issues took front stage; all participants provided verbal permission, and steps were taken to preserve anonymity. Particular care was taken to include low-income households and vulnerable groups like women, in the study.

Water sample collection and laboratory analysis: In January 2024, we collected 23 water samples from various rainwater harvesting systems, comprising 13 individual plastic tanks, 7 community-based shared tanks, and 3 larger RCC tanks to analyse faecal coliform (FC), total coliform (TC), pH, electrical conductivity (EC), total dissolved solids (TDS), iron (Fe), and lead (Pb) contents (Fig. 14.3a). Using standard protocols, we gathered samples in pre-cleaned bottles, directly from tank taps after initial flushing. Each sample was carefully labelled with identification details, including location, tank type, and precise collection time. For microbial analysis (FC and TC), samples were immediately placed in coolers at 4 °C and processed within 6 h to maintain integrity. These samples were analyzed using the membrane filtration method. Each sample was filtered through a 0.45 μm sterile membrane filter, which was then placed on m-Endo agar (for TC) and m-FC agar (for FC) (Rompre et al., 2002). pH was measured using a calibrated digital pH meter, EC was determined using a conductivity meter and TDS was measured gravimetrically after filtration through 1.5 μm glass fiber filters and evaporation at 180 °C (Yan et al., 2024). For heavy metal analysis, samples were acid digested with HNO_3. Fe content was measured using flame atomic absorption spectrophotometry (AAS) at 248.3 nm wavelength, while Pb content was quantified via graphite furnace AAS at 283.3 nm wavelength (Abdelmonem et al., 2025).

Climate data: We obtained mean annual temperature (MAT) data for the period 1990–2024 from the Landsat satellites (5, 7 and 8) of the United States Geological

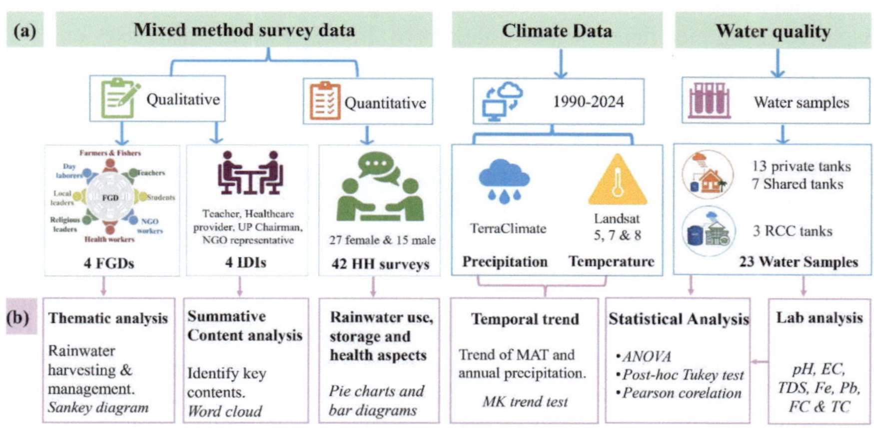

Fig. 14.3 Flowchart showing the data collection steps (**a**), and data processing and analysis (**b**). EC: electrical conductivity; TDS: total dissolved solid; Fe: iron; Pb: lead; FC: faecal coliform; TC: total coliform; FGD: focus group discussion; IDI: in-depth interview; HH: household; RCC: reinforced cement concrete; MAT: mean annual temperature

Survey (USGS) database (Fig. 14.3a). We extracted monthly precipitation data for the same period from the TerraClimate database (Abatzoglou et al., 2018). The spatial resolution of MAT and precipitation data is 30 m and 0.5°, respectively. Monthly precipitation values were summed up to obtain the annual precipitation.

14.2.3 Data Processing and Analysis

Water quality: We employed a one-way analysis of variance (ANOVA; Kaufmann & Schering, 2014) to assess the significance of the differences in the variables (pH, EC, TDS, Fe, Pb, FC and TC) among water supply sources (i.e., private, shared and RCC tanks). A post hoc Tukey's Honest Significant Difference (HSD; Tukey, 1949) test was used to explore the pairwise comparisons of the values between two water sources. We then used Pearson correlation to assess the correlation among the variables (Benesty et al., 2009). A correlation matrix was used to visualize the correlation. In the upper triangle of the correlation matrix, significance level ($p < 0.05$) of the correlations was shown with *asterisk* (*), while insignificant correlations ($p > 0.05$) were left blank. In the lower triangle of the correlation matrix, correlation coefficients were shown. The *p*-values of the correlations were calculated using *cor.mtest()* with 95% confidence. The positive to negative correlations between variables were shown in pie chart symbols using color gradient from slate blue ($r = +1$) to white ($r = 0$) to red ($r = -1$).

Household: Household information was plotted using pie charts and bar diagrams. Specifically, comparisons of the respondents' information on economic status, occupation, education, and rainwater use during monsoon and other seasons were displayed using pie charts, while variations of the respondents' responses on rainwater storage facilities, duration of the stored water usage, water-borne diseases, physical observation of stored water, perceptions of water quality, and water treatment before consumption were displayed using bar diagrams.

FGD: We used thematic analysis of the information derived from FGDs (Ahmed et al., 2025). We implemented a systematic coding approach to classify responses according to recurring themes concerning rainwater harvesting and management practices. The categorized data were systematically organized into a structured spreadsheet to facilitate analytical processing. We documented two core themes (water quality and water contamination). Under these two themes, we then reported three major impacts (health risks, storage challenges, and water quality problems), and seven solutions (rainwater harvesting, water treatment, community awareness, government support, desalinization treatment facility, PSF, and NGO support). The two overarching goals of these solutions are ensuring safe water supply and improving public health. Using a Sankey diagram, we represented the relationships between water-related challenges, their impacts, and potential solutions in the study area.

IDI: We conducted a summative content analysis of the four IDIs (Hsieh & Shannon, 2005). First, all interviews were transcribed verbatim to maintain data accuracy. After reviewing the transcripts to identify preliminary themes, we consolidated them into a single document for systematic analysis. To quantify key terms, we extracted meaningful keywords by filtering out grammatical elements (conjunctions, articles, verbs, prepositions) and technical jargon. From this refined dataset, we calculated the frequency of the top 25 recurring words. These results were then visualized using the *wordcloud* package in statistical software R (R Core Team, 2020; Shapna et al., 2025), where term prominence reflected usage frequency, a methodological approach aligned with established textual analysis practices (DePaolo & Wilkinson, 2014).

Climate: The temporal trend of MAT and annual precipitation of the study area was assessed using the Mann–Kendall trend test (Mann, 1945). We used the Kendall, and ggplot2 packages for this test. The trend line was plotted using a linear method, and *Tau* and p-values of the monotonic trends were derived from the MK test (Fig. 14.3b).

14.3 Results

14.3.1 Water Quality

14.3.1.1 Variation in Water Quality

Water quality analysis revealed significant differences in pH, Fe, Pb, FC and TC among the three different water sources (i.e., private, shared and RCC tanks) (Fig. 14.4, all $p < 0.05$), while no significant differences were observed for EC and TDS among the sources (Fig. 14.4b and c, both $p > 0.05$). Specifically, pH and FC levels were significantly higher in private and RCC tanks compared to shared tanks (Fig. 14.4a and f). Additionally, RCC tanks exhibited significantly higher levels of Fe (Fig. 14.4d), Pb (Fig. 14.4e), and TC (Fig. 14.4g) (all $p < 0.05$). Conversely, no significant differences were observed between private and shared tanks for Fe and Pb levels (Fig. 14.4d and e, both $p > 0.05$).

In terms of water quality standards, pH, EC and TDS in all three water sources fall within the quality standards set by the WHO and the government of Bangladesh (Table 14.1). Conversely, the presence of FC and TC in all three sources exceeded standards for drinking water. Additionally, the level of Pb in the water from private and RCC tanks surpassed both standards. The level of Fe in RCC tanks (0.98 mg L^{-1}) exceeded the standard of WHO (0.3 mg L^{-1}) but remains within the national standard (1 mg L^{-1}) (Table 14.1).

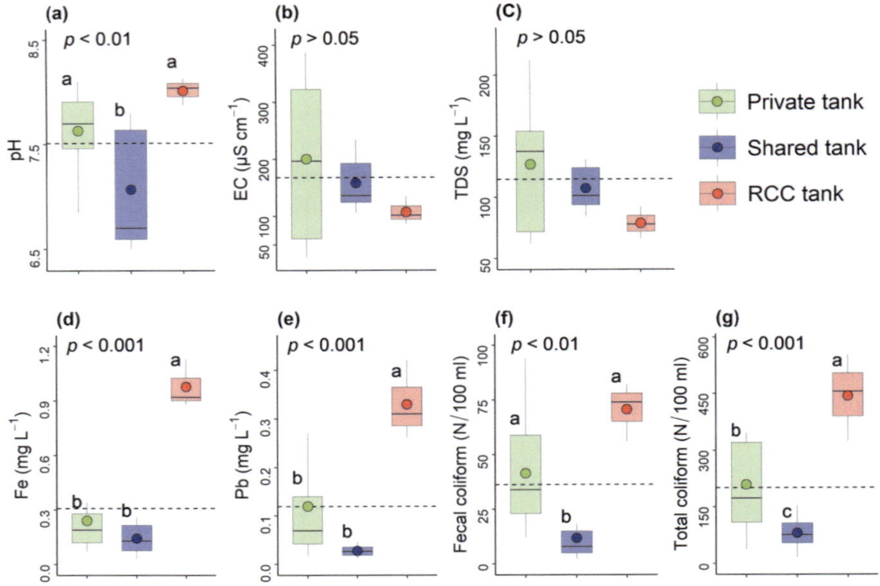

Fig. 14.4 Comparison of water quality parameters: **a** pH levels, **b** electrical conductivity (EC), **c** total dissolved solids (TDS), **d** iron (Fe), **e** lead (Pb), **f** faecal coliform (FC), and **g** total coliform (TC) across three different water sources (private tanks, shared tanks, and RCC tanks). The p values ($p < 0.05$, $p < 0.01$, and $p < 0.001$) indicate the significance of the differences in the levels of water quality parameters among the water sources. The p value ($p > 0.05$) represents the differences of parameters among the water sources is not significant. Different letters (a, b, and c) indicate the significant differences in the levels of water quality parameters between water sources (e.g. private vs. shared tanks). The box plots depict the distribution of the values of water quality parameters. The horizontal lines (solid points) within the boxes indicate the mean (median) of the values of the parameters. The horizontal dashed lines inside the plot depict the mean values of three water sources

14.3.1.2 Correlation of Water Quality Parameters

Pearson correlation analysis of water quality parameters exhibited a significant positive correlation of pH and Fe ($r = 0.43$), Pb ($r = 0.42$), FC ($r = 0.42$) and TC ($r = 0.60$) (Fig. 14.5, all $p < 0.05$). Similarly, significant positive correlations were observed between Fe and Pb ($r = 0.57$), Fe and FC ($r = 0.52$), Fe and TC ($r = 0.67$), Pb and FC ($r = 0.59$), Pb and TC ($r = 0.57$), and FC and TC ($r = 0.65$) (all $p < 0.05$; Fig. 14.5). Conversely, other variables exhibited negative correlations, with a significant negative correlation observed between EC and TC ($r = -0.44$, $p < 0.05$).

Table 14.1 Comparison of water quality of the measured parameters with the standard of World Health Organization (WHO) and national standard of Bangladesh government

Parameter	Unit	Water source			Quality standard	
		Private tank	Shared tank	RCC tank	WHO	National
pH	–	7.6	7.1	8.0	6.5–9.2	6.5–8.5
Electrical conductivity (EC)	$\mu S\ cm^{-1}$	186.85	158.43	107.33	300	800
Total dissolved solid (TDS)	$mg\ L^{-1}$	126.98	107.28	78.67	500	1000
Iron (Fe)	$mg\ L^{-1}$	0.241	0.143	**0.98**	0.3	1
Lead (Pb)	$mg\ L^{-1}$	**0.12**	0.03	**0.33**	0.05	0.05
Faecal coliform (FC)	N/100 ml	**41**	**12**	**71**	0	0
Total coliform (TC)	N/100 ml	**209**	**81**	**444**	0	0

Values in bold font indicate that water quality of the respective parameters exceeded the standards for drinking water

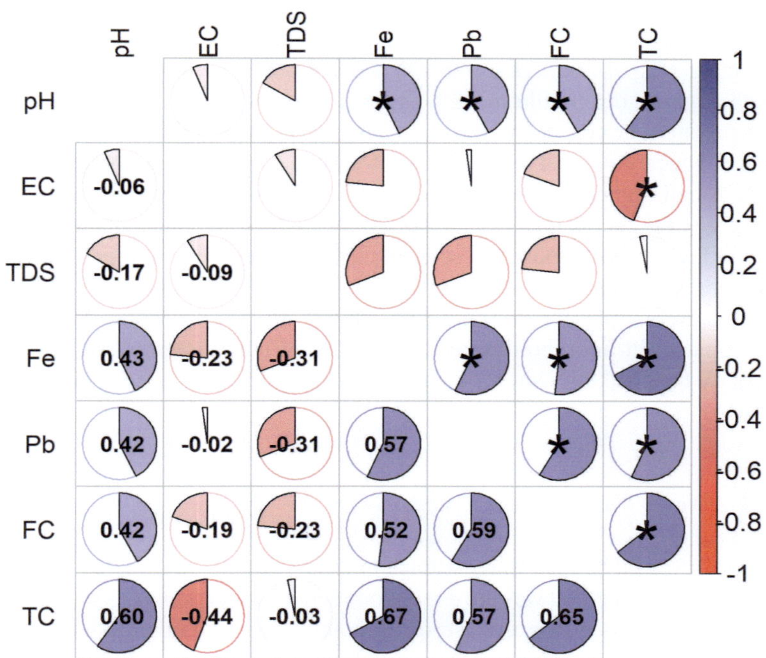

Fig. 14.5 Pearson correlation coefficients among the water quality parameters: pH levels, electrical conductivity (EC), total dissolved solids (TDS), iron (Fe), lead (Pb), faecal coliform (FC), and total coliform (TC). In the upper triangle, significant correlations ($p < 0.05$) are marked with asterisks (*), while insignificant correlations ($p > 0.05$) are left blank. The lower triangle displays the correlation coefficients. Correlations are represented using pie chart symbols with a color gradient ranging from slate blue ($r = +1$, strong positive correlation) to white ($r = 0$, no correlation), to red ($r = -1$, strong negative correlation)

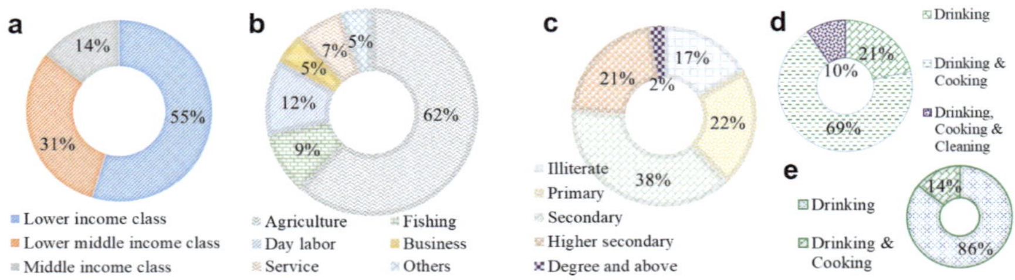

Fig. 14.6 Respondents' economic status (**a**), occupation (**b**), education level (**c**), rainwater use during monsoon (**d**), and rainwater use during other seasons (**e**) in the study areas

14.3.2 Water Usage and Management at the Household Level

14.3.2.1 Description of Socio-economic Status

The results indicate that the majority of respondents (55%) fall into the lower-income class, followed by 31% in the lower middle-income class and 14% in the middle-income class (Fig. 14.6a). Households' occupations showed significant diversity, with 62% engaged in agriculture, 12% as day labourers, 9% in fishing, 7% in government or private services, 5% in business, and 4% in other professions (Fig. 14.6b). In terms of education, most respondents (38%) had completed secondary education, while 21% had higher secondary education, 21% had primary education, and 2% had a degree or higher (Fig. 14.6c). Additionally, 17% of respondents were illiterate (Fig. 14.6c). During the monsoon season, 69% of households rely on rainwater for drinking and cooking, 10% use rainwater for drinking, cooking, and cleaning, and 21% use it exclusively for drinking (Fig. 14.6d). In other seasons, due to limited storage capacity in household water tanks, 86% of households utilize stored rainwater solely for drinking, while 14% use it for both drinking and cooking purposes (Fig. 14.6e).

14.3.2.2 Water Management Practices

The survey revealed diverse water storage practices among households (Fig. 14.7a). The most common method was the use of commercial water tanks (33%). The second most common practice was the use of a combination of earthen pots and plastic drums (26%). Other storage facilities include water tanks and PSFs (14%), PSF (10%), plastic drums (10%) and earthen pots (7%). In terms of the duration of water storage, 45% of households stored water for less than two months, whereas 29% kept water for two to four months. About 17% stored water for four to six months, and 10% for over six months (Fig. 14.7b). Health issues related to water consumption were a significant concern among respondents. Diarrhea was reported by 79% of households, while 67% experienced vomiting. Skin irritation was reported by 57%,

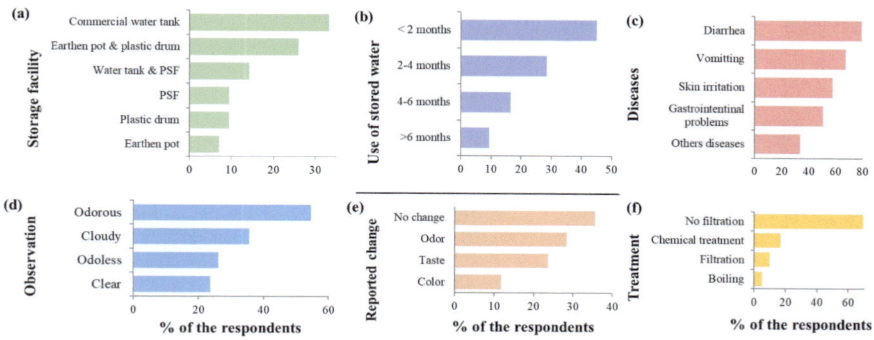

Fig. 14.7 Households' responses on rainwater storage facilities (**a**), the duration of the stored water use (**b**), reported diseases they experienced (**c**), observation of stored water (**d**), changes in water quality (**e**), and treatment before using the stored water (**f**)

and 50% indicated cases of gastrointestinal problems. Additionally, 33% mentioned other illnesses possibly linked to water quality (Fig. 14.7c).

Households also provided insights into the physical characteristics of their water (Fig. 14.7d). While 24% described their water as clear, a much larger group, 55%, reported it as odorous. 36% indicated their water was cloudy, and 26% noted it was odorless. One-third of the respondents reported that their stored water had no change in taste or quality, while 29% reported odor, 24% mentioned a taste change, and 12% stated a color change in their stored water (Fig. 14.7e). The survey also examined water treatment habits before consumption (Fig. 14.7f). A notable proportion of households relied on untreated water, with 69% reporting no filtration or treatment practices. Among those who treated their water, 5% boiled it, 10% employed filtration methods, and 29% used chemical treatments.

14.3.3 Stakeholders' Perceptions in Water Management

14.3.3.1 Perceptions of Focus Groups

We identified two core issues-water quality and water contamination—from the FGD scripts (Fig. 14.8). These two issues were the primary causes of health risks (e.g., diarrhea, skin diseases, pre-eclampsia, and hypertension), storage challenges (e.g., limited rainwater storage facilities), and water quality degradation (e.g., pathogen load and heavy metals). To address these challenges, several solutions were reported by the participants of the FGDs. For example, rainwater harvesting and water treatment emerged as dominant solutions. Other solutions included the establishment of PSFs, desalinization treatment facilities, community awareness, government support (e.g., distribution of rainwater harvesting tanks and water purification systems at the household level), and NGO support (e.g., regular monitoring and maintenance

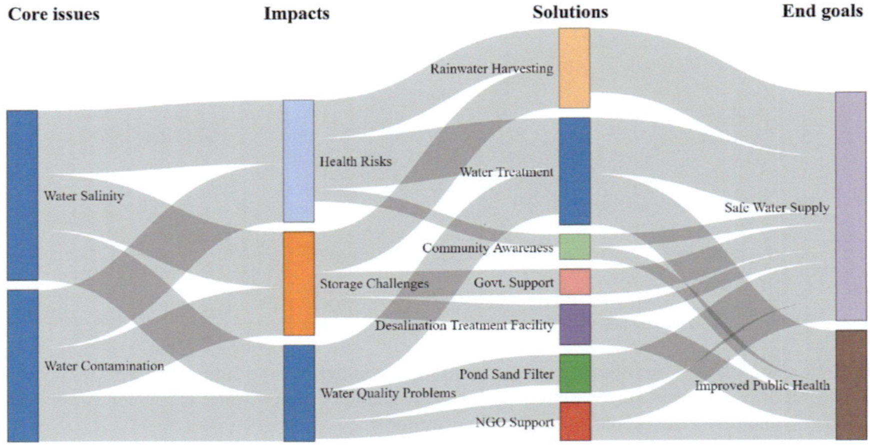

Fig. 14.8 Sankey plot showing the core issues related to water (water salinity and contamination), the impacts of these issues (health risks, storage challenges, and water quality), the viable solutions (e.g., rainwater harvesting and water treatment), and the end goals (safe water supply and improved public health)

support). These major reported solutions are pathways to achieving sustainable outcomes, such as a safe water supply and improved public health (Fig. 14.8).

14.3.3.2 Perceptions of In-depth Interviewees

An analysis of the top 25 keywords identified *Water Salinity* and *Rainwater Harvesting* as the most frequently mentioned issues, each appearing 16 times (Fig. 14.9). Other frequently cited terms included *Water shortage*, *Water Quality*, *Water Tank*, *Rooftop*, and *Pond Sand Filter*, which were mentioned 15, 14, 14, 13 and 13 times, respectively. Keywords such as *Pond Water*, *Disease*, and *Health* were also prominent, appearing 11, 9, and 8 times, respectively. The analysis also highlighted keywords such as *Water Treatment*, *Contaminants*, *Pathogen*, and *Patients*, which suggest active efforts to address water purification and its impact on public health (Fig. 14.9).

Fig. 14.9 Word cloud of the top 25 words obtained from the script of 4 IDIs in the study region. Font size corresponds to the relative frequency of each term, with larger text indicating more prominent issues

14.4 Discussion

14.4.1 Water Quality

Results of the water quality analysis provide important insights into the condition of drinking water from three storage sources: RCC tanks, shared tanks, and private tanks. Among the sources, notable variations in pH, Fe, Pb, FC and TC were noted, highlighting differences in water quality among these sources. Several factors could contribute to the variations in the quality of water stored in these sources, including storage methods, the quality of the catchment surfaces and the length of water storage (Islam et al., 2011, 2019; Rahman et al., 2021; Shapna et al., 2024).

Although pH, EC, and TDS levels across all three water sources comply with the WHO and national drinking water quality standards, the presence of FC and TC in all three sources surpassed the allowed limit for drinking water (Table 14.1). This general bacterial contamination points to systemic problems with sanitation, storage, and water collection methods. In previous studies, the presence of microbial pollutants including FC and TC has been reported the results of inadequate sanitation practices (Islam et al., 2011; Parvin et al., 2022). The observed high levels (in N/100 ml) of FC (TC) in our studied samples: 41 (209) in private tanks, 12 (81) in shared tanks, and 71 (444) in RCC tanks are consistent with the quality of harvested rainwater in different storage facilities in salinisation-affected regions of the country (Islam et al., 2011). For instance, levels of FC found in household-level and community-based rainwater collection systems have been reported as 465 N/100 ml and 856 N/100 ml, respectively in the southwest coastal region of Bangladesh (Islam et al., 2011). The growth of bacteria depends critically on the storage conditions of harvested rainfall.

Especially, if these tanks are not routinely cleaned and maintained, allowing sediment accumulation to harbor pathogens. Dark, warm, and poorly ventilated storage tanks create an ideal environment for bacteria to flourish. Moreover, the first runoff during rain events washes away accumulated pollutants from surfaces (Xu et al., 2021). Without effective first-flush diverters, this contaminated water can unintentionally enter storage tanks. Many households that depend on harvested rainwater also lack access to suitable water treatment options, and the absence of filtration or disinfection processes can further contribute to retaining FC and TC in the water supply (Lepcha et al., 2024).

The measured exceeded levels of Pb in private and RCC tanks highlight even more the need for better infrastructure and thorough water system monitoring. For instance, the presence of Pb in our study may indicate runoff from old plumbing systems, which has been recorded in other studies (Islam et al., 2010, 2019; Palawat et al., 2023), stressing the need of enough infrastructure and water treatment methods. Pb contamination in harvested rainwater may originate from aging infrastructure coated with lead-based paints (Rahman et al., 2021). As these painted surfaces on roofs, gutters, or storage tanks weather and deteriorate over time, rainfall can wash loose lead particles into the water collection system. The higher Pb concentration in the water samples in private and RCC tanks in our study is consistent with similar studies in other coastal areas in the country (Islam et al., 2019; Rahman et al., 2021). For example, Pb concentration in harvested rainwater in RCC tank from a primary school in Mongla sub-district (adjacent to our study area) was reported 0.08 mg L^{-1}, which also exceeds the WHO and national standards (Islam et al., 2019). Similar findings were also reported by Rahman et al. (2021), who found 17% of their samples in the harvested rainwater in private tanks exceeded both national and international standards.

Nonetheless, parameters such as EC and TDS revealed no substantial variations among the sources, implying that salinity and dissolved solids are rather constant in the three sources. Although the raised Fe levels in RCC tanks still fall within Bangladesh's national standard, it exceeded the WHO standard. This implies that although under national guidelines the Fe levels might not be an important hazard, they still compromise water quality according to the international standard. Since they raise the risk of waterborne diseases, the presence of FC and TC in drinking water causes great public health issues (Ghosh et al., 2015; Karim, 2010; Mostafa et al., 2020). Likewise, Pb at levels above recommended limits can cause serious health problems including neurological and developmental abnormalities (Kothapalli, 2021).

Implementing policies including regular cleaning of storage tanks, better sanitation practices, and community awareness campaigns will help to solve these difficulties. First-flush diverters and filtration systems can help to reduce chemical and microbial contamination. Targeted efforts, including the use of water treatment technologies and corrosion-resistant materials are crucial to lower Fe and Pb levels for RCC tanks. Regular water quality monitoring and testing, combined with government and NOG support, can ensure adherence to both national and international standards.

14.4.2 Water-Related Challenges Faced by Households

Our examination of household water usage has revealed a wide range of water storage methods used in different households, ranging from earthen pot, plastic drum, private tank, shared-tank, RCC tank, and PSF (Fig. 14.10). This fluctuation reflects not only personal choices but also local resource availability, socioeconomic background, and financial situation. One-third of our studied households (33%), for example, choose commercial water tanks (private and shared), which are usually made for longevity and simplicity of maintenance. Others (17%), on the other hand, use more readily available and reasonably priced traditional solutions including plastic drums or earthen pots. One-quarter of the households (24%) use private water tanks and PSF. Nonetheless, management strategies and the surrounding environmental conditions greatly affect the efficiency of PFS (Khan & Paul, 2024). While PSF can efficiently lower turbidity and some microbial pollutants (Li et al., 2025), research has indicated that they might not sufficiently remove all harmful substances, especially heavy metals, elevated EC (Hossain & Li, 2024), and FC (Saha et al., 2019).

Nearly 43% of the households we investigated rely on earthen pots, small plastic drums, or a mix of both for storing rainwater, which proves inadequate to meet their drinking water needs during the dry season. For up to two months following the monsoon, for instance, 45% of the households find it difficult to meet their drinking water needs. Households running low on water during the dry season sometimes turn into dangerous sources, especially pond water and PSF. These households ignore treating this unsafe water before using it for cooking and drinking due to financial restrictions and ignorance. Therefore, households affected by the number of diseases are those that had to rely on saline water sources since their household tanks lacked enough stored rainwater. Among the highly cited diseases mentioned by the households are diarrhoea (49%), vomiting (67%), skin irritation (57%), gastrointestinal problems (50%) and other diseases (e.g., eclampsia). These household-level results are in line with empirical research conducted in our study area (Khan et al., 2014), which revealed that compared to filtered pond water and rainwater consumers, saline groundwater consuming women highly affected by pre(eclampsia) and gestational hypertension.

14.4.3 Community Perceptions in Water Management

Like our household survey and water quality findings, the respondents in FGDs and IDIs also reported several concerns regarding harvested water quality. For instance, three key challenges (health risks, storage challenges, and degraded water quality) were highlighted by the FGD respondents. To address these challenges, 7 potential measures were pointed out (Fig. 14.8). Enhancing rainwater harvesting facilities through installing community-based shared water tanks and the establishment

Fig. 14.10 Drinking water sources in the salinization-affected region. © The authors

of water treatment facilities are the most suggested solutions by the FGD respondents (Fig. 14.8), representing actionable strategies for mitigating water quality and contamination issues (Saha et al., 2024). The establishment of technologies like PSFs and desalination facilities highlights the potential for community-based solutions to enhance water accessibility and quality (Khan & Paul, 2024). These technologies, when combined with regular maintenance and monitoring by NGOs, can provide long-term, scalable solutions tailored to local needs. Moreover, the emphasis on government support, including the distribution of rainwater harvesting tanks with higher storage capacity and household water purification systems, underscores the importance of institutional involvement in addressing water security challenges (Hossain & Li, 2024). Community awareness campaigns also play a pivotal role in the success of these interventions. Educating households about safe water storage practices, proper sanitation, and the importance of water treatment can bridge the gap between infrastructure availability and effective utilization (Shapna et al., 2023). A well-informed community is more likely to adopt sustainable practices that improve water quality and reduce health risks, which is also reported in a study in the southwestern district (Satkhira) in Bangladesh (Rahman et al., 2017).

The keyword analysis of the IDIs provides a deeper understanding of the community's priorities and concerns (Fig. 14.9). The frequent mention of '*Water Salinity*', '*Water Quality*', '*Pond Water*' and '*Rainwater Harvesting*' reflects the multiple challenges of saline intrusion into freshwater sources and the reliance on rainwater as a key alternative. This aligns with the broader findings that highlight salinity as a major driver of water quality issues in coastal regions (Rahman et al., 2021; Shapna et al., 2024). The prominence of terms like '*Water Shortage*', '*Water Tank*' and '*PSF*' suggests a collective focus on storage and treatment solutions. These keywords indicate an awareness of the need for infrastructure development to address both the availability and quality of water. These findings are consistent with a recent study on drinking water management in coastal Bangladesh (Hossain & Li, 2024), which reported that proper maintenance of PSFs, rain-fed ponds, and regular cleaning of water tanks minimizes the water shortage challenges in the salinization-affected coastal communities of the country. Similarly, the occurrence of terms such as '*Disease*' and '*Health*' underscores the community's recognition of the direct link between water quality and public health outcomes. Interestingly, terms like '*Pathogen*' and '*Contaminants*' highlight an understanding of the technical aspects of water quality degradation. This suggests that IDI respondents are not only aware of the problems but are also beginning to engage with more scientific concepts, which could facilitate the adoption of advanced treatment methods (Fig. 14.10).

14.4.4 Forward-Looking Strategies

Because of climate-induced catastrophic events in the coastal regions, salinity intrusion in underground aquifers and surface water bodies has been a pressing issue for human health, ecosystems and agriculture (Chen & Mueller, 2018; Hossain & Li, 2024). In our study region, the MAT has shown a significant increase over the past 35 years ($p < 0.001$, $\tau = 0.506$; Fig. 14.11a), while annual precipitation has experienced a substantial but statistically insignificant decline during the same period ($p > 0.05$, $\tau = -0.139$; Fig. 14.11b). The cumulative effects of rising MAT and declining precipitation are expected to result in more frequent hot days, intensified

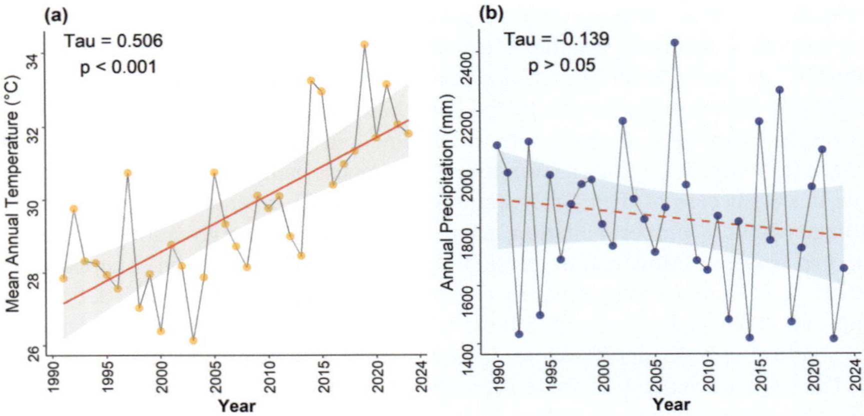

Fig. 14.11 Temporal trends of **a** mean annual temperature (MAT) and **b** annual precipitation in the study region for the period 1990–2024. Points represent MAT and annual precipitation values. Positive (negative) tau (τ) values indicate the strength and direction of the monotonic increase (decrease) in MAT (annual precipitation) over this period. The $p < 0.001$ ($p > 0.05$) values denote statistically significant (insignificant) trends obtained by the Mann–Kendall trend test. The solid red (dashed black) lines represent the trend line of MAT (annual precipitation) obtained by linear regression. The confidence intervals (95%) for fluctuations in MAT and annual precipitation are indicated as bands along the lines

heatwaves and more severe droughts. These conditions further exacerbate the challenge of ensuring access to potable water for the affected communities, underscoring the urgent need for long-term sustainable solutions.

Rainwater harvesting has emerged as a viable solution, as reported by respondents in household surveys, FGDs and IDIs. However, the low purchasing power of these low-income communities limits their ability to invest in large water storage tanks. Thus, support (e.g., large water storage tanks) from the government and donor agencies is highly required. The importance of large water storage tanks at the household and community levels have also been emphasized in recent studies conducted in similarly exposed coastal regions (Hossain & Li, 2024; Rahman et al., 2021; Shapna et al., 2024).

The establishment of water treatment facilities (e.g., reverse osmosis plants) is another potential solution, which could help people to buy water at an affordable price (Fig. 14.12). This type of water treatment facility and low-cost water supply is already operational in other unions and sub-districts in the salinization-affected regions of the country. For example, Islam et al. (2018) reported that reverse osmosis plants in the coastal region of the country supply treated water to the community, albeit serving a relatively small number of people. The water quality (TDS, EC, FC, TC, chloride, bicarbonate and sodium) standard of these plants has been reported as acceptable in the study of Islam et al. (2018). Given that our study union has a population of over 15,000 people (Bangladesh National Portal, 2025), the establishment of some

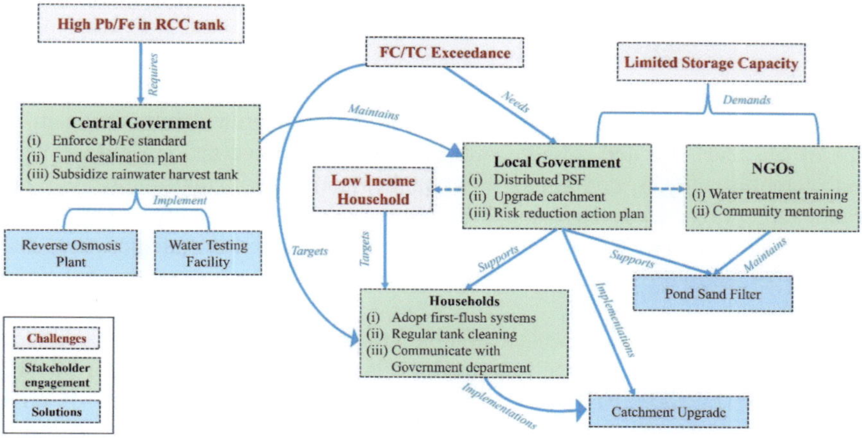

Fig. 14.12 Flowchart showing identified challenges, suggested solutions, and actors' involvement in alleviating the scarcity of safe drinking water in the study area. Pb: Lead, Fe: Iron, RCC: Reinforced Cement Concrete, FC: Faecal coliform, TC: Total coliform; PSF: Pond Sand Filter; NGO: Non-government Organization

reverse osmosis plants could substantially alleviate the ongoing crisis of safe water supply.

If adequate treatment facilities are ensured, health risks caused by microbes, heavy metals, and other chemical contaminants could be avoided. Community awareness, regular cleaning of storage tanks, treatment of unsafe pond water during periods without stored rainwater, maintaining alternative sources (e.g., PSF), and re-excavation of ponds to remove deposited salty silt could help minimize the crisis of safe water supply (Fig. 14.12). To achieve this, the government and NGOs could provide capacity-building training to the community, conduct regular monitoring of water storage facilities, and perform frequent testing of stored water (Fig. 14.12).

14.4.5 Study Limitations

Our study union has 7 RCC tanks, of which 3 were functional, and water samples were collected from these tanks. Collecting water samples from additional RCC tanks could have provided more robust insights into the water quality parameters. Moreover, our study focused solely on the dry season for water quality testing. Variations in water quality across different seasons and sources may exist. In addition, the household sample size was relatively small, with only 42 households surveyed. This limited sample size may not fully represent the broader community's challenges and adaptive practices related to drinking water management. Therefore, caution is advised when interpreting our survey and water quality findings on a broader scale.

14.5 Conclusions and Policy Implications

Examining water quality in a potable water-scarce region and understanding the adaptive mechanisms of the affected community are of critical importance in the face of sea level rise and salinity intrusion into the surface and subsurface water bodies. In this study, by combining water quality analysis with household and stakeholder perceptions, we investigated the quality of rainwater from three types of tanks (private, shared, and RCC), households' challenges and adaptive strategies, and stakeholders' perceptions of drinking water management in a salinization-affected region of Bangladesh.

Water quality analysis revealed notable contamination in all three tank types (private, shared, and RCC), with elevated levels of FC (12–71 N/100 ml), TC (81–444 N/100 ml), and Pb (0.03–0.33 mg L^{-1}) exceeding permissible limits of WHO and national standards, posing severe public health hazards including waterborne diseases and neurological problems. Results of the household survey exhibited that water usage and storage practices vary widely, influenced by socioeconomic conditions, with many relying on inadequate solutions like plastic drums or earthen pots, leading to water scarcity during the dry season. Poor water quality management, including untreated use of unsafe sources like pond water, contributes to widespread health issues such as diarrhea, vomiting, and skin irritation. Community perceptions obtained from FGDs and IDIs highlight key challenges—health risks, storage limitations, and degraded water quality—while advocating for solutions like rainwater harvesting, PSFs, desalination, and government-supported infrastructure. Regular tank cleaning, first-flush diverters, filtration systems, water treatment technologies, and government/NGO support are crucial to ensure safe drinking water and compliance with national and international standards to address these challenges.

Policies should concentrate on improving rainwater harvesting infrastructure with larger, community-based tanks, first- flush diverters, and filtration systems in addition to affordable household water treatment options to improve water quality and access. While replacing ageing infrastructure can help to address Pb and Fe contamination, regular cleaning and maintenance of storage facilities and PSFs are vital to lower FC, TC and heavy metal contamination. Climate-resilient practices and emergency storage solutions can help to lessen water shortage in dry season. Government and NGO collaboration is crucial for subsidies, community awareness campaigns, and regular water quality monitoring to ensure compliance with national and international standards, safeguard public health, and promote sustainable water management.

Acknowledgements The authors express sincere gratitude to the household members and other respondents (FGDs and IDIs) who participated in this research. Our heartfelt appreciation to the enumerators who provided their valuable support throughout this research.

Authors' Contributions **M. L. Hossain**: Conceptualization, methodology, data collection, software, formal analysis, writing-original draft, writing-review & editing, supervision, funding acquisition. **K. J. Shapna**: Methodology, data collection, data processing, software, writing-original draft. **A. Rubaiyat**: Writing-original draft. **M. H. Kabir**: Writing-review & editing. **R. Baidya:** Data collection. **J. Li**: Conceptualization, methodology, writing-review & editing, supervision, funding acquisition, project administration. **D. Y. F. Lai**: Conceptualization, methodology, writing-review & editing, supervision, funding acquisition, project administration.

Funding This work was supported by a research grant from the Research Grants Council (RGC) of the Hong Kong Special Administrative Region (SAR), China [project number RFS2223-2H02]. M. L. Hossain was supported by the RGC of the Hong Kong SAR, China through RGC Postdoctoral Fellowship Scheme 2024/25 [Award Ref. No.: PDFS2425-4S05].

Data Availability Statement Qualitative and quantitative datasets, and summary of IDIs and FGDs are available in Figshare at https://doi.org/10.6084/m9.figshare.29423768.v2.

Ethical consideration The study obtained ethical clearance from the Research Review Board of Centre for Coastal Development (CCD/ResExt/2023/8/011(04). Informed consent was obtained from all participants before conducting the household surveys, focus group discussions and in-depth interviews. Participants were made aware of the study's purpose, their right to withdraw at any time, and the confidentiality of their personal information. The study followed the principles of the Declaration of Helsinki and the guidelines of the International Conference on Harmonization for Good Clinical Practice (ICH-GCP).

Declaration The authors declare that no part of this chapter has been written using AI tools. All content, including texts, figures and supplementary information, has been prepared solely by the authors.

Conflict of Interests The authors declare that they have no competing interests or personal relationships that could appear to influence the work reported in this paper.

References

Abatzoglou, J. T., Dobrowski, S. Z., Parks, S. A., & Hegewisch, K. C. (2018). Terraclimate, a high-resolution global dataset of monthly climate and climatic water balance from 1958–2015. *Scientific Data, 5*, Article 170191. https://doi.org/10.1038/sdata.2017.191

Abdelmonem, B. H., Kamal, L. T., Elbaz, R. M., Khalifa, M., & Abdelnaser, A. (2025). From contamination to detection: The growing threat of heavy metals. *Heliyon, 11*(1), E41713. https://doi.org/10.1016/j.heliyon.2025.e41713

Abdullah, M., Idrak, F., Kabir, P., & Hossain Bhuiyan, M. A. (2024). Suitability of rainwater harvesting in saline and arsenic affected areas of Bangladesh. *Heliyon, 10*(14), Article e34328. https://doi.org/10.1016/j.heliyon.2024.e34328

Afsari, N., Murshed, S. B., Uddin, S. M. N., & Hasan, M. (2022). Opportunities and barriers against successive implementation of rainwater harvesting options to ensure water security in southwestern coastal region of Bangladesh. *Frontiers in Water, 4*, Article 811918. https://doi.org/10.3389/frwa.2022.811918

Ahmed, S. K., Mohammed, R. A., Nashwan, A. J., Ibrahim, R. H., Abdalla, A. Q., Ameen, B. M. M., & Khdhir, R. M. (2025). Using thematic analysis in qualitative research. *Journal of Medicine, Surgery, and Public Health, 6*, Article 100198. https://doi.org/10.1016/j.glmedi.2025.100198

Akhter, S. (2022). Key informants' interviews. In M. R. Islam, N. A. Khan, & R. Baikady (Eds.), *Principles of social research methodology*. Springer, Singapore. https://doi.org/10.1007/978-981-19-5441-2_27

Asma, K. M., & Kotani, K. (2021). Salinity and water-related disease risk in coastal Bangladesh. *EcoHealth, 18*, 61–75. https://doi.org/10.1007/s10393-021-01517-z

Bangladesh National Portal. (2025). Kamarkhola Union Parishad, Dacope Upazila, Khulna District. www.dakop.khulna.gov.bd/en

Barbour, E. J., Adnan, M. S. G., Borgomeo, E., Paprocki, K., Khan, M. S. A., Salehin, M., & Hall, J. W. (2022). The unequal distribution of water risks and adaptation benefits in coastal Bangladesh. *Nature Sustainability, 5*, 294–302. https://doi.org/10.1038/s41893-021-00846-9

Benesty, J., Chen, J., Huang, Y., & Cohen, I. (2009). Pearson correlation coefficient. In *Noise reduction in speech processing*. Springer Topics in Signal Processing (Vol. 2). Springer, Berlin, Heidelberg. https://doi.org/10.1007/978-3-642-00296-0_5

BMD. (2025). Bangladesh Meteorological Department. Climate Data Portal. Government of Bangladesh, Agargoan, Dhaka. https://dataportal.bmd.gov.bd

Boholm, Å., & Prutzer, M. (2017). Experts' understandings of drinking water risk management in a climate change scenario. *Climate Risk Management, 16*, 133–144. https://doi.org/10.1016/j.crm.2017.01.003

Caretta, M. A., Mukherji, A., Arfanuzzaman, M., Betts, R. A., Gelfan, A., Hirabayashi, Y., Lissner, T. K., Liu, J., Lopez Gunn, E., Morgan, R., Mwanga, S., & Supratid, S. (2022). Water. In *Climate change 2022: Impacts, adaptation and vulnerability. contribution of working group II to the sixth assessment report of the IPCC* (pp. 551–712). Cambridge University Press. https://doi.org/10.1017/9781009325844.006

CEGIS. (2006). Impact of sea level rise on land suitability and adaptation options, component 1.4.3 of sustainable environmental management program (BGD/96/007) funded by UNDP. Center for Environmental and Geographic Information Services (CEGIS), Ministry of Environment and Forest, Bangladesh

Chen, J., & Mueller, V. (2018). Coastal climate change, soil salinity and human migration in Bangladesh. *Nature Climate Change, 8*, 981–985. https://doi.org/10.1038/s41558-018-0313-8

DePaolo, C. A., & Wilkinson, K. (2014). Get your head into the clouds: Using word clouds for analyzing qualitative assessment data. *Techtrends Tech Trends, 58*, 38–44. https://doi.org/10.1007/s11528-014-0750-9

Frisbie, S. H., Mitchell, E. J., & Molla, A. R. (2024). Sea level rise from climate change is expected to increase the release of arsenic into Bangladesh's drinking well water by reduction and by the salt effect. *PLoS ONE, 19*(1), Article e0295172. https://doi.org/10.1371/journal.pone.0295172

Ghosh, G.C., Jaha, S., Chakraborty, B., & Akter, A. (2015). Potential of household rainwater harvesting for drinking water supply in Hazard prone coastal area of Bangladesh. *Nature Environment and Pollution Technology, 14*(4), 937. https://neptjournal.com/upload-images/NL-54-33-(31)B-3157.pdf

Ghosh, S., & Ahmed, T. (2022). Assessment of household rainwater harvesting systems in the southwestern coastal region of Bangladesh: Existing practices and household perception. *Water, 14*(21), 3462. https://doi.org/10.3390/w14213462

Hock, R., Rasul, G., Adler, C., Cáceres, B., Gruber, S., Hirabayashi, Y., Jackson, M., Kääb, A., Kang, S., Kutuzov, S., Milner, A., Molau, U., Morin, S., Orlove, B., & Steltzer, H. (2019). High mountain areas. In H. -O. Pörtner, D. C. Roberts, V. Masson-Delmotte, P. Zhai, M. Tignor, E. Poloczanska, K. Mintenbeck, A. Alegría, M. Nicolai, A. Okem, J. Petzold, B. Rama, & N. M. Weyer (Eds.), *IPCC special report on the ocean and cryosphere in a changing climate* (pp. 131–202). Cambridge University Press, Cambridge, UK and New York, NY, USA. https://doi.org/10.1017/9781009157964.004

Hoque, M. A., Scheelbeek, P. F. D., Vineis, P., Khan, A. E., Ahmed, K. M., & Butler, A. P. (2016). Drinking water vulnerability to climate change and alternatives for adaptation in coastal South and South East Asia. *Climatic Change, 136*, 247–263. https://doi.org/10.1007/s10584-016-1617-1

Hossain, M. L., & Hossain, M. K. (2012). *Climate change, sea level rise and coastal vulnerabilities of Bangladesh with adaptation options*. University of Chittagong.

Hossain, M. L., Hossain, M. K., Salam, M. A., & Rubaiyat, A. (2012). Seasonal variation of soil salinity in coastal areas of Bangladesh. *International Journal of Environmental Science, Management and Engineering Research, 1*(4), 172–178. file:///C:/Users/b161886/Downloads/SalinityBD%20(4).pdf

Hossain, M. L., & Li, J. (2024). Salinity challenges and adaptive strategies in salinization-affected coastal Bangladesh: Implications for agricultural sustainability and water resource management. *Environmental Research Letters, 19*, Article 114048. https://doi.org/10.1088/1748-9326/ad7edb

Hossain, M. L., Shapna, K. J., Li, J., Kabir, M. H., Siddika, F., Khandker, S., & Beierkuhnlein, C. (2025). Transforming agriculture with vermicompost: 7-year empirical evidence from drought-prone and salinization-affected regions of Bangladesh. *Journal of Cleaner Production, 508*, Article 145595. https://doi.org/10.1016/j.jclepro.2025.145595

Hsieh, H. F., & Shannon, S. E. (2005). Three approaches to qualitative content analysis. *Qualitative Health Research, 15*(9), 1277–1288. https://doi.org/10.1177/1049732305276687

IPCC. (2023). Summary for policymakers. In H. Lee, & J. Romero (Eds.) *Climate change 2023: Synthesis report. Contribution of Working Groups I, II and III to the Sixth Assessment Report of the Intergovernmental Panel on Climate Change.* IPCC, Geneva, Switzerland, pp. 1–34. https://doi.org/10.59327/IPCC/AR6-9789291691647.001

Islam, M. A., Akber, M. A., Rahman, M. A., Islam, M. A., & Kabir, M. P. (2019). Evaluation of harvested rainwater quality at primary schools of southwest coastal Bangladesh. *Environmental Monitoring and Assessment, 191*, 80. https://doi.org/10.1007/s10661-019-7217-6

Islam, M. A., Akber, M. A., & Ghosh, P. K. (2018). Water quality of small-scale desalination plants in southwest coastal Bangladesh. *Water Supply, 18*(5), 1606–1616. https://doi.org/10.2166/ws.2017.222

Islam, M. A., Sakakibara, H., Karim, M. R., Sekine, M., & Mahmud, Z. H. (2011). Bacteriological assessment of drinking water supply options in coastal areas of Bangladesh. *Journal of Water and Health, 9*(2), 415–428. https://doi.org/10.2166/wh.2011.114

Islam, M. M., Kabir, M. R., & Chou, F. F. (2010). Quality assessment of stored rainwater and its utilization in Dhaka City, Bangladesh. *Water Practice and Technology, 5*(3), wpt2010059. https://doi.org/10.2166/wpt.2010.059

Kabir, A. E., Chakraborty, T. K., & Ghosh, G. C. (2016). Bio-sand filter (BSF): A simple water treatment device for safe drinking water supply and to promote health in Hazard prone hard-to-reach coastal areas of Bangladesh. *American Journal of Environmental Protection, 5*(5), 107–112. https://doi.org/10.11648/j.ajep.20160505.11

Karim, M. R., Rimi, R. A., & Billah, M. S. (2015). Analysis of storage volume and reliability of the rainwater harvesting tanks in the coastal area of Bangladesh. *Desalination and Water Treatment, 54*(13), 3544–3550. https://doi.org/10.1080/19443994.2014.923201

Karim, M. R. (2010). Microbial contamination and associated health burden of rainwater harvesting in Bangladesh. *Water Science and Technology, 61*(8), 2129–2135. https://doi.org/10.2166/wst.2010.031

Kaufmann, J., & Schering, A. G. (2014). *Wily StatsRef: Statistics reference online* (pp. 1–12). https://doi.org/10.1002/9781118445112

Khan, A. E., Scheelbeek, P. F. D., Shilpi, A. B., Chan, Q., Mojumder, S. K., Rahman, A., & Vineis, P. (2014). Salinity in drinking water and the risk of (pre)eclampsia and gestational hypertension in coastal Bangladesh: A case-control study. *PLoS ONE, 9*(9), Article e108715. https://doi.org/10.1371/journal.pone.0108715

Khan, M. S., & Paul, S. K. (2024). Quality and efficiency assessment, health issues and management practices of pond sand filter water in coastal Bangladesh. *International Journal of Energy and Water Resources, 8*(3), 311–325. https://doi.org/10.1007/s42108-023-00264-x

Kothapalli, C. R. (2021). Differential impact of heavy metals on neurotoxicity during development and in aging central nervous system. *Current Opinion in Toxicology, 26*, 33–38. https://doi.org/10.1016/j.cotox.2021.04.003

Lepcha, R., Patra, S. K., Ray, R., Thapa, S., Baral, D., & Saha, S. (2024). Rooftop rainwater harvesting a solution to water scarcity: A review. *Groundwater for Sustainable Development, 26*, Article 101305. https://doi.org/10.1016/j.gsd.2024.101305

Leveque, B., Burnet, J. B., Dorner, S., & Bichai, F. (2021). Impact of climate change on the vulnerability of drinking water intakes in a northern region. *Sustainable Cities and Society, 66*, Article 102656. https://doi.org/10.1016/j.scs.2020.102656

Li, N., Li, X., Zhao, L., Lu, Z. D., Liu, Y. W., & Wang, N. (2025). Slow sand filters with variable filtration rates for rainwater purification: Microecological differences between biofilm and water phases. *Journal of Environmental Management, 375*, Article 124210. https://doi.org/10.1016/j.jenvman.2025.124210

Ma, B., Hu, C., Zhang, J., Ulbricht, M., & Panglisch, S. (2022). Impact of climate change on drinking water safety. *ACS ES&T Water, 2*(2), 259–261. https://doi.org/10.1021/acsestwater.2c00004

Mamun, M. A. A., Li, J., Cui, A., Chowdhury, R., & Hossain, M. L. (2024). Climate-adaptive strategies for enhancing agricultural resilience in southeastern coastal Bangladesh: Insights from farmers and stakeholders. *PLoS ONE, 19*(6), Article e0305609. https://doi.org/10.1371/journal.pone.0305609

Mann, H. B. (1945). Nonparametric test against trend. *Econometrica, 13*, 245–259. https://doi.org/10.2307/1907187

Mekonnen, M. M., & Hoekstra, A. Y. (2016). Four billion people facing severe water scarcity. *Science Advances, 2*(2), Article e1500323. https://doi.org/10.1126/sciadv.1500323

Meredith, M., Sommerkorn, M., Cassotta, S., Derksen, C., Ekaykin, A., Hollowed, A., & Schuur, E. A. G. (2022). Polar regions. In H. O. Pörtner, D. C. Roberts, V. Masson-Delmotte, P. Zhai, M. Tignor, E. Poloczanska, K. Mintenbeck, A. Alegría, M. Nicolai, A. Okem, J. Petzold, B. Rama, & N. M. Weyer (Eds.), *IPCC special report on the ocean and cryosphere in a changing climate* (pp. 203–320). Cambridge University Press. https://doi.org/10.1017/9781009157964.005

Mostafa, A. I., Samina, Z., Tapos Kumar, C., & Gopal Chandra, G. (2020). Assessment of rainwater quality from harvested rainwater in the coastal region of Bangladesh. In A. Haque, & A. Chowdhury (Eds.), *Water, flood management and water security under a changing climate*. Springer. https://doi.org/10.1007/978-3-030-47786-8_12

Mukarram, M. M. T., Kafy, A. A., Mukarram, M. M. T., Rukiya, Q. U., Almulhim, A. I., Das, A., & Chowdhury, M. A. (2023). Perception of coastal citizens on the prospect of community-based rainwater harvesting system for sustainable water resource management. *Resources, Conservation and Recycling, 198*, Article 107196. https://doi.org/10.1016/j.resconrec.2023.107196

Nyumba, T. O., Wilson, K., Derrick, C. J., & Mukherjee, N. (2018). The use of focus group discussion methodology: Insights from two decades of application in conservation. *Methods in Ecology and Evolution, 9*(1), 20–32. https://doi.org/10.1111/2041-210X.12860

Palawat, K., Root, R. A., Cortez, L. I., Foley, T., Carella, V., Beck, C., & Ramírez-Andreotta, M. D. (2023). Patterns of contamination and burden of lead and arsenic in rooftop harvested rainwater collected in Arizona environmental justice communities. *Journal of Environmental Management, 337*, Article 117747. https://doi.org/10.1016/j.jenvman.2023.117747

Parvin, F., Haque, M. M., & Tareq, S. M. (2022). Recent status of water quality in Bangladesh: A systematic review, meta-analysis and health risk assessment. *Environmental Challenges, 6*, Article 100416. https://doi.org/10.1016/j.envc.2021.100416

R Core Team. (2020). *R: A language and environment for statistical computing* (R Foundation for Statistical Computing). (Available at: www.R-project.org/).

Rahman, M. A., Hossain, M. L., Foysal, M. A., & Asma Akter, A. A. (2014). Impact of climate change on food security in south-west coastal region of Bangladesh. *American Journal of Experimental Agriculture, 4*(12), 1916–1934. https://doi.org/10.9734/AJEA/2014/6165

Rahman, M. A., Hashem, M. A., Sheikh, M. H. R., & Fazle Bari, A. S. M. (2021). Quality assessment of harvested rainwater and seasonal variations in the southwest coastal area, Bangladesh. *Environmental Earth Sciences, 80*, 1–12. https://doi.org/10.1007/s12665-021-09622-6

Rahman, M. T. U., Rasheduzzaman, M., Habib, M. A., Ahmed, A., Tareq, S. M., & Muniruzzaman, S. M. (2017). Assessment of fresh water security in coastal Bangladesh: An insight from salinity, community perception and adaptation. *Ocean & Coastal Management, 137*, 68–81. https://doi.org/10.1016/j.ocecoaman.2016.12.005

Raimondi, A., Quinn, R., Abhijith, G. R., Becciu, G., & Ostfeld, A. (2023). Rainwater harvesting and treatment: State of the art and perspectives. *Water, 15*(8), 1518. https://doi.org/10.3390/w15081518

Rakib, M. A., Sasaki, J., Matsuda, H., & Fukunaga, M. (2019). Severe salinity contamination in drinking water and associated human health hazards increase migration risk in the southwestern coastal part of Bangladesh. *Journal of Environmental Management, 240*, 238–248. https://doi.org/10.1016/j.jenvman.2019.03.101

Rakib, M. A., Sasaki, J., Matsuda, H., Quraishi, S. B., Mahmud, M. J., Bodrud-Doza, M., & Bhuiyan, M. A. (2020). Groundwater salinization and associated co-contamination risk increase severe drinking water vulnerabilities in the southwestern coast of Bangladesh. *Chemosphere, 246*, Article 125646. https://doi.org/10.1016/j.chemosphere.2019.125646

Rodrigues de Sá Silva, A. C., Bimbato, A. M., Balestieri, J. A. P., & Vilanova, M. R. N. (2022). Exploring environmental, economic and social aspects of rainwater harvesting systems: A review. *Sustainable Cities and Society, 76*, 103475. https://doi.org/10.1016/j.scs.2021.103475

Rompre, A., Servais, P., Baudart, J., De-Roubin, M. R., & Laurent, P. (2002). Detection and enumeration of coliforms in drinking water: Current methods and emerging approaches. *Journal of Microbiological Methods, 49*(1), 31–54. https://doi.org/10.1016/S0167-7012(01)00351-7

Ruiz Martínez, H. G., & Cornejo Tueros, J. V. (2022). Rainwater harvesting system as a strategy for adaptation on climate change: A review. *IOP Conference Series: Earth and Environmental Science, 1121*(1), Article 012007. https://doi.org/10.1088/1755-1315/1121/1/012007

Saha, A., Setu, S., Das, S., Hossain, M. I., Rahman, A. K., & Rahman, M. M. (2024). Decision support system for community managed rainwater harvesting: A case study in the salinity-prone coastal region of Bangladesh. *Heliyon, 10*, Article e30455. https://doi.org/10.1016/j.heliyon.2024.e30455

Saha, R., Dey, N. C., Rahman, M., Bhattacharya, P., & Rabbani, G. H. (2019). Geogenic arsenic and microbial contamination in drinking water sources: Exposure risks to the coastal population in Bangladesh. *Frontiers in Environmental Science, 7*, 57. https://doi.org/10.3389/fenvs.2019.00057

Shahid, S. (2010). Rainfall variability and the trends of wet and dry periods in Bangladesh. *International Journal of Climatology, 30*, 2299–2313. https://doi.org/10.1002/joc.2053

Shaibur, M. R., Parvin, S., Ahmmed, I., Rahaman, M. H., Das, T. K., & Sarwar, S. (2021). Gradients of salinity in water sources of Batiaghata, Dacope and Koyra Upazila of coastal Khulna district, Bangladesh. *Environmental Challenges, 4*, Article 100152. https://doi.org/10.1016/j.envc.2021.100152

Shapna, K. J., Li, J., & Hossain, M. L. (2025). Solar energy integration in off-grid communities: Empowering remote areas in Bangladesh for sustainable development. *Discover Sustainability, 6*, 371. https://doi.org/10.1007/s43621-025-01214-9

Shapna, K. J., Li, J., Kabir, M. H., Salam, M. A., Khandker, S., & Hossain, M. L. (2024). Strengthening adaptation in coastal Bangladesh: Community-based approaches for sustainable agriculture and water management. *Disaster Prevention and Resilience, 3*(5). https://doi.org/10.20517/dpr.2023.41

Shapna, K. J., Hasan, K., Kabir, K. H., Li, J., & Hossain, M. L. (2023). Water, sanitation and hygiene challenges of forcibly displaced Myanmar nationals in Rohingya camps in Bangladesh. *Journal of Water and Health, 21*(10), 1385–1403. https://doi.org/10.2166/wh.2023.045

SRDI. (2019). *Annual report 2018–2019*. Soil Research Development Institute, Ministry of Agriculture, Dhaka. Available at: 2020-08-16-15-18-0eec130cc27a84d0b61fa3e4d3e13b0d.pdf

Tsai, C., Hoque, M. A., Vineis, P., Ahmed, K. M., & Butler, A. P. (2024). Salinisation of drinking water ponds and groundwater in coastal Bangladesh linked to tropical cyclones. *Scientific Reports, 14*, 5211. https://doi.org/10.1038/s41598-024-54446-6

Tukey, J. (1949). Comparing individual means in the analysis of variance. *Biometrics, 5*(2), 99–114. https://pubmed.ncbi.nlm.nih.gov/18151955/

Ward, P. J., Jongman, B., Weiland, F. S., Bouwman, A., van Beek, R., Bierkens, M. F. P., Ligtvoet, W., & Winsemius, H. C. (2013). Assessing flood risk at the global scale: Model setup, results, and sensitivity. *Environmental Research Letters, 8*(4), Article 044019. https://doi.org/10.1088/1748-9326/8/4/044019

WHO. (2011). *Guidelines for drinking water quality* (4th ed.). World Health Organization, Geneva. https://www.paho.org/en/documents/guidelines-drinking-water-quality-4o-ed-2011

Xu, J., Wu, X., Ge, X., Tian, Y., Ma, X., Li, Y., Xu, X., & Li, Z. (2021). Variations of concentration characteristics of rainfall runoff pollutants in typical urban living areas. *Bulletin of Environmental Contamination and Toxicology, 106*, 608–613. https://doi.org/10.1007/s00128-021-03110-0

Yan, S., Zhang, T., Zhang, B., Feng, H., & Siddique, K. H. (2024). Calibration of saline water quality assessment standard based on EC and CROSS considering soil water-salt transport and crack formation. *Journal of Hydrology, 633*, Article 130975. https://doi.org/10.1016/j.jhydrol.2024.130975

Chapter 15
Cyclone Fani and Rural Well-Being: A Community-Centered Analysis from Puri District, Odisha

Swosti Subhrajyoti Sahoo and Sukanya Das

Abstract This paper explores the multi-dimensional impacts of Cyclone Fani (2019) on rural households in Puri district, Odisha, using a social well-being framework that integrates economic, relational, and subjective dimensions. Employing a community-based approach informed by the Community Toolbox, data were collected through stakeholder interviews in eleven affected villages and were analysed thematically. Results show extensive housing, livelihood, infrastructure, and environmental damage, with agriculture and fishing most affected. Economic well-being declined due to income loss and inadequate recovery support. Relational well-being was reinforced by strong community networks, Self-Help Groups, and informal institutions enabling mutual aid and collective action. Subjective well-being was sustained through trust, solidarity, and shared identity. The findings highlight interlinked feedback loops between well-being, disaster preparedness, coping strategies, and environmental challenges, underscoring the need for policies that integrate social, cultural, and psychological factors alongside economic recovery to strengthen resilience.

Keywords Cyclone Fani · Social well-being · Disaster resilience · Odisha · Rural communities

15.1 Introduction

Climate change affecting economic activity and human well-being is an old idea, even featuring in the writings of the Ancient Greeks (Dell et al., 2014). Climatic variables—such as temperature, precipitation, and wind patterns—exert significant

S. S. Sahoo (✉) · S. Das
Department of Policy and Management Studies, TERI School of Advanced Studies, New Delhi, India
e-mail: swosti.sahoo@terisas.ac.in

S. Das
e-mail: sukanya.das@terisas.ac.in

P. Datta et al. (eds.), *Climate Change and Human Well-being*, Advances in Global Change Research 83, https://doi.org/10.1007/978-3-032-25479-5_15

influence on economic activities and outcomes. This impact is increasingly critical for effective policy design, governance, and resource management. Recent literature examining climate change effects has documented how variations in these variables shape diverse economic dimensions, including agricultural output (Schlenker & Lobell, 2010), energy demand (Auffhammer & Aroonruengsawat, 2011), labor productivity, mortality rates (Barreca, 2012), crime and conflict levels (Rotton & Cohn, 2004; Sekhri & Storeygard, 2014; Theisen et al., 2012), industrial production (Dell et al., 2014), migration patterns (Feng et al., 2010), and overall economic growth (Sandhani et al., 2020). But the literature on the impact of climate is currently restricted to economic impact, which is only one aspect of the overall social well-being of the people and households (Armitage et al., 2012). Most of the existing studies use pre-defined indicators to assess the impact of climate change and extreme climatic events. But this only offers a partial and incomplete picture. As the impact of climate change and extreme climatic events extends beyond the conventional indicators like economic productivity, crime rates, and mortality. It shapes households' everyday lives—impacting their interactions, food habits (Muzerengi et al., 2023), sense of loss (Charlson et al., 2021), and recovery strategies. Adopting a well-being approach helps in moving beyond established conventional indicators (Costanza et al., 2007). Human activities are deeply interdependent with natural processes, and recent evidence shows that this relationship is bidirectional, i.e., humans influence nature just as nature shapes human lives. Hence, this study adopts a social well-being approach, which goes beyond material indicators by emphasizing collective and relative processes over individualistic notions (Gough, 2004).

It moves beyond material/asset-based and basic needs conceptions, acknowledging the critical importance of social, psychological, and cultural needs of humans to not just exist but to thrive (McGregor et al., 2009; White, 2010). The concept of well-being, especially in development, helps in understanding and capturing the context-specific needs in times of distress (Copestake, 2008). Using this concept of well-being, we look into how a natural disaster/extreme climatic event cyclone can affect households and communities as integrating well-being with human need frameworks is vital for effective climate policy, as it directly addresses the neglected demand-side of mitigation strategies required for extreme climatic events (Lamb & Steinberger, 2017). The objective of this chapter is to explore the multi-dimensional effect of cyclone Fani on rural households in the Puri district of Odisha.

Thinking from a well-being perspective, we have to look into the relationship between the natural processes and human lives, how the natural processes affect human lives, their day-to-day activities, and how humans adapt to changes created by the natural processes. Natural disasters like cyclones aren't just natural processes, but they are extreme climatic events that have a varied impact on human lives and communities (Ebi et al., 2021; Honda et al., 2012). Current literature has looked into the well-being impact of natural disasters and extreme climatic events through well-being indicators, which are mostly economic in nature. Cissé and Barrett (2018) used the livestock holding of the households belonging to pastoral communities in

Kenya (Tropical Livestock Unit[1]) as a measure of well-being or Yang et al. (2018) use changes in livelihood or Tamuly and Mukhopadhyay (2022) use monthly consumption expenditure adjusted for healthcare expenditure as an indicator for well-being against natural disasters and extreme climatic events. But the issue with these indicators is that they don't capture the exact needs of the households after a natural disaster and hence don't give a clear picture of the effect of a natural disaster on a community. Additionally, the origin of the natural disasters may be natural, but their effects are not entirely natural in the sense that they depend on how the households interact with their surrounding (Yang et al., 2018). Hence, based on this motivation, we engage in the exercise of taking a community-based approach to understanding the effect of Cyclone Fani on households in the Puri district of Odisha. We use the community toolbox to design our questionnaire and carry out a thematic analysis of the transcripts to find out the impact of cyclone on households and the relevant indicators for the well-being of the households.

15.2 Study Sites

Odisha, being a coastal state, is highly vulnerable to cyclones, storm surges, and the associated flooding (Mishra & Kar, 2016). Climate change and global warming are expected to further increase both the frequency and intensity of cyclones in the coming years (IPCC, 2019). With growing infrastructure and population in the region, the scale of economic losses and damages from such events is also projected to rise (Chittibabu et al., 2004; Crowell et al., 2007). Within the state, the Puri and Jagatsinghpur districts are considered the most susceptible to cyclones and storm surges (Kumar et al., 2010). Historically, Odisha has experienced the highest number of cyclones among all coastal states in India, facing 95 cyclones between 1891 and 2017, followed by Andhra Pradesh with 77 during the same period (IMD, 2018). The destructive potential of these events is evident from the 1999 super cyclone, which claimed approximately 10,000 lives and brought wind speeds of 250–260 km/h (Chhotray & Few, 2012). Historical analyses suggest that at least three such super cyclones occurred in the nineteenth century in this region, with an estimated recurrence cycle of around 50 years (Chittibabu et al., 2004). Given this high vulnerability, the present study seeks to identify key contributing factors and, most importantly, determine the minimum household resilience threshold needed to prevent collapse during extreme events.

Cyclone Fani, one of the most powerful cyclones in recent history, struck Odisha in May 2019, causing widespread devastation. With wind speeds reaching up to 175 km/h (Chauhan et al., 2021) and storm surges comparable to the Super Cyclone of 1999, Fani impacted fourteen districts across the state. The cyclone resulted in total damages and losses estimated at USD 3454 million, with the housing sector

[1] 1 Tropical Livestock Unit—Cattle (adult, 250 kg live weight): 1.0, Sheep/Goats: 5–6 animals, Camels: 0.7–0.8 animals, Pigs: 3–5 animals, Chickens: 50–100 birds.

bearing a substantial share of the destruction. In total, 3,61,743 houses were either partially or completely damaged, leaving a lasting impact on the affected communities (OSDMA et al., 2019). Cyclone Fani made landfall in Puri district, Odisha, before moving north-east and affecting all eleven blocks of the district. To capture its diverse impacts, our survey covered eight blocks of the district and one village from each block. At the village level, a snowball sampling strategy was adopted to identify stakeholders engaged in disaster management, including both formal actors and community-based groups. This method ensured the inclusion of a wide range of perspectives on preparedness, response, and recovery. The following section details these stakeholders and the methodological steps undertaken to gather and analyse data on their roles and experiences.

15.3 Research Methods

This paper has a two-step approach to solve the research question. We started by looking into the methods of data collection and then into the methods of data analysis. For data collection it was pertinent to look into existing methodologies that can capture the essence of community in the whole disaster management process. As evident from the literature on social resilience, community and the feeling of belongingness to the community are important for households to rebuild and restart after a cyclone (Ahmed et al., 2016; Choudhury et al., 2021).

15.3.1 Questionnaire Design

To capture the essence of community, we used the Community Toolbox to design our questionnaire. The Toolbox is a freely available resource that provides practical guidance for understanding communities.[2] Chapter 3, in particular, offers a framework for assessing community assets and needs, which can then be used to set priorities for development programs. Table 15.1 presents the various sections of the questionnaire and their description, designed using Community Toolbox.

The first step in this process is describing the makeup and history of the community, which helps situate data collection in context. This includes information on demographics, geography, infrastructure, settlement patterns, industries, and existing groups or institutions. Such details can be obtained through interviews, surveys, public forums, records, and archival sources. Researchers must carefully evaluate credibility to ensure accuracy.

---

[2] Source—https://ctb.ku.edu/en/table-of-contents/assessment/assessing-community-needs-and-resources.

The second step is to understand what matters to people in the community. This involves forums, concern surveys, or interviews to identify pressing problems. Problems are assessed by frequency, duration, severity, the number of households affected, and the level of deprivation they cause.

The third step is to capture the perspective of stakeholders, identified according to their relation to the issue.[3]

- Primary stakeholders are directly affected, either positively or negatively.
- Secondary stakeholders are indirectly affected.
- Key stakeholders are those with influence—positively or negatively—over the effort or institution.

Engaging stakeholders through interviews and surveys provides insights into how they plan to benefit communities and which subgroups are prioritized.

The fourth step is to set the priority level of identified problems using community-level indicators, such as the number of people affected, duration, severity, feasibility of addressing the issue, and potential consequences of intervention.

Finally, the process requires identifying barriers and resources to address the issues. Plans should aim to overcome challenges while drawing upon community assets—resources, skills, institutions, and relationships that already exist locally. By being inclusive and asset-based, this approach ensures that strategies are grounded in the strengths of the community while addressing its most urgent needs.

In sum, the Community Toolbox framework allows researchers to move beyond a narrow needs-based view, incorporating both the challenges and assets of communities into disaster recovery and resilience planning (Table 15.1).

15.3.2 Data Analysis

After implementing the methodology outlined above, we gathered valuable data through questionnaires and interviews with selected stakeholders at various levels. Once the data was collected, we meticulously prepared transcripts of the stakeholders' responses. From these transcripts, we conducted a thematic analysis, drawing on themes identified in both academic and non-academic literature. Furthermore, we constructed thematic networks to illustrate the relationships between identified themes and overlapping sub-themes, enhancing our understanding of the data. This structured approach allows us to build a more comprehensive picture of the insights gained from our stakeholders.

[3] Source—https://ctb.ku.edu/en/table-of-contents/participation/encouraging-involvement/identify-stakeholders/main.

Table 15.1 Questionnaire design for stakeholder data collection

Step	Input	Output
Describing the community	Demographics of the community, geographic location of the community, infrastructure, patterns of settlements and industries, and existing groups and institutions	Description of the community
Describe what matters to the people of the community	Issues in the community that people care about and, importance of the issues	Problems faced by the community
Describe what matters to stakeholders	Understanding stakeholders' interests, what stakeholders want to know, and what populations and subgroups are prioritized by the stakeholders	Targets of change
Evidence indicating whether the problem/goal should be a priority issue	Community level indicators related to the issue, frequency of problem, number of people affected, severity of problem, feasibility to address the problem, and possible impact/ consequences of addressing the problem	Relative priority to address the problem
Barriers and resources for addressing the identified issue	Identifying who cares about the problem, community assets and needs, effect of solving the problem on different subgroups in the community	Barriers and resource constraints that can affect the solving of the problem

15.4 Results

In the thematic analysis, we were able to identify three broad categories of themes from the response, which were related to three broad categories of well-being of the households—economic well-being, relational well-being, and subjective well-being. Taking this approach of identifying the themes on the basis of social well-being part is our method of going beyond the narrow measures of material assets and basic needs, as it is vital to recognize the social, psychological, subjective, relational, and cultural dimensions essential for human flourishing. Economic well-being refers to the material component of well-being, which encompasses physical requirements like income, assets, physical health, and environmental services (Armitage et al., 2012). Relational well-being refers to the households' social interaction, and collective actions (Armitage et al., 2012). These include the neighbourhood relations, the relations with the state, and the informal social structures (Armitage et al., 2012). And lastly, the subjective well-being component encompasses cultural values, norms, and belief systems, while also reflecting individuals' sense of self.

It includes both personal and collective hopes, fears, and aspirations, as well as perceived levels of satisfaction or dissatisfaction, trust, and confidence (Armitage et al., 2012; White, 2009, 2010). Subjective well-being is the personal, affective component of the broader well-being construct. It is described as "the feeling-good side" of wellbeing, encompassing happiness, contentment, enjoyment, curiosity and engagement (White, 2009).

15.4.1 Impact of Cyclone Fani

Cyclone Fani had devastating and far-reaching impacts on housing, livelihoods, infrastructure, livestock, and community well-being in Odisha. Housing damage was extensive, especially in thatched and kutcha houses in villages like Badrikilo, Kakatpur, Khandimangalpur, Suanda, and Arakhakuda. Many homes lost roofs, collapsed, or became uninhabitable due to flooding and structural damage, forcing families into school buildings or neighbours' pucca houses. This led to mass displacement and prolonged recovery.

"There was major loss in terms of loss of houses. Many houses were damaged and some were damaged beyond repair. Most of the households had thatched houses, with roof either made of tiles or hay." (Villager from Kakatpur)

Agriculture and livelihoods suffered severe setbacks. Standing crops such as paddy and Muga were destroyed across multiple villages, while fishing and poultry-based livelihoods were disrupted by damaged equipment and environmental degradation. Tourism-dependent artist households also lost income, and many income-generating activities remained stalled for months.

Infrastructure breakdowns further delayed recovery. Electricity outages lasting one to three months affected water supply, communication, and daily life. Fallen trees blocked roads, restricting mobility and delaying relief. Damage to public infrastructure, such as schools and veterinary centres, compounded the crisis.

Losses in livestock and inadequate veterinary services increased vulnerability. Many animals died from falling trees, rain, or disease. In some cases, animals were sold prematurely due to a lack of fodder and shelter. The absence of functioning veterinary services, such as in Dandamakundapur, highlighted critical weaknesses in disaster-time animal care.

Health, education, and emotional well-being were also severely impacted. Schooling was disrupted for weeks to months in villages like Badrikilo and Suanda. Health issues such as diarrhoea, skin diseases, and viral infections emerged, particularly in flooded or overcrowded shelters. The psychological toll was significant, with residents reporting distress from witnessing destruction, loss, and displacement. For the most vulnerable, recovery remains incomplete.

15.4.2 Economic Well-Being

In the category of economic well-being, as per the definition given earlier, we were able to identify two sub-categories in this section: livelihood structures and impact from the cyclone, and the infrastructure gaps in the study sites.

15.4.2.1 Livelihood Structures and Impact

The livelihood structure in the cyclone-affected villages of Odisha is primarily agrarian, with agriculture—especially paddy and vegetable cultivation—serving as the main income source in areas like Suanda, Khandimangalpur, Dandamakundpur, and Malada. This is often supplemented by livestock rearing or wage labour. Some villages, such as Dandamakundpur and Maltipatpur, have diversified into seasonal activities like mushroom farming, and brick kilns. Along with that in Maltipatpur Patachitra art is another way of income for many households. However, these ventures remain fragile and insufficient for year-round stability. Fishing is the dominant livelihood in Arakhakuda, particularly in Chilika Lake, but dependence on this single activity increases vulnerability to cyclones and environmental shocks. Women play an important economic role through Self-Help Groups (SHGs), engaging in handicrafts, tailoring, and informal credit activities. In Malada and Dandamakundpur, migration and wage labor in nearby cities indicate a growing dependence on external income sources, marking a shift away from purely agrarian livelihoods.

"After Cyclone Fani, many people got into labour work for house building. Major number of people are going into that. Its better paying than working in the agriculture and its less tyring than working in the agricultural field. Currently people are giving that more importance." (A Villager from Dandamakundapur)

Long-term livelihood insecurity is a persistent challenge. In villages such as Badrikilo, Khandimangalpur, and Suanda, recovery after disasters has been largely self-driven, with limited institutional aid. Seasonal and irregular income patterns—like those in Dandamakundpur—or the loss of tourism income in Maltipatpur reflect economic fragility. The pre-emptive sale of livestock in Malada and the destruction of fishing boats in Arakhakuda have caused prolonged livelihood losses. These problems with income sources also led to a reduction in expenditure by the households as mentioned by respondents in Khandimangalapur, Malada, and Badrikilo.

"Nearly for 2 months. Our boats were broken, so we had to repair them. Also, people had to dig up the river so that the boats could travel to the sea. Due to loss of income and fishing activities, nearly 1000 people from our village are working in some other places to support their families and households." (Villager from Arakhakuda, a fishing village on the banks of Chilika Lake)

"Actually, after the cyclone we kind of reduced our expenditures in all aspects of life. Expenditure will reduce as the we lost our crop. Before, we were spending 10 rupees easily, and after the cyclone, you will naturally reduce it." (A Villager from Khandimangalapur)

Small businesses and wage labor were severely disrupted in Sirei and Suanda, causing cascading financial distress that, for some, delayed resumption of income generation for months.

"Loss of standing crop, leading total breaking of their financial stability and resources. As the cyclone was in the month of May. Many people did not receive any compensation for the loss of their standing crops" (A Panchayat Member from Suanda Village)

Government compensation has often been inadequate or delayed, as seen particularly in Suanda, deepening household debt and prolonging recovery timelines. This has reinforced poverty cycles, pushing vulnerable families further into financial precarity. Overall, the livelihoods in these regions are marked by high dependency on climate-sensitive sectors, limited diversification, and systemic vulnerabilities, which are exacerbated by inadequate recovery support. These factors contribute to prolonged economic instability, forcing many households to adopt coping strategies such as asset liquidation, migration, and reliance on informal income sources, all of which have long-term implications for resilience and well-being.

15.4.2.2 Infrastructure Gaps

Infrastructure gaps were reported to be majorly affecting what households do after the cyclone, especially in terms of arranging basic necessities—food, shelter, and water. Unavailability of cyclone shelters in the villages of Arakhakuda, Dandamakundpur, Khandimangalpur, and Serei led to households staying in the school buildings, and, in some cases, even depending on their neighbours' as was the case in Kakatpur village. This dependence on neighbours' houses for shelter post cyclone is an example of neighbourhood resiliency, which arises from social structures, i.e., in this case, the neighbour-neighbour relationship (Breton, 2001).

Infrastructure gaps affected service delivery in the case of healthcare services for households, as well as for the livestock of households. Unavailability of veterinary health care services is one of the reasons mentioned by the households in Malada about the pre-emptive sale of livestock before the cyclone. In Badrikilo, Maltipatpur, and Dandamakundpur, the respondents reported about non-functional primary healthcare centres. This led to dependence of the households on healthcare services from outside the village, thus adding an additional cost component to the healthcare services required for the households in the form of transport costs to outside the village.

"Due to damaged roads, people with illness couldn't go to the hospital. We either have to carry them or use tubes." (Villager from Sirei village on healthcare facilities after the cyclone)

Even where health centres exist, they are either poorly equipped or lack personnel, causing households to travel far for treatment, especially during emergencies. Service delivery issues due to infrastructure gaps was also visible in delayed electricity and communication restoration. In our study sites we find that the average time to restore electricity ranged from 30 to 45 days.

"Additionally, the government repair and relief work gets slowed down due to the washed away roads. This posed a major hinderance to the recovery of the households in the village." (Villager from Sirei village).

"No, nothing like that was available. Roads were blocked due to fallen trees. Machines and people removed the debris and that resumed the communication. Due to blocked roads, people weren't able to get out of their houses" (Villager from Malada on healthcare facilities after cyclone)

Additionally, as households relied on electricity for communication and to operate water pumps, power outages also disrupted water security. With a limited number of hand pumps in the villages, households had to manually get water from the hand pumps. Despite causing an additional burden on households for arranging water, getting water from handpumps ensured that the households weren't using contaminated surface water sources like ponds and rivers.

Relief and recovery systems are slow and uneven, with delays or partial distribution of essential food, financial aid, and recovery support. Contributing factors include the absence of pre-positioned relief materials, unclear compensation mechanisms, and a lack of formal credit facilities. These gaps disproportionately affect poorer households and those reliant on livestock, whose recovery is further slowed by inadequate institutional support.

15.4.3 Relational Well-Being

Relational well-being is the social dimension of the "social conception of well-being," focusing on the quality, diversity, and structure of people's interactions, collective actions, and relationships (Armitage et al., 2012). It highlights how individuals and groups are embedded in, and derive meaning from, connections with families, communities, institutions, and broader cultural and political networks (White, 2015). Distinct from material well-being (income, assets, health) and subjective well-being (beliefs, aspirations, satisfaction), it encompasses three key dimensions: close relationships—intimate bonds with family, friends, and partners that provide emotional support, identity, and care; social connections—broader networks such as community ties, neighbourhood relations, and group memberships that foster trust, belonging, and collective action; and relationships with institutions—interactions with governance structures, markets, religious bodies, and service providers that influence access to resources, rights, and opportunities. But the relational well-being is not limited to social human-to-human connections but is beyond human connections, to local environments too. These connections or relations of humans to other humans and their surroundings make people who and what they are (White, 2010, p. 164). Hence, under this theme, we have looked at our respondents' relations with their surroundings, consisting of their neighbours and ecosystems.

15.4.3.1 Social Structures

Social structures refer to relatively stable forms of social relations which are not flexible and are abstract organization forms within the social setting (Martin & Lee, 2015). The social structures across the surveyed villages reveal strong community networks, inclusive practices, and adaptive institutional arrangements. Women's Self-Help Groups (SHGs) are notably active in villages like Dandamakundpur, Sirei, Suanda, and Maltipatpur, often organized around neighborhood clusters (Sahis), providing access to credit and fostering women-led micro-enterprises.

"My wife and my sister – in – law received some money from their SHG. We used the money to buy some rice and firewood after the cyclone, I think, after 10 – 12 days. Before that, we were dependent on cooked food given in the village by the panchayat at the school building"(Villager from Sirei)

The SHGs are rule-based social structures but are ingrained in the lives of the households in the study sites, forming a dependable source of income and credit as well as neighbourhood bonding. Caste diversity is navigated differently—villages like Suanda demonstrate inter-caste cooperation without residential segregation, while others, such as Khandimangalpur, maintain caste-based Sahis but still encourage inclusion.

"We have different Sahis in the village, like Harijan Sahi and Behera Sahi, but during cyclone everyone cooperates with each other." (Villager from Saunda)

In areas where formal credit access is limited, such as Arakhakuda and Malada, informal institutions, including women-led cooperatives and lending groups, bridge financial gaps, proving crucial in post-disaster recovery. However, several villages, including Suanda, Badrikilo, and Khandimangalpur, face restricted access to essential formal services like healthcare and secondary education, despite having strong social cohesion and shared community spaces. Hence, in some cases, it can be said that social cohesion and shared community feeling are not enough for the respondents to access better public services.

Cooperative societies and credit networks further strengthen economic resilience. Artist groups in Maltipatpur and cooperatives in Dandamakundpur and Sirei provide financial support and livelihood-specific assistance, demonstrating localized institutional innovation. Overall, these social structures reflect a blend of traditional networks and adaptive strategies, enabling communities to address economic needs, maintain social cohesion, and build resilience despite gaps in formal service provision.

15.4.3.2 Informal Institutions

Informal institutions are the unwritten, socially shared rules, norms, values, and conventions that guide behavior and expectations within a community or society, but are not codified in laws, regulations, or official policies (Barbalet, 2023). Informal institutions played a pivotal role in community resilience and recovery during Cyclone Fani. Across villages, strong community solidarity formed the backbone of coping strategies, with families sharing shelter, food, water, and essential resources to protect the most vulnerable. Women-led Self-Help Groups (SHGs) were instrumental in post-disaster relief, providing small but critical items like mosquito nets and sarees, facilitating loans for income diversification, and offering emotional support. In places lacking formal credit systems, such as Arakhakuda, residents relied on local moneylenders (mahajans and sahukars) to finance recovery, highlighting both the debt burden and the significance of informal financial networks.

"We had to take loans from Mahajan (informal credit). We had to depend on such sources to maintain necessary expenditures after the cyclones or any other problems. We have to use whatever is left from the necessary expenditure to pay back the Sahukar. We don't have farms or agricultural lands or anything on which we can depend to generate income or get food." (Villager from Arakhakuda)

Collective action also drove infrastructure recovery, with villagers voluntarily clearing roads, transporting materials, and aiding electricity restoration in the absence of a timely institutional response. Mutual aid and trust reinforced social stability, with no significant conflicts reported. Cooperation extended beyond neighbours to strangers, and in some cases, like Suanda, relief distribution was perceived as fair and need-based. Overall, these informal systems—rooted in trust, reciprocity, and shared responsibility—proved vital in bridging institutional gaps, sustaining livelihoods, and fostering equitable recovery. They demonstrated that in disaster-prone areas, community-driven mechanisms can serve as critical complements to formal governance and emergency response systems.

15.4.3.3 Environmental Challenges

Cyclone Fani triggered significant environmental challenges that disrupted livelihoods and strained village ecosystems. In Arakhakuda, the drying of Chilika Lake and siltation of boat routes severely impacted fishing, the primary livelihood, with no formal forecasting systems to aid adaptation. Widespread tree loss in villages like Badrikilo, Maltipatpur, and Suanda altered local ecology, reduced social spaces, and heightened emotional distress.

"We lost two very old and large Banyan trees in our village. They were a gathering spot for many people in our village, especially in the evening. During most parts of the day you will find someone there. The place now feels empty, like 'charm' is missing" (Villager from Badrikilo)

In Sirei, recurring flooding and saline intrusion degraded farmland, creating long-term uncertainty for agriculture. Post-cyclone, stagnant water and debris in Badrikilo and Suanda contaminated water sources, leading to disease outbreaks and poor sanitation, underlining serious environmental health risks. Livestock also suffered from environmental exposure, particularly in Sirei and Malada, where cold, wet conditions and inadequate shelter caused animal deaths, intensifying rural ecological stress. These combined impacts not only undermined immediate recovery but also threatened the long-term resilience of communities dependent on natural resources.

15.4.4 Subjective Well-Being

The subjective well-being component encompasses cultural values, norms, and belief systems, while also reflecting individuals' sense of self. It includes both personal and collective hopes, fears, and aspirations, as well as perceived levels of satisfaction or dissatisfaction, trust, and confidence (Armitage et al., 2012; White, 2009, 2010). Subjective well-being refers to the component of social well-being that is constructed from the mutual trust, shared values, and community engagement. This is the outcome of collective actions the community members take, not for themselves but for the overall good of their community. Here, we have explored the collective actions the respondents have taken or have described in the pre-cyclone and post-cyclone activities as indicators of subjective well-being. Here, we revisit the themes that we have already identified in the previous sections, but with a lens of subjective well-being.

Informal community networks played a vital role in helping villages cope with Cyclone Fani's aftermath, ensuring protection and support for even the most vulnerable. Across affected areas such as Badrikilo, Suanda, Malada, and Khandimangalpur, families extended shelter to others and shared essential resources, including food, water, firewood, and cooking materials. This solidarity formed the backbone of survival and early recovery. This reflects the household's engagement with each other in sharing resources. Grassroots collective action was central to restoring damaged infrastructure. In villages like Dandamakundapur and Khandimangalpur, residents organized themselves to clear roads, transport poles, and assist linemen in repairing electricity lines. Such voluntary labor substituted for delayed or insufficient formal institutional assistance. Mutual aid and trust reinforced community stability.

"Yes, all the people took responsibility for post cyclone works. This helped in increasing the work speed and it took one month to completely repair power lines and other things. If people wouldn't have been involved it might have taken 3 months or more than that for department workers to repair and rebuild. People realised that everyone requires current for day to day activities and that motivated people to work. Of course, not everyone was involved, but many were. They even used their own personal vehicles, like bikes and tractors, to transport things from one place to another." (A Panchayat Member from Dandamakunpur)

Villagers cooperated not only with their neighbors but also with strangers, as seen in Malada and Arakhakuda, displaying strong inter-household trust and social cohesion.

"During the cyclone, people helped each other. At that, no one will back out. If a single household also faces a problem, everyone will come to help them out. If you are close by, you will help each other." (A Villager from Arakhakuda, showing a strong sense of community when asked about co-operation among villagers after Cyclone Fani)

In Suanda, relief distribution was reported as fair and need-based, reflecting transparent and accountable local governance practices. Community mobilization was particularly effective in infrastructure repair. Villagers across multiple settlements, including Suanda, took proactive roles in clearing debris, transporting materials, and managing logistics during delays in official response. These coordinated efforts showcased high levels of resilience, self-organization, and the capacity of informal institutions to fill critical recovery gaps.

15.4.5 Disaster Management

From the responses of the stakeholders, we find that there are two phases of disaster management. The first phase is disaster preparedness. The disaster preparedness consists of early warning systems, the arrangement of necessary resources, and moving to cyclone resilient shelters. But the level of preparedness varies from village to village as well as in terms of different aspects of preparedness.

"With the first weather reports about cyclone formation, we start by spreading the news regarding the incoming cyclone using loudspeakers on auto-rickshaws and police vehicles in all the coastal villages. Along with the weather warning, fishing boats and trawlers are advised not to venture into the ocean until the system fully subsides." (Officer from District Disaster Management Office in Puri).

Disaster preparedness across the surveyed villages showed both progress and persistent gaps. Early warning systems have improved, with most villages like Khandimangalpur, Sirei, and Suanda receiving alerts via TV, mobile phones, or local announcements 3–4 days before the cyclone. However, follow-up actions and evacuation protocols were inconsistent, especially in Badrikilo and Malada. Shelter capacity was critically inadequate.

"Most of the families went to the school buildings. There are two school buildings nearby and an Anganwadi center. Most people stayed in those places. Also, some people went and stayed in their relatives' places too, who had pucca houses." (Villager from Badrikilo)

Villages such as Arakhakuda and Sirei had limited shelters that could accommodate only a fraction of residents, while Malada and Maltipatpur lacked dedicated shelters entirely, relying instead on schools or neighbors' homes. Water, food, and electricity backup systems were poorly integrated. While Suanda and Khandimangalpur managed some early arrangements, areas like Dandamakundapur and Badrikilo suffered water shortages due to prolonged electricity outages and limited pump access. Few villages had pre-positioned relief supplies or reliable power backups. Health and veterinary preparedness were weak. Dandamakundapur had no veterinary doctor for over a decade, and Maltipatpur's PHC response was delayed, increasing vulnerability to post-cyclone health crises. Community awareness and local initiatives, however, have strengthened compared to the 1999 Super Cyclone. Villages like Khandimangalpur and Suanda demonstrated proactive measures—securing food, livestock, and shelter—reflecting improved disaster awareness, stronger social cohesion, and quicker recovery capacities.

"The panchayat was active during the cyclone. They distributed cooked food to the poor households in the village. Food was also provided" (Villager from Khandimangalapur Village)

And the second aspect of disaster management is coping strategies. This is employed by the respondents after the cyclone. Coping strategies in the cyclone-affected villages relied heavily on community solidarity and informal systems.

"Now in the village we people will labour and then we can earn to eat. Not everyone has land for farming. Here it's a Behera family, they have buffalo. Now, if for 15–20 days they don't take out the milk from the buffalos, they won't be able to eat as their income depends on it. Hence they were able to arrange food from other households in the village. people help each other in the village" (Villager from Khandimangalapur)

Informal social networks played a central role, with households sheltering neighbours, sharing resources, and jointly clearing debris. In the absence of formal cyclone shelters, school buildings and Anganwadi centres became vital emergency shelters, offering temporary housing and communal cooking spaces. Women-led Self-Help Groups (SHGs) in villages like Dandamakundapur, Maltipatpur, and Suanda provided small relief items, facilitated access to loans, and supported both economic recovery and social well-being.

Community mobilization was key in infrastructure rebuilding and restoration, with villagers restoring electricity, clearing debris, and managing the movement of necessary materials, helping the government departments. This came out both as a need for faster restoration as well as a frustration for slow administrative response. In areas like Arakhakuda, limited formal aid led to reliance on local moneylenders and collective petitions to authorities. Families in Malada and Sirei hosted strangers, coordinated shared food distribution, and provided emergency transport, highlighting the resilience and adaptability of these communities during crisis recovery.

15.5 Interlinkages Between Well-Being Categories

Figure 15.1 illustrates the dynamic interconnections between economic, relational, and subjective well-being, environmental challenges, disaster preparedness, and coping mechanisms in cyclone-affected villages. Economic well-being is closely tied to relational well-being, as livelihood losses increase dependence on social networks, Self-Help Groups (SHGs), and informal credit, and strong community ties enable faster recovery. Environmental challenges, such as ecosystem degradation, directly affect climate-sensitive livelihoods, creating financial instability and lowering morale, thus linking economic and subjective well-being. Relational well-being influences disaster preparedness by facilitating trust-based early warning dissemination and collective action, while in turn, preparedness activities strengthen community cohesion. Similarly, strong relationships enhance coping mechanisms through resource sharing, voluntary labour, and mutual support, which reinforces trust and informal governance. Coping mechanisms contribute to restoring economic

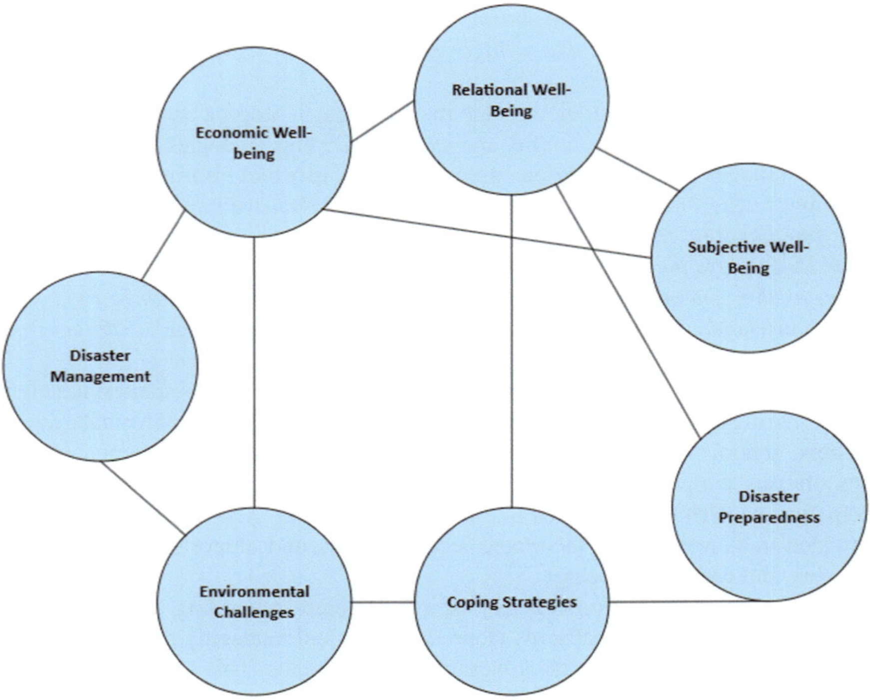

Fig. 15.1 Interlinkages between well-being themes and sub-themes; Authors' creation

stability and improving subjective well-being by boosting morale and a sense of agency. Environmental challenges also shape preparedness behaviours and affect community identity when natural spaces are lost, impacting subjective well-being. Preparedness capacity determines the speed and efficiency of coping actions, while high subjective well-being—rooted in optimism, trust, and belonging—encourages proactive disaster readiness. Overall, these interlinked systems show that resilience depends on a continuous feedback loop between social, economic, environmental, and subjective dimensions.

15.6 Discussions

This study highlights the multi-dimensional nature of disaster impacts and recovery, moving beyond narrow economic indicators to encompass relational and subjective well-being. The case of Cyclone Fani in Puri district, Odisha, illustrates that disaster effects are embedded within interconnected social, economic, environmental, and psychological systems. Further papers like Dung et al. (2024) take a multidimensional impact of droughts, floods, and storms on poverty approach, where the impact is studied on education, health, housing, and access to information, but our findings demonstrate that relational networks and community-driven coping strategies are equally critical to resilience. Additionally, taking multi-dimensional well-being approach can be beneficial in terms of recovery strategy too. McClelland et al. (2023) showed through a study on the impact of the COVID-19 disaster that a multidimensional approach to disaster recovery can produce a recovery strategy more suitable to local needs.

Economic well-being in the study villages remains highly dependent on climate-sensitive livelihoods such as agriculture and fishing, leaving households vulnerable to environmental shocks. Storm surges due to cyclone cause increase in soil salinity thus leading to change in cropping patterns as well as crop production (Ghosh & Mistri, 2023). Further, the Arakhkuda village in our sample is a fishing dependent village, and existing evidence has shown that without proper adaptive strategies, the impact of cyclones, and ensuing climate change is highly negative for such communities (Islam et al., 2014). The inadequacy of formal recovery mechanisms—delayed compensation, insufficient infrastructure, and limited credit access—forces reliance on informal institutions. Here, relational well-being plays a pivotal role. Women's Self-Help Groups, neighbourhood ties, and informal credit systems bridged institutional gaps, providing both material support and emotional stability. Uddin et al (2020) found from Bangladesh that community networks like family and friends and local informal organizations are an important part of recovery post cyclone, as they are the first to assist after the cyclone. Hence, we find that interaction between different well-being aspects helps in the recovery of the households.

Subjective well-being emerged as both an outcome and driver of resilience. It has been found that coastal communities have strong social values like trust and a sense of identity from their place (Berkes & Ross, 2013). This trust in community institutions and other households brings together the community as one and thus makes the recovery path easier. Activities like repairing electricity lines, or clearing debris and fallen trees, are not just carried out by people for their own convenience but for the whole village. There is a sense of belongingness among the people in the villages studied. Environmental challenges—such as ecosystem degradation, siltation, and saline intrusion—acted as compounding stressors, linking ecological resilience directly to social well-being. The interplay between these dimensions reinforces the argument that disaster risk reduction must integrate social capital, cultural norms, and emotional security alongside physical and economic recovery.

The thematic interlinkages identified in this study show that resilience is sustained through feedback loops: relational cohesion improves coping capacity, coping actions restore economic stability, and enhanced subjective well-being positively affects preparedness. However, these loops can also reverse—economic fragility, environmental decline, and institutional neglect can erode trust and collective action, undermining long-term resilience. Hence, there are both positive and negative interlinkages between different well-being.

15.7 Conclusions

Taking a well-being approach based on qualitative methods for studying the impacts of cyclone Fani has given us not just material/economic impact and needs arising from the cyclone but also highlighted the social aspects of the impact too. We find that material needs are required for individual well-being and recovery after a cyclone. But if we incorporate the relational connections the households have in a community and how they tap into it, led to overall community recovery after a cyclone. This highlights the interlinkages that are present between the different aspect of cyclone impact and post impact activities. The collective actions address not just the collective needs but also individual needs of the people too. The findings from the analysis highlights that by applying a social well-being framework, this study expands the understanding of disaster impacts beyond asset loss to include relational and subjective dimensions. Cyclone Fani's aftermath revealed that while infrastructure and livelihoods can be rebuilt, the strength of social connections and the community's psychological state often determine the pace and equity of recovery.

Conflict of Interest The authors declare no conflict of interest.

Declaration We declare that no part of the paper is generated using AI.

References

Ahmed, B., Kelman, I., Fehr, H. K., & Saha, M. (2016). Community resilience to cyclone disasters in coastal Bangladesh. *Sustainability (Switzerland), 8*(8), 1–29. https://doi.org/10.3390/su8080805

Armitage, D., Béné, C., Charles, A. T., Johnson, D., & Allison, E. H. (2012). The interplay of well-being and resilience in applying a social-ecological perspective. *Ecology and Society, 17*(4). https://doi.org/10.5751/ES-04940-170415

Auffhammer, M., & Aroonruengsawat, A. (2011). Simulating the impacts of climate change, prices and population on California's residential electricity consumption. *Climatic Change, 109*(S1), 191–210. https://doi.org/10.1007/s10584-011-0299-y

Barbalet, J. (2023). Conceptualising informal institutions: Drawing on the case of guanxi. *The British Journal of Sociology, 74*(1), 70–82. https://doi.org/10.1111/1468-4446.12984

Barreca, A. I. (2012). Climate change, humidity, and mortality in the United States. *Journal of Environmental Economics and Management, 63*(1), 19–34. https://doi.org/10.1016/j.jeem.2011.07.004

Berkes, F., & Ross, H. (2013). Community resilience: Toward an integrated approach. *Society & Natural Resources, 26*(1), 5–20. https://doi.org/10.1080/08941920.2012.736605

Breton, M. (2001). Neighborhood resiliency. *Journal of Community Practice, 9*(1), 21–36. https://doi.org/10.1300/J125v09n01_02

Chauhan, A., Singh, R. P., Dash, P., & Kumar, R. (2021). Impact of tropical cyclone "Fani" on land, ocean, atmospheric and meteorological parameters. *Marine Pollution Bulletin, 162*, Article 111844. https://doi.org/10.1016/j.marpolbul.2020.111844

Charlson, F., Ali, S., Benmarhnia, T., Pearl, M., Massazza, A., Augustinavicius, J., & Scott, J. G. (2021). Climate change and mental health: A scoping review. *International Journal of Environmental Research and Public Health, 18*(9), 4486. https://doi.org/10.3390/ijerph18094486

Chhotray, V., & Few, R. (2012). Post-disaster recovery and ongoing vulnerability: Ten years after the super-cyclone of 1999 in Orissa, India. *Global Environmental Change, 22*(3), 695–702. https://doi.org/10.1016/j.gloenvcha.2012.05.001

Chittibabu, P., Dube, S. K., Macnabb, J. B., Murty, T. S., Rao, A. D., Mohanty, U. C., & Sinha, P. C. (2004). Mitigation of flooding and cyclone hazard in Orissa, India. In *Natural hazards* (Vol. 31).

Choudhury, M. U. I., Emdad Haque, C., Nishat, A., & Byrne, S. (2021). Social learning for building community resilience to cyclones: Role of indigenous and local knowledge, power, and institutions in coastal Bangladesh. *Ecology and Society, 26*(1). https://doi.org/10.5751/ES-12107-260105

Cissé, J. D., & Barrett, C. B. (2018). Estimating development resilience: A conditional moments-based approach. *Journal of Development Economics, 135*, 272–284. https://doi.org/10.1016/j.jdeveco.2018.04.002

Copestake, J. (2008). Wellbeing in international development: What's new? *Journal of International Development, 20*(5), 577–597. https://doi.org/10.1002/jid.1431

Costanza, R., Fisher, B., Ali, S., Beer, C., Bond, L., Boumans, R., Danigelis, N. L., Dickinson, J., Elliott, C., Farley, J., Gayer, D. E., Glenn, L. M., Hudspeth, T., Mahoney, D., McCahill, L., McIntosh, B., Reed, B., Rizvi, S. A. T., Rizzo, D. M., … Snapp, R. (2007). Quality of life: An approach integrating opportunities, human needs, and subjective well-being. *Ecological Economics, 61*(2–3), 267–276. https://doi.org/10.1016/j.ecolecon.2006.02.023

Crowell, M., Edelman, S., Coulton, K., & McAfee, S. (2007). How many people live in coastal areas? *Journal of Coastal Research, 23*(5). https://doi.org/10.2112/07A-0017.1

Dell, M., Jones, B. F., & Olken, B. A. (2014). What do we learn from the weather? The new climate-economy literature. *Journal of Economic Literature, 52*(3), 740–798. https://doi.org/10.1257/jel.52.3.740

Dung, N. Q., Thi Hue, H., & Thanh, T. P. (2024). The impact of natural disaster on multidimensional poverty of rural households in Vietnam: The regulating role of social assistance. *Poverty & Public Policy, 16*(3), 213–231. https://doi.org/10.1002/pop4.412

Ebi, K. L., Vanos, J., Baldwin, J. W., Bell, J. E., Hondula, D. M., Errett, N. A., Hayes, K., Reid, C. E., Saha, S., Spector, J., & Berry, P. (2021). Extreme weather and climate change: Population health and health system implications. *Annual Review of Public Health, 42*(1), 293–315. https://doi.org/10.1146/annurev-publhealth-012420-105026

Feng, S., Krueger, A. B., & Oppenheimer, M. (2010). Linkages among climate change, crop yields and Mexico–US cross-border migration. *Proceedings of the National Academy of Sciences, 107*(32), 14257–14262. https://doi.org/10.1073/pnas.1002632107

Ghosh, S., & Mistri, B. (2023). Cyclone-induced coastal vulnerability, livelihood challenges and mitigation measures of Matla-Bidya inter-estuarine area, Indian Sundarban. *Natural Hazards, 116*(3), 3857–3878. https://doi.org/10.1007/s11069-023-05840-2

Gough, I. (2004). Human well-being and social structures. *Global Social Policy, 4*(3), 289–311. https://doi.org/10.1177/1468018104047489

Honda, J. Y., Kundzewicz, Z. W., Arnell, N., Benito, G., Hatfield, J., Mohammed, I. F., Peduzzi, P., Wu, S., Sherstyukov, B., Takahashi, K., & Yan, Z. (2012). Changes in impacts of climate extremes: human systems and ecosystems. In C. B. Field, V. Barros, T. F. Stocker, D. Qin, D. J. Dokken, K. L. Ebi, M. D. Mastrandrea, K. J. Mach, G.-K. Plattner, S. K. Allen, M. Tignor, & P. M. Midgley (Eds.), *Managing the risks of extreme events and disasters to advance climate change adaptation* (pp. 231–290). A Special Report of Working Groups I and II of the Intergovernmental Panel on Climate Change (IPCC), Cambridge University Press.

IMD. (2018). Annual Frequency of landfalling cyclonic disturbances, cyclones & severe cyclones. In *Regional Specialized Meteorological Centre for Tropical Cyclones over North Indian Ocean, Indian Meteorological Department, Ministry Earth Sciences, Government of India.*

IPCC. (2019). Summary for policymakers. In *Climate change and land: An IPCC special report on climate change, desertification, land degradation, sustainable land management, food security, and greenhouse gas fluxes in terrestrial ecosystems.*

Islam, Md. M., Sallu, S., Hubacek, K., & Paavola, J. (2014). Vulnerability of fishery-based livelihoods to the impacts of climate variability and change: Insights from coastal Bangladesh. *Regional Environmental Change, 14*(1), 281–294. https://doi.org/10.1007/s10113-013-0487-6

Kumar, T. S., Mahendra, R. S., Nayak, S., Radhakrishnan, K., & Sahu, K. C. (2010). Coastal vulnerability assessment for Orissa State, East Coast of India. *Journal of Coastal Research, 26*(3), 523–534. https://doi.org/10.2112/09-1186.1

Lamb, W. F., & Steinberger, J. K. (2017). Human well-being and climate change mitigation. *WIREs Climate Change, 8*(6). https://doi.org/10.1002/wcc.485

Martin, J. L., & Lee, M. (2015). Social structure. In *International encyclopedia of the social & behavioral sciences* (pp. 713–718). Elsevier. https://doi.org/10.1016/B978-0-08-097086-8.32154-7

McClelland, A. G., Shaw, D., O'Grady, N., & Fattoum, A. (2023). Recovery for development: A multi-dimensional, practice-oriented framework for transformative change post-disaster. *The Journal of Development Studies, 59*(1), 1–20. https://doi.org/10.1080/00220388.2022.2130055

McGregor, J. A., Camfield, L., & Woodcock, A. (2009). Needs, wants and goals: Wellbeing, quality of life and public policy. *Applied Research in Quality of Life, 4*(2), 135–154. https://doi.org/10.1007/s11482-009-9069-7

Mishra, S., & Kar, D. (2016). Cyclonic hazards in Odisha and its mitigation. *Odisha Review*, 37–41.

Muzerengi, F., Gandidzanwa, C. P., & Chirubvu, L. (2023). Impacts of climate change on household food security in Matande communal lands, Mwenezi district in Zimbabwe. *Jàmbá Journal of Disaster Risk Studies, 15*(1). https://doi.org/10.4102/jamba.v15i1.1499

OSDMA, G. of O. O. S. D. M. A., Agencies, U. N., Bank, T. W., & Bank., A. D. (2019). Cyclone Fani : Damage, loss and needs assessment. In *United Nations India ADB the World Bank.*

Rotton, J., & Cohn, E. G. (2004). Outdoor temperature, climate control, and criminal assault. *Environment and Behavior, 36*(2), 276–306. https://doi.org/10.1177/0013916503259515

Sandhani, M., Pattanayak, A., & Kavi Kumar, K. S. (2020). *Impact of climate change on economic growth : A case study of India* (2020–204).

Schlenker, W., & Lobell, D. B. (2010). Robust negative impacts of climate change on African agriculture. *Environmental Research Letters, 5*(1), Article 014010. https://doi.org/10.1088/1748-9326/5/1/014010

Sekhri, S., & Storeygard, A. (2014). Dowry deaths: Response to weather variability in India. *Journal of Development Economics, 111*, 212–223. https://doi.org/10.1016/j.jdeveco.2014.09.001

Tamuly, R., & Mukhopadhyay, P. (2022). Natural disasters and well-being in India: A household-level panel data analysis. *International Journal of Disaster Risk Reduction, 79*. https://doi.org/10.1016/j.ijdrr.2022.103158

Theisen, O. M., Holtermann, H., & Buhaug, H. (2012). Climate wars? Assessing the claim that drought breeds conflict. *International Security, 36*(3), 79–106. https://doi.org/10.1162/ISEC_a_00065

Uddin, M. S., Haque, C. E., Walker, D., & Choudhury, M. U. I. (2020). Community resilience to cyclone and storm surge disasters: Evidence from coastal communities of Bangladesh. *Journal of Environmental Management, 264*(March), Article 110457. https://doi.org/10.1016/j.jenvman.2020.110457

White, S. C. (2009). *Bringing wellbeing into development practice* (09/50; Wellbeing in Developing Countries (WeD) Working Papers).

White, S. C. (2010). Analysing wellbeing: A framework for development practice. *Development in Practice, 20*(2), 158–172. https://doi.org/10.1080/09614520903564199

White, S. (2015). *Relational wellbeing: A Theoretical and operational approach* (43; Bath Papers in International Development and Wellbeing).

Yang, H., Dietz, T., Yang, W., Zhang, J., & Liu, J. (2018). Changes in human well-being and rural livelihoods under natural disasters. *Ecological Economics, 151*, 184–194. https://doi.org/10.1016/j.ecolecon.2018.05.008

Part VI
Policy, Law, and Governance

Chapter 16
The Effectiveness of Climate Finance in Temperature Control: Implications for Monetary Policy in India and Its Neighbouring Countries

Rizwan Qasim, Irshad Ahmad, Dastgir Alam, and Abid Hussain

Abstract This study investigates the impact of climate-specific finance on extreme temperature patterns. It uses annual average maximum temperature as the dependent variable, with adaptation finance, CO_2 emissions, and inflation as explanatory variables. The analysis focuses on India, Bangladesh, and Nepal over the period 2011 to 2020 and applied a panel Autoregressive Distributed Lag model to capture both short- and long-run relationships. In the long run, the results reveal that adaptation finance has a negative association with annual maximum temperature, whereas CO_2 emissions and inflation show positive associations. This negative relationship reflects the role of adaptation finance in the development of infrastructure, the transfer of technology, and the enhancement of institutional capacity to counter rising temperatures. However, their influence is weak. In contrast, the positive association of CO_2 emissions with temperature suggests that rising emissions remain a key driver of climate stress in the region. Similarly, the positive relationship between inflation and temperature shows that high inflation levels reduce the capacity to allocate resources for climate adaptation and lead to greater exposure to climate risks. Furthermore, the error correction term (− 1.18), which is significant at the 10% level, indicates a relatively fast and stable return to long-run equilibrium. Hence, the study suggests that policymakers of the respective countries must prioritize emission reduction, macroeconomic stability, and sustained adaptation finance to address temperature-related climate risks.

Keywords Climate finance · Temperature variation · South Asia and panel ARDL

R. Qasim · I. Ahmad (✉) · D. Alam · A. Hussain
Department of Economics, Aligarh Muslim University, Aligarh, Uttar Pradesh, India

P. Datta et al. (eds.), *Climate Change and Human Well-being*, Advances in Global Change Research 83, https://doi.org/10.1007/978-3-032-25479-5_16

16.1 Introduction

Climate change remains one of the most complex and far-reaching challenges that confront policymakers in the twenty-first century (Maruf et al., 2024). In South Asia, particularly in developing economies such as India, Bangladesh, and Nepal, the adverse effects of climate change extend beyond environmental degradation and significantly threaten macroeconomic stability, food security, and long-term economic growth (Bhandari, 2018). These effects create risks for food security, household income, and long-term economic growth. Moreover, these risks directly undermine human wellbeing, particularly in regions where climate vulnerability coincides with weak economic conditions. Increases in temperature reduce labour productivity, heighten health risks, and intensify heat-related mortality. At the same time, inflationary pressure and food system disruptions erode purchasing power and food access, exacerbating poverty and social stress. As a result, the consequences of climate change extend beyond environmental degradation, affecting fundamental dimensions of wellbeing such as health security, income stability, and overall quality of life. The rise in temperature, irregular monsoon patterns, retreat of Himalayan glaciers, and severe climate-related disasters now place heavy pressure on public finances and central bank operations (Santos & Araújo, 2024). In this context, climate finance from both domestic and international sources plays a critical role in the reduction of vulnerability to climate-related shocks across India, Bangladesh, and Nepal, and in the control of temperature escalation. Climate finance receives strong attention in global climate agreements such as the Paris Accord. However, its actual role in reducing temperature levels across South Asia remains insufficiently explored in empirical studies. The link between climate finance and key economic factors such as monetary policy, inflation expectations, and price stability does not follow a clearly defined path in this region. In light of the above, this paper focuses on India, Bangladesh, and Nepal to assess whether climate finance inflows have led to control over rising temperatures. It also examines how these outcomes influence the monetary policy frameworks in these countries. This study builds its foundation on environmental macroeconomics and the evolving role of central banks within climate-linked financial systems. Traditional macroeconomic models now incorporate climate shocks, which affect inflation, output gaps, and interest rates through both supply-side and demand-side channels. In South Asia, central banks usually function with a primary focus on inflation control and monetary stability. However, the rise of climate finance introduces new areas that require coordination between monetary tools and regulatory oversight. This shift holds particular importance for economies where climate-sensitive sectors such as agriculture and hydropower affect the consumer price index and shape external trade outcomes. By applying panel data regression methods, this analysis explores the relationship between climate finance flows and annual temperature changes across India, Bangladesh, and Nepal over the past ten years. In doing so, it addresses a significant gap by linking the role of climate finance not only to its environmental objective of temperature regulation but also to broader macroeconomic outcomes. Consequently, when adequate climate finance helps reduce climate instability, central banks benefit from greater price stability and face fewer fiscal pressures linked to climate-related disruptions. This contributes to a more predictable economic environment and supports long-term monetary and financial stability. These results can provide room for more flexible monetary measures. On the other hand, if climate finance shows limited effect, central banks may require more targeted approaches such as green credit schemes, sustainable refinancing plans, or selective liquidity instruments. India, Bangladesh, and Nepal reflect diverse economic structures but face common regional climate risks, which provide a sound basis for comparative analysis. India holds a leading position due to its advanced green finance architecture and the proactive regulatory measures undertaken by the Reserve Bank of India. Bangladesh, on the other hand, reflects a climate-vulnerable economy, where the central bank adopted green refinance schemes at an early stage to address environmental challenges. Nepal, in comparison, emphasizes the critical importance of international climate finance in a country that relies heavily on hydropower and agriculture. These three countries collectively represent key features of the Global South, where climate finance acts not only as an urgent requirement but also as a potential catalyst for innovation in monetary governance. These findings contribute to the broader global discourse on resilient macroeconomic structures and evolving responsibilities of central banks under climate-related stress.

16.2 Research Methodology

This study employs a panel data approach to examine the relationship between climate-specific finance and extreme temperature outcomes. The dependent variable is annual maximum temperature (TMAX), which serves as a proxy for extreme climatic events. The independent variables include adaptation finance (ATF), carbon dioxide (CO_2) emissions, and consumer price index (CPI) as a proxy for inflation (INF) as independent variables. The selection of variables is based on their direct relevance to the research objectives, focusing on the relationship between climate-specific finance and extreme temperature outcomes. Adaptation finance serves as a crucial variable because it represents the financial resources allocated to mitigate and adapt to the effects of climate change, including extreme temperature fluctuations. As such, adaptation finance plays a critical role in understanding how investments in climate resilience affect the severity and frequency of extreme climatic events, such as heatwaves (Lam & Delina, 2024). Carbon dioxide (CO_2) emissions are a critical factor in global warming, significantly contributing to the increase in extreme temperature events. The rise in atmospheric CO_2 levels, primarily driven by anthropogenic activities such as fossil fuel combustion, deforestation, and industrial processes, has led to a substantial increase in global temperatures, which in turn exacerbates climate extremes (Anwar et al., 2019; Nunes, 2023; Kanna et al., 2024). On the other hand, the inclusion of inflation is necessary, as it sheds light on broader economic conditions. Moreover, inflationary pressures can potentially alter energy demand, which, in turn, leads to a rise in temperature (Li et al., 2023). By selecting these three variables, the study captures both environmental and economic factors critical to shaping climate outcomes.

16.2.1 Descriptive Trends in Climate and Economic Indicators (2011–2020)

This section presents the annual trends in maximum temperature, adaptation finance, CO_2 emissions, and inflation across India, Bangladesh, and Nepal for the period 2011–2020. These patterns provide context for understanding the climate and economic dynamics across the sample countries and help clarify the relevance of the variables used in the analysis.

Figure 16.1 presents the trend in annual maximum temperature across the three countries. India consistently records the highest temperature levels, ranging between 44 and 46 °C. A slight increase is visible up to 2015, followed by relative stability. These consistently elevated temperatures reflect long-term heat stress, possibly due to industrial activity, urban expansion, and delayed adaptation measures. Bangladesh shows more variation. Maximum temperature increases until 2014, then declines sharply in 2016 and 2018, with partial rebounds in the intervening years. Nepal maintains the lowest values, between 17 and 20 °C, with minor fluctuations over the period. These patterns confirm the diverse thermal exposure profiles of the three countries.

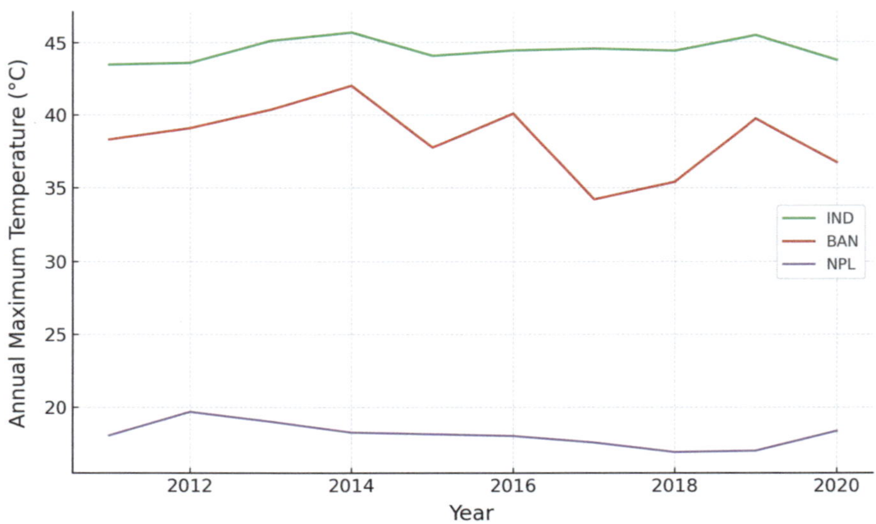

Fig. 16.1 Maximum temperature

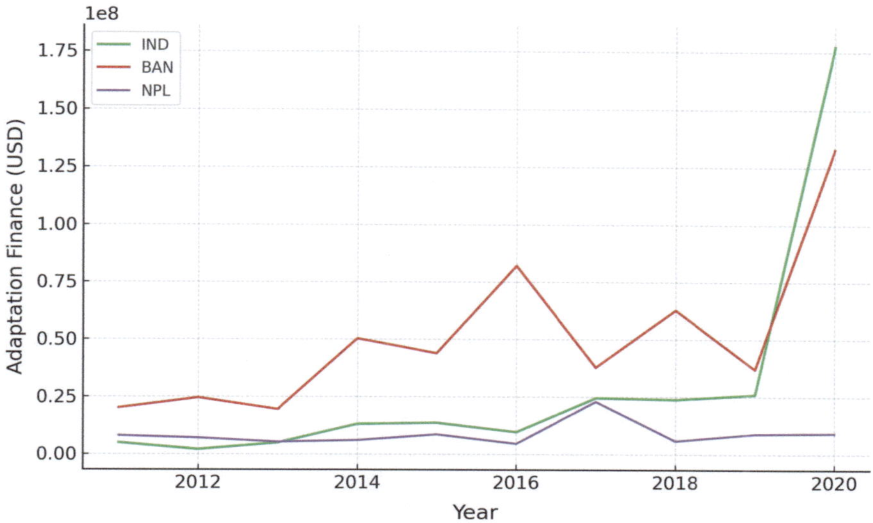

Fig. 16.2 Adaptation finance

Figure 16.2 illustrates the annual flow of adaptation finance. Bangladesh receives the highest funding in most years, peaking above USD 80 million in 2016 and USD 130 million in 2020. India shows a sharp increase in 2020, exceeding USD 175 million, after relatively moderate flows throughout the previous years. Nepal's adaptation finance remains modest but steady, rarely surpassing USD 25 million. The rapid expansion of funding in 2020 for both India and Bangladesh may reflect heightened international attention or a response to climate vulnerabilities accentuated by the pandemic. Overall, the figures highlight uneven access to adaptation finance and suggest the presence of institutional or absorptive capacity constraints.

Figure 16.3 tracks per capita CO_2 emissions. India exhibits a clear upward trend, rising from approximately 1.4–1.8 metric tons by 2018, before declining slightly in 2020. Bangladesh follows with a gradual increase, reaching nearly 0.58 metric tons per capita. Nepal shows a similar trajectory, although the growth appears more abrupt between 2015 and 2018. These trends correspond with each country's stage of economic development and energy consumption patterns. The deceleration in 2020 may reflect economic contraction or reduced industrial activity.

Figure 16.4 displays the inflation rate for each country. All three begin the period with relatively high inflation—above 9% in India and Nepal, and above 11% in Bangladesh. From 2012 to 2017, a downward trend dominates, with all countries reaching lows between 3 and 5%. This decline may reflect fiscal consolidation, improved monetary coordination, or lower commodity prices. After 2018, inflation begins to rise again in India and Nepal, suggesting renewed macroeconomic pressures. Bangladesh maintains a comparatively stable rate between 5 and 7% throughout the period.

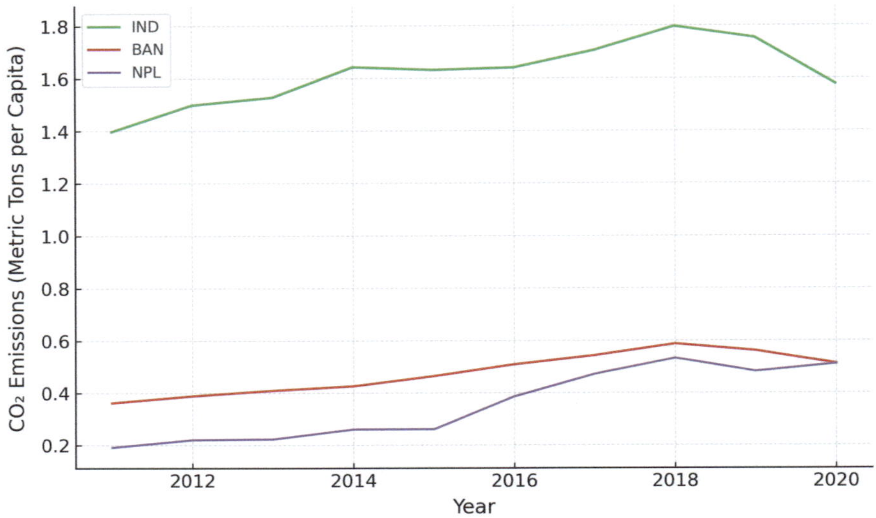

Fig. 16.3 CO_2 emissions

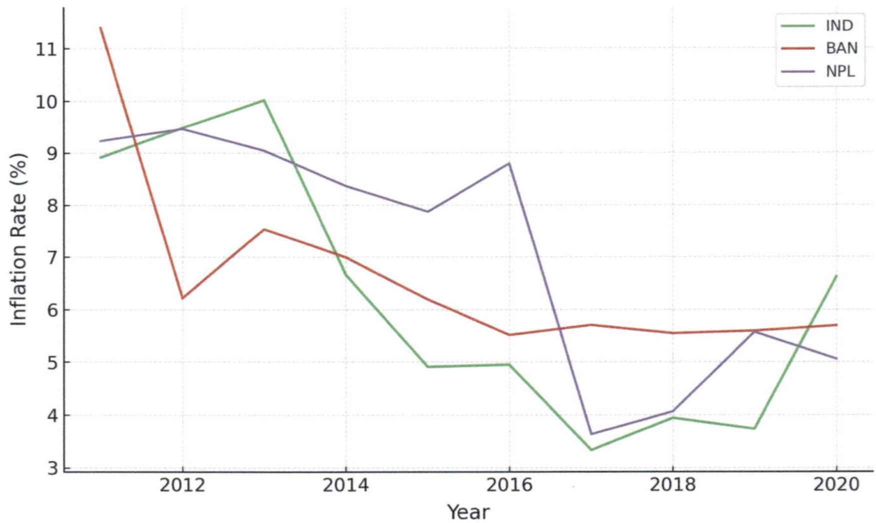

Fig. 16.4 Inflation

Overall, the graphical trends show notable variation across both countries and variables over the 2011–2020 periods. India consistently records higher values in temperature and CO_2 emissions, while Bangladesh shows relatively larger adaptation finance flows and more stable inflation. Nepal presents lower levels across most indicators, though with visible fluctuations in inflation and finance. These differences across variables and countries justify the application of panel data models that capture both temporal and cross-sectional variation. This selection ensures the analysis remains focused and theoretically grounded, while avoiding complexity and data limitations that might arise from including additional variables. Moreover, the information regarding climate finance has been taken from the United Nations Framework Convention on Climate Change portal. This variable reflects the annual financial disbursements reported under adaptation-related projects and is measured in constant US dollars. In addition, the dependent variable, annual maximum temperature, is derived from the NASA Power Data Bank and represents the annual average of daily maximum surface air temperatures, expressed in degrees Celsius. Furthermore, data on CO_2 emissions and inflation (INF) are collected from the World Development Indicators published by the World Bank. CO_2 emissions are measured in metric tons per capita and reflect environmental pressure. Inflation, on the other hand, is reported as the annual percentage change in consumer prices and captures macroeconomic volatility, which may influence a country's capacity to manage climate risks. The selected variables provide an integrated view of environmental stress, financial response capacity, and macroeconomic dynamics in the sample countries.

The dependent variable, Annual Maximum Temperature, is obtained from the NASA Power Data Bank, while CO_2 and INF data are sourced from the World Bank. The study covers a ten-year period from 2011 to 2020 and includes three South Asian countries: India, Bangladesh, and Nepal. The selection of these countries is based on the availability and consistency of climate finance data. This analysis highlights regional disparities.

16.2.2 Model Selection

After collection of data, the stationarity properties of the variables are examined. To achieve this, four standard unit root tests are applied. These tests were selected based on their suitability for the sample size and asymptotic characteristics (da Silva et al., 2018). The panel unit root tests operate under the null hypothesis that all variables in the panel contain a unit root, as outlined in studies by (Choi, 2001; Im et al., 2003; Levin et al., 2002; Maddala & Wu, 1999). These tests are particularly useful in panel data contexts, where they assess the stationarity properties of multiple time series or cross-sectional units simultaneously. The results indicate that the variables exhibit a mixed order of integration, which means some are stationary at level, while others are stationary after their first difference. This justifies the application of the Panel ARDL model (Pesaran & Smith, 1995). After the stationarity tests, the study also includes the cointegration test. Although the presence of cointegration is not a strict requirement, its existence allows the Panel ARDL model to take the form of an error correction model. This provides stronger evidence that the long-run coefficients remain consistent across all countries, which supports the use of the Pooled Mean Group (PMG) estimator (da Silva et al., 2018). The Panel ARDL method holds several advantages over other dynamic panel estimation techniques, such as fixed effects, instrumental variables, and GMM estimators (Pesaran et al., 1999; Anderson & Hsiao, 1982).

Further, conventional methods often fail to provide consistent estimates when slope coefficients differ across cross-sectional units. In this context, the panel ARDL model emerges as a more appropriate and robust estimation method (Arellano & Bover, 1995). This method allows the coefficients of the variables in levels to reflect the long-run impact on the dependent variable, as shown in the following equation.

$$\Delta TMAX_{t-1} = \rho^0 + \sum_{i=0}^{n} \rho^1 \Delta TMAX_{t-1} + \sum_{i=0}^{n} \rho^2 \Delta ATF_{t-1} + \sum_{i=0}^{n} \rho^3 \Delta CO2_{t-1} + \sum_{i=0}^{n} \rho^4 \Delta INF_{t-1} + \delta^1 TMAX_{t-1} + \delta^2 ATF_{t-1} + \delta^3 CO2_{t-1} + \delta^4 CPI_{t-1} + \varepsilon_t \quad (1)$$

where ρ and σ are the white-noise, ε_t is the error term, $TMAX_{t-1}$, ATF_{t-1}, $CO2_{t-1}$, INF_{t-1} are the short run and the long run coefficients of the model, respectively, and Δ is the first difference operator; 't' denotes time period, and 'n' is the maximum number of lags in the model, based on the Akaike Information Criterion.

16.3 Results Analysis

This section presents the results of the study, with key findings aligned with the research objectives and hypotheses. The data reveal important trends, patterns, and variations observed during the investigation. The next section presents detailed interpretation and discussion of these results within the context of the study.

The descriptive statistics in Table 16.1 present a detailed summary of the behaviour of the four variables: TMXI, ATF, CO_2 emissions, and INF, based on a dataset of 30 observations. First, the comparison between mean and median values reveals that TMXI and ATF exhibit a fairly symmetrical distribution. The mean of TMXI stands at 3.44, closely aligned with the median of 3.65. Similarly, ATF shows a mean of 16.61 and a median of 16.60. However, CO_2 and INF reflect some degree of asymmetry. The mean value of CO_2 (0.81) surpasses the median (0.51), indicating the presence of higher values that raise the average. A similar trend appears in the INF data, where the mean (6.66) remains slightly above the median (6.20). Moreover, the examination of standard deviation confirms that INF exhibits the highest variability (2.16), followed by ATF (1.08), and CO_2 (0.59). In contrast, TMXI appears relatively stable with the lowest standard deviation of 0.40. This observation implies that inflation and adaptation finance fluctuate more widely than temperature and CO_2 levels.

In addition, the skewness values indicate the direction of the data distribution. TMXI holds a negative skewness (– 0.63), which points to a longer left tail and a

Table 16.1 Descriptive analysis

	TMXI	ATF	CO_2	CPI
Mean	3.444823	16.61938	0.814939	6.666017
Median	3.656178	16.60852	0.510007	6.205892
Maximum	3.822098	18.99390	1.795595	11.39517
Minimum	2.827314	14.50866	0.190682	3.328173
Std. dev.	0.402794	1.083647	0.589129	2.160584
Skewness	– 0.625001	0.329005	0.652414	0.330810
Kurtosis	1.524469	2.492473	1.618425	2.127563
Jarque-Bera	4.674617	0.863203	4.514157	1.498609
Probability	0.096587	0.649468	0.104656	0.472695
Sum Sq. dev.	4.705055	34.05442	10.06510	135.3756
Observations	30	30	30	30

concentration of higher values. In contrast, ATF, CO_2, and INF show positive skewness, with CO_2 (0.65) indicating a notable tilt toward lower values and occasional spikes on the higher end. Kurtosis values for all variables remain below the benchmark of 3, which suggests a platykurtic distribution. This outcome denotes a flatter distribution than the normal curve, with fewer extreme observations. The Jarque-Bera test provides further insight into the normality of each variable. All four variables fail to reject the null hypothesis of normal distribution at the 5% level. Although the test values for TMXI (JB = 4.67, p = 0.09) and CO_2 (JB = 4.51, p = 0.10) approach marginal significance, they do not cross the conventional threshold. Therefore, the data used in this study appears suitable for further econometric analysis. Overall, the variables exhibit strong statistical properties. TMXI, as the dependent variable, shows moderate variation and near-normal distribution, while the explanatory variables, particularly ATF and INF, display patterns that enable a thorough exploration of their relationship with temperature levels. In order to examine the trends of these variables, trend analysis through graphical representation is carried out. Figure 16.1 illustrates the temporal evolution of key climate and macroeconomic variables across the selected South Asian economies. The observed trends reflect considerable variation in adaptation finance inflows, gradual changes in CO_2 emissions, and distinct inflationary pressures across the three countries. While temperature maximum shows limited short-term fluctuation, its overall stability contrasts with the dynamic nature of the other indicators. These differences across countries and variables support the use of panel data models to capture both cross-sectional and temporal heterogeneity.

The results of the stationarity tests based on four different methods are reported in Table 16.2. In the context of panel data analysis, a variable attains the status of stationarity when at least three out of the four tests produce significant results at the 5% level. This criterion provides a more reliable basis to determine the integration order of the variables. Based on this approach, the variables TMAX and INF satisfy the condition of stationarity at level, which confirms their integration at I (0). On the other hand, the variables ATF and CO_2 achieve stationarity at their first difference, which confirms their integration at I (1). The existence of both I (0) and I (1) variables in the model confirms the presence of a mixed order of integration across variables. This condition justifies the selection of the ARDL model for estimation because the ARDL approach remains suitable when variables possess different orders of integration.

Table 16.3 presents the outcomes of the Pedroni panel cointegration test, which examines whether a long-run equilibrium exists among the selected variables. The test consists of two dimensions: within-dimension (panel statistics) and between-dimension (group statistics). In the panel statistics, both the Panel PP-Statistic and the Weighted Panel PP-Statistic reject the null hypothesis of no cointegration at the 1% level, confirming the presence of a long-run relationship among the variables. The Panel ADF-Statistic and its weighted counterpart also show statistically significant evidence of cointegration at the 5% level. In the group statistics, the Group PP-Statistic rejects the null hypothesis at the 1% level, while the Group ADF-Statistic provides weaker support for cointegration at the 10% level. Overall, the test statistics show strong significance, confirming the existence of a stable long-run relationship among the variables. This evidence justifies the application of long-run estimation techniques, such as the panel ARDL model, in the next phase of analysis.

Table 16.4 presents the long-run estimates from the panel ARDL model, with Average Maximum Temperature as the dependent variable. The results reveal statistically significant relationships between the explanatory variables and temperature. The coefficient for adaptation finance holds statistical significance which indicates that an increase in adaptation finance leads to a reduction in maximum temperature.

Table 16.2 Stationarity tests of the variables

	LLC	IPS	ADF—Fisher	PP—Fisher
Level: Constant				
TMAX	– 3.16821 **(0.0008)**	– 1.43755 **(0.0753)**	12.8435 **(0.0456)**	15.2237 **(0.0186)**
ATF	1.03198 (0.8490)	0.69078 (0.7551)	2.91984 (0.8188)	10.4308 (0.1076)
CO_2	– 1.68425 (0.0461)	0.24841 (0.5981)	4.01076 (0.6752)	5.37227 (0.4970)
INF	– 4.78734 **(0.0000)**	– 2.18525 **(0.0144)**	16.2297 **(0.0126)**	14.4300 **(0.0252)**
First difference: Constant				
ATF	– 1.02997 (0.1515)	**– 1.31652** **(0.0940)**	**12.5959** **(0.0499)**	**38.5161** **(0.0000)**
CO_2	– 2.27647 **(0.0114)**	+	**11.4529** **(0.0753)**	**11.9440** **(0.0632)**

Note Probability values are given in parentheses. Bold values indicate that the variable is stationary at the 1, 5 and 10% significance level
+ Not available due to insufficient number of observations

Table 16.3 Pedroni cointegration results

			Weighted	
	Statistic	Prob.	Statistic	Prob.
Panel PP-statistic	− 5.157186	0.0000***	− 4.965353	**0.0000*** **
Panel ADF-statistic	− 1.956697	0.0252**	− 1.835405	**0.0332** **
Alternative hypothesis: individual AR coefficients (between-dimension)				
	Statistic	Prob.		
Group PP-statistic	− 6.088523	0.0000***		
Group ADF-statistic	− 1.431178	0.0762*		

Note ***denotes significance at the 1% level, **at the 5% level, and *at the 10% level

This result implies that financial measures aimed at climate adaptation contribute effectively to temperature control over the long term. In contrast, the coefficient for CO_2 emissions demonstrates a strong positive association with maximum temperature. This indicates that an increase in CO_2 emissions causes a rise in temperature, which highlights the adverse climatic impact of higher carbon concentrations (Pustišek, 2025). Similarly, the coefficient for INF also holds statistical significance, reflecting a positive relationship with temperature levels. This outcome suggests that inflationary trends may correlate with higher temperatures (Kotz et al., 2023) possibly due to economic pressures that coincidsse with environmental degradation or increased energy consumption. The error correction term of − 1.18 suggests a damping fluctuation, which indicates that the error correction (ECT) process fluctuates around the long-run value in a diminishing manner, rather than reaching the equilibrium trajectory monotonically (Narayan & Smyth, 2006). Once the process finishes, convergence to equilibrium is fast (Zafar & Tarique, 2023).

Table 16.4 Long-run coefficients

Dependent variable: maximum temperature				
Variable	Coefficient	Std. error	t-Statistic	Prob.
ATF	− 0.028144	0.010971	− 2.565350	0.0248
CO2	0.207655	0.040696	5.102565	0.0003
INF	0.030848	0.003261	9.459553	0.0000
Constant	3.8986	1.985	1.964	0.0731
ECT(− 1)	− 1.18	0.6622	− 1.782	0.1000

16.4 Discussion

The long-run coefficients reported in Table 16.4 provide key insights into the relationship between climate-specific financial flows and temperature levels across the countries—India, Bangladesh, and Nepal. Each of these countries faces frequent exposure to extreme climatic events, limited fiscal space, and increasing environmental pressure. The inclusion of ATF, CO_2 emissions, and INF provides a comprehensive understanding of the environmental and economic factors that influence temperature trends in South Asia.

The negative coefficient of ATF is significance at the 5% level. This value confirms that increases in climate adaptation finance correlate with reductions in maximum temperature. The result aligns with theoretical expectations, as adaptation funding often supports projects that enhance climate resilience through infrastructure, ecosystem protection, and water resource management.

In the context of India, this outcome reflects the effect of state-led missions under the National Action Plan on Climate Change (NAPCC) which directs climate finance towards critical areas such as infrastructure, ecosystem protection, and water resource management, all essential for mitigating the impacts of climate change (Dey & Bhattacharya, 2024). In Bangladesh, international funds such as the Green Climate Fund and the Bangladesh Climate Change Trust Fund have supported coastal and flood-prone zones (Pervin et al., 2019). Similarly, Nepal, with its mountainous terrain, has utilized adaptation finance for glacial lake outburst risk reduction and forest-based carbon sequestration (Agrawal, 2022). Despite variation in administrative capacity and resource absorption across the three countries, the negative coefficient affirms the climate-stabilizing potential of adaptation measures in the long term.

In contrast, the coefficient for CO_2 emissions appears positive and statistically significant at the 1% level. This positive relationship confirms that increased CO_2 emissions result in a corresponding rise in maximum temperature. Among the three countries, India contributes the highest volume of emissions due to its energy-intensive industrial and transportation sectors (Philip & Kuriakose, 2025). Bangladesh, although lower in absolute emissions, experiences rapid urban expansion and deforestation, which reduce natural carbon sinks. Nepal shows a relatively smaller emission footprint but still registers upward trends due to increasing dependence on diesel generators and cross-border energy imports. The strong and direct impact of CO_2 on temperature highlights the urgent need for emission control policies, transition to cleaner fuels, and greater enforcement of environmental regulations.

The INF, which acts as a control variable in this model, shows a positive coefficient and significant. It does not directly relate to environmental factors but captures macroeconomic volatility and policy inefficiency. In India, inflation often links with fuel costs and food supply shocks due to climatic extremes, thereby indirectly influencing energy use and urban heat. In Bangladesh, food price inflation ties closely with seasonal flood cycles. In Nepal, inflation reflects fuel imports and agricultural dependency. The positive coefficient of INF therefore supports the hypothesis that unstable macroeconomic conditions may coincide with greater exposure to temperature anomalies, possibly through weak climate governance or inadequate public investment.

Overall, the long-run model confirms three critical points. First, ATF emerges as a meaningful tool for climate mitigation, although its effect appears moderate. Second, CO_2 emissions exert a strong influence on temperature, thereby calling for strict emission control strategies. Third, INF, despite serving as a control variable, also shows a significant association with temperature outcomes, which indicates the role of macroeconomic stability in climate resilience. For India, Bangladesh, and Nepal, these findings provide useful suggestions. Reducing emissions, improving adaptation finance, and maintaining economic discipline can lower climate risks and protect vulnerable populations from rising heat extremes. These findings carry significant implications for human wellbeing in the region. Rising temperatures, which are associated with increased emissions and inflation, pose risks to public health, reduce labour productivity, and disrupt food and water security, particularly among low-income and climate-vulnerable populations. In contrast, adaptation finance holds potential to reduce these risks by supporting resilient infrastructure, improving water access, and enabling ecosystem-based interventions. Although the regression model focuses on temperature as the outcome, the underlying mechanisms are closely linked to wellbeing dimensions such as health protection, economic stability, and disaster preparedness. The results, therefore, underscore the importance of understanding climate finance not only as a tool for environmental adaptation but also as a mechanism for the protection of wellbeing in South Asia.

16.5 Conclusion and Policy Implications

This study examines the role of climate-specific finance in addressing extreme temperature patterns across India, Bangladesh, and Nepal. The findings reveal significant relationships between adaptation finance, CO_2 emissions, inflation, and temperature trends. Adaptation finance, although negatively correlated with efforts to mitigate rising temperatures, faces limitations due to variations in resources, governance, and institutional capacity. CO_2 emissions and inflation are key drivers of temperature increases, emphasizing the urgent need for effective climate policies.

To mitigate the impacts of rising temperatures, it is crucial to enhance adaptation finance, implement emission control policies, and stabilize inflation, as these factors emerged as the primary drivers of temperature increases. Moreover, Inflation, often seen as an economic issue, must be managed carefully, as it impacts the availability of resources for climate adaptation and increases vulnerability to extreme temperatures. Stabilizing inflation will ensure that financial resources are allocated efficiently to address climate risks.

The cross-country differences in climate exposure, financial flows, and macroeconomic indicators point to the need for differentiated policy responses. Specifically, in the case of India, sustained high temperatures and rising emissions, alongside fluctuations in adaptation finance, indicate a need to expand climate investment, enhance policy coherence across sectors, and integrate climate considerations into fiscal and monetary frameworks. By contrast, in Bangladesh, relatively high levels of adaptation finance do not correspond to consistent moderation in temperature trends. This outcome suggests the importance of strengthening institutional capacity, improving the efficiency of fund allocation, and prioritizing adaptation in geographically vulnerable regions. Meanwhile, Nepal exhibits low levels of adaptation finance and emissions yet remains highly exposed to climate risks. As such, national efforts should focus on mobilizing additional international support, particularly for hydropower development, forest-based mitigation, and decentralized adaptation planning. Together, these policy directions reflect the structural and environmental distinctions among the three countries and may enhance the long-term effectiveness of climate finance in building resilience. Moreover, enhancing climate finance mechanisms has the potential to reduce climate vulnerability and indirectly improve wellbeing by limiting exposure to extreme temperatures, preserving economic stability, and reducing risks to health and livelihoods.

Limitation of the Study

In examining the effectiveness of climate finance in temperature control, this study encountered several limitations. The most significant constraint was the unavailability of comprehensive and consistent data across all South Asian countries. As a result, the analysis was restricted to India, Bangladesh, and Nepal. While these countries provide useful case studies, the findings may not fully capture the regional variation in climate finance mechanisms or policy frameworks across South Asia. Additionally, the model included only three explanatory variables—adaptation finance, CO_2 emissions, and inflation. Although these variables are relevant and supported by literature, the exclusion of other potentially important factors such as renewable energy adoption, forest cover, urbanization, institutional quality, and external aid may limit the explanatory power of the model. This raises the possibility of omitted variable bias. The study also uses a relatively short time frame (2011–2020), which may not reflect the long-term effects of climate finance. Climate adaptation measures often produce results over longer periods, and short-run analysis may underestimate or misrepresent their true impact. Lastly, while this study focuses on temperature-related outcomes, it does not assess how these climates impacts translate into social or economic wellbeing. The broader literature on climate finance emphasizes links to livelihoods, health, and long-term development, which remain outside the scope of this analysis. Future studies should address these limitations by using longer time series, expanding the country sample, and including additional explanatory variables that capture economic, institutional, and environmental factors. Moreover, considering both development and environmental dimensions of climate finance may strengthen the contribution of such research to policy design.

Acknowledgements We express our sincere gratitude to the Indian Council of Social Science Research (ICSSR) for the Doctoral Fellowship (Award Letter No.: ICSSR/RFD/24-25/ECO/GEN/32). We are also thankful to the Department of Economics, Aligarh Muslim University, for providing the necessary infrastructure and academic support throughout the course of this research.

Conflict of Interest The authors declared that there are no conflicts of interest.

Author Declaration The authors confirm that the content of this manuscript is the result of original academic work conducted by the authors. No part of the paper has been generated using artificial intelligence. All ideas, analyses, arguments, and interpretations remain the intellectual contribution of the authors. The final content has been thoroughly reviewed and approved by the authors to ensure academic integrity and accuracy.

References

Agrawal, D. (2022). Climate adaptation financing in Nepal. *Armed Police Force Command and Staff College Journal, 5*(1), 131–146. https://doi.org/10.3126/japfcsc.v5i1.49353

Anderson, T. W., & Hsiao, C. (1982). Formulation and estimation of dynamic models using panel data. *Journal of Econometrics, 18*(1), 47–82.

Anwar, M. N., Iftikhar, M., Bakhat, B. K., Sohail, N. F., Baqar, M., Yasir, A., & Nizami, A. S. (2019). *Sources of carbon dioxide and environmental issues* (pp. 13–36). Springer. https://doi.org/10.1007/978-3-030-29298-0_2

Arellano, M., & Bover, O. (1995). Another look at the instrumental variable estimation of error-components models. *Journal of Econometrics, 68*(1), 29–51.

Bhandari, M. P. (2018). The climate change induced problems in South Asia, a case study of Bangladesh, India, Nepal, and Pakistan. *Advances in Agriculture and Environmental Science Open Access (AAEOA), 1*(1), 39–41.

Choi, I. (2001). Unit root tests for panel data. *Journal of International Money and Finance, 20*(2), 249–272.

da Silva, P. P., Cerqueira, P. A., & Ogbe, W. (2018). Determinants of renewable energy growth in sub-Saharan Africa: Evidence from panel ARDL. *Energy, 156*, 45–54.

Dey, B., & Bhattacharya, K. (2024). Climate financing: As an economic approach towards sustainable environment. *International Journal of Research Publication and Reviews, 5*(5), 8448–8460. https://doi.org/10.55248/gengpi.5.0524.1339

Im, K. S., Pesaran, M. H., & Shin, Y. (2003). Testing for unit roots in heterogeneous panels. *Journal of Econometrics, 115*(1), 53–74.

Kanna, I. V., Roseline, S., Balamurugan, K., Jeeva, S., & Santhiyagu, I. A. (2024). *The effects of greenhouse gas emissions on global warming*. Elsevier BV. https://doi.org/10.1016/b978-0-323-93940-9.00216-4

Kotz, M., Kuik, F., Lis, E., & Nickel, C. (2023). The impact of global warming on inflation: Averages, seasonality and extremes. *Social Science Research Network*. https://doi.org/10.2139/ssrn.4457821

Lam, R. Y. H., & Delina, L. L. (2024). Financing structural and non-structural extreme heat adaptation measures in southeast Asian cities: Statuses and prospects. *Environmental Research Communications*. https://doi.org/10.1088/2515-7620/ad9f10

Levin, A., Lin, C. F., & Chu, C. S. J. (2002). Unit root tests in panel data: Asymptotic and finite-sample properties. *Journal of Econometrics, 108*(1), 1–24.

Li, C., Zhang, X., & He, J. (2023). Impact of climate change on inflation in 26 selected countries. *Sustainability*. https://doi.org/10.3390/su151713108

Maddala, G. S., & Wu, S. (1999). A comparative study of unit root tests with panel data and a new simple test. *Oxford Bulletin of Economics and Statistics, 61*(S1), 631–652.

Maruf, I. R., Sari, A. L., Saptayuda, A., Purwadi, M. A., & Susilawaty, A. (2024). Navigating the escalating climate crisis: Comprehensive policy responses and collaborative global action for a sustainable future. *International Journal of Science and Society*. https://doi.org/10.54783/ijsoc.v6i1.1097

Narayan, P. K., & Smyth, R. (2006). What determines migration flows from low-income to high-income countries? An empirical investigation of Fiji–us migration 1972–2001. *Contemporary Economic Policy, 24*(2), 332–342.

Nunes, L. J. R. (2023). The rising threat of atmospheric CO_2: A review on the causes, impacts, and mitigation strategies. *Environments, 10*(4), 66. https://doi.org/10.3390/environments10040066

Pervin, M., Barua, P., Imam, N., Haque, M. M., & Hossain, N. J. (2019). *Climate governance and finance in Bangladesh* (pp. 65–83). Springer. https://doi.org/10.1007/978-3-030-05237-9_6

Pesaran, M. H., Shin, Y., & Smith, R. P. (1999). Pooled mean group estimation of dynamic heterogeneous panels. *Journal of the American Statistical Association, 94*(446), 621–634.

Pesaran, M. H., & Smith, R. (1995). Estimating long-run relationships from dynamic heterogeneous panels. *Journal of Econometrics, 68*(1), 79–113.

Philip, A., & Kuriakose, N. E. (2025). Trends and drivers of greenhouse gas emissions in India: A decadal analysis (2010-2020). *Current World Environment, 19*(3), 1355–1364. https://doi.org/10.12944/cwe.19.3.26

Pustišek, A. (2025). *CO_2 concentration and global temperature: Missing experimental evidence* (pp. 111–122). https://doi.org/10.9734/bpi/geserh/v3/3633

Santos, L. H. B., & Araújo, E. (2024). Financial instability and climate change: New challenges for central banks. *Brazilian Journal of Political Economy, 44*(4). https://doi.org/10.1590/0101-31572024-3617

Zafar, S., & Tarique, M. (2023). Efficacy of public spending for agricultural development in India: A disaggregate analysis contextualizing subsidies vs investment debate. *International Journal of Social Economics, 50*(7), 925–940.

Chapter 17
Governance Challenges in the Age of Climate Induced Migration: Re-conceptualizing Vulnerabilities in Trafficking-Migration Continuum Through Indexing

Yu Kojima and Chiranjeevi Tallapragada

Abstract Whereas climate-induced migration is increasingly recognized as a viable survival means globally, the failure of global migration governance in recent years suggests limited impacts of public agency's efforts to mitigate extreme policy trade-offs between measures to counter human trafficking and promote international labour migration. Drawing upon the political economic landscape of multi-dimensional vulnerabilities adhered to women in migration, this paper recognizes that human trafficking as a systemic problem. As exhibited in the gendered and ethno-racial normative framework that promotes labour migration and prevents human trafficking, the failure of conventional policy measures to incorporate its subjective nature of domestic labour's value is therefore identified as a major policy blind spot. Under this premise, this paper intends to advance discussion on building greater institutional capacity and resilience of migration governance. We explore how indexing could offer improved understanding of interactions between institutional and migration outcomes in public policy planning through blind spots and trade-off analysis. We argue that through this approach, more comprehensive assessment of and progress monitoring of policy measures may be feasible, claiming this would enhance response and learning capacity of a public agency to effectively address challenges and monitor counter trafficking initiatives.

Keywords Climate induced migration · Trafficking in persons · Migration policy · Gender and racism · Vulnerabilities in migration · Institutional trajectory analysis · Trade-offs · Indexing · Feminist political theory · Domestic labour · Boundary science

---

Y. Kojima (✉)
The Climate Panel, State College, Pennsylvania, USA
e-mail: yucita2003@yahoo.com

C. Tallapragada
Livelihoods and Natural Resources Management Institute, Hyderabad, India
e-mail: tc.lnrmi@gmail.com

P. Datta et al. (eds.), *Climate Change and Human Well-being*, Advances in Global Change Research 83, https://doi.org/10.1007/978-3-032-25479-5_17

17.1 Introduction

Evidentially, climate-induced disasters, environmental degradation, and the growing scarcity of resources are emerging as serious drivers of migration through displacement and loss of livelihoods (Furlan & Rosado, 2023; Vince, 2022). As reported by the International Organization for Migration (2017), these factors exacerbate pre-existing structural socio-economic vulnerabilities, deepening exclusion, marginalization, and violence, driving displacement and migration, and heightening the risk of modern slavery and trafficking in persons, especially women, youth, and children (Molinari, 2017). In communities without adequate pre-disaster preparation strategies and post-disaster recovery mechanisms, climate change-related hazards lead to conditions of multi-dimensional vulnerability. Most notably, hazard-affected communities are driven into unemployment and poverty, displacement and migration, food insecurity, health insecurity, lack of education, and psychological distress (Srihari et al., 2024). Thus, we stress that to better understand the rooted drivers of vulnerability to human trafficking, key areas to account for in the evidence of vulnerability factors relate particularly to climate change and its linkages to environmental degradation, livelihood stress, impoverishment, and forced migration.

Against this background, as seen through a recent introduction of the Modern Slavery Bill (2013) by the British government, a contemporary use of the term 'modern slavery' (Walk Free, 2018) reflects a shift in the global consensus in the area of anti-human trafficking policy initiatives. It may provide an improved interventional framework, offering a clarity and more nuanced understanding of the status and condition of the individuals (debt bondage, forced labour, forced marriage, slavery and slavery-like practices) and constituting processes (human trafficking) in which enslave people for sexual exploitation, forced labour and forced marriage purposes. Nonetheless, a scarcity of reliable and methodologically sound data on modern slavery has been yet recognized as a major challenge[1] while a closer gaze into individual governments responses triggers urgent scrutiny into the effectiveness of current migration governance to alleviate modern slavery.

For instance, the Organization for Economic Co-operation and Development (OECD) donors have been contributing anti-slavery funds both locally and internationally. One study conducted by Ucnikova (2014) reveals that OECD donors spend an average of USD 124 million annually in the form of ODA through funding over 2000 individual initiatives to counter modern slavery during the period of 2003–2012. This means that public funds of the total amount of USD 1.2 billion were disbursed through a country's development agency, foreign affairs and directly by police, immigration and interior departments to recipient governments over ten years.

---

[1] One landmark attempt to overcome this challenge engaged by Walk Free Foundation in collaboration with International Labour Organization (ILO) and the International Organization for Migration (IOM) has estimated that 40.3 million people were held in modern slavery in 2016 in which 15.4 million in forced marriage while 24.9 million in forced labour. It also confirms that modern slavery is a gender parity issue—78% female to 29% male held in slavery and slavery-like practices and human trafficking (Walk Free, 2018).

Nevertheless, there is an increasing concern voiced by rights advocates over the lack of transparency regarding actual spending of the anti-slavery funds and confusion, inconsistency and harm entailing the spending (Dottridge, 2014).

A key question here is how can we further reinforce government accountability, particularly enhancing public agency's efforts to mitigate extreme policy trade-offs in the context of countering human trafficking? Debates on the current anti-slavery public funds spending miss a critical point in this respect by disconnecting the discussion from the budget process and its significant role in shaping behaviour of public agencies through which political commitments for donor-led interventions are easily subverted. We must thus learn more about the features of norms and organizational behaviour of public agencies and the institutional mechanisms that facilitate implementation of anti-trafficking initiatives, their inherent challenges and scope of change for more sustainable and effective implementation.

Under this premise, this paper intends to advance theoretical discussion on the migration governance in respect to building greater institutional capacity and resilience in addressing counter-trafficking initiatives. Identifying the political economic landscape of multi-dimensional vulnerabilities entailing human trafficking as background, we demonstrate how indexing offers improved understanding of interactions between institutional and migration outcomes in public policy planning through blind spots and trade-off analysis. We argue that through this approach, more comprehensive assessment of and progress monitoring of policy measures be feasible that enhance response and learning capacity of a public agency in effectively address challenges and monitor counter trafficking initiatives.

In this connection, we recognize the Boundary Science (Kurian & Kojima, 2021) as a resourceful framework in examination of migration governance for this paper. Boundary science is a versatile framework which offers new ways to improve our undertaking in development and scientific practice in relation to: (1) pathways for integrating scientific inputs into migration policy and planning; (2) integrative modelling that combines feminist political theory and institutional perspectives and (3) refining the scope for data and statistics for the purpose of more effective monitoring of development outcomes and impacts. Boundary Science has its conceptual component of blind spots examination, complimented by a methodological construct of trade-off analysis. Arising from the theoretical discussion on environmental governance (Daher et al., 2018), analysis on the role of policy blind spot is also useful to unpack the mechanism that prevents the institutional accountability and synergies in the context of migration governance, pointing out dominant norms and assumptions in defining research designing and monitoring, incentives for organizational capacity building and feedback.

To this end, the paper begins with introduction of key theoretical debates on vulnerabilities in trafficking-migration nexus, teasing out the characteristics and extent of interconnectivity between these blind spots and the policy trade-off intensity. Our discussion then will focus on the conceptual and methodological aspects of the institutional trajectory analysis through identifying policy shortcomings related to counter-trafficking within Sustainable Development Goals (SDGs) framework as a starting point. We argue for a rationale for indexing and demonstrate how typology of trade-off analysis is useful to systematically analyze key policy challenges related to public agency's enrolment in counter-trafficking initiatives. By summation of these findings, we finally propose a framework for cross-compatible composite index through which we conclude that more comprehensive assessment of policy measures be feasible that enhance response and learning capacity of a public agency in effectively addressing challenges and monitor and evaluate impacts of counter trafficking initiatives.

17.2 Key Definitions and the Study Scope

In this paper, we define vulnerabilities related to trafficking in persons in a larger scale by incorporating both the relationship between individuals' decisions and changes in their socio-economic environment (agency) and state policies identified in the form of policy norms, choices and patterns of state practices to regulate and protect individuals in trafficking and migration (structure). In this vein, with rising climate change impacts, the line between displacement and migration, the latter generally regarded as voluntary, is increasingly blurring. Climate change intersections with inequality and marginalization are also driving movements across borders. As a result, those forced to migrate can become victims of exploitation and modern slavery (Climate Refugees, 2022). Having these emerging forces in mind, this study identifies five policy blind spots in relation to migration governance analysis, with an explicit focus on dynamics related to policy decision-making processes in the context of climate induced vulnerabilities, trafficking, migration and modern slavery nexus as below;

Institutional gaps emerge as policy decisions divert depending on the scale and impact of: (1) guidance offered by global versus regional normative model (as seen as legal framework) to counter-human trafficking; (2) guidance inspired by knowledge of science versus policy (political economy of domestic labour as work); (3) guidance offered by victim-criminal framework in relation to law enforcement models and options in judicial approach (4) assumptions about data availability and attitudes toward data sharing that guide the design of Management Information System (MIS)[2] of the public agency (5) assumptions that define characteristic of feedback loop between MIS and decision-making that supports the prediction, response and recovery from extreme migration trade-offs. More detailed discussion on these points will be spared in the following section.

It is also important to note here that this paper recognizes that human trafficking occurs in continuum of migration (Kojima, 2011). Here, migration is understood

[2] One example of a mainstream guiding assumption could be that data collected from stakeholder consultations will lead to coherent decisions (Daher et al., 2018) without taking into consideration of how group politics behind could shape the dynamics of such information sharing activity.

in a larger process of movement of people involving documented migrants, undocumented migrants, refugees and other persons of concern[3] (asylum seekers, returnees, stateless and internally displaced persons).[4] Climate change and/or climate-induced migration intersects with severe forms of exploitation along at least three pathways: slow-onset disasters (droughts, crop failures), rapid-onset disasters (floods, cyclones) and an amalgamation of conflicts and climate change events. Yet policymakers have rarely considered climate change as a driver of human trafficking (Bharadwaj et al., 2022). Climate change acts as a push factor for human migration and providing an easier access to a large pool of human resources susceptible to trafficking. Harsher environments and fragile livelihood conditions have traditionally been the grounds for exploitation and trafficking. The numbers were limited and the process was demand driven. With climate change induced migration, the geographical spread and the pool of human resources forced to migrate becomes very large. This type of trafficking will be in a way is supply driven, which can make the process more exploitative, riskier as well as more dangerous. While trafficking itself is exploitative, the factors driving it will shape the extent, type as well as the sections of society affected. Governance issues in addressing the trafficking problem becomes even more complicated. Institutional set up, resources required and strategies to protect the migrants will require a broader approach cutting across borders.

Mindful that the issues arising from the nexus of climate induced migration-trafficking are complex and require tedious investigation, this study proposes rather an explorative methodology that aims to draw out a systematic understanding in this connection and thus promising. As growing evidence suggest that climate change related impacts drive communities such as the farmers, fishing communities and others who depend on such environmentally sensitive activities into poverty. Poverty with compounded vulnerabilities further reinforce people susceptible to trafficking. Women and children are the most affected in any adverse situation. Furthermore, we focus our discussion on the form of human trafficking that involves women and girls who engage in private care services (PCS) and commercial sexual services (CSS)[5] for the following reasons: (1) theoretical discussions are most advanced in these categories, (2) empirical information are most available compared to that of involving men and boys and (3) all available statistical projection indicates that the majority of those in modern slavery are female.

---

[3] Movement of refugees and other persons of concern take place within mixed migratory flows. State-prosecution of the ethnic minority groups in Burma led incidents of stateless Rohingya women refugees trafficked for marriage (Kojima, 2015) and Shan girls and women trafficked for sexual exploitation (SWAN, 2002) in Thailand, Malaysia and beyond for instance.

[4] This is to reflect what UNHCR (2009) repeatedly urges in recent years that refugees and other persons of concern who travel through mixed migratory flows increasingly have fallen victim of human trafficking thereto and/or are vulnerable to becoming victims of human trafficking during the process of their movement.

[5] Literature on female migration suggest that there are three types of migrant labour engaged in PCS and CSS. They are person in legal migration for oversea employment, those in undocumented migration and those in migration associated with conjugal relations (mail-order brides, refugee women seeking asylum overseas through marriage arrangement) or study overseas (au pairs).

17.3 Vulnerabilities in Migration-Trafficking Continuum-Theoretical Debates

This paper recognizes that human trafficking for PCS and CSS as a systemic problem drawing upon the theoretical account that views the political economic structure and entailing cultural valuing mechanism related to domestic labour[6] as the key determinant of exploitation and discrimination against migrant women. Against this background, review of existing theoretical discussions denotes that conventional policy measures have failed to recognize the compound vulnerabilities of migrants in PCS and CSS that are attributed to: (1) ambivalent value and status of domestic labour as work (knowledge), (2) narrowly defined victimhood adopted in the law enforcement and judicial system that does not reflect emerging complex material and subjective conditions that women face in migration (instrumental and management models) and (3) androcentric and racist normative basis identified within migration policy intervention and market and labour management practices related to PCS and CSS (norms). From the perspective of institutional analysis, we recognize these vulnerabilities are cross-cutting features of the five policy blind spots we have previously described that prevent effective mitigation of policy trade-offs between trafficking of prevention and labour migration promotion. In principle, the lesser the government engage with blind spots, the extent of efforts required to mitigate policy trade-off be more laborious.

Let us begin our discussion by asking one simple question—what is the moral behind the current international female migration policy? Are they for the equality of opportunity or equality of outcome for women? Familiarity in theoretical discussion on the role of migration and gender equality is crucial as it provides the larger context in which a nuanced understanding on female migrants' vulnerabilities in relation to the scope of women's agency and its policy implication should be grounded. In this connection, gender racism (Essed, 1991) is a useful concept to navigate our discussion on the inequities produced by both racist and sexist policy conflated. Migrant labour precarity (Landolt & Goldring, 2015; Silvey & Parreñas, 2019) in the context of PCS and CSS is first and foremost attributed to the fact that institutionalization of PCS and CSS as migratory work relies on two sub-sets of norm hierarchies in which cultural valuing of domestic labour is infused by the understanding of differences in gender and race.

[6] The concept of "domestic labour" arose through the historical struggles of the feminist scholars in 1960s to challenge this essentialist assumptions on women's engagement with work in the home. Domestic labour is mainly composed of two separate subsystems of care and sex. The Care system represents the labour required to care for young and old people and for the social maintenance of labour that is predominantly adult male. Sex system on the other hand, fulfils the biological needs of human procreation and is shaped and ruled by a particular set of mechanisms and arrangements that should be satisfied in a socially consensual manner (Rubin, 1975). Drawing on Marxist analysis, domestic labour as a concept wage for revaluing women's labour performed in the home as value-producing in a capitalist economy (Smith, 1978; Windebank, 2001).

Trajectory of feminist political theories in this respect offer rich insights on how ambivalent value and status of domestic labour is a product of knowledge construction and on politics evolved around the public /private dichotomy in which the family as institution was given the least political significance in theoretical discussion (Okin, 1998, 1990; Pateman, 1983; Squires, 1999). Historically, domestic labour performed by women has been marginalized within the liberalism discussions, leaving it as a biological characteristic of femininity and therefore unrecognized- an orphan without sufficient theoretical references and political debates within the policy domain (Agustin, 2003; Himmelweit, 2002). This conceptual marginalization in political economic theory has been translated into the state policy to exclude domestic labour from the public work category.

Under the male-defined political economic landscape, gendered norms dominant in the labour market justifies devaluing of women's labour (Lee, 1996; Macklin, 2003; Yea, 2004), resulting in domestic labour being systematically underappreciated and unregulated in the process of gender capitalism. Propositions by Hooper (2000) and Hearn and Collinson (1994) further offer some intriguing insights in this connection by pointing out how gender norms combined with other social traits reinforce certain human behaviour and practices masculine or feminine while such processes define entailing cultural values and the degrees of power associated with activities performed by both men and women. It may be then fair to assume that domestic labour is identified as the central element of what shapes feminine identity while the cultural value of domestic labour may shift according to who performs such labour (Kojima, 2001). In this vein, the promotion of domestic labour through international labour migration policy is viewed as global restructuring processes of gender capitalism and racial capitalism (Kendi, 2019) combined, in which gauging of its value is determined by working through perceptions on gender, class and historically defined ethno-racial hierarchy. Consequently, the associated cultural valuing mechanism that not only undervalues domestic labour but promotes the practice of determining its labour value based on the racialized identity of the person who performs domestic labour offers a justifiable ground for the institutionally based gender and racial discriminatory practice.

In commercial relations, female migrants from less advanced economies are strategically placed in a niche market of PCS and CSS that promises satisfaction of emerging expectations that cannot be met by conventional service suppliers. This is a common strategy adopted by the agents to beat the competition in the market whereas compromises on working conditions and remuneration have limits. What is presented as an added value is often gratification of suggestive racialized traits related to ethnicity, behaviour, body and colour. Racist and sexist assumptions arise from ethnicity/nationality determines the rank of female migrants and they are translated into unequal working conditions among female migrant workers in PCS and CSS. These racist ideas are engineered by colourism[7] which is powerful in defining perceptions of what is a "desirable" employee (Tyner, 1996) or woman in large.

[7] Kendi (2019) defines "colourism is a collection of racist policies that cause inequities between Light people and Dark people, and these inequities are substantiated by racist ideas about Light

These racist ideas are fictional, but they have become powerful norms to sustain the institutional racial inequities among female migrants in PCS and CSS.

Contribution by contemporary feminist theories has advanced our understanding further by incorporating rich empirical findings to demonstrate how the domestic labour requires a constant reinforcement of gendered and racialized processes of othering to overcome its legal ambiguity as work. These studies also offer sound evidence that othering as the idea has played a major catalyst role in forging varied links between the neoliberal marketing and the formation of women's agency. For instance, rationalization of otherness of female migrants is a valid and widely adopted concept in labour management regulations and strategies (Wong, 1996; Yeoh & Huang, 1998) and recruitment agencies (Cheng, 2003; Tyner, 1996). The idea also defines the scope and nature of female employers' agency in maintaining the migrant labour. Often migrant women find themselves not regarded as an equal peer by female employers in whose interactions are driven by the idea of maternalism (Anderson, 2001; Constable, 2002; Human Rights Watch, 2004) or their personal desire to materialize a sense of social upward mobility through hiring a maid (Sabban, 2002; Suzuki, 2003). Further it impacts the self-image of female migrants themselves. A study conducted in Hong Kong found out that more than 60% of foreign domestic workers felt they were discriminated due to their occupation. The idea of othering distorts the self-image of foreign migrant women and forces them to accept abusive treatment by employers and clients (Asian Migration Centre, 2001).

These literatures offer complimentary evidence to an idea that support gender racism centred around domestic labour is indeed a prerequisite for the business principle and the main thread of migrant labour management for PCS and CSS. This structural flaw however has been disguised by the rhetoric of women's advancement where the role of women is elevated as an international wage earner to support their family and contribute to the overall welfare of the country through remittance. This seemingly an egalitarian idea could be traced to what body of feminist theory depicts as an 'equality' approach which sees achieving the gender-neutrality is the key to overcome gender inequality. In this respect, migration could be regarded as a device for female migrants to pursuit what Squires (1999) points out the importance of a procedural approach in attaining the equality. Equality feminist theorists stress that a just society prioritizes equality of opportunity rather that outcome of equality. In this vein, individuals, regardless to gender, can equally enjoy their full potential to pursuit their projects while everyone's outcome of the project does not necessarily be the same (ibid, p.120).

Under the bilateral agreements, however, governments of both labour sending and importing countries must accommodate paradoxical circumstance where the foreign migrant labour must be regulated, sometimes by creating the double labour standards for their own national labour while keeping the status of domestic labour as unregulated work (Kojima, 2011). Series of weak protection schemes and bureaucratic and often unrealistic legal redress and compensation procedures administered by the

and Dark People. Colourist ideas are also assimilationist ideas, encouraging assimilation into- or transformation into something close to—the White body" (ibid, p.110).

labour-sending governments suggests that official labour migration schemes often suffer from a charade of legality where once women are given a chance to migrate overseas, they are left alone to strive for recognition in society where it is designed for those who could assimilate with male qualities could only survive (Perry & Gesualdi-Facteau, 2021). This is because the normative basis of the conventional migration policy by default is not designed to appreciate women's authentic qualities that are prerequisite skills to perform domestic labour. When exclusion of women is the default, it is no longer gender-neutral but, androcentric. Policy for migrant work for domestic labour has been disastrous precisely because much of focus has been spared for ensuring the equal opportunity to migrate for women but without careful attention being paid to scenarios when gender and sector-specific vulnerabilities heighten to the point where migrant women are subject to trafficking episodes. Further theoretical and policy implications can be drawn from this proposition to highlight serious conflation of migrant women's agency and policy norms.

While repeated emphasis on global effective implementation of a victim-oriented approach continues as evident in the form of Trafficking in Persons Annual Report (U.S. Department of State, 2020), gender and racial bias and social stigmatization has been yet internalized in law enforcement authorities and judicial system, rendering repressive implications on migrants in PCS and CSS as often manifest in the process of migrant screening to identify a trafficking survivor. Failure to recognizing emerging rights violation of migrant women in PCS and CSS must be recognized in the context of gaps in normative basis related to standards and institutional practices ranging from the global (UN conventions) versus national, central and reginal within the national framework. In this premise, the dominant policy instruments known as a victim-criminal paradigm framework and a victim-oriented judicial approach have been currently serving mainly its instrumental function attributed to the human rights strategy- in provision of legal resources and political instruments for legitimacy (Fortman, 2003). Continued reporting on malpractice of law enforcement authorities and the judiciary i.e. corruption and arbitrary implementation of rights-protection measures in litigation procedures points out that these instruments are ineffective to mitigate structural injustice and thus transformational function of human rights strategy is yet fully realized (Kojima, 2007).

This shortcoming is attributed to the instrumentalized idea of victimhood which has been driven by over emphasis on the prosecution of traffickers and failing to embrace material and subjective conditions of migration that trafficked women undergo. Current migration screening system neglects empirical knowledge that women in documented and undocumented migration in fact undergo a process of migration that is not very different in substance (Chantavanich et al., 2001; Hondagneu-Sotelo, 2001; Thorbek & Pattanaik, 2002). Further, current patterns of trafficking demonstrate complex characteristics which increasingly denies conventional understanding that trafficking is one fixed process, and it features throughout the entire migration experience.[8] Present forms of trafficking adopt and synchronize

[8] Following the sequential flow of migratory experience of women and girls in Asia for instance, reveals that due to fluid and unregulated employment patterns, management practices combined

with women's life cycle and formation of their agency. The migrant vulnerability in PCS and CSS is an outcome of structural (law and policy) and subjective (identity) social forces intersected where the position of individual migrant is constantly redefined by decisions she makes to leave or remain in a situation at each migratory stage in which self-image (how each migrant perceives herself) and the threshold of tolerance (each migrant's standards as to what is morally acceptable to settle during the migration experience) play key roles (Kojima, 2011). The scope of their agency, choices and preferences[9] therefore play a key role in defining them either as a victim of trafficking or a criminal of violating migration law (Kojima, 2007; Pearson, 2002a; Roces, 2009).

In sum, shortcomings of current counter-trafficking policy measures identified in the safeguard measures within the official labour migration schemes (prevention and protection) and victim-criminal framework that underpins migrant screening procedures (prosecution and protection) must be addressed through a critical review of current dominant public–private divide paradigm that discourages a fuller policy understanding regarding the relationship between women's agency and migration, its connectivity with trafficking and its overall implication on gender equality. In this respect, a lack of substantial theoretical discussion on blind spots related to MIS denotes that there is an urgent need for the framework which ensures assessment of the present status and monitor the progress of counter-trafficking efforts that reflect three areas of blind spots representing gaps in norms, knowledge and management models. We argue that the index construction through typology analysis of trade-offs would appropriate this need.

17.4 Institutional Trajectory Analysis in the Context of Migration Governance

Migration governance as an analytical concept enables us to map out the institutional functions among members of stakeholders and their interplay with individual policy initiatives and thus more resourceful. Mindful that the maintenance of the current form of migration governance relies significantly on the contribution provided by the private service providers i.e. NGOs should not replace the government's prime role

with marginalized socio-economic status adhered to migrants engaged in PSC and CSS, growing numbers of female migrants find themselves in situations where they are trapped in trafficking episode among all other employment that they undergo in overall migration process. It hence points out that trafficking occurs in the continuum of the overall migration process (Kojima, 2011).

[9] As much as it matters to secure financial gains, migrant women spend their time away from home, anxiously exploring how migration could provide with autonomous, independent and sometimes adventurous experience for their enhanced identity and capacity building without legal ambiguity (Agustin, 2003; Roces, 2009). Increasingly recent empirical evidence also suggest that migrant women could be both victims and agents while their prime concern is their migrant status in relation to the institutional surrounding, not organizational concerns related to prostitution (Agustin, 2003; Outshoorn, 2004; Roces, 2009).

in making transparent policy choices to mediate trade-offs in migration-trafficking policy nexus. Hence, the government accountability should come to the fore. This is the reason why our analysis primarily focuses on the role and function of the public agency. An inquiry into the characteristics and functions of institutions necessitates us to ask a fundamental question: What is a normative framework for policy and planning in general term? In principle, norms are shared values while institutions are the set of rules and organizations are public agencies that implement rules which are shaped by norms. Further, institutions have two faces: one is related to legislation and the policy directives - we call this institutional environment, and the other is related to institutional arrangements-to enhance the capacity of government to effectively translate legislation and policy directives via a focus on institutional arrangements.

Let us then examine how this mechanism works in the context of trafficking in women and girls, with a simple positive norm. Generally speaking, many agree that trafficking in women and girls is inhumane and should not be tolerated; it should be eliminated by the government's efforts. These shared values are considered "norms" of the population. This may be reflected in the central legislature passing a Human Trafficking Bill which is composed of anti-prostitution law, Violence Against Women Act together with an issue of the directive on inter-agency collaboration to eliminate human trafficking (institutional environment). Further, the establishment of a government-owned women's shelter and focal point within relevant ministries are some examples of "institutional arrangements" that can promote an institutional environment that discourages human trafficking. This set of rules is "institutions" and should not be mistaken for the brick-and-mortar government offices. Finally, law enforcement authorities and government mandated NGOs are the ones who are involved in implementing the government policy directives and legislation. They are referred to as "organizations".

Evidentially, few dispute to the idea that the norm is regarded as the most crucial aspect of the overall process of institutionalization. Norms are translated into legislation, regulations, policy and operational norms which define how people involved in migration/trafficking should be treated institutionally. Norms shape the behaviour and attitudes of organizations (government offices, judicial/law enforcement authorities, employer Associations, clients, recruitment agencies etc.) which interact with migrants/trafficked persons. Perceptions and behaviours of people concerned themselves are also shaped by these norms. Whereas the public agency normally sets forth rules and the guidance in place, institutional norms are yet open for different interpretation and implementation by agents (people who work in organizations) as evident in the common use of discretion by officers in decision-making process.

Existing literature further suggests that there is a gulf between a universal normative framework (as represented in the form of international conventions) and dominant social norms which surface as the policy challenges in a wide spectrum involving the counter-trafficking measures engaged by the public agencies at national level. These norms often stem from stereotypical gender values and racist assumptions. Table 17.1 summarizes these gaps from the two contrasting world perspectives. Recent study conducted by Walk Free (2019) provides the assessment of government progress on SDG 8.7 offers insightful findings in this connection. Overall, 47 countries out of 183 have not yet criminalized human trafficking in line with the definitions provided by the UN Trafficking Protocol while 96 countries have not criminalized forced labour (ibid, p.1).

Table 17.1 Summary of dominant social norms in human trafficking policy environment

Policy world view	TIPs as victim versus criminal	Real world scenario
Status as victim of human trafficking		Status as criminal of illegal migration
Dominant social/working norms supported and practiced in the area of		
People should migrate cross country only legally	Migration	Foreign migrants who engage in unregularized industries/ certain nationality have migrated illegally
Migrants rights should be protected	Labour	Foreign migrants are willing victims of labour exploitative working conditions
Screening of trafficked persons should be carried out in a gender sensitized manner, and they should not be mistreated under the police custody or detained	Law enforcement authorities (immigration and police)	– All foreign migrants are criminal and do not have rights and should be punished and detained – All female trafficked victims worked in sex industry and require no respect
– Any foreign migrants, including victims of trafficking must be protected their basic rights and dignity in line with domestic law and international guidelines; – Boys and men working in non-sex industry could be victimized by trafficking	Judicial system	– All foreign migrants are criminal or those exploited are willing victims; only those who can provide hard evidence of trafficking should be protected by law – Only women and girls in sex industry are trafficked victims
Women are vulnerable to risky migration, and they should be provided support during post-trafficking phase	Health/social welfare	– Sex trafficking victims require rehabilitation – Women in need should be prioritized due to limited resources (minor citizen of origin first)
Women are vulnerable to risky migration, and they should be protected from human trafficking and should receive support for reintegration	Community	– Women should be held responsible for engaging in risky migration and victimized by trafficking – Female victims are fallen women and should be stigmatized

TIPs Trafficking in persons
Source ACIL (2004), Anti-Slavery International (2006), Human Rights Watch (2004), Field notes and interviews (1994–2022), Pearson (2002b), PHR (2004)

Trafficking survivors are often subject to repressive treatment even in the case where key conventions are duly ratified by individual governments due to lack of capacity, resources and foremost the political commitment. This trend is particularly evident in the areas of migration screening process, implementation of anti-trafficking legislations in judicial system (GAATW, STV & International Human Rights Law Group, 1999; Huda, 2006; Kojima, 2011; Pearson, 2002b) and regulations that define limited accessibility and eligibility of programme interventions involving donor funding practices (Clancey et al., 2014; Hoff, 2014; Surtees & De Kerchove, 2014). In sum, gender and racial injustice is a cross-cutting structural feature of human trafficking and it is prevalent at the levels of norms, institutions and organizations. We argue thus this understanding must be incorporated as the central component of our index construction.

17.4.1 SDG Targets and Indicators on Human Trafficking—Why Indexing Helps?

In this section, we will demonstrate conceptual framework of the institutional trajectory analysis, which involves policy trade-offs and typology analysis and indexing. To this end, we discuss a rationale for the proposed Index development in reference to our analysis on challenges related to U.N. SDG targets and their indicators—5 (gender equality), 8(decent work),10 (migration) and 16 (justice) which address pressing policy blind spots related to human trafficking- in three-fold. Table 17.2 provides a snapshot of these findings. First, there is a lack of reliable and basic statistical overview of the problem in general. Some selection of indicators reflects organizational need to grasp the magnitude of the current status and establish mapping of the status as the priority rather than monitoring of the progress of the changes itself.

Further, there is a strong instrumentalist view that is translated into the selection of indicators that neither realistic nor substantive to measure the magnitude of the policy challenges. For instance, access the level of protection of migrants' labour rights through the number of occupational injuries cases (8.8.1) and national legislation (8.8.2) is not only limited but also not sufficient to capture the dynamics entailing the nature of the industries (entertainment, care work and agrofishery industries) that many migrants engage in such industries. Mindful of ideological strife that has been lingering for decades around the issue (Huda, 2006), furthermore, a political choice to select indicators only for the form of child trafficking while excluding trafficking as the core indicators for VAW (Jansen, 2012, p.6) is highly discouraging. It dilutes the political willingness to effectively address the long-lasting ideological disputes on the agency-victimhood debates on prostitution. Further, 'safe' migration is a contested concept as evidence suggest that the government-led safe migration scheme de facto encourages trafficking of legally recruited migrant workers (Target 10.7.2).

Table 17.2 SDG targets and indicators on human trafficking: underlying assumptions (Source: UN SDGs with authors'original analysis)

SDGs number	Thematic Goal	Specifically mentioning trafficking related to	Indicators	Underlying assumptions	Aspect of trafficking vulnerability related to	Remarks
5.2	Gender equality and empowerment of women and girls	Women	5.2.1 Proportion of ever-partnered women and girls aged 15 years and older subjected to physical, sexual and psychological violence by a current/former intimate partner in the previous 12 months, by form of violence and by age	DV as the most common VAW form across the regions	Women's rights/ VAW	– Officially, trafficking is excluded from the core indicator selection – Conceptual unclarity: lack of recognition that trafficking takes forms along the life cycle of and in association with gender roles assigned to girls and women
			5.2.2. Proportion of ever-partnered women and girls aged 15 years and older subjected to sexual violence by persons other than an intimate partner in the previous 12 months, by age and place of occurrence	Sex work/ victim-agency debate is too controversial and thus be avoided	Women's rights/ VAW	
5.3		N.A	5.3.1 Proportion of women aged 20–24 years who were married or in a union before age 15 and before age 18	Early, forced and child marriage is not linked with trafficking	Women's rights/ VAW	
5.4		N.A	5.4.1 Proportion of time spent on unpaid domestic and care work, by sex, age and location	(Economic) re-valuation of reproductive work (CSS/PCS) without link between migrant women's contribution to reproductive work	Gendered traditional roles and values	

(continued)

Table 17.2 (continued)

SDGs number	Thematic Goal	Specifically mentioning trafficking related to	Indicators	Underlying assumptions	Aspect of trafficking vulnerability related to	Remarks
8.7	Decent work for all	General description (human trafficking)	8.7.1 Proportion and number of children aged 5–17 years engaged in child labour, by sex and age	Elimination of child labour Child labour is the least disputed form of modern slavery (no agency debate involved) and the easiest to address politically	Child protection/ child rights	Political decision to select child trafficking as a primary indicator costs the SDG exercise unable to address the core policy challenges highly relevant to human trafficking
8.8	Decent work for all	N.A	8.8.1 Fatal and non-fatal occupational injuries per 100,000 workers, by sex and migrant status	Migrant would seek medical treatment for their occupational injuries as normalcy	Labour rights of migrants, particularly women migrants	The assumption is not realistic given the nature of work sectors where many undocumented and female migrants engage in, and thus the relevancy of the selected indicator is questionable
		N.A	8.8.2 level of national compliance with labour rights (freedom of association and collective bargaining) based on ILO textual sources and national legislation, by sex and migrant status	Labour exploitation could be eliminated through ensuring national legislation	Labour rights/ elimination of unregulated work categories	Selected indicator alone does not fully reflect the reality arising from instrument versus practice gaps

(continued)

Table 17.2 (continued)

SDGs number	Thematic Goal	Specifically mentioning trafficking related to	Indicators	Underlying assumptions	Aspect of trafficking vulnerability related to	Remarks
10.7	Reduce inequality within and among countries	N.A	10.7.2 Number of countries with migration policies that facilitate orderly, safe, regular and responsible migration and mobility of people	Safe migration policy equates with elimination of trafficking phenomenon	Migration regimes/ institutional policy	Conceptual review is required from the empirical perspective that safe/ government-led migration schemes de facto encourage trafficking of legally recruited migrants
16.2	Justice for all	Children	16.2.2 Number of victims of human trafficking per 100,000 population, by sex, age and form of exploitation	Overlaps with 8.7 child labour	Child protection /child rights	Same as 8.7
			16.2.3. Proportion of young women and men aged 18–29 years who experienced sexual violence by age 18	Overlaps with 8.7 child labour	Child protection /child rights	Same as 8.7

Selection of narrowly focused and insufficient indicators may be underlined by a larger misconception by policy makers that refuse to embrace emerging realities of human trafficking in present time. Limited understanding resulted from failure to recognize: (1) the material and subjective contexts and processes of present migration that nexus with human trafficking (2) marginalized position of domestic labour as migrant work in the current neoliberal global policy. In this connection, our second point stresses that there is an acute fragmentation and compartmentalization of trafficking-migration policy initiatives. This may have occurred due to: (1) out of sync knowledge gaps combined with (2) political decision to maintain an interventionist approach to keep the territoriality of custodiam agencies responsible for individual targets intact.

Subsequently, policy challenges related to human trafficking have been dismantled along with thematic focuses in the area of gender, labour, migration and justice as evident in the way SDG targets and intervention strategies are organized. It thus lacks a holistic approach. Given that under valuation of reproductive work serves as the basis for labour exploitation of women migrant workers in domestic and care services, for instance, it would make much sharper and solid intervention should the Target 5.4 (revaluation of unpaid care and domestic work/UNWOMEN) be combined with the Target 8.8.2 (ILO as a custodian agency) which call for consolidation of migrant labour rights.

Third, there is a lack of reliable and effective data management system. Despite of the series of high-level consultations and research initiatives involving the scientific community and international organizations for several decades, given the sensitivity of the problem, empirical data collection has been yet facing serious ethical, methodological and financial challenges. Its mundane policy dialogue environment also has failed to overcome the technical issue of lack of reliable statistical data through a more cost-effective, inclusive and creative research methods.

Without having able to establish benchmark to fairly grasp the magnitude of the problem, it explains why the conventional programme interventions by the international organizations and donors have been ineffective in mitigating human trafficking/exploitative migration. Patterns and characteristics of past interventions also hint that there is an urgent need for a paradigmatic change in how we engage with the problem. These shortcomings have offered a solid starting point for our exploration of the indexing initiative. Our proposed index construction is based on the idea that citizen science is a mathematical expression of valuable relationships. It is a revolutionary tool which promises to overcome these challenges as it is drawn from the data-light approach while it is inclusive and cost-effective to carry out compared to conventional data collection method. More details will be discussed in the subsequent section.

17.5 Typology of Policy Trade-Offs in the Context of Human Trafficking and Migration Nexus

To begin, migration typology analysis bears some characteristics and analytical limitations. Current human trafficking involving women in PCS and CSS point to patterns that women undergo combination of more than one migration system as the result of business management strategies while such transfer may also occur in attempts to overcome precarious visa status through the marriage of convenience and long-term settlement of other kinds. Female migration systems are thus a hybrid and highly subject to change in response to the political economic environment (Lee, 1996) and the typology analysis should be mindful of this point. Further, there is a need to devise the new way of analysis to overcome the conceptual limitation attributed to the conventional migration system framework.

One striking feature we could recognize immediately from the SDG target analysis may be the acute compartmentalization of issues surrounding migration governance. Table 17.3 summarizes typology analysis of seven key areas of public policy/programming interventions by general intervention options. It also identifies dominant trade-offs that each sector is driven by. A quick review of individual policy trade-offs informs that they are split between a general idea to support either judicial/ institutional efficiency or gender/ethno-racial equity of trafficked persons. What does it mean in a more policy-friendly term? A closer look into each policy trade-off makes one realize overlapping nature of some trade-offs that could be clustered into two dominant groups of policy trade-offs. They are: (a) prevention of human trafficking versus promotion of migration and (b) prosecution of traffickers versus protection of rights of trafficked persons.

Furthermore, Table 17.4 illustrates an analysis on the intensity of trade-offs which shows the levels of effective policy mitigation of human trafficking is contingent on the balancing of trade-offs between these two dominant trade-offs. At a glance, one may consider what we define high or low intensity here is disputable. However, what we intent to underscore here is the systematic framework to compile the analysis of trade of intensity -the level of tension or dilemma that the individual government may undergo between two selected policy options and how much such tension stresses the institutional function in terms of revenue and expenses. Under this premise, we argue that counter-trafficking policy that costs the individuals' right to migrate has a repressive implication for the government as it must forego a vital revenue source-the remittances. It also violates the global normative consensus set forth by the UN Migrant Convention. Policy status as such, therefore, is considered high intensity.

From visualizing trade-off analysis by timeline of the policy goal, whether it is a short or long-term, empirical evidence suggests that the present common policy strategies adopted by the individual governments are primarily the short-term strategy in which human trafficking incidence is high. Through comparing scenario between short and long term-strategies, one could also realize that short-term migration strategies (based on prioritization of promotion of migration over prevention of trafficking) goes hand in hand with prosecution of traffickers over protection of trafficked person's

Table 17.3 Typology of policy trade-offs by sector

Areas of policy/programming intervention	General intervention option	Example of trade-off	Typology considerations between
Law	Technical support for legal instruments/judicial system: counter-trafficking/victim-oriented legal measures	Judicial efficiency versus equity	Urban–rural/labour exploitation—sexual exploitation
Law enforcement authorities	Gender sensitization and pro-migrant rights advocacy training for government officers/professionals	National security versus gender/ethno-racial equity	Native survivors—foreign survivors/native majority—thnic minority/female-male/adults-minors
Labour	Technical assistance for migrant labour protection/anti-unregulated work categories measures: legal instruments and sensitization training/Migrants Resource Centre	Economic advancement versus labour rights protection	Urban–rural/native labour—foreign labour
Migration	Technical assistance for government-led Safe migration scheme/programmes	Economic development versus migrant rights protection	Urban–rural/female-oriented migrant scheme-male-oriented migrant scheme
Health/social welfare	Technical support for institutional updates: women's shelters/referral systems	Women's socio-economic advancement versus women's well-being and safety	Urban–rural/adult women-minors/sex work-non sex work
Livelihoods	Income/employment-generating projects for vulnerable communities	Limited economic sustainability/life options versus migration	Urban–rural/technical-nontechnical job categories/single-married status/gender, ethno-racial and class
Public advocacy	Government-led advocacy campaign and activities through media and arts/online resource platforms for networking and collaboration	Economic development versus anti-migration	Urban–rural/

Source Compiled from UN and donor reports; Field notes and interviews (1994–2022)

Table 17.4 Policy trade-off dynamics on human trafficking (HT) mitigation (Source: Filed notes and interviews 1994-2022)

Overall policy trade-off in counter-trafficking in Women				
Judicial and institutional efficiency versus gender/ethno-racial equity of trafficked				
Prevention of human trafficking versus promotion of migration		Trade-off between	Prosecution of traffickers versus protection of the rights of trafficked persons	
Trade-off intensity (general scenario)				
HT is highly prevented so that migration opportunity for women is highly restricted		High	Prosecution of traffickers is highly successful, using trafficked persons as a tool only (no protection/benefits for trafficked/ criminalization of trafficked occur due to gender/ethno-racial discrimination)	
HT is moderately prevented while women migrate through both risky and not risky paths		Medium	Modest prosecution of traffickers while trafficked persons receive moderate protection	
HT is not at all prevented and women are encouraged to migrate at any costs, resulting in mostly through risky migration paths		Low	Prosecution of traffickers occurs seldom; trafficked persons receive the minimum protection	
– Counter-trafficking legal measures/ procedures – Availability of resources (staff and budget) – Policy norm that recognizes migration as formal livelihoods strategy – Gender sensitive migration policy/pathways		Factors that shape trade-off intensity (Typology)	– Profile of trafficked (nationality—native versus foreign migrant, age-adult versus minor, job category- sex versus non-sex work) – Nationality of traffickers	
Prevention of trafficking < promotion of migration		Characteristics of short-term strategy	Prosecution of traffickers > protection of trafficked persons	
Pros	Cons		Pros	Cons
– Remittances as government revenue – High demand of labour (more employment opportunity)	– More migrant labour abuse – More incidence of HT – No market advantage to negotiate terms		– Effective and high prosecution of traffickers	– Rights of trafficked not respected – Victimization/ stigmatization of trafficked reinforced – Poor recognition of trends of HT/needs of trafficked by judiciary resulting in trafficked could be wrongly prosecuted as criminal
Prevention of trafficking > promotion of migration		Characteristics of long-term strategy	Prosecution of traffickers < protection of trafficked persons	
Pros	Cons		Pros	Cons
– Migrant labour protected in destination – Less incidence of HT	– Limited remittances – Low demand of labour (less employment opportunities)		– Rights and dignity of trafficked restored – Better recognition of needs/trends of HT by the judiciary sector	– Proceedings may take longer time/ require more resources

rights while the maximization of development impacts of labour migration identified as the background policy intent. From this perspective, long-term strategies on the other hand, demand more human and financial resources, raising a question over the fundamental idea of labour migration in PCS and CSS that may not be economically valid any longer. Hence, the conflicting aspect of the current dominant labour migration policy (Lim & Oishi, 1996) is exposed.

We must stress here that trade-off analysis be best carried out in the context of functioning states rather than non-functioning states due to civil war or conflict. While trade-off analysis would give you a general picture, typology analysis compliments it by providing contextual characteristics. Typology analysis would yield more enriching results when it is taken into the consideration of what we call seven boundary conditions. Features of migration governance would differ depending on norms that embedded within the politico-administrative regimes (Pollitt & Bouckeart, 2000): (1) state structure, (2) the degree of autonomy, (3) extent of coordination, (4) level of accountability (5) form of fiscal mechanisms (6) nature of electoral systems and (7) consistency of party voting behaviour on a range of policy options (Kurian & Kojima, 2021, 30). Analysis mindful of these conditions may provide: (1) a refreshing way to re-examine fundamental features of migration governance beyond the political ideology debates and (2) rich contexts and nuanced understanding on contrasting world views that enhance indicators selection and weights allocation discussion in indexing process, and thus pragmatic and more resourceful to policy making and programming processes.

17.6 Delineating Institutional Failure via Typology Analysis: Characteristics of the Policy Synergies in the Context of Counter-Trafficking Interventions

Managing critical trade-offs is crucial to maintain resilience of the migration governance to effectively and timely responds to emerging trends of and changes in labour markets and needs of migrants. Through acute compartmentalization of institutional functions, migration governance is however characterized by sectoral siloes leading to overlapping mandates and functions among members of public agencies. Uncoordinated planning and implementation combined with limited human and financial resources (Dottridge, 2014;) suggests there are several missing links within the mechanism of government's efforts to mitigate such trade-offs. Table 17.5 below summarizes key characteristics in this regard.

Borrowing from the institutional analysis of environmental governance (Batley, 2004; Gregory, 1997; Kurian & Kojima, 2021), key components of (1) information and coordination (2) human resource management and (3) technology and finance are equally highly relevant to elaborate more on this inter-connectivity between the level of mitigation of trade-offs and a sectoral-bound resource distribution of

Table 17.5 From institutional failure to synergies pathways

Current policy world view (uni-dimensional, linear, recursive)			Conditions	Real World scenario (multi-dimensional, non-linear, non-recursive)		
Finance and technology	Human resources	Information and coordination		Finance and technology	Human resources	Information and coordination
One policy option, one outcome – Victim versus criminal policy framework – Simple victimhood: HT victims are girls in sex industry only – Legal solution (prosecution of traffickers) is the foremost priority	*Skills and roles synchronized* – Well cross-coordination of operations among the police, the public welfare office and private organizations	*Pre-determined, uni-dimensional information required only* – Information to effectively prosecute traffickers required only	(Uni)/multi-dimensional policy framework	*Financing and policy option not synchronized (Access to budget with no/wrong policy choice; policy identified but no access to budget)* – Migrant screening instruments do not embrace well trafficking for forced labour/boys and men as victims	*Skills and roles not synchronized* – Limited capacity to implement pro-victim approach strategies due to gap of understanding on the standardized measures for cooperation	*Multi-faceted information not collected* – Information on HT is overly focused on sex trafficking involving women and girls – Information is limited on integration/post HT phases – Hub governments focus on information collection on HT involving foreigners (national security)

(continued)

Table 17.5 (continued)

Current policy world view (uni-dimensional, linear, recursive)			Conditions	Real World scenario (multi-dimensional, non-linear, non-recursive)		
Finance and technology	Human resources	Information and coordination		Finance and technology	Human resources	Information and coordination
No budget constraint for most advanced policy choice Non-classic trafficked survivors are also included as beneficiary	*Incentive and budget structure aligned* Police are secured with adequate facilities and funds to investigate any suspicious case Shelters are well equipped with both financial and human resources	*Project/budget cycle aligned* New procedures on non-conventional trafficking cases i.e. fishery involving men and boys, introduced by a donor project be smoothly integrated into the existing operational system	(Non)linear Time/phase bound sequencing in budgeting/ financing	*Budget constraints* Use of own's salary for preliminary investigation before it is officially filed as a common police practice	*Incentive and budget structure not aligned* – No effective impact/ commitment of government funding on sensitization trainings (100% funded by foreign aid) – Rent-seeking behaviour by low level functionaries in LEA and SP – Low level functionaries are not consulted with introduction of new procedures	*Project/budget cycle not aligned* – Government fails to respond to emerging policy challenges due to unavailability of foreign aid funded budget

(continued)

Table 17.5 (continued)

Current policy world view (uni-dimensional, linear, recursive)			Conditions	Real World scenario (multi-dimensional, non-linear, non-recursive)		
Finance and technology	Human resources	Information and coordination		Finance and technology	Human resources	Information and coordination
Feedback loop between expenditure and outcomes are robust – Budget allocation is updated in response to emerging HT patterns and policy challenges	*Unintended effects of policy mitigated* – Staff endowed with rich resources to handle unexpected policy effects	*Interventions are responsive to feedback* – Counter-trafficking interventions reflect the emerging needs of trafficked	(Non)recursive Organizational characteristics related to information updating and coordination	*Feedback loop between expenditure and outcomes non-existence/weak* – Weak accountability mechanism on government infrastructure for counter-trafficking related expenditure	*Unintended effects of policy exacerbated* – Lack of policy recognition on effects of challenges related to trafficking-migration continuum in both unregulated and safe migration path	*Interventions are not responsive to feedback* – Budget allocation is not updated in response to emerging HT patterns and policy challenges – Inefficiency in redress procedure for HT survivors: too many channels for grievance mechanism by adding new systems

Source Adopted from Kurian & Kojima (2021, p. 104), Field notes and interviews (1994–2022), The US Trafficking in Persons Report (2001–2024) public agency in the case of counter-trafficking interventions. Siloes in information and knowledge sharing refers to an institutional state where no sharing information results in preventing cooperation among public agencies and their staff and thus networking and coordination across and among public agencies is must. Synchronization of skills and roles within public agencies reflect how different functionaries are effectively committed to an instructional environment that promotes adequate staffing, capacity building and policy support that coordinate decision-making. Further, when the public agency can afford to choose right-fit technology/policy option that ensures desired social outcome by utilizing the available financial resources, robust institutional synergies are realized (ibid, p. 6–7).

Table 17.5 offers an experimental analysis in this connection. It illustrates a simplified but yet common status of two contrasting conditions related to the institutional synergies identified using conventional assumptions in policy realm and a real-world scenario. Therein, three elements of institutional functions (technology and finance; human resources and information and coordination) are assessed in relation to conditions in the form of uni/multi-dimensional, (non)linear and (non)recursive to highlight the significant aspect of each institutional function in respect to its (1) diversity (i.e. extent and scope of policy instruments); (2) time/phase bound sequencing (in relation to budget cycle for instance) and (3) information updating and coordination of organizational characteristics (ibid, p. 6–7/103). In sum, effective implementation of counter-trafficking measures is hindered by three institutional specific disfunctions: (1) financing, information and policy outcome not synchronized (2) skills, roles and incentive of staff not synchronized (3) lack of sequencing in project implementation and donor funding due to donor driven initiatives. Lack of transparency in allocation of anti-trafficking fundings and expenditure (Nwogu, 2014) further impose a serious challenge in achieving institutional synergies.

17.7 Designing a Composite Index—The Concept and Methodological Challenges

Whereas there is ample qualitative material is available on the state of its changing trend, shaping of women's agency in migratory experience and limitation of the current judicial approach to effectively prosecute traffickers, it is alarming that there is hardly any effective indicator that helps us fully understand how different factors compounded and contributing to the growing and more diversified forms of human trafficking. This trend precisely explains why the public agency have failed to mitigate human trafficking phenomenon effectively so far. Given the limited statistical data combined with most available information is based on qualitative in nature, it is not easy to construct a composite index to represent highly contextual issue such as human trafficking using the regular statistical methods and tools.

In general, statistical indexes suffer from two major problems: (1) high abstraction leading to compromising on ground reality and (2) high dependency on 'perfect' data alone leading to inferior quality of holistic analysis. Because of these challenges, too often composite indices may lose some of their relevance to the cause, leaving the development of indices themselves become a purpose of its own. We must be cautious that statistical tools applied to sparse or inferior quality data could dilute the results to mere numbers with little help in explaining the complex issues. It is fair to argue that studies in human trafficking have perhaps refrained from index construction mainly because of both problems. There are limited number of indices therefore used for studying the trends in human trafficking.

Furthermore, extensive efforts with experimental spirits will have to go in exploring the issue to find a direct route to the crux of the problem in the environment where additional political economic challenges are in play. It is also not easy to not get overwhelmed by the conventional understanding that dictates what is considered appropriate evidence in the policy process which reveals a strife divide among the scientific community. Complex political dynamics that evolve around the difference in the ideological standing on agency-victim divide in prostitution debates further mar the path. Thus, what is being explored here is a small step in the direction of reaching the crux of the matter. We have identified our proposed framework of Boundary Science (Kurian & Kojima, 2021) is particularly resourceful in the context of construction of composite index to overcome some of these challenges through its key functions of: (1) multi-disciplinary, (2) connection to policy and (3) scaling up.

First, challenges related to lack of common understanding about what constitutes evidence for policy and decision-making can be overcome by multi-disciplinary and multi-dimensional approach that guide through a composite construction, Second, a composite index which accommodates key institutional dimension of the political economy (i.e. power, justice, diversity and etc.) can better inform the design, monitoring and evaluation of policy instruments and thus more policy effective. Finally, the third is that a composite index constructed mindful of point 1 and 2 would offer more solid conceptual framework and it can be ready to inform policy that has relevance and impacts on a larger scale through utilizing a big data analytic platform. This means it does away from a need to repetitive data collection therefore low cost, data light and high impact.

Under this premise, we propose the design of Counter-trafficking Preparedness Composite Index (CPI). This index is intended to measure the degree of the individual government's capacity to effectively engage in counter-trafficking measures involving women for PCS and CSS while anchoring vulnerabilities attribute to areas of knowledge, instrumental and management models and norms as the key policy blind spots. To this end, 'counter-trafficking' is defined as a composite of three dimensions- prevention of trafficking, prosecution of traffickers and protection of trafficked persons.[10] Similarly, 'preparedness' is defined as a composite of

[10] These include: (1) The degree of trafficking prevented through the safe migration path, (2) The level of use of trafficking specific criminal procedures to prosecute traffickers, (3) The extent of organizational formalization of protection of trafficked persons.

three dimensions; normative, institutional and behaviour and attitude of government institutions/organizations.[11] Component for the 'counter-trafficking' dimension would demonstrate the level, degree and scope of adoption and effectiveness of measures whereas components for the 'preparedness' dimension would illustrate institutional settings—environment and arrangement regarding norms, legal and operational instruments and organizational practice related to counter-trafficking measures respectively. Further, our CPI is grounded on the co-designing and cross-fertilization learning experience drawing upon citizen science[12] and cost-effective research methods. This means that the proposed methodology envisions the co-curation of research process- selection of data and the scaling of weights for indicators would be carried out through the consultation with the local panel experts consists of key stakeholders of government officials, academics, NGO executives and activists of the selected country or region.

Overview design framework of CPI (see Fig. 17.1) is following. First, CPI is comprised of multi-faceted components of Gender and Political Economy (GPE), Institutional Responses (IR) and Migration-Trafficking Threshold (MTT). Second, construction of CPI will be carried out through these three parallel sub-indices development paths while these sub-indices correspond to the individual blind spot areas of vulnerabilities identified earlier as norms (GPE), institutional and management models (IR) and knowledge (MTT) respectively. These sub-indices are tentatively identified as Trafficking Intensity Index, Synergy Pathway Meld Index and Trafficking Propensity Index. Third, indicators for individual sub-indices will be identified through typology analysis utilizing different methods. Whereas indicators for GPE component will be sought through the advanced analysis of typology of policy trade-off dynamics as presented in Table 17.3, MTT component will rely on typology of case studies of human trafficking to delineate relevant factors and patterns that victimize migrant women. Selection of indicators for IR will be identified based on analysis on policy synergy pathways matrix (Table 17.5) combined with typology analysis of counter-trafficking institutional characteristics in prevention, protection and prosecution policy initiatives. Finally, these sub-indices be summated into one composite index (CPI) to evaluate what we are up against and monitor the progress of the extent building of institutional capacity and resistance against human trafficking.

---

[11] These include: (1) The level of use/adoption of an institutional norm that views human trafficking should not be tolerated and trafficked persons as right-holders, (2) The degree of institutional environment and arrangement that promotes effective prosecution of traffickers and protection of trafficked persons, (3) The extent of organizational discriminatory practice against women and foreigners.

[12] Citizen science offers a new pathway to critical inquire into challenges related to scientific objectivity and theory of knowledge—positivism (Harding, 1987)—through expanding the definition of scientific knowledge by incorporating richer perspectives on relationships between knowledge and experience (Harding, 1987). Citizen science is particularly resourceful in redefining the public policy knowledge and behaviour in the context of public service delivery from the perspective of representation (Kurian & Kojima, 2021). See more Kabeer (2005), for instance.

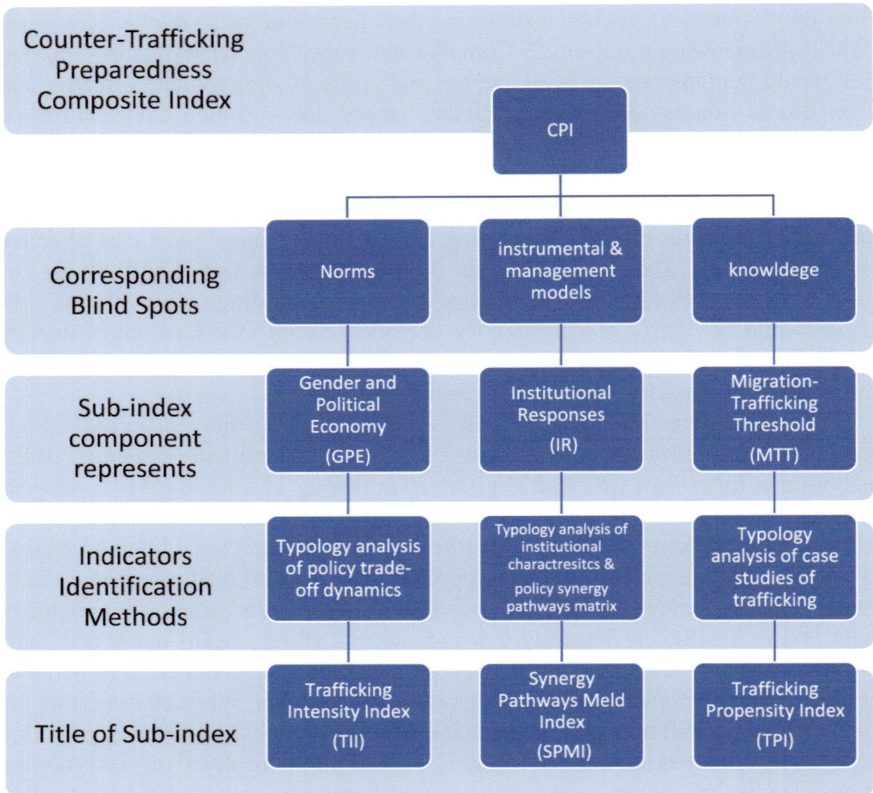

Fig. 17.1 CPI—framework of composite index design and methodology

17.8 Conclusion

This paper re-examined the policy challenges of the present migration governance from the perspective of building greater institutional capacity and resilience in addressing counter-trafficking initiatives involving women in PCS and CSS. We argue that there is a hard lesson learned by all from the decades of policy experiments involving female migration in domestic labour. It points out serious blind spots in how we do science related to human trafficking—starting from identification of policy challenges, process of the scientific knowledge production (in academia), and translation of knowledge into policy that match the real needs of migrant women. It also urges for immediate needs to devise innovative ways to convince policy makers better. As demonstrated in this paper, our proposed analytical framework and the method of indexing combined with blind spots and typology of trade-off analysis is promising in this respect. Through this paper, we foresee to pre-empt further discussion on creative ways to achieve improved understanding of interactions between the institutional and migration outcomes in public policy planning.

Declarations We hereby confirm that there is no conflict-of-interest in producing/publishing this manuscript while the content of this manuscript is not AI-generated.

References

ACIL (2004). *Asia regional cooperation to prevent people trafficking: The role of victim support agencies in the criminal justice response to human trafficking—Lessons from Thailand, Cambodia, Lao PDR and Myanmar*. AusAID.

Agustin, L. (2003). A migrant world of services. *Social Politics-International Studies in Gender, State & Society, 10*(3), 377–396.

Anderson, B. (2001). Why madam has so many bathrobes: Demand for migrant domestic workers in the EU". *Tijdschrift Voor Economische En Sociale Geographie (Journal of Economic and Human Geography), 92*(1), 18–96.

Anti-Slavery International. (2006). *Trafficking for forced labour UK country report*. Anti-Slavery International.

Asian Migration Centre. (2001). *Baseline research on racial and gender discrimination towards Filipino, Indonesian and Thai domestic helpers in Hong Kong*. Asian Migration Centre.

Batley, R. (2004). The politics of service delivery reform. *Development and Change, 35*(1), 31–56.

Bharadwaj, R., Chakravarti, D., Karthikeyan, N., Hazra., S., Daniel, U., Topno J., & Abhilashi, R. (2022). Climate change, migration and vulnerability to trafficking. *IIED Working Paper.* IIED. http://pubs.iied.org/20936IIED

Chantavanich, S., Nittayananta, S., Ratanaolan-Mix, P., Ruenkaew, P., Khemkrut, A., Kanchai, S. and Stern, A. (2001). *The migration of Thai women to Germany: Causes, living conditions and impacts for Thailand and Germany*. Asian Research Centre for Migration Chulalongkorn University and International Organization for Migration.

Cheng, S. A. (2003). Rethinking the globalization of domestic service: Foreign domestics, state control and the politics of identity in Taiwan. *Gender and Society, 17*(2), 166–186.

Clancey, A., Khushrushahi, N., & Ham, J. (2014). Do evidence-based approaches alienate Canadian anti-trafficking funders? *Anti-trafficking Review, 3*, 87–108.

Climate Refugees (2022, January 27). To mitigate climate change, combat slavery and protect climate migrants & refugees. https://www.climate-refugees.org/spotlight/01/27/22/climate-migration-slavery

Constable, N. (2002). Filipina women in Hong Kong homes: Household rules and relations. In B. Ehrenreich & A. Hochschild (Eds.), *Global woman-nannies, maids and sex workers in the new economy* (pp. 115–141). Granta Publications.

Daher, B., Mohtar, R., Pistikopoulus, P., Portney, K., Kaiser, R., & Saad, W. (2018). Developing socio-techno-economic-political solutions for addressing resource nexus hotspots. *Sustainability, 10*(512), 1–14.

Dottridge, M. (2014). Editorial: How is the money to combat human trafficking spent? *Anti-trafficking Review, 3*, 3–14.

Essed, P. (1991). *Understanding everyday racism: An interdisciplinary theory*. Sage Publications.

Fortman, B. (2003, March 26–28). *Persistent poverty and inequality in an era of globalisation: Opportunities and limitations of a rights approach.* Paper presented at Tilburg University Lustrum Conference—Sustainable Ties in the Information Society. Tilburg.

Furlan, M., & Rosado, L. (2023). *Human trafficking and climate change in the Philippines: Understanding intersections and strengthening responses in the Philippines*. Free the Slaves.

Global Alliance against Traffic in Women (GAATW), STV and International Human Rights Law Group. (1999). *Human rights standards for treatment of trafficked persons*. GAATW.

Gregory, R. (1997). Political rationality or incrementalism? In M. Hill (Ed.), *The policy process a reader* (pp.175–191). Prentice Hill.

Harding, S. (Ed.). (1987). *Feminism and methodology*. Indiana University Press.

Hearn, J., & Collinson, D. L. (1994). Theorizing unities and differences between men and between masculinities. In H. Brod & M. Kaufman (Eds.), *Theorizing masculinities* (pp. 97–118). Sage Publications.

Himmelweit, S. (2002). Making visible the hidden economy: The case for gender-impact analysis of economic policy. *Feminist Economics, 8*(1), 49–70.

Hoff, S. (2014). Where is the funding for anti-trafficking work? A look at donor funds, policies and practices in Europe. *Anti-trafficking Review, 3*, 109–132.

Hondagneu-Sotelo, P. (2001). *Domestica-immigrant workers cleaning and caring in the shadows of affluence*. University of California Press.

Hooper, C. (2000). Masculinities in transition: The case of globalization. In M. Marchand & A. S. Runyan (Eds.), *Gender and global restructuring: Sightings, sites and resistance* (pp. 59–73). Routledge.

Huda, S. (2006). *Integration of the human rights of women and a gender perspective report of the special rapporteur on the human rights aspects of the victims of trafficking in persons, especially women and children*. United Nations Commission on Human Rights. Sixty-Second Session 20 February. (E/CN.4/2006/62).

Human Rights Watch. (2004). *Bad dreams: Exploitation and abuse of migrant workers in Saudi Arabia*. Human Rights Watch.

Jansen, H. A. F. M. (2012). *Prevalence surveys on violence against women—Challenges around indicators, data collection and use*. Expert Paper presented at U.N. Expert Group Meeting Prevention of Violence Against Women and Girls, 17–20 September, Bangkok. (EGM/PVAWG/INF.9).

Kabeer, N. (Ed.). (2005). *Inclusive citizenship: Meanings and expression*. Zubaan Publisher.

Kendi, I. X. (2019). *How to be an Antiracist*. One World.

Kojima, Y. (2001). In the business of cultural reproduction: theoretical implication of the mail-order bride phenomenon. *Women's Studies International Forum, 24*(2), 199–210.

Kojima, Y. (2007). *Women in trafficking-migration continuum: From perspective of human rights and social justice*. Maastricht.

Kojima, Y. (2011). Migrant women and their vulnerability—Evidence from Asia. In T. D. Truong & D. Gasper (Eds.), *Transnational Migration and Human Security: The Migration-Development-Security Nexus (Hexagon Series)* (pp.147–158). Springer.

Kojima, Y. (2015, December 8). Rohingya women in migration: Lost voices. *Our World—Online Journal of United Nations University*. http://ourworld.unu.edu/en/rohingya-refugee-and-migrant-women-shadowed-by-sexual-and-gender-based-violence

Kurian, M., & Kojima, Y. (2021). *Boundary science: Reimagining water-energy-food interactions in the context of a data light approach to monitoring the environment-development Nexus*. Elsevier.

Landolt, P., & Goldring, L. (2015). Assembling non-citizenship through the work of conditionality. *Citizenship Studies, 19*(8), 853–869.

Lee, S. (1996). Issues in research on women, international migration and labor. *Asian and Pacific Migration Journal, 5*(1), 5–26.

Lim, L., & Oishi, N. (1996). International labour migration of Asian women: Distinctive characteristics and policy concerns. *Asian and Pacific Migration Journal, 5*(1), 85–115.

Macklin, A. (2003). Dancing across borders: Exotic dancers, trafficking and Canadian immigration policy. *International Migration Review, 37*(2), 464–500.

Molinari, N. (2017). Intensifying Insecurities: The impact of climate change on vulnerability to human trafficking in the Indian Sundarbans. *Anti-trafficking Review, 8*, 50–69.

Nwogu, V. (2014). Anti-trafficking interventions in Nigeria and the principal-agent aid model. *Anti-trafficking Review, 3*, 41–63.

Okin, S. M. (1990). Feminism, the individual, and contract theory (Review of *the sexual contract*, by C. Pateman). *Ethics, 100*(3), 658–669. Retrieved May 6, 2025, from http://www.jstor.org/stable/2381814

Okin, S. M. (1998). Feminism, Women's Human Rights, and Cultural Differences. *Hypatia, 13*(2), 32–52.

Outshoorn, J. (Ed.). (2004). *The politics of prostitution: Women's movements, democratic states and the globalisation of sex commerce*. Cambridge University Press.

Pateman, C. (1983). Feminist critiques of the public/private dichotomy. In S. I. Benn & G. F. Gaus (Eds.), *Public and private in social life* (pp. 281–303). Croom Helm.

Pearson, E. (2002a). *Human traffic human rights: Redefining victim protection.* Anti-Slavery International.

Pearson, E. (2002b). Half-hearted Protection: What does victim protection really mean for victims of trafficking in Europe? *Gender and* Development, *10*(1), 56–59.

Perry, J. A., & Gesualdi-Fecteau, D. (2021). Contending with precarity, temporariness and life aspirations in the context of international labour migration. *International Migration, 59*(2), 3–7.

PHR. (2004). *No status: Migration, trafficking and exploitation of women in Thailand*. Physicians for Human Rights.

Pollitt, C., & Bouckeart, G. (2000). *Public sector management: A comparative analysis*. University Press, Oxford.

Roces, M. (2009). Prostitution, women's movements and the victim narrative in the Philippines. *Women's Studies International Forum, 32*(4), 270–280.

Rubin, G. (1975). The traffic in women: Notes on the political economy of sex. In R. R. Reiter (Ed.), *Towards an anthropology of women* (pp. 157–210). Monthly Review Press.

Sabban, R. (2002). United Arab Emirates: Migrant women in the United Arab Emirates- The case of female domestic workers. *GENPROM Working Paper* No. 10, Gender Promotion Programme. Geneve: International Labour Organization.

Shan Women's Action Network (SWAN). (2002). *License to rape*. SWAN.

Silvey, R., & Parreñas, R. (2019). Precarity chains: Cycles of domestic migration from Southeast Asia to the Middle East. *Journal of Ethic and Migration Studies, 46*(1), 1–15.

Smith, P. (1978). Domestic labour and Marx's theory of value. In A. Kuhn & A. Wolpe (Eds.), *Feminism and materialism: Women and modes of production* (pp. 198–219). Routledge and Kegan Paul.

Squires, J. (1999). *Gender in political theory*. Polity Press.

Srihari, G., Ishaan, S., Beth, S., Benslimane, Y., Imen, E. A., Venté Coha, F., Kulesza, V., Sanghrajka A., Luchs, A., & Ho, C. (2024). Climate change, modern slavery, and its impact on health—A youth perspective and global call to action. *The Journal of Climate Change and Health, 20,* 2667–2782.

Surtees, R., & De Kerchove, F. (2014). Who funds RE/integration? Ensuring sustainable services for trafficking victims. *Anti-trafficking Review, 3*, 64–86.

Suzuki, N. (2003). Transgressing 'victims' reading narratives of Filipina brides in Japan. *Critical Asian Studies, 35*(3), 399–420.

Thorbek, S., & Pattanaik, B. (Eds.). (2002). *Transnational prostitution: Changing global patterns.* Zed Books.

Tyner, J. (1996). Construction of Filipina migrant entertainers. *Gender, Place and Culture, 33*(1), 77–93.

U.K. the Secretary of State for the Home Department (2013). *Draft modern slavery bill.* Retrieved 6 May 2025, from https://assets.Publishing.service.gov.uk/media/5a7c8afae5274a0bb7cb7bb6/Draft_Modern_Slavery_Bill.pdf

U.S. Department of State Government (2020). *Trafficking in persons report.* Retrieved 4 May 2025, from https://www.state.gov/reports/2020-trafficking-in-persons-report/

Ucnikova, M. (2014). OECD and modern slavery: How much aid money is spent to tackle the issue? *Anti-trafficking Review, 3*, 133–151.

UNHCR (2009). *Human trafficking and refugee protection: UNHCR's perspective*. Paper presented at the Ministerial Conference on Towards Global EU Action Against Trafficking in Human Beings. Brussels, 19–20 October.

Vince, G. (2022). *Nomad Century: How climate migration will reshape our world*. Flatiron Books.

Walk Free (2018). *The global slavery Index 2018*. Minderoo Foundation.

Walk Free (2019). *Measurement action freedom—An independent assessment of government progress towards achieving UN sustainable development goal 8.7*. Minderoo Foundation.

Windebank, J. (2001). Dual-earner couples in Britain and France: Gender divisions of domestic labour and parenting work in different welfare states. *Work, Employment & Society, 15*(2), 269–290.

Wong, D. (1996). Foreign domestic workers in Singapore. *Asia Pacific Journal, 5*(1), 117–137.

Yea, S. (2004). Runaway brides: Anxieties of identity among trafficked Filipina entertainers in South Korea. *Singapore Journal of Tropical Geography, 25*(2), 180–197.

Yeoh, B. S. A., & Huang, S. (1998). Negotiating public spaces: Strategies and styles of migrant female domestic workers in Singapore. *Urban Studies, 35*(3), 583–602.

Chapter 18
Climate Change and Environmental Rule of Law: Limitations from Law—Making to the Adjudication Process

Ramandeep Singh Sidhu

Abstract Climate change has threatened the ecological balance and become a challenge to social and legal systems around the globe. Climate change has pushed humanity to the edge of planetary sustainability and become one of the greatest challenges including social, political and economic. The Greenhouse gases are increasing though international endeavours to curtail and bring the temperature in required Celsius through climate change policymaking and judicial bodies are striving to implement those policies and also contributing through judicial activism to control the consequences of climate change. The rule of law is an essential element of policymaking to fairly consider the GHG contributions and imposing policies for consequences. This paper delves into discussion on how environmental rule of law in policy making and climate adjudication assists to deal with these matters fairly. There are gaps in climate change policymaking which must be addressed at the adjudication stage such as considering vulnerability, human rights etc.

Keywords Climate change · Rule of law · Adjudication · Policymaking · Decision-making

18.1 Introduction

Climate change has threatened the ecological balance and become a challenge to social and legal systems around the globe. Climate change has pushed humanity to the edge of planetary sustainability and become one of the greatest challenges including social, political and economic. When any system thinks to enact a law for climate change then it requires to be considered from every aspect from right to life, health biodiversity, land use, natural resources and basic human rights and human security (Cinnamon Carlarne, 2020). Scientific evidence reveals that anthropogenic Green House Gases (GHG) having significant impact on the climate change and

R. S. Sidhu (✉)
Bathinda College of Law, Punjab, India
e-mail: singh.ramansidhu@gmail.com

P. Datta et al. (eds.), *Climate Change and Human Well-being*, Advances in Global Change Research 83, https://doi.org/10.1007/978-3-032-25479-5_18

it is decided that it should not be more than 2 Celsius at all (IPCC, 2023). The global warming will continue to increase from rule of law helps to govern the system fairly and requires considerations from democracy and human rights perspective. International climate change law has started evolving in 1992 from the United Nations Framework Convention on Climate Change, and the Kyoto Protocol, 1997 and Paris Agreement 2015 are legally binding in nature. But when it comes to environmental rule of law in adjudication of climate change is a significant loophole in the system. COP 26 has urged the parties to reduce the GHG emissions and keep the global warming below 1.5 °C (COP 26). When it comes to enforcement of climate change adjudication, is essentially a herculean task it varies according to the jurisdictions and regional standards vary too.

This chapter delves into invoking the environmental rule of law in climate change adjudication which is significantly includes human rights, democracy, rule of law as core concepts of environmental rule of law and its application at international level decision making by international courts and tribunals. Human rights and rule of law required to be integrated while adjudication of environmental matters including climate change takes place but to what an extent international climate change law has addressed human rights and rule of law in its treaties and conventions is a matter of consideration. Though John Knox and other Special Rapporteurs have evidently established climate change and human rights relationship and its impact not only on human beings but on the biodiversity and ecosystem. Environment has been recognized as a human right in the Reports and also in UNGA resolution in 2022 (UNGA, 2022). This chapter contends that how environmental rule of law is essential to adjudicate climate change issues and its application to protect the future generations. Section 18.1 of the chapter provides an insight into relationship between climate change, human rights and rule of Law by invoking human rights instruments and impacts of climate change on the planet. Section 18.2 illustrates the rule of law and its transformation into environmental rule of law while addressing the environmental concerns. Section 18.3 of the chapter signifies the concept of environmental rule of law, its contents and features of the concept. Section 18.4 evaluates the environmental rule of law addresses full range of environmental challenges including climate change, water scarcity, biodiversity loss, pollution and soil degradation. The major threat, as of now, to humanity is climate change, that transcends the boundaries.

18.2 Climate Change, Human Rights and Rule of Law: An Integration from International Perspective

Global governance has been constantly changing and it is enhancing its ambit. While talking about global governance international politics is in constant change and it requires international community should work in solidarity and should examine the assumptions regarding the content and legitimacy of rule of law (Nicola Lacey, 2019; Amy Oloo, 2019). To address the climate change; stable, accountable and transparent laws are required. Rule of law is an essentially contested concept (Bingham, 2010). While considering the normative account of rule of law in its thick version, elements such as justice, human rights, equity and respect for international law demand greater attention. The concept of cooperation is imbibed in the international rule of law. General Secretary Kofi Anan in 2004 stated "rule of law is at the very heart of the UN" representing "principle of governance in which all persons, institutions and entities, public and private, including the State itself, are accountable to laws that are publicly promulgated, equally enforced and independently adjudicated, and which are consistent with international human rights norms and standards" (Annan, 2004). In 2012 in UN Meeting on Rule of Law recognized rule of law, democracy and human rights are inter-linked and mutually reinforce (UN, 2012).

The concept of "global community" or 'international community" has remained an elusive, and requires states and their citizens to cooperate to shared values at global level and it also helped to develop the international law. When it comes to cooperation, is least realized concept in certain areas of international law. International climate change signifies a classic example of assumption that we are a collective human community and prepared to cooperate (Carlarne & Helal, 2018). Concept of 'international community' concedes collective initiative, shared common interests and normative common ground for cooperation required to progressive realize humankind. As the models of climate change have exacerbated, endeavours to control have faltered, climate change have caused inequities which are intrinsic to human rights (John Knox, 2016). The linkage between human rights and climate change has been developed fully by various studies. The development of human rights and environment has started with the adoption of UN Conference on the Human Environment (1972) and then Rio Declaration (1992). In 1992 UNFCCC (1992) was adopted and then in 1994 Fatma Zohra Ksentini was appointed as a special rapporteur to study the relationship between human rights and environment, presented a Draft Declaration on Principles of Human Rights and the Environment, which recognized right to healthy environment as a human right (Zohra Ksentini, 1994), though the new right to healthy environment is right belongs to solidarity rights and still has to be considered from various aspects. After first report on the same, then in 2012 another special rapporteur was appointed to continue the study and to examine "the human rights obligations relating to the enjoyment of a safe, clean, healthy and sustainable environment," and "promote best practices of the use of human rights in environmental policymaking" (UN, 2012). John Knox was appointed as an independent expert to study human rights and environment relationship and in 2015 was converted his position as special rapporteur, in his 2016 report on climate change. At the end of his term, he presented a framework principle for states to ensure the enjoyment of a safe, clean, healthy, and sustainable environment within the context of human rights, 2018. John Knox recognized human rights and environment are interdependent. David Boyd, special rapporteur in his 2019 report on safe climate described it as "global climate emergency" on the human rights enjoyment and it needs to be addressed on human rights-based approach. Further in his 2021 Report, in "Healthy Biosphere, rapporteur considered climate change is not affecting the human beings, it has significant impact on the entire ecosystem and for the planetary sustainability (Boyd, 2021).

United Nations has taking strides to strengthen the climate change regime especially its enforcement mechanism. Various international environmental organizations are contributing in the development of human rights congruent with rights of nature. For example, IUCN World Commission on Environmental Law, drafted a global pact for the environment to ensure substantive and procedural rights and aspects which are essential for environmental protection and preservation of ecosystem, that entails environmental rule of law which will help to adjudicate climate change matters fostering integrity of ecological system (IUCN, 2017). Then Inuit Petition filed before the Inter-American Commission of Human Rights, (discussed in detail in judicial aspect part) for the adjudication of climate change effects on the indigenous communities in the region. In 2007, a vulnerable group of small islands met in Maldives and adopted Male' Declaration on the Human Dimension of Climate Change, which expressly recognized the "climate change immediate effects on the enjoyment of human rights and urged the UN to address the issue as matter of urgency" (Male Declaration 2007). In 2009, UN Human Rights Council, adopted a resolution affirming the international Bill of Rights, UNFCCC, Rio Declaration and other instruments signifying the impact of climate change on human rights as "Noting that climate change-related impacts have a range of implications, both direct and indirect, for the effective enjoyment of human rights including, inter alia, the right to life, the right to adequate food, the right to the highest attainable standard of health, the right to adequate housing, the right to self-determination and human rights obligations related to access to safe drinking water and sanitation, and recalling that in no case may a people be deprived of its own means of subsistence" (UNHRC, 2009) and further affirmed that human rights obligations have capability to development international and national level policies to explore relationship between human rights and climate change also to help the adjudication process. The litigation process in climate change matters have been taking place at another level and national courts have started recognizing adverse effects of climate change on human rights. Climate change law can be able to strengthen the rule of law only if human rights instruments such as Universal Declaration of Human Rights, 1948 (UN, 1948), International Covenant on Civil and Political Rights, 1966, International Covenant on Social, Economic and Cultural Rights and Child Rights Convention 1990 are invoked with these human rights instruments as these instruments establish human rights committee and other institutions for the violation at national level.

18.3 Concept and Characteristics of Environmental Rule of Law (EROL): From Rule of Law to Environmental Rule of Law

Rule of Law as a Tool of Governance

Good governance is a meta concept and can easily be connected with other meta concepts, rule of law, is quintessential for functioning of democracy. Rule of law is an essentially contested concept. The traditional rule of law entails application of abstract principles and limited state discretion in governance system, and government is subject to same as individual citizens (Addink, 2019). Rule of law can be formal, procedural or substantive in nature. Formal elements of rule of law include generality, publicity, clarity etc., these elements show that laws are enacted within this framework and applied accordingly. Lon Fuller's 'inner morality of law' contains eight elements of rule of law, ensuring that government should do whatever it wants to do and it must publish these to give advance notice to its citizens (Lon Fuller, 1963). Fuller's principles are as following:

A. Generality: the element of generality ensures that laws should be general in nature and should not be made for specific individuals. If the laws are general then, there will be equal consequences for violating the law for everyone irrespective of class, gender etc. and guarantees equity, fairness and justice (Lon Fuller, 46–48).
B. Publicity or promulgation: Fuller contended that it is impossible to educate every citizen about the full meaning of every law but laws must be promulgated and publicized so that can give enough notice about the existence of law.
C. Non-retroactivity: it is essential to enable law to guide human conduct, to prescribe rules prospectively and not retrospectively. The retroactive rules can do more harm than benefit and will not be able to achieve the purpose of actually guiding human conduct.
D. Clarity: element of clarity is most essential element of legality. If laws are ambiguous then promulgation will be futile and it will lead to arbitrariness. Laws must be clear in its context to convey its effect to citizens.
E. Non-contradictory laws: explaining this principle fuller stated law should not be contradictory itself. Every law must be consistent though unseen contradictions are possible.
F. Practicable: Fuller, considers that law must be practicable and easy to be comprehensible. Practicality aspect cannot be ignored and if laws are impossible then these will not be able to guide human conduct perfectly.
G. Stability: stability is considered an essential ingredient of law, so that law can achieve its purposes, whatever they are. If changes in law are frequent then it will be impracticable for the people to follow the law. This principle emerges partly from that many of our actions that the law purports to regulate require advance planning, preparation and a certain level of guaranteed expectations about the future normative environment.
H. Congruence: a law can guide human conduct well only if it maintains congruence between the rules promulgated and their actual application to specific cases. This is the most complex requirement of law which may include whole range of principles and practices. In general terms, it necessitates that agencies dealing with applying and enforcing law to specific categories actually apply rules promulgated by the law.

These formal aspects are required to be present in the rules to restrict the arbitrariness of the government and its officials.

Procedural aspects of rule of law: mere formal aspects of rule of law cannot ensure good governance and democracy if these are not complemented by procedural aspects. Procedural aspect requires no one should have any penalty stigma, serious loss imposed upon them by the government except following the rules of natural justice. These include:

a. A right to fair hearing by an independent and impartial tribunal.
b. A right to representation by a counsel at such hearing
c. Right to be present, to confront and questions witness, that is to argue the case fairly; and
d. A right to hear reasons from the tribunal when it reaches to the conclusion of the trial (Lord Bingham, 2010).

Joseph Raz in his works considered procedural/formal aspects of rule of law enough though there are certain jurists who considers that substantive aspects are equally essential for the rule of law. However, a careful consideration is required to distinguish between substantive requirements of rule of law and specificities of the deeper values that underlie and motivate the ideal even in its formal and procedural necessities. Ronald Cass states "critical aspect for the commitment of the commitment to the rule of law is the definition and protection of property rights" (Ronald Cass, 2004). Richard Epstein considers rule of law is separate concept from property rights, but there is a significant connection between rule of law and property rights as property rights protection is sought under the rule of law (Epstein, 2011). Josph Raz's proposition that rule of law is just one of the virtues of the law and it should not be read into other considerations such as democracy, human rights and social justice (Joseph Raz, 1979) but this proposition was rejected by Bingham expressly, as it seems he wants to fill the gap between procedural and formal aspects of rule of law by invoking human rights component. As liberals associate rule of law with human dignity. Democratic values allow protection of human rights and participation as a component of rule of law.

The analysis of these three aspects of rule of law encompasses the climate change and human rights are intertwined and environmental rule of law is required to protect the environmental rights of human and other beings including the rights of nature. There is a vast scholarly literature on international rule of law and its relevance. Research reveals that international rule of law has significantly evolved from the concepts of national practices. The international rule of law evolution is evident as formulated by United Nations, International courts such as ICJ and scholars: such concepts include accountability, certainty predictability and non-arbitrariness Next part of the chapter charts the outlines of environmental rule of law with its significant features.

18.4 Concept of Environmental Rule of Law

United Nations Environmental Programme defined environmental rule of law by considering three components: fundamental rights, accountability and inclusivity developed fairly effectuated. The law should be in harmony with fundamental rights; law should be fairly and inclusively developed and must bring forth accountability not just to show on paper but in actual so that compliance process can be affected fairly and evenly and generate respect in the community (UNDP, 2019). The environmental rule of law integrates critical environmental needs with the pivotal components of the rule of law and provides the premise for reforming environmental governance. Environmental rule of law is significant to achieve environmentally sustainable development and it stems from the Sustainable Development Goal 16 which commits to "rule of law at the national and international level" in respect to "promote peaceful and inclusive societies for sustainable development, provide access to justice for all and build effect, accountable and inclusive institutions at all level" (SDG, 16) and when considering EROL in context of climate change, then SDGs provide "take urgent action to combat climate change and its impacts" (SDG, 13). (UNDESA, 2015).

Rule of law is considered as quintessential for governance but when it comes to Environmental governance rule of law is lacking. Although, all countries of the world have some kind of environmental rules and regulations, but the implementation and enforcement mechanisms are not as good as for other fields which result into mounting crises of biodiversity loss, pollution and climate change (UNEP, 2019).

The environmental rule of law and environmental governance are two distinct matters. Environmental governance includes broad range of approaches and principles relating to how decisions are made and implementation in regard to the environment. Environmental rule of law deals with implementation, enforcement and compliance process. Environmental rule of law ensures public participation, transparency, accountability and protection of environmental human rights. environmental rule of law is significant for proper functioning of the environmental governance (UNEP, 2019).

Environmental rule of law is significant to realise environmental goals for effective legal control of environmental issues. Apart from environmental and climatic benefits, social and economic benefits are provided. Environmental rule of law protects natural resources and public health, as both are economically valuable and intertwined. The absence of environmental rule of law impedes the environmental governance in many ways. Fair and effective implementation of environmental laws bolsters legitimacy of government action and strengthens the public confidence in public institutions (Bruch et al., 2016).

There are seven core elements of environmental rule of law: (a) Fair, clear and implementable environmental laws, (b) access to information, public participation and access to justice, (c) accountability and integrity of institutions and decision makers, (d) clear and coordinated mandates and roles, across and within institutions, (e) recognition of the relationship with rights, (f) accessible, fair, impartial and timely adjudication and (g) specific criterion for interpretation of environmental laws (UNEP, 2015).

First Global report on Environmental rule of law has identified seven characteristics for environmental and climate change regime law. These are as following: *firstly*, EROL is critical to human health and welfare. It requires adherence to rules and regulations and approaches set forth in laws related to environment and health laws. It requires peoples' right to land, health, justice, water, forests and natural resources to advance the livelihood of the people be strengthened (Bosselmann, 2014). *Secondly*, political, social and economic dynamics particular to natural resources are required to be managed under the environmental rule of law as if these resources will be depleted then earth's regenerative capacity will not be capable to bear the burden and human will have to bear the brunt of climate change. *Thirdly*, environmental rule of law is multidimensional. It includes various forms of laws including customary. Social and statutory laws. The environmental governance at local level such as indigenous peoples and other national and sub national communities to ensure public participation (Epremian et al., 2016). *Fourthly*, moral and ethical duties humans towards non-human species and resources. All species are equal in nature and ecocentric approach it required to maintain the balance of the universe and to prevent existing threats to nature caused by climate change and it is only possibly if environmental rule of law is made part of management of the environment. *Fifth*, human communities depend on natural resources for livelihood. When the management and decision making related to water, and clean air, requires public involvement in environmental decision making (Beierle, 2010). Local populations and indigenous communities which are affected by environment and their life is based on nature or forests where they live. *Sixth*, environmental rule of law must consider the uncommon timescales. The entire universe including human and non-human species are dependent on nature and future generations will be dependent on resources left by their predecessors. So environmental rule of law must include management of resources in such timeless scales so that future generations can enjoy their lives and it is only possible if environmental rule of law integrates the elements of sustainable development. (Brown Weiss, 1983, 2007, 2011). *Seventh*, environmental rule of law based on decision making in the face of uncertainty. Scientific community has raised more questions than answers and those answers are quite uncertain in the changing climate and other environmental issues. What precautions and measures can be taken and what can be avoided is a question of certainty and how this can be fixed. Can environmentally rule of law assist those tackle these questions?

18.5 Prominence of Rule of Law in International Environmental and Climate Change Law

Climate change policy-making started taking place in 1990s, but the Environmental law making started in 1970s only. First part of this section contains international and regional climate change instruments and second part discusses the problems persists in climate change policymaking. There are More than 1000 bilateral and multilateral environmental agreements, which either protect the ecosystems or regulate the natural resources, polluting and hazardous activities. There is number of MEA on climate change and its regulation to curb the carbon dioxide and other greenhouse gases. The rule of law in environmental law is still in infancy because there is no express mention of rule of law in Stockholm Declaration 1972 and Rio Declaration 1992. Moreover, the IEL emphasises on the protection of environment that is human value oriented (Nicolas de Sadeleer, 2020). Rule of law is anthropocentric in nature like other principle such as human rights and democracy. principle of separation of power is not germane to IEL, as the MEAs focus on fostering international cooperation rather than structuring the manner in which states integrate. MEAs are not self-executing treaties, which can directly be enforced, rather states are required to enact legislation at national level to give effect to these MEAs.

Taking legality aspect of IEL into consideration reveals that majority of the MEAs establish procedural safeguards such as reporting. Consultation, transparency, impact assessment, scientific cooperation and assistance, but hardly provide any substantive standards for these procedural aspects. A few MEAs have developed uncanny mechanisms such as emissions allowances trading schemes established within the ambit of the Kyoto Protocol and payment schemes for CO_2 emissions reductions to reduce deforestation within the ambit of the Paris Agreement (Art. 5 of REDD+ (2015)).

The 1972 United Nations Conference on Human Environment, under Principle 21 provides state responsibility for transboundary harm which is essential to realize the environmental rule of law in climate change adjudication. The Rio Declaration, 1992 (UN, 1992) in its principle 10 promotes public participation and access to justice are significant to realize environmental democracy which aligns with AARHUS convention (UNECE, 1998), and Principle 11 effective environmental legislation and 15 provides precautionary principle. But when it comes to effectiveness, it lacks binding impact. Principle 7 of the Rio Declaration contains common but differentiated responsibility essential for climate change but it is contentious in climate negotiation (Rajmani, 2011). 1998 AARHUS Convention (UNECE, 1998), is significant milestone in developing the environmental rule of law, contains access to information (Articles 4–5) Public participation in decision making (Articles 6–8) and access to justice in Article 9 of the Convention. This convention is a central to environmental governance and strengthening the rule of law. Escazú Agreement in the Latine America and the Caribbean (2018) is in line AARHUS Convention but a broader perspective has been outlined.

United Nations Framework Convention on Climate Change (UNFCC, 1992), the 1985 Convention for the Protection of the Ozone Layer or the 1979 Convention on Long-Range Transboundary Air Pollution (CLRATP) or the Bonn Convention on Conservation of Migratory Species of Wild Animals (1979), set out vague objectives, declare general principles and institutional framework and obliges states to take "all practicable measures" in the form of additional agreements. There are many regional instruments adopted for climate change, EU Climate and Energy Framework, 2014 entails legally binding targets such as 55% reduction in GHG by 20230, 32.5% enhancement in energy efficiency. EU has adopted European Climate law (EU2021/1119) mandates the EU to become climate neutral by 2050 which sets net zero GHG emission by 2050. Court of Justice of European Union deals with the violation of these binding treaties. African Union (AU) has always taken pivotal for environmental and climate issues. In 2003, the Maputo Declaration on Environment and Sustainable Development adopted by AU to manage the environment. AU Climate Change Policy Framework (AUCCPF) adopted in 2014 focuses on implementation of international commitments for climate change. It is non-legally binding in nature (AU Climate Change Policy Framework (African Union, 2022). There are various other international and regional instruments deal with the climate change and adaptation and mitigation enforcement mechanism which are invoked by the international and regional court to adjudicate climate change and its impact and somehow furthering climate justice. Next section analyses the role of international and regional courts in climate change and human rights relationships. Majority of the agreements are soft law which are not legally binding impede the compliance and adjudication process in climate and environmental instruments.

Climate change policymaking consists of four aspects: (I) substantive alignment with science-based gaols; (II) procedural rights reflected in Aarhus Convention and Escazu agreement such as: access to information, justice and participation; (III) institutional check and balances by courts and regulatory authorities and (IV) Enforcement mechanism (Boyd, 2012). There are various challenges that persist in the aligning the rule of law with policymaking, enlists as following: There is ambition and implementation gap in policymaking and enforcement like Art. 2 of the Paris agreement sets the temperature goals, emission reductions, but National Determined Contributions in domestic policies are insufficient cause of political factors, legal frameworks and institutional capacity constraints etc. Moreover, climate framework goals are long term without any interim obligation creating implementation hindrances, whereas from rule of law perspective, these goals become vague and non-transparent (Averchenkova et al., 2018). International instruments and policies set the normative ambitions, but operational reality depends on state laws such as compliance mechanism, budget allocation and political will.

Principle of CBDR is based on equity norms which designs prospects adaptation, mitigation, finance and technology transfer, but burden sharing remains vexed, both at international and domestic level, as just transition measures at absent from the policies, highly impacting the vulnerable sections of the society. The rule of law requires fairness and foreseeability, socially acceptable legislation and regional development and to bolster legitimacy (Sachs et al., 2019). Article 6 of the Paris agreement requires environmental and social integrity as essential elements of rule of law in climate change policymaking for international cooperation to bring transparency in carbon counting, bolster grievance mechanism. Huge investment is required for climate scaling, but there are certain gaps such as inadequate pledges for assessed needs, specifically for adaptation and damage, vulnerable states do not receive timely disbarments due to complex fiduciary standards and disintegrated funds, debt and micro-stability restraints and technology transfer issues hinder climate change policymaking at domestic level (Sachs et al., 2019). Rule of law must ensure domestic policy to include clarify fiduciary standards, protection of social and environment safeguards, enable blended finance with accountability. The most significant challenge which is premised on the procedural aspects of climate change policy is impact of climate change on vulnerable sections of society such as children, women, indigenous peoples and minorities. Climate policymaking must include intersectionality as an essential facet for targeting policy support to do away with discriminatory practices prevalent in the policymaking. Climate change policymaking must be recognitional and distributive and procedural (Schlosberg, 2007). Distributive justice requires fair allocation of burdens and benefits and financing for climate change funds and technology transfer, whereas recognitional justice entails recognition of identities, livelihood and knowledge systems of indigenous and local communities. recognitional justice will facilities procedural justice to entail access to information, participation in decision making and remedies (Sovacool et al., 2017). The rule of law ensures non-arbitrary enforcement and human and ecological rights protection (UN, 2012). Climate change policymaking needs to be designed for adaptation and mitigation, with consideration of above-mentioned challenges though not illustrated in detail. These challenges can be overwhelmed, if rule of law, is considered as essential of climate change policy-making.

18.6 International Climate Change Adjudication: International, Regional and National Perspective

International, regional and national courts are receiving matters on climate change. when it comes to international court of justice has limited jurisdiction and significantly impacts the adjudication process. In international adjudication compulsory jurisdiction is lacked and without which effective enforcement cannot be sought. Moreover, states when agree to something are more likely to cooperate than accepting the imposition of rules from outside authority (Phillipe Sands, 2016). Another problem states least contributed to the climate change are equally or more than contributors bearing the consequences of climate change and for that purpose a fair adjudication process is required to deal with transboundary harms and effectively impose international obligations stemming from the climate change and human rights law with the assistance of rule of law.

In *Certain Phosphate Lands in Nauru* (Nauru v. Australia, 1989), In May 1989 Naura filed the case against Australia before ICJ for violation of Article 76 of UN

Charter along with Articles 3 and 5 of Trusteeship Agreement, including violation of general principles of international law such as self-determination, permanent sovereignty over natural resources. The court considered the environmental obligations of state. It held that it is state's duty to rehabilitate land damaged by the extraction of resources and it established an outline for post-colonial environmental justice. ICJ in Advisory opinion in nuclear case considered rule of law to strengthen the protection of environment. It acknowledged that environment is under daily threat. The court has invoked various principles of environmental law and humanitarian law based on rule of law include principle of proportionality, doctrine of necessity, precautionary principle and principle of inter-generational justice. In Gabčíkovo-Nagymaros Project (1993) judgment was based on international rule of law and environmental legal procedure; aspect was included in the issue. Budapest Treaty 1977 was dealing Dams and other related matters and violation of rules not only led to Vienna Convention on law of treaties but also principles of sustainable development which require economic development must commensurate with the sustainable development (Pulp Mills case, 2007). Recently international courts have started adjudicating climate change matters though not as binding judgments but most of them are advisory opinions issued by the ICJ, ITLOS etc. the United Nations General Assembly, on 29th March 2023 (ICJ, 2025) requested an advisory opinion from the ICJ on states obligations under international law to combat climate change. The key questions before the ICJ are whether court should confine its judgments to UNFCCC, Kyoto Protocol and Paris Agreement or apply laws with wider ambit including customary international law, humanitarian law and principles of international environmental law considering future generations? IUCN and others have argued that principle of due diligence urges the states to limit the global warming to 1.5 °C including Common but Differential Responsibility for the proportionality for the consequences of the global warming. ITLOS in advisory opinion delivered in 2024, COSIS asked two questions under part XII of UNCLOS that, whether states have obligations to reduce, prevent and control marine pollution from the GHG emissions causing acidification of sea, sea level rise and ocean warming? Whether states have an obligation to protect marine environment from GHG emissions? ITLOS adjudicating under articles 194 due diligence duty and transboundary harm, article 212, 211, 213 217, 223 197 asking the states to adopt domestic measures to address the land-based pollution and emissions, bilaterally cooperate, assist the vulnerable states and Environmental Impact Assessment for monitoring the marine environment. But the ITOL could not precisely define the stringent obligations related to due diligence. Advisory opinions given by the ICJ and other international courts and tribunals are not binding on the states just state how states can go ahead with their obligations which help in adjudicating at domestic level by invoking international climate change and human rights law (ICJ, 2025).

At regional level, European Court of Human Rights and Inter-American Court of Human Rights have dealt with various cases. First subpart of this chapter deals with European Court of Human Rights (Strasbourg Court) extensively dealt with climate change issues through environmental matters came before it. Risk of climate harms and greenhouses gases is acute and it is required to be brought less than 2 °C. ECHR in KlimaSeniorinnen v. Switzerland (2021) a petition came before the court invoking Articles 2, 8, 6 of the European Convention on Human Rights. a third-party intervention containing observation on the effects of climate change on right to life and privacy and what are the positive obligations of the states resulting from these rights and applying international climate change instruments such as Paris Agreement 2015. Two experts warned that court 1.1 °C is harmful for the vulnerable humans, animals and natural systems. Emissions in the past and present not only will impact today but will impact future times as well. While considering the positive obligations under the ECHR, it is required that protection against the convention violations of which the state "knew or ought to have known" by considering actual or putative risks. Critical questions arise while considering the relationship between attribution, causation and attributing state knowledge, which is made specifically serious by the denial of the court to develop a general theory of positive obligations which may be inferred from the Convention. In Duarte Agostinho and Others v. Portugal and Others (2022), a complaint was filed against the members of Council of Europe as well as Norway Russia, Turkey, Switzerland United Kingdom. The complaint was raised that states parties failed to implement their human rights obligations by failing to keep emission reductions less than 1.5 °C as contained in the Paris Agreement. The court dealt with this case in a fast-track way. The European Court declared the case as inadmissible, while considering the jurisdiction question, the court held "climate change has special features including global impact, state control over emissions and existential threat, applicant should justify the expansion of jurisdiction (para 174). The court considered that national remedies must be exhausted which are not taken by the Portugal so case cannot be accepted by the court. In Carême v. France (2024), Damien Carême, former mayor of Grande Synthe filed a case before the France and European Court of Human Rights for violation of not adopting the measures for reduction of greenhouses gases, targeted at 40% cut by 2030 in harmony with the Paris Agreement. The adopted the victim status in climate related cases approach and considered Carême is not direct victim and dismissed the case. In European Court of Human Rights climate change litigation is taking place but court is inclined to inadmissibility of cases before it as it finds climate change is not expressly mentioned in the European Convention on Human Rights, 1950. Though somewhere it is accepting the cases but unable to lay down general theory.

Inter-American Court of Human Rights and Commission has significantly dealt with the climate change adjudication. In 2005, Sheila Walt-Cloutier, Chair of Inuit Circumpolar conference made a statement at Conference of Parties of UNFCCC at 2005, mentioning the effects of climate change in arctic regions. This petition opened the door for indigenous communities to seek environmental rule of law for protecting the cultural environment. Climate change threatens individual as well collective rights of the Inuit people such as right to life, right to health, right to property and cultural identity of the peoples (IACtHR, 2005). In 2017 in an advisory opinion recognizing environment as an autonomous human right having direct linkage with right to life, health and culture. The IAHRC considered the specific obligations of states regarding prevention of effects of the climate change on human rights enjoyment and specific violation of American Declaration of Rights and its

Protocol San Salvodor. The court considered the the World Declaration on the Environmental Rule of Law of the International Union for Conservation of Nature adopted at the IUCN World Environmental Law Congress, held in Rio de Janeiro from April 26–29, 2016, Principles 1 and 2 (IACtHR, 2017 para. 164). In 2019, a petition before the Inter-American Commission was filed to recognize climate crisis as a priority that threatens ecosystems and human rights. the claimants requested the commission to urge the states to stop activities which aggravate the climate change and adopt energy model which commensurate with the ecological system and guarantee social, economic, political and right to vulnerable communities. In 2025 the Inter-American Court of Human Rights, issued an advisory opinion invoking environmental rule of law from the climate change perspective (IACtHR, 2025). The democratic principles should be considered for the protection of environment, "it must be based on the strengthening of the democratic rule of law as an indispensable scenario for the protection of human rights. Therefore, neither the urgency of effective measures, nor the seriousness of the climate impacts, nor the complexity of the required response can be invoked to justify the weakening of democratic systems or the disregard of the guarantees inherent to them. In this regard, the Court recalls that, according to Principle 10 of the Rio Declaration, environmental problems are best solved through broad public participation, adequate access to environmental information and effective access to judicial and administrative procedures" (IACtHR, 2025 para 214). Climate emergency has arisen to the extent especially to the vulnerable sections of society that environmental rule of law, democracy and human rights must pave the way to mitigate the effects of climate change.

The African Court of Human and Peoples' Rights decided various cases on the effects of climate change on vulnerable communities. *African Commission vs Republic of Kenya* (Ogiek Case, 2012), application was filed by the Ogiek, an indigenous community against the eviction notice issued by the Kenyana Forst Service authority to leave the Mau Forest. They contended that Mau Forest is their home and livelihood of the peoples, having cultural identity. This decision has become basis in various other judgments of the court and commission on which climate change litigation has been expanded. In Africa Climate Alliance et al., v. Minister of Mineral Resources & Energy et al. (Cancel Coal Case, 2021) South African high court delivered the judgment on the issue whether the government power to procure new coal-fired power is unconstitutional? The court invoked human rights and held "the rights to life, dignity, and equality. They contended that coal-fired power, a major contributor to climate change and environmental degradation, posed severe risks to public health, particularly for children and future generations" (Cancel Coal Case, 2021).

Sacchi, et al. v. Argentina, et al. (2021), the child Rights committee received an application from 16 children from various countries alleging the violation of UN Convention on Rights of Child caused by emission of greenhouse gases and their reduction has not been controlled by the countries. Though CRC did not admit the application but found that states are responsible for controlling the emissions and prevent the transboundary harm causes by such emissions, and asserted that climate change requires are global response and individual states are responsible for their actions and inactions regarding climate change to for their contributions.

Supreme court of India in MK Ranjit Singh v. Union of India (2024) the court addressed the adverse effects of climate change on the ecosystem said that right to healthy environment is a fundamental right and right to be free from adverse effects of climate change is a constitutional right. the court emphasising on the extraterritorial obligations under international law which are essential for the development of environmental rule of law by invoking UNFCCC, Kyoto protocol and Paris Agreement, legally binding instruments and other international human rights treaties because without invoking human rights obligations. The court asked the state to look for renewable energy sources for the mitigation of climate change and reduction of GHG gases required to be kept under 2 °C. The court held that sustainable development goals need to be implemented as well. In State of the Netherlands v. Urgenda Foundation (2015) respondent requested the supreme court Netherlands to issue directions to the state to reduce emission of GHG. The court considered exceeding 2° causes global warming which threatens right to life and family which are human rights under the European Convention of human rights. The court held that substantive rights can be invoked for environmental and climate change concerns.

The international, regional and national courts imbibing human rights obligation for climate change concerns along with UNFCC, Paris Agreement and other climate change policies which entails emission reduction of GHG and fostering laws of nature and preservation of planet through planetary ethics by signifying nature has rights too and courts need to adopt ecocentric approach to save the ecosystem from the scourge of climate change.

18.7 Conclusion

Climate change adjudication necessities human rights obligations implementation. Environmental governance includes governance of environmental management and environmental rule of law is a principle which can be exercises through democratic principles. Climate change litigation has significantly enhanced in the last few years though majority of the matters are coming to the courts for violation of human rights. Emission of green house gases are provided under legally binding agreements such as UNFCCC, and Paris Agreement. But how these agreements are interpreted by the national and international courts, matter of consideration. Climate change policymaking is having certain challenges those can be remedies through the rule of law, considering the climate justice as major aspect for policy development. So far as policymaking at domestic level is concerned, will of a nation and how far they are implementing the international carbon reduction targets is essential question because contribution and consequences are entirely different for contributors and non-contributors, rather non-contributors are bearing most of the brunt of the climate change. Questions arise about sovereignty of nations when it comes to CBDR and other principles which can potentially help to reduce and control global warming caused by the climate change. Climate change policymaking at international level is not fair in distribution allocation of budgets and making major contributors to pay more. Non-contributor nations are suffering more in consequences than contributors. Judicial mechanism and enforcement require a stringent compliance process so that climate change matters can be effectively dealt through environmental rule of law. Judiciaries at national level are significantly invoking the human rights jurisprudence to deal with adverse effects of climate change, not only on humans but also on ecological system.

References

Addink, H. (2019). *Good governance: Concept and context*. Oxford University Press.

Africa Climate Alliance et al. v. Minister of Mineral Resources & Energy et al. (Cancel Coal Case 2021), South Africa High Court, Case No. 56907/21.

African Commission on Human and Peoples' Rights v. Republic of Kenya, African Court on Human and Peoples' Rights, Application No. 006/2012 Ogiek Case (2012).

African Union (2022, June 28). *Climate change and resilient development strategy and action plan 2022–2032.*

Annan, K. (2004). Report of the Secretary-General on the work of the organization. *The Yearbook of the United Nations,* 3–50.

Averchenkova, A., Fankhauser, S., & Finnegan, J. (Eds.). (2018). *Climate change legislation: Global review.* Edward Elgar.

Beierle, T. C. (2010). *Democracy in practice: Public participation in environmental decisions.* Routledge.

Bingham, T. (2010). *The rule of law*. Penguin.

Bosselmann, K. (2014). The rule of law grounded in the Earth: Ecological integrity as a Grundnorm. In L. Westra & M. Vilela (Eds.), *The Earth Charter, ecological integrity and social movements.* Routledge.

Boyd, D. R. (2012). The constitutional right to a healthy environment. *Environment: Science and Policy for Sustainable Development, 54*(4), 3–14.

Boyd, D. R. (2020, July 15). *Human rights depend on a healthy biosphere* (Report of the Special Rapporteur on human rights obligations relating to the enjoyment of a safe, clean, healthy and sustainable environment, A/75/161). United Nations. https://undocs.org/A/75/161

Brown Weiss, E. (1983). The planetary trust: Conservation and intergenerational equity. *Ecology Law Quarterly, 11*(4), 495–582.

Brown Weiss, E. (2007). Climate change, intergenerational equity, and international law. *Vermont Journal of Environmental Law, 9*, 615–627.

Brown Weiss, E. (2011). The evolution of international environmental law. *Japanese Yearbook of International Law, 54*, 1–27.

Bruch, C., Muffett, C., & Nichols, S. (2016). *Governance, natural resources and post-conflict peacebuilding*. Routledge.

Carême v. France, European Court of Human Rights, Application No. 7189/21 (2024).

Carlarne, C. (2020). Climate change, human rights and the rule of law. *UCLA Journal of Law and International Affairs, 25*(1), 11–40.

Carlarne, C. P., & Helal, M. S. (2018). A conversation about climate change law and the 'international community.' *Climate Law, 8*(3–4), 229–243.

Caribbean. (2018). Regional Agreement on access to information, public participation and justice in environmental matters in Latin America and the Caribbean. Santiago. United Nations.

Cass, R. (2004). Property rights systems and the rule of law. In E. Colombatto (Ed.), *The Elgar companion to the economics of property right* (pp. 131–163). Edward Elgar.
Convention on the Conservation of Migratory Species of Wild Animals (1979, November 23), 1651 U.N.T.S. 333.
Duarte Agostinho and Others v. Portugal and Others, European Court of Human Rights, Application No. 39371/20 (2022).
Epremian, L., Lujala, P., & Bruch, C. (2016). *High-value natural resources revenues and transparency: Accounting for revenues and peace*. Oxford University Press.
Epstein, R. A. (2011). *Property rights and the rule of law: Classical liberalism confronts the modern administrative state*. Harvard University Press.
European Climate Law, Regulation (EU) 2021/1119.
Fuller, L. L. (1963). *The morality of law*. Yale University Press.
Human Rights Council (2009). *Human rights and climate change* (10th Session, Res. 10/4). https://ap.ohchr.org/documents/E/HRC/resolutions/A_HRC_RES_10_4.pdf
Inter-American Court of Human Rights. (2005). Petition to the Inter-American commission on human rights seeking relief from violations resulting from global warming caused by acts and omissions of the United States.
Inter-American Court of Human Rights (2017). Advisory Opinion OC-23/17, requested by the Republic of Colombia.
Inter-American Court of Human Rights (2025). Advisory Opinion OC-32/25, requested by the Republic of Chile and the Republic of Colombia on climate emergency and human rights.
Intergovernmental Panel on Climate Change (2023) Climate Change 2023, Synthesis Report.
International Court of Justice (1989). Certain phosphate lands in Nauru (Nauru v. Australia).
International Court of Justice (1993). Gabčíkovo-Nagymaros Project (Hungary/Slovakia).
International Court of Justice (2007). Pulp mills on the River Uruguay (Argentina v. Uruguay).
International Court of Justice (2025). *Advisory opinion on obligations of States on climate change* (to be delivered July 23, 2025).
ITLOS Opinion. (2025, 23 July). Obligations of states in respect of climate change, advisory opinion. ITLOS.
IUCN World Commission on Environmental Law (2017). *Global Pact for the Environment*. IUCN. Retrieved from https://www.iucn.org/our-union/commissions/world-commission-environmental-law/our-work/history/foundational-documents-2
Knox, J. H. (2016). Human rights principles and climate change. In C. P. Carlarne, K. R. Gray, & R. G. Tarasofsky (Eds.), *International climate change law* (pp. 213–231). Oxford University Press.
Ksentini, F. Z. (1994). *Human rights and the environment* (E/CN.4/Sub.2/1994/9). United Nations Commission on Human Rights.
Lacey, N. (2019). Populism and the rule of law (London Sch. of Econ. and Pol. Sci., Working Paper No. 28). https://eprints.lse.ac.uk/101867/1/Lacey_populism_and_the_rule_of_law_wp28.pdf
Sacchi et al. v. Argentina et al., United Nations Committee on the Rights of the Child, Comm. No. 104/2019 & Comm. No. 105/2019 (2021).
M. K. Ranjitsinh v. Union of India, 2024 INSC 280. Petition to the Inter-American Commission on Human Rights seeking relief from violations resulting from global warming caused by acts and omissions of the United States (2005).
Male' Declaration on the Human Dimension of Global Climate Change (2007, November 2007).
Oloo, A. (2019). Global rule of law continues to decline, but resistance is gaining strength. *Democracy Without Borders*. https://www.democracywithoutborders.org/
Rajamani, L. (2011). The Cancun climate agreements: Reading the text, subtext and tea leaves. *International and Comparative Law Quarterly, 60*(2), 499–519.
Raz, J. (1979). *The authority of law*. Oxford University Press.
Sachs, J., et al. (2019). Six transformations to achieve the SDGs. *Nature Sustainability, 2*, 805–814.
de Sadeleer, N. (2020). *Environmental principles: From political slogans to legal rules*. Oxford University Press.

Sands, P. (2016). Climate change and the rule of law: Adjudicating the future in international law. *Journal of Environmental Law, 28*(1), 19–35.
Schlosberg, D. (2007). *Defining environmental justice: Theories, movements, and nature*. Oxford University Press.
Verein KlimaSeniorinnen Schweiz and Others v. Switzerland, European Court of Human Rights, Application No. 53600/20 (2021).
Sovacool, B. K., Burke, M., Baker, L., Kotikalapudi, C. K., & Wlokas, H. (2017). New frontiers and conceptual frameworks for energy justice. *Energy Policy, 105*, 677–691.
State of the Netherlands v. Urgenda Foundation (2015) HAZA C/09/00456689.
United Nations (1948). *Universal Declaration of Human Rights*. United Nations.
United Nationssss Conference on the Human Environment, Stockholm (1972 June 5–16).
United Nations Framework Convention on Climate Change (1992, May 9), 1771 U.N.T.S. 107.
United Nations. (1992, June 14). *Rio Declaration on Environment and Development*. United Nations Conference on Environment and Development, Rio de Janeiro.
United Nations Economic Commission for Europe (1998, June 25). *Convention on access to information, public participation in decision-making and access to justice in environmental matters (Aarhus Convention)*
United Nations General Assembly (2012, November 30). *Declaration of the high-level meeting of the General Assembly on the rule of law at the national and international levels* (A/RES/67/1). United Nations Digital Library. https://digitallibrary.un.org/record/739199
United Nations Environment Programme (2012, May 7–18). *Report of the United Nations Environment Programme (UNEP)*. United Nations.
United Nations (2015). *Paris Agreement*. United Nations Treaty Collection. https://treaties.un.org/doc/Treaties/2015/12/20151215%2003-34%20PM/Ch_XXVII-7-d.pdf
United Nations Department of Economic and Social Affairs (2015). *Sustainable Development Goals*.
United Nations Environment Programme (2015). *Environmental rule of law: Critical to sustainable development* (Issue brief). https://www.unep.org/zh-hans/node/15094
United Nations Environment Programme (2019). *First global report on environmental rule of law*. https://environmentalmigration.iom.int/sites/g/files/tmzbdl1411/files/documents/2023-11/environmental_rule_of_law.pdf
United Nations General Assembly (2022, July 28). *The human right to a clean, healthy and sustainable environment: Resolution adopted by the General Assembly, A/RES/76/L.75*. https://press.un.org/en/2022/ga12437.doc.htm
World Declaration on the Environmental Rule of Law, International Union for Conservation of Nature (2016).

Chapter 19
Assessing the Impact of Low Emission Zone Policies on Small and Medium Enterprises in Delhi: Insights and Recommendations

Ananya

Abstract Low Emission Zone (LEZ) is an emerging global policy intervention aimed at reducing vehicular pollution. Many countries have successfully implemented LEZs, resulting in notable improvements in air quality and enhanced public spaces. This study explores the perspectives and preparedness of small and medium-sized businesses regarding LEZ implementation in Delhi's Connaught Place (CP), located within the New Delhi Municipal Corporation (NDMC) area. Using a mixed-method approach, the research engages directly with business owners through structured interviews to examine their dependence on vehicles, willingness to pay potential entry fees, and the challenges they anticipate in transitioning to cleaner alternatives. By focusing on independent retail, wholesale, and order-based enterprises, rather than large branded outlets, the study captures the experiences of those most vulnerable to LEZ restrictions. These conversations reveal a strong environmental awareness among business owners, alongside significant concerns about practical issues such as increased costs and limited infrastructure. The findings suggest that while many are open to adapting, a successful transition will depend on supportive policies, targeted incentives, and improved infrastructure. Ultimately, this study aims to contribute to a more equitable and effective LEZ implementation strategy that balances environmental objectives with the economic sustainability of Delhi's crucial commercial hubs.

Keywords Low emission zone (LEZ) · Commercial hubs · Vehicular emissions · Delhi air pollution · Commercial vehicle transition

Ananya (✉)
Department of Policy and Management Studies, TERI School of Advanced Studies, New Delhi, India
e-mail: mananya729@gmail.com

P. Datta et al. (eds.), *Climate Change and Human Well-being*, Advances in Global Change Research 83, https://doi.org/10.1007/978-3-032-25479-5_19

19.1 Introduction

Vehicular emissions contribute significantly to overall air pollution in an urban landscape (Margaryan, 2019). Numerous case studies have established that traffic congestion and tailpipe pollution have a high share in deteriorating air quality (Moreno et al., 2022). According to the Intergovernmental Panel on Climate Change (IPCC) Sixth Assessment Report of 2021, the transportation sector accounted for approximately 23–28% of global energy-related CO_2 emissions in 2019. The World Health Organisation (WHO) reports that road transport is responsible for a substantial share of particulate matter (PM) emissions globally, particularly due to diesel traffic. In cities, vehicular emissions typically contribute 20–30% of $PM_{2.5}$ at the breathing level. According to the International Council on Clean Transportation (ICCT), vehicle tailpipe emissions were linked to approximately 361,000 premature deaths from ambient $PM_{2.5}$ and ozone worldwide in 2010 and 385,000 in 2015. This represents about 11.4–11.7% of global ambient $PM_{2.5}$ and ozone premature deaths. Additionally, the United Nations Environment Programme (UNEP), in 2024, stated that vehicles are major contributors to nitrogen oxide (NOx) emissions, with heavy-duty vehicles accounting for over 40% of on-road NOx emissions in urban areas.

Prolonged exposure to vehicular pollutants is linked to severe health outcomes, including increased risks of heart attacks, strokes, cancer, and developmental issues in children, with vulnerable populations such as children and those living near busy roads facing even greater risks (Boogaard et al., 2012; Gehrsitz, 2017; Margaryan, 2019). Therefore, various policies are implemented globally to combat the problem of vehicular pollution and reduce risks related to public health (Gehrsitz, 2017). Low Emission Zone (LEZ) is one of the policies that has emerged in response to increasing vehicular emissions across the globe, in order to improve urban air quality. There is no standard definition, understanding, or application of LEZs; however, it can be defined as a designated area in a city restricting the entry of high-polluting vehicles (CSE, 2024a, 2024b; ITDP, 2023). As per reports by ITDP, ICCT, and CSE, this zone is a manmade boundary within a city or an urban region legitimised and authorised by the government to reduce tailpipe emissions from older and high-polluting vehicles. These LEZs aim to cause a shift from older vehicles to vehicles with country-specific approved standards or Electric Vehicles (EVs). Unlike traditional traffic management measures focused primarily on congestion, LEZs specifically target air quality improvement, uniquely positioning them to address public health concerns that disproportionately affect children, elderly individuals, and those with existing health conditions.

LEZs are primarily imposed by two particular policy scenarios: (1) a complete ban on high-polluting vehicles with harsh penalties on defaulters, or (2) an imposition of a restriction fee to enter the LEZ for the high-polluting vehicles (CSE, 2024a, 2024b; ITDP, 2023). Hence, the goal of LEZ is threefold: Avoid, Shift, and Improve; this strategy is called the ASI-approach (CRDF, 2025). This ASI approach talks about avoiding or reducing freight movement by managing it efficiently for logistic operations; shifting to cleaner alternatives and public transportation; and improving vehicle technology for easier adaptation to LEZs (CRDF, 2025; ITDP, 2025). Hence, this highlights the core idea of the implementation of LEZ in an urban region.

19.2 Literature Review

19.2.1 Global Case Studies of LEZs

According to ITDP, by 2022, there were about 320 LEZs within Europe. There are also a few LEZs in Asia, such as China, Japan, and South Korea. It has been a successful policy intervention in the city of London over the past 2 decades, and in recent years, the city has come up with an Ultra Low Emission Zone (ULEZ) (Browne et al., 2005; Ellison et al, 2013; Zhai & Wolff, 2021). Similarly, for cities like Brussels, Amsterdam, Antwerp, Lisbon, Rome, Madrid, Paris, and Milan, a policy intervention like LEZ has turned out to be a decent attempt at cutting significant levels of air pollution (ITDP, 2023). Table 19.1 shows some of the most successful cases of LEZs interventions worldwide. The primary reason behind listing these case studies is their great success rate and future estimates to reduce emissions even more.

Initiatives, like LEZ, have set air quality benchmarks that require both national authorities and cities to take steps to cut down on emissions from vehicles (Dablanc & Montenon, 2016).

Table 19.1 shows that **London's ULEZ** saw significant outcomes (Ellison et al., 2013; ITDP, 2023). It was implemented in phases since 2019 and expanded city-wide in 2023, charging for non-compliant vehicles (pre-Euro IV petrol and pre-Euro VI diesel) was £12.50 daily. To address equity, a £110 million scrappage scheme supported low-income residents and SMEs transitioning to electric vehicles (EVs).

Table 19.1 Examples of successful LEZs

Location	Policy	Outcome
London	Introduction of the Ultra Low Emission Zone (ULEZ): daily charge for non-compliant vehicles	• 44% NO_2 reduction • 27% $PM_{2.5}$ reduction • 13,500 fewer high-polluting vehicles entering the zone daily • 4% reduction in CO_2 emissions
Brussels	Low Emission Zone (LEZ): progressive ban on older diesel and petrol vehicles	• 38% reduction in $PM_{2.5}$ • 9% reduction in NOx • 62% of diesel vehicles in 2018, and 50% of diesel vehicles in 2020, were replaced by hybrid vehicles
Oslo	Planned Zero Emission Zone (ZEZ): only zero-emission vehicles allowed in the city centre	• Estimated reduction of 8000–31,000 tonnes CO_2, depending on zone size

Source ITDP and ICCT

Challenges included limited $PM_{2.5}$ improvements and compliance costs. It reduced $PM_{2.5}$ emissions by 20% and NOx by 13% in outer London, accelerating the adoption of electric vehicles (EVs) through scrappage schemes (McLoughlin, 2023).

Similarly, **Brussels, Belgium,** implemented a city-wide LEZ in 2018, banning pre-Euro one petrol and pre-Euro five diesel vehicles. Brussels' LEZ cut $PM_{2.5}$ by 38% and NOx by 9% within a year, demonstrating the scalability of such measures (ITDP, 2023).

Oslo's comprehensive LEZ strategy exemplifies how Nordic cities can transform urban mobility while prioritizing human health and environmental sustainability through bold yet pragmatic policy implementation (Cui et al., 2021). ICCT recognizes Oslo's approach as contributing to the city's goal of avoiding a minimum of 5000 tons of CO_2 emissions annually. Beyond environmental benefits, Oslo's human-centered approach has fundamentally transformed urban life by creating spaces where pedestrians and cyclists take precedence over vehicles with massive public transport investments that have increased cycling modal share targets from 8.3 to 25% by 2025 (Hartmann & Abel, 2020).

According to a study by the Institute of European Environmental Policy (IEEP), in 2024, in **Milan, Italy,** the Area C LEZ (established in 2012) combines a €5 daily fee for non-residents with phased restrictions. The complementary Move-In Project limits annual mileage for polluting commercial vehicles (e.g., 10,000 km/year for Euro 4 diesel). Diesel vehicles in Milan's sub-zones fell from 40 to 32% within a year, while EV adoption rose to 2%. Equity measures, including €27 million in grants for SMEs and 50 free annual entries for residents, improved social acceptance. Exemptions for street vendors during initial resistance further eased implementation (Marchetti & Antonelli, 2024).

Similarly, LEZs also drive systemic shifts: **Shenzhen's zero-emission freight zones** spurred a 40% reduction in logistics-related $PM_{2.5}$, aided by subsidies for e-trucks and charging infrastructure. Shenzhen, China, pioneered zero-emission freight zones in 2020, mandating electric trucks in logistics hubs alongside subsidies of up to ¥150,000 (~ $21,000) for e-trucks and 50% grants for charging infrastructure. This policy reduced logistics-related $PM_{2.5}$ by 40% (2020–2023) and electrified 99% of buses and 85% of taxis. However, SMEs faced high upfront costs, and grid instability during peak charging periods posed operational challenges (ITDP, 2023).

Seoul, South Korea, adopted a 3.5 km^2 Green Transport Zone with toll discounts for EVs and hybrid buses, paired with pedestrianization and expanded bike lanes. $PM_{2.5}$ levels fell by 16% (2021–2023), while retail footfall rose by 20% due to improved walkability. The success underscored the efficacy of small, strictly regulated zones (ITDP, 2023).

Amsterdam's LEZ represents a comprehensive policy framework designed to improve urban air quality and achieve climate neutrality by progressively restricting access to polluting vehicles within designated city areas (Wappelhorst & Cui, 2022; Kok, 2023). The city's Clean Air Action Plan, adopted in 2019, established clear objectives, including meeting World Health Organization air quality guidelines by

2030, reducing CO_2 emissions from mobility to zero by 2030, and ultimately transforming the entire city into a zero-emission zone covering all vehicle types (Wappelhorst & Cui, 2022). The LEZ implementation follows a phased approach, beginning with restrictions on heavy-duty vehicles and gradually expanding to include all motorized transport, with the current environmental zone within the A10 ring road serving as the foundation for the future city-wide zero-emission zone by 2030 (Kok, 2023). The ICCT reports that Amsterdam's approach has successfully reduced traffic contributions to air pollution, with NOx particles decreasing by over 30% following LEZ implementation.

19.2.2 Key Trends and Lessons

Milan's Area C LEZ, which reduced traffic by 28%, paired access fees with EUR 27 million in grants for low-income groups and SMEs to adopt EVs (Marchetti & Antonelli, 2024). Conversely, poorly designed LEZs risk exacerbating inequalities; for example, Madrid's initial LEZ proposal faced backlash from retailers and logistics operators, necessitating revised fee structures and nighttime freight corridors (Tarriño-Ortiz et al., 2022; Marchetti & Antonelli, 2024). The efficacy of LEZs is based on equitable design and complementary policies (Holman et al., 2015). This means that any such zone has to be created while keeping all the aspects in mind.

The expansion of LEZs in significant cities of developing nations, such as **Jakarta**, has been driven by alarming levels of air pollution, with transportation identified as a primary contributor (Rizki et al., 2022). The government implemented a pilot LEZ in the Kota Tua Tourism Area, focusing on emission-based vehicle restrictions without congestion charges, and subsequently explored public acceptance and preferences for expanding LEZs to other locations. This research, by Rizki et al., in this context underscores the significance of socio-psychological factors—such as trust in government, environmental concern, and personal and social norms—in shaping public support for LEZ policies. Younger and wealthier individuals were found to be more critical of LEZ effectiveness, and overall acceptability was higher among those who perceived greater environmental benefits and trusted governmental intentions. Notably, acceptability is not only a function of policy design but also local context, including the scale of implementation and the types of restrictions.

Parallel findings emerge from studies (Jiménez-Espada et al., 2023) in European contexts, such as Spain, where mandatory LEZs in medium-sized cities like Cáceres have prompted investigations into citizen perceptions. Surveys conducted before LEZ implementation reveal that while citizens are generally receptive to measures that enhance accessibility, subsidize public transport, and promote pedestrianization, there is considerable apprehension about restrictions on vehicle access and parking.

Globally, phased implementation (e.g., London's incremental expansion) and equity mechanisms (e.g., Milan's SME grants) emerge as critical for compliance and social acceptance. Complementary policies like EV incentives (Shenzhen) and pedestrianization (Seoul) amplify environmental and economic benefits.

The literature points to a critical need for public administrations to engage in information campaigns and participatory processes to increase awareness of LEZ benefits and to address the knowledge gap surrounding the anticipated effects of such policies. In Cáceres, for instance, citizens expressed support for sustainable mobility initiatives but also voiced concerns about the potential reduction in personal mobility and the adequacy of public transport infrastructure (Jiménez-Espada et al., 2023). The findings suggest that successful LEZ implementation requires not only technical and regulatory measures but also robust stakeholder engagement and tailored communication strategies to foster public buy-in and compliance.

19.2.3 India's Approach to LEZs

In the past 4–5 years, conversations around implementing LEZs have started, primarily in cities like Ahmedabad, Delhi, Jaipur, and Pimpri-Chinchwad (CSE, 2024a, 2024b; ITDP, 2024). Table 19.2 shows some of the emerging cases of LEZs in India (CRDF, 2025; CSE, 2024a, 2024b; ITDP, 2025).

A comprehensive study by CEPT Research Development Foundation (CRDF), in 2025, has established various aspects of implementing restrictive policies of LEZs in and around the **Walled City of Ahmedabad**, which is of prime importance for UNESCO Heritage Sites. This research sought to address the complex challenge of reducing vehicular emissions in a densely populated UNESCO World Heritage Site that houses 328,000 residents and accommodates significant commercial and cultural activities. The expected outcomes focus on achieving a 40% reduction in vehicular pollutants through integrated interventions, including electrification of three-wheelers and light commercial vehicles, rationalized public transport services, and enhanced pedestrian infrastructure. The framework prioritizes community engagement and phased implementation, recognizing that successful LEZ deployment requires balancing environmental protection with social equity while maintaining the area's economic vitality and cultural heritage significance.

Similarly, ITDP and ICCT have made a collaborative effort in creating a potentially successful LEZ in the Pimpri Chinchwad Municipal Corporation **(PCMC)**.

Table 19.2 Indian Cities that are planning for LEZs

City	Location proposed	Expected outcome
Ahmedabad	The Walled City of Ahmedabad	• Electrification of freight • Reduce vehicular emissions by up to 40% in Ahmedabad • Enhance public spaces and livability • Reduce traffic congestion • Increased use of public transport • Pedestrianization around key markets
Pimpri-Chinchwad	Pimpri-Chinchwad Municipal Corporation (PCMC)	
Jaipur	Pink City Heritage Area	
New Delhi	New Delhi Municipal Corporation (NDMC)	

Source CSE, ICCT, and CRDF

In Pimpri-Chinchwad, this study by ICCT 2024 piloted LEZ restricted BS-III and older vehicles in a 12 km^2 commercial area, projecting an 18% $PM_{2.5}$ reduction. ICCT's technical assessment aimed to provide scientific evidence for LEZ effectiveness through sophisticated emissions modeling and spatial analysis of Pimpri-Chinchwad, a major industrial city in Maharashtra. The research objective focused on identifying optimal geographic boundaries and restriction strategies to deliver maximum emission reductions while considering local fleet characteristics and industrial activities. Expected outcomes demonstrate substantial environmental benefits, with projections showing up to 79% reduction in particulate matter and 67% decrease in nitrogen oxides by 2030 under age-based vehicle restrictions. The study anticipates that implementing LEZ measures could reduce total city-wide emissions by 3–15% depending on zone size and restriction scenarios, with the most significant benefits achieved when non-compliant vehicles are replaced with zero-emission alternatives rather than merely upgraded to newer internal combustion engines. Challenges include resistance from informal logistics networks (e.g., three-wheeler operators lacking EV financing) and sparse charging infrastructure (only four public stations in the pilot zone) (ICCT, 2024).

The CSE's focused investigation of **Jaipur's Pink City heritage area** aimed to assess LEZ implementation potential in a UNESCO World Heritage Site context, recognizing the unique challenges faced by smaller cities with significant cultural and tourist importance. This study sought to understand how heritage conservation priorities could align with air quality improvement goals while maintaining the area's economic viability as a significant commercial and tourist destination. The research methodology incorporated detailed surveys of various vehicle users, from two-wheeler owners to goods service providers, analyzing their willingness to adapt to potential restrictions and their capacity for transitioning to cleaner alternatives. Expected outcomes include meaningful emission reductions through age-based vehicle restrictions, with projections showing 90% compliance achievable at access fees of Rs 200–350 for two-wheelers and Rs 600+ for cars.

CSE conducted an extensive perception-based impact assessment in 2024 to understand how LEZ could be strategically implemented in **NDMC**, focusing on community acceptance and behavioral adaptation. This study aimed to evaluate the potential effects of LEZ policies on different income groups and vehicle user segments through comprehensive surveys covering over 1,000 respondents across various transportation modes. The research examined two policy scenarios—complete vehicle bans versus fee-based access charges—to determine India's capital region's most socially acceptable and environmentally effective approach. Expected outcomes include significant modal shifts toward public transportation, with projections showing 90% compliance rates achievable through carefully calibrated access fees ranging from Rs 79 for two-wheelers to Rs 120 for cars. The study anticipates substantial health benefits, including potential prevention of thousands of premature deaths annually, while emphasizing the critical need for enhanced public transport services and targeted financial support for economically vulnerable communities to ensure equitable implementation.

19.2.4 Delhi Air Pollution and Application of LEZ

Delhi's air pollution crisis, driven significantly by vehicular emissions, has become a year-round challenge (WRI, 2025a, 2025b). This poses environmental issues and significant public health concerns in the short and long term. Since 2021, the Commission for Air Quality Management (CAQM) has been implementing the Graded Response Action Plan (GRAP), which aims at reducing the emission levels in Delhi and the National Capital Region (NCR), especially in winter. In 2024, CAQM reported that only 209 days fell under the "Good to Moderate" air quality category, despite stringent measures under GRAP and Bharat Stage VI norms (PIB, 2024).

Real-world emissions testing in Delhi reveals a stark disparity: BS-VI-compliant vehicles emit 81–95% less NOx in labs, but on-road emissions remain alarmingly high due to traffic congestion, outdated fleets, and poor enforcement (ICCT, 2024). According to the report by the Comptroller and Auditor General (CAG) of India on Performance Audit of 'Prevention and Mitigation of Vehicular Air Pollution in Delhi', vehicular sources contribute 39% of $PM_{2.5}$, 81% of NOx, and 84% of CO in Delhi's air, underscoring the urgent need for targeted interventions. Hence, the assessment study of LEZ in Delhi is essential.

Despite global evidence, LEZ research in India remains narrowly focused on air quality and modal shifts, neglecting business-centric impacts. For instance, the 2024 CSE study on Delhi's NDMC area analyzed vehicle-user behavior but omitted commercial enterprises, despite 67% of traffic originating from logistics and delivery vehicles (ICCT, 2024). Similarly, ITDP's LEZ guidelines for Chennai projected a 20% reduction in NO2 but offered no insights into small-business adaptation costs (ITDP, 2025). This gap is critical: Delhi's commercial hubs, like Connaught Place (CP), rely on non-branded SMEs and informal logistics networks, which face unique challenges in transitioning to cleaner technologies.

19.2.5 Knowledge Gaps

This study focuses on the potential impact of restrictive policy scenarios on businesses operating within a potential LEZ. The goal is to include the perception and preparedness of commercial hubs to upcoming policies. The research also highlights the disparity in vehicle usage between the types of businesses and their level of operations. Delhi's LEZ design could integrate phased restrictions (starting with Connaught Place as a pilot), targeted subsidies for SMEs (modeled after Milan), and nighttime freight corridors (tested in Madrid, Tarriño-Ortiz, 2022) to balance environmental and economic priorities. As highlighted in Pimpri-Chinchwad's pilot, these measures must address informal logistics dependencies and charging infrastructure gaps (Tarriño-Ortiz et al., 2022; ICCT, 2024).

An impact assessment study conducted by CSE in Delhi in 2024 has highlighted the importance of LEZ in the urban Indian landscape. On similar lines, various think tanks and research organisations like the World Resource Institute (WRI), ICCT, ITDP, CEPT Research and Development Foundation have discussed the application of LEZs in India. These discussions have been around reducing air pollution and its associated health risks. Their idea is to encourage freight electrification and enhanced public spaces for sustainable mobility. Interestingly, they have highlighted that the commercial vehicles have the highest share in the vehicular emissions; therefore, a phased implementation of policies like LEZs would help in improving the Air Quality Index (AQI) of the cities in India (ICCT, 2024). However, they have not talked about the perception and preparedness of the commercial hubs or the businesses towards the restrictive policies of LEZs.

As mentioned earlier, LEZs can be implemented by two policy scenarios: a complete ban or a restriction fee. Now, these policy scenarios of LEZ require careful consideration of the perception of the stakeholders (Rizki et al., 2022; Sfendonis et al., 2017). Hence, this study, in extension to the proposed LEZ in NDMC by CSE in 2024, focuses on the perception and preparedness of businesses towards the restrictive policies of LEZ. The study tries to cover this gap by a comprehensive methodology of assessing the vehicle dependency and cost implications on the logistic operations of businesses within the potential LEZ. The focus area is a commercial hub in the proposed LEZ, and the assessment is done based on the opinions of business owners operating in NDMC. The goal is to bring an equitable framework of recommendations that can help in providing an informed decision while implementing a policy like LEZ in a city like Delhi.

19.3 Objectives

The objectives of this study are purely based on the recognised knowledge gaps of the existing literature. As mentioned, the commercial fleet significantly contributes to Delhi's overall vehicular emissions (ICCT, 2024). However, there is a lack of perspectives of the business owners taken into account while framing the restrictive policies of LEZ. These business owners rely on commercial vehicles for their business and logistics operations. Hence, it is important to assess the potential impact of LEZ restrictions on them. The idea is to gain insights into businesses operating inside LEZ. Therefore, the main objectives are:

1. To analyse the logistic operational characteristics and vehicle dependency of SMEs located within the proposed LEZ of NDMC in Connaught Place, Delhi.
2. To assess the SMEs' willingness to pay potential entry fees and the expected impact on business operations in CP.
3. To evaluate the preparedness of SMEs to transition to cleaner vehicle alternatives, such as electric vehicles (EVs) or BS-VI compliant models, and to identify key factors influencing this transition.

The first objective is primarily based on the idea of how the business units in the proposed LEZ are operating. The motive is to first understand their reliance on vehicles on a daily basis and then to gain their perspective on the willingness to pay for a potential entry fee, which is the second objective. Since the ultimate aim of LEZ policies is to cause a shift to a cleaner mode of transport (ITDP, 2023), it is also necessary to keep in mind how willing these SME owners are to shift to cleaner alternatives if there is a strict imposition of LEZ policies. The target is to have a strong and equitable framework of recommendations that informs better policy decisions (Holman et al., 2015). In regard to acceptance of LEZ policies rather than backlash or any form of resistance (Tarriño-Ortiz et al., 2022; Marchetti & Antonelli, 2024), it is essential to first analyse the daily needs of the SMEs and then their willingness to comply.

19.4 Data and Methodology

In order to achieve valuable results for the objectives, the methodology for the study employs a mixed-method approach to systematically assess the potential impact of LEZ in Delhi with a focus on commercial hubs. This study is based on both qualitative and quantitative techniques to capture the perception and preparedness of business owners towards the restrictive policies of LEZ. The idea is to converge an environmentally resilient policy with a socioeconomic effect to make more informed policy decisions. The methodology addresses the research objectives of understanding business profiles and vehicle dependency, analyzing willingness to pay for restrictions, and examining the transition potential to cleaner alternatives.

19.4.1 Study Area and Sampling

The impact assessment study done by the Centre for Science and Environment (CSE) in 2024 considers NDMC as a potential LEZ. It has extensively highlighted the importance and rationale behind creating NDMC as LEZ. Additionally, according to this assessment, the Government of the National Capital Territory of Delhi's (GNCTD) Transport Department is actively exploring creating an LEZ in the NDMC region. Hence, it would make more sense that the assessment of the potential impacts of LEZ restrictions on commercial hubs should be done on the businesses operating inside the NDMC. Figure 19.1 depicts the ward map of Delhi, in which NDMC consists of 9 wards. This area is strategically and environmentally important (CSE, 2024a, 2024b). The area of NDMC is about 42.7 sq. km, with proper metro and other public transport connections. This area holds both administrative and corporate importance as it consists of major government offices and corporate buildings that have a regular movement of government and private vehicles (CSE, 2024a, 2024b; Sinha, 2025).

According to NDMC's official website, this area covers Rashtrapati Bhavan (President's Estate), Parliament House, India Gate, Connaught Place, Chanakyapuri (Diplomatic Enclave), Lodhi Estate, Gole Market, Moti Bagh, Sarojini Nagar,

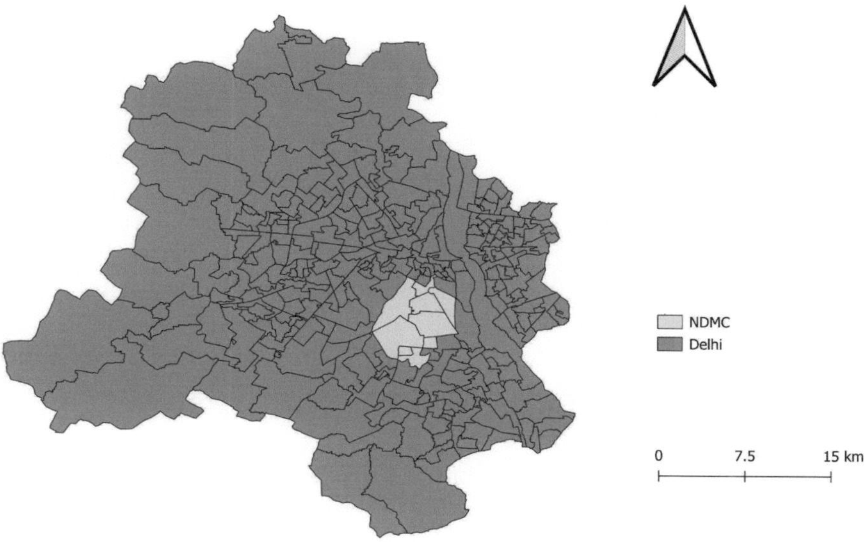

Fig. 19.1 Map of Delhi, NCT wards. *Source* QGIS and OpenStreetMap

Laxmibai Nagar, Kidwai Nagar, Sardar Patel Marg, Ashoka Road, Janpath, Shantipath, Tilak Marg, Barakhamba Road, Sunder Nagar, Golf Links, Prithviraj Road, Aurangzeb Road (now Dr. APJ Abdul Kalam Road), Safdarjung Airport, Talkatora Road, and Mandir Marg.

Given the strategic importance of NDMC, it is evident that Connaught Place (CP) is the iconic commercial hub of this area (Sinha, 2025). Thus, this study consists of an on-ground survey of CP. Figure 19.2 depicts the specific survey area of CP. The blue dots represent the businesses surveyed, the orange highlighted area represents the commercial land, the yellow region shows the nearby residential area, and the light red thin lines depict the roads of NDMC. This map shows that businesses heavily populate CP, a commercially essential region of Delhi's National Capital Territory (NCT).

After selecting CP as the research focus, a pilot survey was conducted to understand local business operations and their reliance on vehicles for logistics. The pilot survey revealed that CP is dominated by retail businesses, with a mix of branded outlets, restaurants, and numerous small and medium enterprises operating alongside them. Even though the study aims to assess how businesses are prepared for potential LEZ restrictions, it became clear that big branded stores would be largely unaffected by such measures, as their logistics are managed internally and they can absorb additional costs, such as entry fees or transitioning to cleaner vehicles. A shop owner from Janpath Lane highlighted:

> Only bigger companies like Westside, H&M, and even Maruti Suzuki and Honda have the necessary passes for large vehicle entry, which is already restricted during the day.

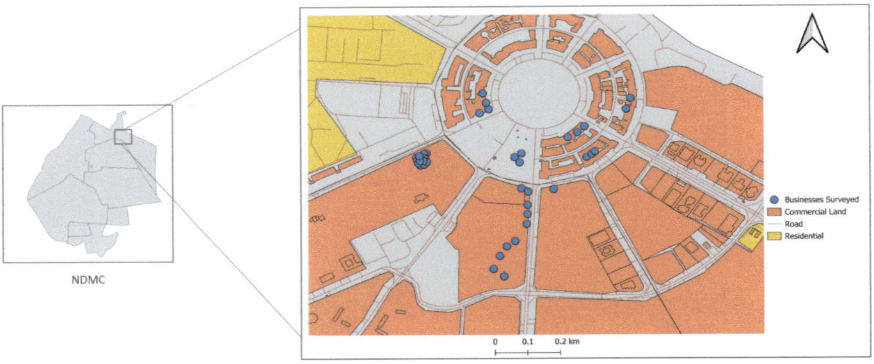

Fig. 19.2 NDMC (Left) and surveyed area, CP (Right). *Source* QGIS and OpenStreetMap

It meant that branded chains of shops and businesses require large commercial vehicles that are already permitted to enter CP with only a valid license. This was confirmed by a 2010 Delhi Transport Department notice outlining regular limits on commercial vehicle access in CP to manage congestion and pollution. As a result, focusing on large enterprises for this study was deemed unnecessary, since they are well-equipped to handle these restrictions.

Instead, the study concentrated on small and medium businesses, which are more vulnerable to the uncertainties of LEZ rules. CP has over 1,200 shops, including 250 restaurants and more than 800 informal vendors, according to NDMC data. The research targeted these smaller operations, especially those relying on commercial vehicles like cars, vans, autos, and bikes for logistics, since they lack the resources of larger brands.

The survey identified three main business types in CP: retail, wholesale, and order-based, with locations such as Palika Bazar, Janpath Market, Mohan Singh Place, and the inner and outer circles of CP. Using convenient random sampling, perspectives were gathered from 50 business owners: 35 from retail, six from wholesale, and nine from order-based businesses, ensuring representation from each category. Only small and medium enterprises were included, as they are most likely to be impacted by future LEZ restrictions (Fig. 19.3).

After defining that this study will only take the perspectives of the medium and small business owners, it was necessary to develop strong data collection methods. The pilot survey suggested that collecting the data through interview schedules would be more appropriate. Since LEZ is a relatively new concept, the respondents needed to understand its goals and implications. Thus, this study relies on data collected through interviews with structured, closed-ended, and open-ended questions. Each interview lasted for 15–20 min. The idea was to gain insights from the business owners regarding the emerging concept of LEZ. The data collected was then analysed using Excel and Flourish to see if there was any visual pattern concerning the defined objectives of the study.

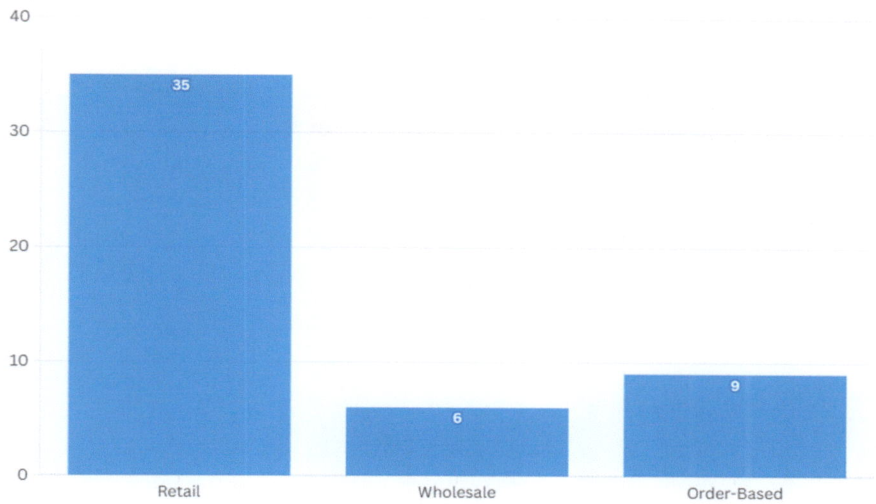

Fig. 19.3 Types of businesses surveyed in CP. *Source* Survey Conducted at CP by Author

19.5 Results and Discussion

19.5.1 Business Profile and Vehicle Dependency

As discussed in previous sections, the commercial ecosystem of CP, situated within NDMC, is characterized by a diverse mix of medium-sized, non-branded businesses. The surveyed businesses span retail outlets, including clothing, apparel, showpieces, sports goods, home decor, and electronic gadgets, alongside wholesale and order-based enterprises. According to NDMC records, there are approximately 800 such shops in CP, making it a representative commercial hub for assessing the potential impacts of LEZ policies. Within the surveyed group, retail businesses constitute the majority, reflecting the broader commercial profile of CP. Wholesale and order-based businesses, while fewer in number, play a crucial role in the supply chain dynamics of the area. The business operations in CP are predominantly small to medium-scale, with owners often managing daily logistics and inventory personally. This demographic is particularly relevant for LEZ policy analysis, as these businesses are susceptible to changes in operational costs and regulatory requirements. A central finding from the survey is the high degree of vehicle dependency among CP businesses. Each business type relies on various vehicles to source goods, fulfill customer orders, and manage logistics.

These were mainly three types of vehicles: two-wheeler, three-wheeler, and four-wheeler, as depicted in Fig. 19.4.

The data reveals that **Retail businesses** are the most vehicle-dependent, with the majority requiring commercial vehicles at least twice a week, and some relying on

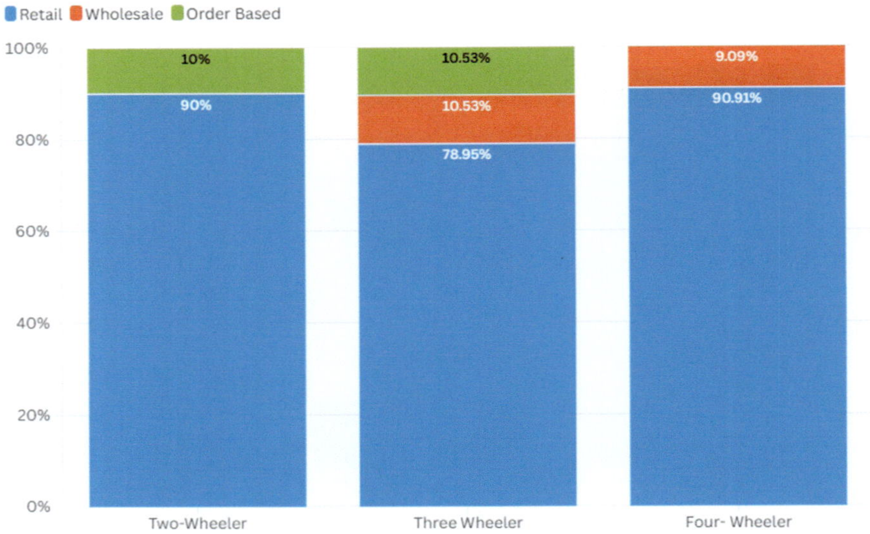

Fig. 19.4 Types of businesses and types of vehicles used by them. *Source* Survey Conducted at CP by Author

them daily. The need for frequent restocking and moving goods from warehouses or suppliers to storefronts drives this dependency. **Like retail, wholesale businesses** also need vehicles multiple times weekly, primarily for bulk procurement and distribution to smaller retailers. **Order-based businesses** have less frequent vehicle use, typically requiring transportation at least once a month, often aligned with larger, scheduled consignments rather than daily operations, as shown in Table 19.3.

The types of vehicles employed vary depending on the business scale and operational needs. Two-wheelers (such as mopeds and scooters) are commonly used for small, quick deliveries, while three-wheelers (autos) and four-wheelers (cars, vans, taxis) are preferred for larger consignments or bulk movement of goods. Notably, 90.91% of surveyed businesses reported using two-wheelers, 10.53% used three-wheelers, and 78.95% relied on four-wheelers at some point in their operations, highlighting the multi-modal nature of commercial logistics in CP.

The operational rhythm of CP businesses is closely tied to their vehicle usage patterns. Most businesses operate between 9 a.m. and 9 p.m., with most commercial activity occurring during these hours. However, the loading and unloading of

Table 19.3 Types of businesses and frequency of vehicle use

Type of business	Frequency of requirement for vehicles
Retail and wholesale	Twice a week
Order-based	Once a month

Source Compiled by Author

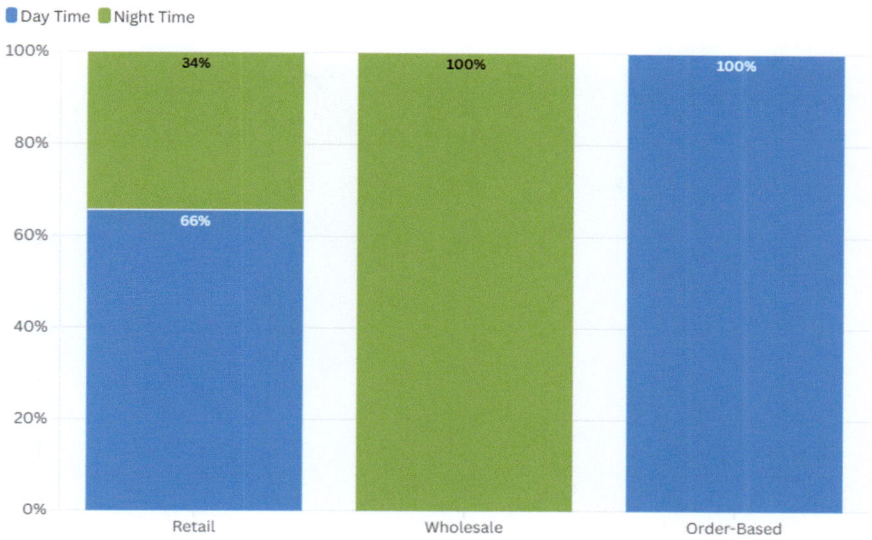

Fig. 19.5 Preferred time of LEZ restrictions by the businesses. *Source* Survey Conducted at CP by Author

goods—a critical logistics function—typically occurs at night, outside peak business hours. This nocturnal logistics window is essential for minimizing disruptions to customer-facing operations and navigating the high congestion levels typical of central Delhi during the day. Therefore, on being asked about the preferred time for LEZ restrictions, 66% of retailers and 100% of order-based business owners said that daytime restrictions would make more sense for them. While 100% of wholesalers and 34% of retailers said that nighttime restrictions would be preferred, many of their business operations do not operate at night, and a larger part of their businesses also occur during the day (Fig. 19.5).

The frequency of vehicle use is also influenced by business type; retail and wholesale businesses require vehicles on a weekly or even daily basis; Order-based businesses, in contrast, have less frequent but larger-scale transportation needs, often tied to monthly cycles or specific client orders. This temporal and operational dependency on vehicles underscores the potential challenges that LEZ restrictions could pose, particularly if entry bans or fees are not aligned with business logistics cycles.

19.5.2 Willingness to Pay and Economic Implications

The survey conducted on CP businesses reveals a complex and varied landscape regarding willingness to pay an entry fee for accessing LEZ. Retailers, wholesalers,

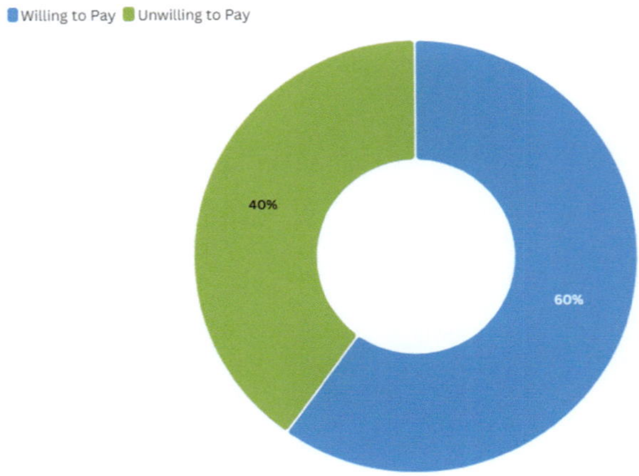

Fig. 19.6 Businesses willing to pay for LEZ entry fee. *Source* Survey Conducted at CP by Author

and order-based businesses display distinct attitudes shaped by operational needs and financial flexibility (Fig. 19.6).

Among retailers, only 40% indicated a willingness to pay an LEZ entry fee, while a notable 33.33% were unwilling, reflecting concerns about additional costs eating into already tight profit margins. In contrast, all surveyed wholesale businesses (100%) expressed readiness to pay, likely due to their greater reliance on frequent and large-scale logistics, which makes alternative arrangements more challenging. Order-based businesses fell in between, with 66.67% willing to pay, as their less frequent but bulk transportation needs allow for some adaptability (Fig. 19.7).

The trend becomes even more pronounced when examining willingness to pay by vehicle type. Operators of four-wheelers—such as vans and taxis, typically used for bulk or high-value deliveries—showed the highest readiness, with 77.2% prepared to pay the fee rather than immediately transition to cleaner vehicles. This starkly contrasts with two-wheeler users, where only 50% were willing to pay, and the remainder either preferred to seek alternatives or expressed reluctance to absorb additional costs. Three-wheeler operators were nearly evenly split, with 52.7% willing to pay. These figures suggest that larger businesses or those with greater logistical demands are more likely to absorb new costs in the short term. At the same time, smaller operators may be pushed to reconsider their transport strategies (Fig. 19.8).

A closer look at the amounts businesses are prepared to pay reveals further nuances. Regardless of type, most businesses strongly preferred keeping the entry fee below ₹100. This threshold is particularly significant for two-wheeler operators, with only 10.53% willing to consider fees up to ₹500. Four-wheeler users, however, showed more flexibility: 52.63% were comfortable with fees between ₹100 and ₹400, and 17.14% could tolerate charges up to ₹500. These differences highlight the varying financial capacities and operational priorities across business types and vehicle classes (Fig. 19.9).

Retail and order-based businesses, which often operate on thinner margins and with less frequent transport needs, are more sensitive to increased costs. At the same

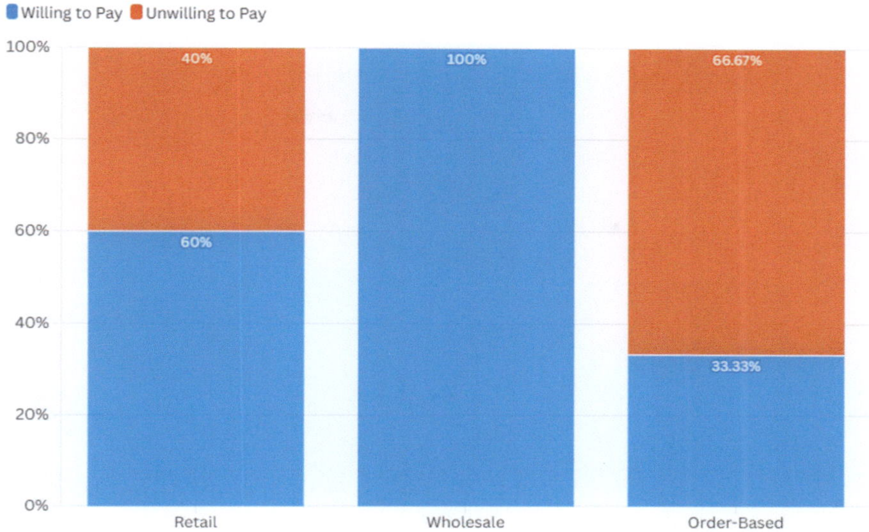

Fig. 19.7 Types of businesses and their willingness to pay. *Source* Survey Conducted at CP by Author

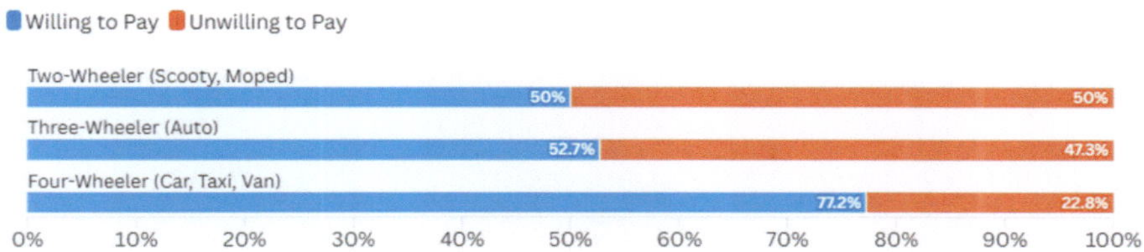

Fig. 19.8 Types of vehicles used and willingness to pay. *Source* Survey Conducted at CP by Author

time, wholesalers and those using larger vehicles may have more room to manoeuvre (Fig. 19.10).

The expectation of increased operational costs is nearly universal among CP businesses. Eighty percent of all respondents foresee a rise in expenses due to LEZ restrictions, and this concern is most acute among four-wheeler operators, with 100% predicting a significant impact on their operational budgets (Fig. 19.11).

Retailers, in particular, worry that added costs will either be passed on to customers, potentially reducing competitiveness, or absorbed internally, squeezing already narrow profit margins. Wholesalers, while more willing to pay, also anticipate disruptions to their supply chains if fees become prohibitive or if alternative logistics arrangements are not feasible. Given their less frequent but larger shipments, order-based businesses are concerned about the cumulative effect of repeated fees over time. There is also apprehension about indirect effects, such as reduced sales if delivery costs rise and are reflected in product prices. Nearly half of the

Fig. 19.9 Types of vehicles and the amount they are willing to pay. *Source* Survey Conducted at CP by Author

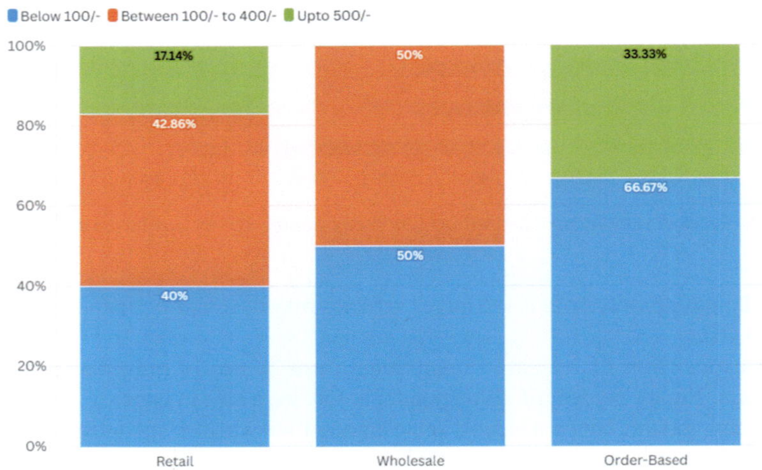

Fig. 19.10 Types of businesses and the amount they are willing to pay. *Source* Survey Conducted at CP by Author

surveyed businesses (48.57%) expressed concern that increased logistics costs could dampen customer demand, especially in a competitive retail environment like CP, where price sensitivity is high. This highlights the broader economic ripple effects that LEZ policies could have beyond direct transport costs (Fig. 19.12).

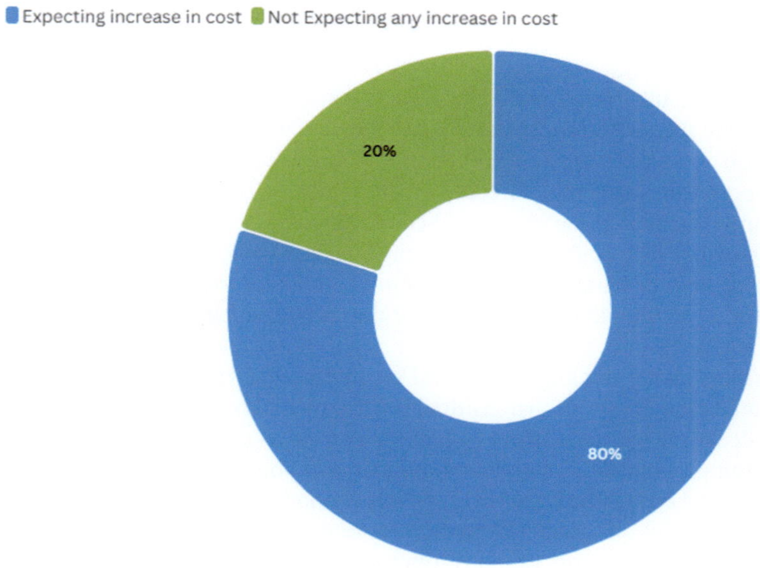

Fig. 19.11 Businesses expecting cost increase. *Source* Survey Conducted at CP by Author

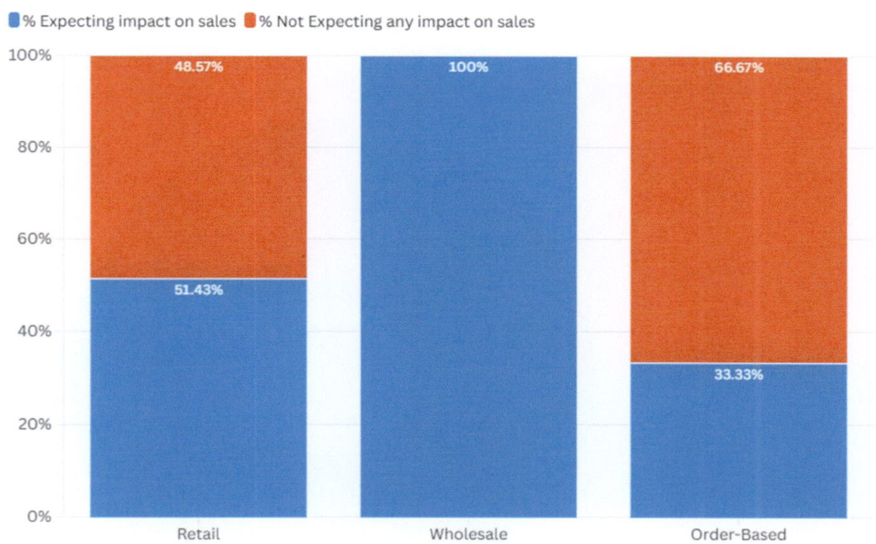

Fig. 19.12 Types of businesses expecting an increase in costs. *Source* Survey Conducted at CP by Author

19.5.3 Transition to Cleaner Alternatives

The survey results from Connaught Place businesses show a clear and encouraging trend: more than 90% of respondents say they are willing to switch to cleaner vehicles, such as electric vehicles (EVs) or BS-VI compliant models, if strict restrictions are imposed on conventional commercial vehicles. This willingness, however, is not simply a matter of environmental consciousness—a mix of practical, economic, and regulatory factors shapes it. For many business owners, the prospect of government mandates or harsh penalties is a strong motivator to consider cleaner options. However, switching depends on whether the transition feels feasible and supported (Fig. 19.13).

Business owners highlighted several key factors when asked about the main influences on their willingness to transition. Environmental benefits were cited by 63% of respondents, showing that there is genuine awareness and concern about air quality and pollution impacts. A significant number also pointed to government incentives (16.67%) and cost-effectiveness (17.39%) as important motivators. Some mentioned parking subsidies and clear government restrictions could tip the balance in favor of cleaner vehicles. This mix of motivations suggests that while there is a strong sense of social responsibility, practical considerations—especially financial ones—play an equally crucial role in decision-making (Fig. 19.14).

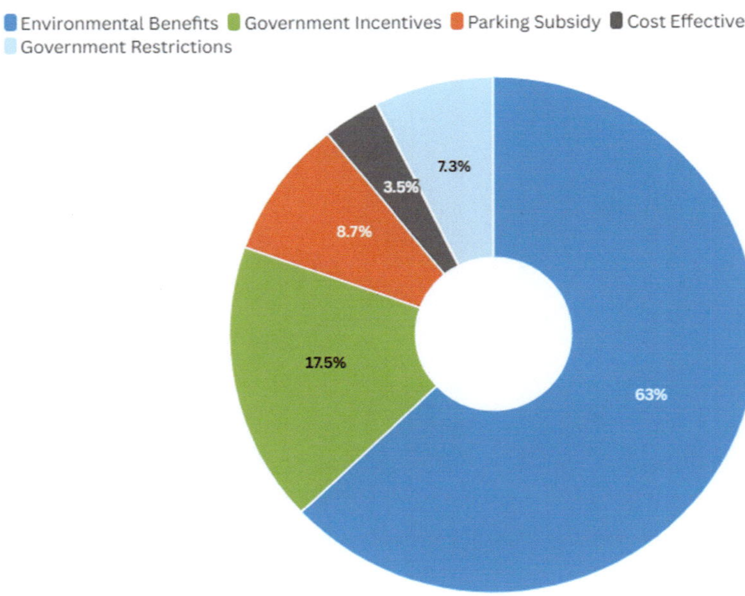

Fig. 19.13 Factors influencing businesses in transitioning to cleaner alternatives. *Source* Survey Conducted at CP by Author

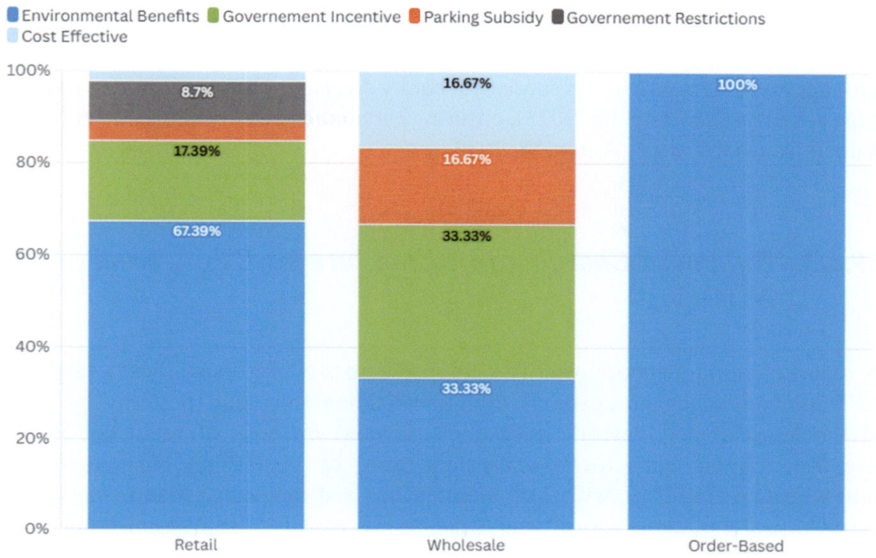

Fig. 19.14 Types of businesses and influencing factors in transitioning to cleaner alternatives. *Source* Survey Conducted at CP by Author

Despite the positive attitude toward cleaner alternatives, businesses face real and immediate obstacles. The most frequently mentioned barrier is the high upfront cost of purchasing EVs or upgrading to BS-VI vehicles. For small and medium-sized enterprises, which comprise most of Connaught Place's commercial scene, such investments can be daunting and sometimes out of reach without external support. In addition to cost, many respondents expressed concerns about the lack of reliable charging infrastructure in and around Connaught Place. Several business owners shared frustrations about non-functional or inaccessible charging stations, making switching to EVs less attractive and riskier for daily operations.

19.6 Policy Recommendations

19.6.1 Financial Incentive Framework for Cleaner Vehicle Transition

Delhi must establish a tiered financial support system tailored to SMEs, prioritizing higher subsidies for two- and three-wheeler operators, with only 10% and 10.53% willing to pay fees exceeding ₹500, respectively. This includes direct grants for EV/BS-VI vehicle purchases, low-interest loans with extended repayment periods, and temporary operational subsidies to offset transition costs. Such measures address the

80% of businesses anticipating cost hikes while leveraging the 16.67% of respondents who identified government incentives as a key motivator for transitioning. This approach mirrors Milan's successful equity-focused grants and London's scrappage schemes (McLoughlin, 2023), ensuring economic viability alongside emission reductions (Marchetti & Antonelli, 2024).

19.6.2 Enhanced Charging Infrastructure for EVs Around NDMC

A robust charging network must be deployed across NDMC, specifically targeting CP's commercial logistics patterns: nighttime loading/unloading (preferred by 100% of wholesalers) and daytime operations (favored by 66% of retailers). Prioritize fast-charging hubs for two-wheelers (used by 90.91% of businesses) and four-wheelers (78.95%), with 24/7 maintenance and innovative load-management systems. Collaborations with private operators can expedite deployment, while preferential commercial-user rates would incentivize adoption. This directly addresses the primary barrier cited by businesses, non-functional or sparse charging stations, and aligns with Shenzhen's model of infrastructure grants for logistics electrification (ITDP, 2023).

19.6.3 Inclusive Policy Framework for Small and Medium Businesses

Delhi's LEZ design must adopt a differentiated approach that accounts for business heterogeneity, implementing tailored restrictions based on operational patterns: daytime limits for retailers and order-based businesses, and nighttime restrictions for wholesalers. The framework should offer flexible compliance pathways while incorporating phased implementation with grace periods and exemptions for SMEs facing transitional challenges. This structure directly addresses the disparities revealed in the study, including wholesale willingness to pay fees (100%) versus retailer resistance (33.33%). It integrates lessons from Madrid's revised fee structures and temporal freight corridors (Tarriño-Ortiz, 2022). Regular stakeholder consultations, modeled after Cáceres, Spain, will ensure equity and address informal logistics dependencies, balancing environmental goals with commercial viability.

19.7 Conclusion

This comprehensive assessment of business perspectives towards LEZ implementation in NDMC's CP reveals a complex landscape where environmental awareness meets practical operational challenges. The study demonstrates that while over 90% of small and medium business owners express willingness to transition to cleaner vehicles when faced with strict restrictions, their capacity remains constrained by significant financial barriers and inadequate infrastructure support. The findings show distinct patterns across business types, with wholesale operations displaying greater readiness to absorb entry fees compared to retail establishments, which operate on tighter margins and show more resistance to additional costs. Most businesses prefer keeping LEZ entry fees below ₹100, with 80% anticipating increased operational expenses that could potentially impact their competitiveness and customer demand. Despite these challenges, the strong environmental consciousness among 63% of respondents, coupled with their recognition of government incentives as motivating factors, suggests that carefully designed policy interventions could successfully balance environmental goals with business sustainability. The research underscores that effective LEZ implementation in Delhi requires a nuanced approach that addresses the unique needs of small and medium enterprises while providing adequate support mechanisms for their transition to cleaner alternatives.

References

Boogaard, H., Janssen, N. A. H., Fischer, P. H., Kos, G. P. A., Weijers, E. P., Cassee, F. R., van der Zee, S. C., de Hartog, J. J., Brunekreef, B., & Hoek, G. (2012). Impact of low emission zones and local traffic policies on ambient air pollution concentrations. *Science of The Total Environment, 435–436*, 132–140. https://www.sciencedirect.com/science/article/abs/pii/S0048969712009229

Browne, M., Allen, J., & Anderson, S. (2005). Low emission zones: The likely effects on the freight transport sector. *International Journal of Logistics: Research and Applications, 8*(4), 269–281. https://doi.org/10.1080/13675560500405899

CEPT Research Development Foundation (CRDF) (2025). *Planning framework for low emission zone (LEZ) in core areas of Indian cities*. https://crdf.org.in/project/planning-framework-for-low-emission-zone-lez-in-core-areas-of-indian-cities

Centre for Science and Environment (CSE) (2023). *Policy brief: Low emission zones in Indian cities—What is needed*. https://www.cseindia.org/low-emission-zones-in-indian-cities-11811

Centre for Science and Environment (CSE) (2024a). *Low-emission zones: The case of Delhi*. https://www.cseindia.org/low-emission-zones-in-indian-cities-11811

Centre for Science and Environment (CSE) (2024b). *Low-emission zones: Opportunities and impacts, Jaipur*. https://www.cseindia.org/low-emission-zone-opportunities-and-impacts-the-case-of-jaipur-12472

Cui, H., Gode, P., & Wappelhorst, S. (2021). *A global overview of zero-emission zones in cities and their development progress*. International Council on Clean Transportation. https://coilink.org/20.500.12592/d6h4wd

Dablanc, L. & Montenon, A. (2016). Impacts of environmental access restrictions on freight delivery activities: The example of Low Emission Zones in Europe. *Transportation Research Procedia, 12*, 360–369. https://doi.org/10.3141/2478-02

Ellison, R. B., Greaves, S. P., & Hensher, D. A. (2013). Five years of London's low emission zone: Effects on vehicle fleet composition and air quality. *Transportation Research Part D: Transport and Environment, 23*, 25–33. https://www.sciencedirect.com/science/article/abs/pii/S136192091300059X

Gehrsitz, S. (2017). The effect of low-emission zones on air pollution and infant health. *Journal of Environmental Economics and Management, 81*, 121–144. https://www.sciencedirect.com/science/article/abs/pii/S0095069617300736

Government of National Capital Territory of Delhi, Transport Department (2010, September 27). *Notification No. F. 19 (76)/Tpt/Sectt./2010/3979: Restrictions on entry, plying, and idle parking of goods vehicles in Delhi*. https://transport.delhi.gov.in/sites/default/files/transport_data/CWGas+goods-115.pdf

Hartmann, A., & Abel, S. (2020, June 18). *How Oslo achieved zero pedestrian and bicycle fatalities, and how others can apply what worked*. The City Fix. https://thecityfix.com/blog/how-oslo-achieved-zero-pedestrian-and-bicycle-fatalities-and-how-others-can-apply-what-worked/

Holman, C., Harrison, R., & Querol, X. (2015). Review the efficacy of low-emission zones to improve urban air quality in European cities. *Atmospheric Environment, 111*, 161–169. https://www.sciencedirect.com/science/article/abs/pii/S1352231015300145

International Council on Clean Transportation (ICCT) (2021a). *A global overview of zero-emission zones in cities and their development progress*. https://theicct.org/publication/a-global-overview-of-zero-emission-zones-in-cities-and-their-development-progress/

International Council on Clean Transportation (ICCT) (2021b). *Evaluation of real-world vehicle emissions in Brussels*. https://theicct.org/sites/default/files/publications/TRUE-Brussels-report-2021-11-19.pdf

International Council on Clean Transportation (ICCT) (2021c). *Planning and implementation of low- and zero-emission zones in cities*. https://theicct.org/wp-content/uploads/2023/09/Planning-ZEZs-working-paper-21-A4-v4.pdf

International Council on Clean Transportation (ICCT) (2024). *Real-world motor vehicle exhaust emissions in Delhi and Gurugram using remote sensing*. https://theicct.org/publication/rw-motor-vehicle-exhause-emissions-delhi-gurugram-remote-sensing-aug24/

Institute for Transportation and Development Policy (ITDP) (2023). *The opportunity of low emission zones: A taming traffic deep dive report*. https://itdp.org/publication/the-opportunity-of-low-emission-zones-a-taming-traffic-deep-dive-report/

Institute for Transportation and Development Policy (ITDP) (2024). *Low emission zone (LEZ) basics: A step-by-step guide for cities*. https://itdp.in/resource/lez-basics/

Institute for Transportation and Development Policy (ITDP) (2025). *Pimpri-Chinchwad's low-emission zone strategy explained*. https://itdp.in/pimpri-chinchwads-low-emission-zone-strategy-explained/

Intergovernmental Panel on Climate Change (IPCC) (2021). *Climate change 2021: The physical science basis*. Contribution of Working Group I to the Sixth Assessment Report of the Intergovernmental Panel on Climate Change. https://www.ipcc.ch/assessment-report/ar6/

International Council on Clean Transportation (ICCT) (2019). *A global snapshot of the air pollution-related health impacts of transportation sector emissions in 2010 and 2015*. https://theicct.org/publication/global-snapshot-health-impacts-transportation-emissions-2010-2015/

International Council on Clean Transportation (ICCT) (2022). *Impact of a low-emission zone on air pollutants: A case study of Pimpri-Chinchwad, India*. https://theicct.org/publication/impact-of-a-lez-on-air-pollutants-pimpri-chinchwad-india-oct24/

Jiménez-Espada, M., García, F. M. M., & González-Escobar, R. (2023). Citizen perception and ex ante acceptance of a low-emission zone implementation in a medium-sized Spanish city. *Buildings, 13*(1), 249. https://www.mdpi.com/2075-5309/13/1/249

Li, W., Long, R., Chen, H., Chen, F., Zheng, X., & Yang, M. (2019). Effect of policy incentives on the uptake of electric vehicles in China. *Sustainability, 11*(12), 3323. https://www.mdpi.com/2071-1050/11/12/3323

Kok, I. (2023). Planning and implementation of low-and zero-emission zones in cities. *International Council on Clean Transportation.* Pobrano z https://theicct.org/wp-content/uploads/2023/09/Planning-ZEZs-working-paper-21-A4-v3.pdf

Marchetti, E., & Antonelli, C. (2024). *Social aspects of low emission zones: Milan case study.* Institute for European Environmental Policy. https://ieep.eu/wp-content/uploads/2024/06/Social-aspects-of-low-emission-zones-Milan-case-study-IEEP-2024.pdf

Margaryan, S. (2019). *Low-emission zones and population health.* Research Paper. https://www.sciencedirect.com/science/article/abs/pii/S0167629620310481

McLoughlin, J. (2023, June 2). *TfL scrappage scheme encourages small companies to go electric ahead of ULEZ expansion.* Logistics Manager. https://www.logisticsmanager.com/tfl-scrappage-scheme-encourages-small-companies-to-go-electric-ahead-of-ulez-expansion/

Moreno, E., Schwarz, L., Host, S., Chanel, O., & Benmarhnia, T. (2022). The environmental justice implications of the Paris low emission zone: A health and economic impact assessment. *Air Quality, Atmosphere & Health, 15*(12), 2171–2184. https://doi.org/10.1007/s11869-022-01243-7

Press Information Bureau (PIB) (2024). Air quality perspective in Delhi. *Government of India.* https://www.pib.gov.in/PressReleaseDetail.aspx?PRID=2210305&reg=1&lang=1

Rizki, M., Irawan, M. Z., Dirgahayani, P., Belgiawan, P. F., & Wihanesta, R. (2022). Low emission zone (LEZ) expansion in Jakarta: Acceptability and restriction preference. *Sustainability, 14*(19), 12334. https://www.mdpi.com/2071-1050/14/19/12334

Sfendonis, N., Basbas, S., Mintsis, G., Taxiltaris, C., & Politis, I. (2017). Investigation of the user's acceptance concerning a low emission zone in the center of Thessaloniki, Greece. *Transportation research procedia, 24*, 280–287. https://www.sciencedirect.com/science/article/pii/S2352146517304027

Sinha, S. (2025, February 22). *NDMC to roll out smart integrated parking system by June.* Hindustan Times. https://www.hindustantimes.com/cities/delhi-news/ndmc-to-roll-out-smart-integrated-parking-system-by-june-101740163593530.html

Tarriño-Ortiz, J., Gómez, J., Soria-Lara, J. A., & Vassallo, J. M. (2022). Analyzing the impact of low emission zones on modal shift. *Sustainable Cities and Society, 77*, 103562. https://www.sciencedirect.com/science/article/pii/S2210670721008283

United Nations Environment Programme (UNEP) (2024). *Used heavy-duty vehicles and the environment: A global overview of used heavy-duty vehicles: Flow, scale and regulation.* https://wedocs.unep.org/20.500.11822/44826

Wappelhorst, P. S., & Cui, H. (2022). *Update on zero-emission zone development progress in cities.* International Council on Clean Transportation. https://theicct.org/wp-content/uploads/2022/09/Global-ZEZs-update-09292022.pdf

World Resources Institute (WRI) (2025a, January 1). *Implementing low-emission zones for cleaner air in India [LinkedIn post].* LinkedIn. https://www.linkedin.com/posts/wri-india_lez-cp-activity-7310897027024916480-1DXq/

World Resources Institute (WRI) (2025b). *Low-emission zones: A blueprint for Delhi's urban transformation.* https://wri-india.org/perspectives/low-emission-zones-blueprint-delhis-urban-transformation

Zhai, M., & Wolff, H. (2021). Air pollution and urban road transport: Evidence from the world's largest low-emission zone in London. *Environmental Economics and Policy Studies, 23*(4), 721–748. https://link.springer.com/article/10.1007/s10018-021-00307-9

MIX
Papier aus verantwortungsvollen Quellen
Paper from responsible sources
FSC® C105338

If you have any concerns about our products,
you can contact us on
ProductSafety@springernature.com

In case Publisher is established outside the EU,
the EU authorized representative is:
Springer Nature Customer Service Center GmbH
Europaplatz 3, 69115 Heidelberg, Germany

Printed by Libri Plureos GmbH
in Hamburg, Germany